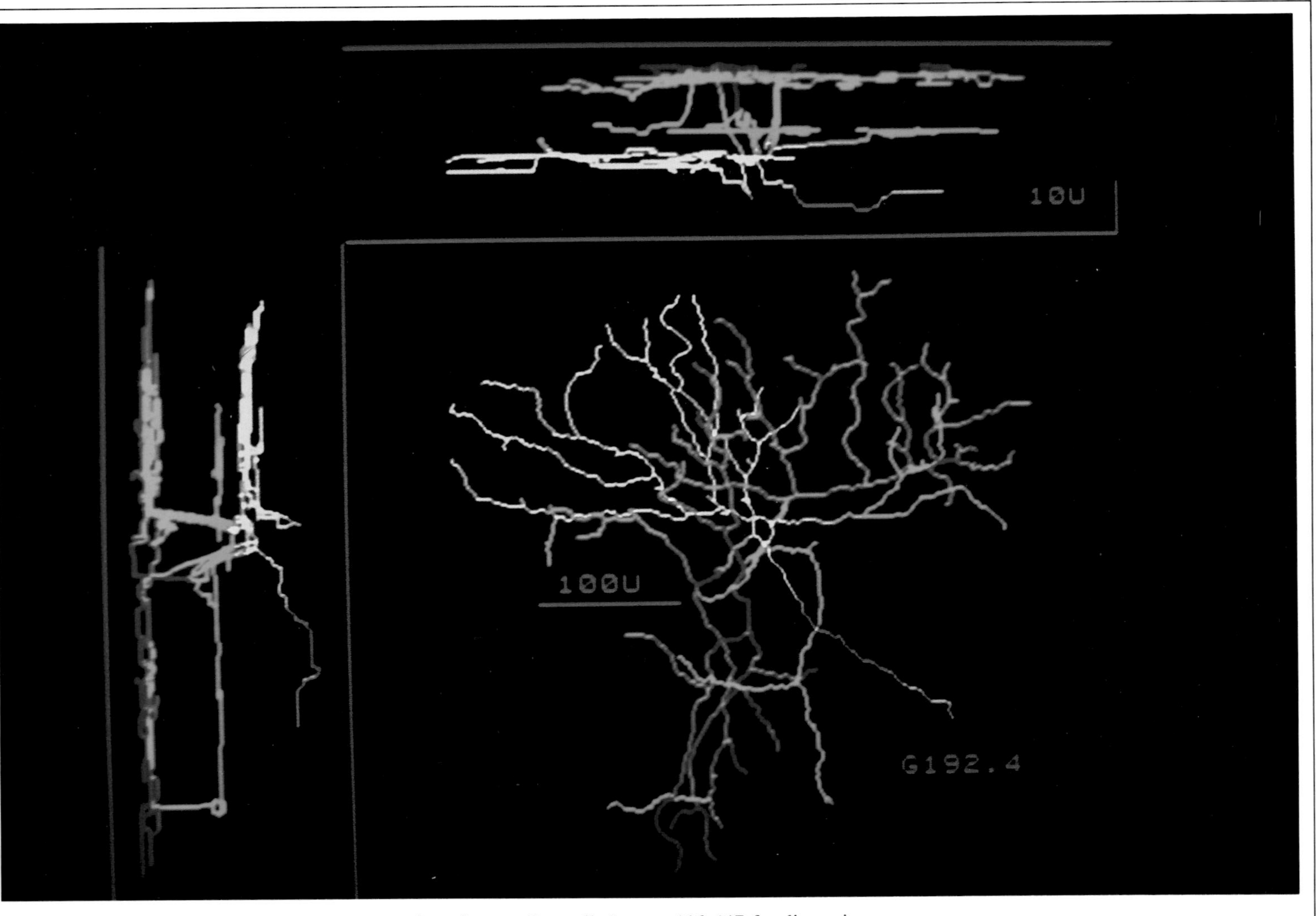

Color-coded 3-dimensional computer reconstruction of a ganglion cell. See pp. 146–147 for discussion.

THE MICROCOMPUTER IN CELL AND NEUROBIOLOGY RESEARCH

Edited by

R. Ranney Mize, Ph.D.

Associate Professor, Department of Anatomy
and Division of Neuroscience
College of Medicine, University of Tennessee
Center for the Health Sciences, Memphis

ELSEVIER

New York · Amsterdam · Oxford

Elsevier Science Publishing Co., Inc.
52 Vanderbilt Avenue, New York, New York 10017

Sole distributors outside the United States and Canada:

Elsevier Science Publishers B.V.
P.O. Box 211, 1000 AE Amsterdam, The Netherlands

Library of Congress Cataloging in Publication Data

Main entry under title:

The Microcomputer in cell and neurobiology research.

Includes bibliographical references and index.
1. Cytology—Data processing. 2. Neurobiology—Data processing. 3. Microcomputers. I. Mize, R. Ranney.
QH585.5.D38M53 1985 574.87′02854 84-18637
ISBN 0-444-00842-X

Manufactured in the United States of America

In memory of John Lawrence,
without whose inspiration
this book would not have been completed

Contents

Preface

The computer revolution has arrived in cell and neurobiology research. The advent of the microcomputer makes measurement, control, and analysis of data both simpler and far faster than seemed imaginable a decade ago. In the mid-70s, computer analysis of cell and neurobiology experiments was generally limited to a few "wealthy" laboratories that could afford high-priced mainframe or minicomputers. Today, inexpensive microprocessors and microcomputers make computer analysis accessible to virtually every scientist. Spurred by the popularity of the personal computer and the availability of low cost computers like the Apple and IBM-XT, many cell and neurobiologists have "computerized" their laboratories. In the near future, the microcomputer will become an indispensable general purpose research tool in the laboratory.

In a recent review of microcomputer applications in cell and neurobiology (Mize, 1984), I became aware of a number of microcomputer systems that had been developed for use in those disciplines. Descriptions of these systems, however, either had not been published or were scattered in a wide variety of journals. Those that were published varied greatly in technical detail and research use. Often they were too technical to be understood by the computer novice. Just as frequently they included too little detail to be of use to those with a background in computers. The articles often lacked a comprehensive description of the research application for which they were designed. Most importantly, I found very few books that collected such information in a single volume. This book is an effort in that direction.

Although the book is designed primarily for cell biologists, neurobiologists, and others who analyze the structure and function of tissue, scientists in related fields will also find it useful for understanding microcomputer architecture and the potential of microcomputers in a laboratory setting. Graduate and advanced undergraduate students should be able to use it as a supplementary text in biologically oriented computer courses.

The book is divided into six sections: the first section includes chapters describing the components of a microcomputer, programming languages, and hardware and software selection. The remaining sections describe microcomputer systems used in Light and Electron Microscopy, Morphometry, Serial Section Reconstruction, Imaging and Densitometry, and Electrophysiological Recording. These sections cover a broad range of topics of interest to most scientists in the rapidly growing fields of cell and neurobiology. The chapters are written for those with experience in research but without computer expertise. Each chapter includes sufficient technical detail, however, to be useful to those with extensive backgrounds in computer science.

I have chosen an introductory chapter for each section, which will introduce the reader to principles of hardware and software design for that type of research. The introductory chapter is followed by chapters that describe specific systems used for particular research applications. I have chosen systems that should interest a large group of biologists. Several criteria were used in choosing these chapters. First, I looked for systems that used commercially available desk-top or microcomputers. (A microcomputer was defined as a portable self-contained unit with 8- or 16-bit microprocessors, memory, and standard peripherals such as printers, plotters, and a CRT.) Second, I tried to choose systems that were just that—systems. Programs developed to collect or analyze data without the necessary design of appropriate interfaces to laboratory instruments or other hardware were not considered. The computer systems also had to include specially designed software of special value to the biological community. Third, I tried to choose systems in which the software was written at least partially in a high-level programming language such as BASIC, FORTRAN, or PASCAL. This should allow the software to be transported to a variety of microcomputers. Finally, I chose computer systems that have been used extensively to collect data in a laboratory environment. Most of the application chapters in the book include sections that describe actual applications in cell and neurobiology research.

There are a number of people I wish to thank for helping put this book together. The late John Lawrence of Elsevier was extremely cooperative in all phases of the publication. His enthusiasm for the idea and his expert publishing skills were indispensable to the completion of the book. Louise Calabro Gruendel, Dorita James, and Jane Licht of Elsevier were of great assistance in preparing and editing the book. Their pleasant and cooperative spirit is gratefully acknowledged. Mary King Givens and Ellen McDonell of the Medical Library of the University of Tennessee Center for the Health Sciences assisted considerably in the library searches that brought some of the contributors to my attention. My research assistant, Linda Horner, was most helpful in proofing the text. I also want to thank several colleagues who first exposed me to computer science. They include Dr. John Stevens, Dr. Larry Palmer, and Cameron Street, who taught me to appreciate structured programming. Finally, I want to thank my wife, Dr. Emel Songu-Mize, who has tolerated and often encouraged my obsession, both with computers and the completion of this book.

The computer revolution has arrived. I hope this book helps the reader join that revolution more easily. Happy computing.

Reference

Mize, R. R. (1984) Computer applications in cell and neurobiology: A review. Int. Rev. Cytol. 90:83–124.

Contributors

Steven Akeson
Department of Biochemistry, University of California, San Francisco, California

Franklin R. Amthor
Department of Physiological Optics, School of Optometry, University of Alabama in Birmingham, Birmingham, Alabama

Michael Breton
Scheie Eye Institute, University of Pennsylvania, Philadelphia, Pennsylvania

J. J. Capowski
Department of Physiology, University of North Carolina Medical School, Chapel Hill, North Carolina

Vivien A. Casagrande
Department of Anatomy, Vanderbilt University, Nashville, Tennessee

Thomas L. Davis
Department of Anatomy, University of Pennsylvania School of Medicine, Philadelphia, Pennsylvania

Bruce Drum
Department of Ophthalmology, George Washington University, Washington, D.C.

Daniel DuVarney
Emory University, Atlanta, Georgia

Raymond C. DuVarney
Department of Physics, Emory University, Atlanta, Georgia

Lennart Enerbäck
Department of Pathology II, University of Goteborg, Goteborg, Sweden

Dale G. Flaming
Department of Physiology, University of California, San Francisco, California

C. R. Gallistel
Department of Psychology, University of Pennsylvania, Philadelphia, Pennsylvania

Ellen M. Johnson
Department of Physiology, University of North Carolina Medical School, Chapel Hill, North Carolina

David C. Joy
AT&T Bell Laboratories, 600 Mountain Avenue, Murray Hill, New Jersey

W. Daniel Kornegay III
Department of Anatomy, East Carolina University School of Medicine, Greenville, North Carolina

Noel Kropf
Department of Biological Sciences, Columbia University, New York, New York

Jeffrey H. Kulick
Department of Computing and Information Sciences, Queen's University, Kingston, Ontario, Canada

Cyrus Levinthal
Department of Biological Sciences, Columbia University, New York, New York

James A. McKanna
Department of Anatomy, Vanderbilt University, Nashville, Tennessee

Cary N. Mariash
Division of Endocrinology and Metabolism, Department of Medicine, University of Minnesota, Minneapolis, Minnesota

Robert Massof
Wilmer Eye Institute, Johns Hopkins University, Baltimore, Maryland

R. Ranney Mize
Department of Anatomy and Division of Neuroscience, University of Tennessee Center for the Health Sciences, Memphis, Tennessee

Walter H. Mullikin
Department of Anatomy, University of Pennsylvania School of Medicine, Philadelphia, Pennsylvania

Melburn R. Park
Department of Anatomy and Division of Neuroscience, University of Tennessee Center for the Health Sciences, Memphis, Tennessee

S. Mark Poler
Departments of Anesthesia and Medicine, University of California, San Francisco, California
Present address: Washington University, St. Louis, Missouri

Max C. Poole
Department of Anatomy, East Carolina University School of Medicine, Greenville, North Carolina

Peter Ramm
Department of Psychology, Queen's University, Kingston, Ontario, Canada

Ingemar Rundquist
Department of Pathology II, University of Linköping, Linköping, Sweden

Eric D. Salin
Department of Chemistry, McGill University, Montreal, Quebec, Canada

Steven Seelig
Division of Endocrinology and Metabolism, Department of Pediatrics, University of Minnesota, Minneapolis, Minnesota

R. L. A. Sing
Department of Chemistry, McGill University, Montreal, Quebec, Canada

Irwin Sobel
Department of Biological Sciences, Columbia University, Fairchild Center, New York, New York

Cameron H. Street
Department of Anatomy, University of Pennsylvania School of Medicine, Philadelphia, Pennsylvania

Oleh Tretiak
Department of Computer Engineering, Drexel University, Philadelphia, Pennsylvania

THE MICROCOMPUTER AS A RESEARCH TOOL

1

The Components and Architecture of the Microcomputer

Raymond C. DuVarney

Introduction

It is difficult these days to define what we mean by the term *microcomputer*. I shall, however, take as a working definition those computers whose arithmetic functions and logical decision-making processes are performed using a microprocessor, which is a single, very large-scale integrated (VLSI) circuit. Dramatic improvements in very large-scale integration technology, particularly the ability to mass-produce circuits containing about 50,000 transistors on a silicon chip less than a square centimeter in area, have revolutionized the computer industry. This miniaturization is largely responsible for the increases in microprocessor operating speed and complexity and decreases in cost that we have witnessed during the past decade. As the development of integrated circuits continues, there is every indication that today's microcomputers will be considered toys in comparison to what will become available before the end of this decade.

These small wonders have proliferated into our homes, offices, and laboratories. They provide us, at affordable prices, tools for research that will continue to revolutionize the way in which science is conducted. In this chapter I want to introduce the reader to the way in which the microcomputer functions. In particular, I want to emphasize what is happening at the hardware or electronic level, not with regard to electric currents and charges, but rather the flow, processing, and storage of information in its most fundamental form within the microcomputer.

Microcomputer Components

Figure 1 shows a block diagram of the basic architecture of a microcomputer. Depicted are the central processing unit (CPU) or microprocessor, a block of random access memory (RAM, here labeled read/write memory), a section of read only memory (ROM), and two input/output interfaces (parallel and serial ports are illustrated in the figure). Each of these components of the microcomputer is interconnected by three types of bus: the address, data, and control buses. The term *bus* refers to a parallel set of conducting lines used collectively to transmit data or control signals. An electronic oscillator, called the *clock,* supplies timing pulses to the microprocessor, triggering each step in the operating sequence of the microcomputer. The microprocessor performs the arithmetic and logic functions that one usually associates with the term *computing*. In addition, it controls the three buses that interconnect all the components of the microcomputer. Microcomputers usually contain two types of electronic memory. RAM may be written to as well as read by the microprocessor. The information it contains can be changed under the control of the microprocessor. On the other hand, ROM may only be read by the CPU. The information it contains never changes. The input/output (I/O) interfaces interconnect the internal components of the microcomputer with external peripheral components such as the keyboard or the printer. Mass storage devices such as floppy disk drives, fixed hard disk drives, or cassette tape recorders also receive and transmit information through an I/O interface.

Information Storage

To understand the operation of a microcomputer (or any computer, for that matter), it is important to have a clear idea of the way in which information is coded and stored within the computer. A computer memory consists of a vast collection

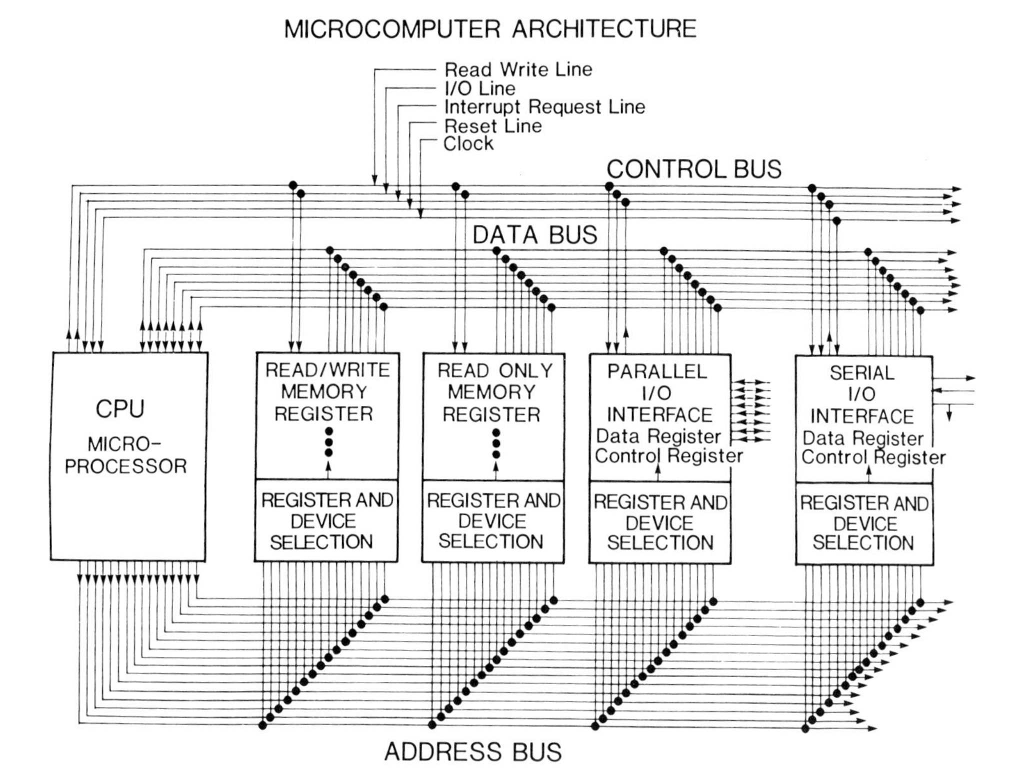

Figure 1. Block diagram illustrating the internal architecture of a typical 8-bit microcomputer. The microprocessor is interconnected to the other system components via the address, data, and control buses. The system components shown include read/write memory (RAM), read only memory (ROM), and a parallel and serial I/O interface. Each component is connected to the address bus through an address decoder, which selects and activates the addressed register and connects it to the data bus so that the microprocessor may read the register's contents or change the register's contents.

of electronic switches that can be opened or closed (written to) by the central processing unit, or read by the central processing unit to "see" which are open and which are closed. Each switch can store one piece of information, which is called a *bit.* There are only two possible states for a switch. The information stored in one bit could therefore represent Yes or No, True or False, On or Off, Plus or Minus, the numbers 0 or 1, or any other quantity that can be represented as one of two possible states.

To be able to represent more complex sets of things, such as letters of the alphabet (26 states) or the decimal integers 0–9 (10 states), the switches are organized into groups called *registers.* Registers contain multiple bits. Most microcomputers available today contain either 8- or 16-bit registers. An 8-bit register is a group of eight switches that functions as if it were one 256-position switch. (Eight two-position switches have 2^8 or 256 possible configurations.) These can range from all eight switches open to all eight switches closed. An 8-bit register holds one *byte* of information. This means it contains one out of 256 possible codes. The types of data that can be represented by one byte of information are practically unlimited; they must only be members of a set of 256 elements or less. Here are some examples.

We shall use 1s and 0s to indicate the state of each bit or switch in the register.

A 1 will represent the on-state and a 0 the off-state. Let us suppose that we stored the pattern 10101010 at some memory register. The 1s and 0s indicate that a bit is set or not set, respectively. This pattern could represent the decimal number 170, which is $(1 \times 2^7) + (0 \times 2^6) + (1 \times 2^5) + (0 \times 2^4) + (1 \times 2^3) + (0 \times 2^2) + (1 \times 2^1) + (0 \times 2^0)$ in *binary code.* Here each bit shows the presence or absence of the successive powers of 2 that comprise a number, and in this fashion all the integers from 0 to 255 may be represented. Of course, the information stored in a register can also represent things other than numbers. For example, 01010000 (binary for 80) represents the letter P using the American Standard Code for Information Interchange (ASCII) or could represent one of 256 possible instructions for the microprocessor. The meaning lies in the interpretation of the code.

Binary, Octal, and Hexadecimal Numbers

All numbering or counting systems have a set of symbols to represent the number of items in a group. The number of symbols one uses is called the *base.* In the familiar decimal or base 10 system there are ten symbols (0–9). If we wish to express a number greater than the number of symbols in our counting system, we do so by counting the number of times each power of the base appears as a constituent of that number. For example, in base 10 the number 111 means $(1 \times 10^2) + (1 \times 10^1) + (1 \times 10^0)$. The same 111 in base 2 or binary means $(1 \times 2^2) + (1 \times 2^1) + (1 \times 2^0) = 7$. In *octal* (base 8) the same 111 would mean $(1 \times 8^2) + (1 \times 8^1) + (1 \times 8^0) = 73$, and in *hexadecimal* (base 16) it means $(1 \times 16^2) + (1 \times 16^1) + (1 \times 16^0) = 273$. The hexadecimal system has a base of 16 and thus requires 16 symbols to express all possible integers. Any 16 symbols would suffice but the convention is to use the ten decimal symbols 0–9 plus the first six letters of the alphabet. As an instructive example we will convert a decimal representation of a number into each of the other representations mentioned above. We begin with the decimal number 205 and first convert this into binary. The first question we must ask is what is the largest power of 2 that is less than or equal to 205? The answer is 128, which is 2^7. In binary this is 10000000. Subtracting 128 from 205 leaves 77. Now we find the largest power of 2 that is less than or equal to 77. The answer this time is 64, which is 2^6 (1000000 in binary). Subtracting 64 from 77 leaves 13. The greatest power of 2 contained in 13 is 8 or 2^3 (1000 in binary). Subtracting 8 from 13 leaves 5. The greatest power of 2 contained in 5 is 4 or 2^2 (100 in binary). Subtracting 4 from 5 leaves 1. The greatest power of 2 less than or equal to 1 is 2^0. (Any number to the power of 0 equals 1; thus, in binary this is 1.) Combining all of the binary constituents of the decimal number 205 yields the following:

Binary Constituents

Decimal Representation	Binary Representation
1 × 128 = 128	10000000
1 × 64 = 64	1000000
1 × 8 = 8	1000
1 × 4 = 4	100
1 × 1 = 1	1
205	11001101

Had we converted this decimal number 205 into an octal or base 8 representation, we would have obtained the following octal constituents.

Octal Constituents

Decimal Representation	Octal Representation
$3 \times 64 = 192$	300
$1 \times 8 = 8$	10
$5 \times 1 = 5$	5
205	315

Similarly the hexadecimal conversion yields the following constituent table.

Hexadecimal Constituents

Decimal Representation	Hexadecimal Constituents
$12 \times 16 = 192$	C0
$13 \times 1 = 13$	D
205	CD

Examining this hexadecimal conversion we see that the greatest power of 16 that is less than 205 is 16^1. Sixteen may be subtracted from 205 twelve times leaving a remainder of 13. The hexadecimal symbol for 12 is C, which appears in the "16s" column of the hexadecimal representation. The remainder of 13 is represented by a D in the units (16^0) column.

It should be clear from our discussion of two-state information storage that the binary representation of a number is the most natural way of storing that number within the microcomputer. However, it is not the easiest way for us mortals to express a number. Base 10 numbers are easy to use but are difficult to convert into binary. Base 8 or 16 numbers, however, may be converted to and from binary representation quite simply. This is because 8 and 16 are powers of 2 whereas 10 is not. We can think of a base 8 or base 16 number as a binary number expressed as groups of 3 or 4 bits, respectively. Let us return briefly to the previous example. The decimal number 205 has the binary representation.

11001101

Let us write this in groups of 3 bits (starting from the right),

11 001 101

and then convert each group into one of the eight octal symbols. The leftmost or most significant group becomes a 3, the middle group a 1, and the rightmost or least significant group a 5. The octal representation is thus

315

Let us now write the binary number in groups of four bits,

1100 1101

and then convert each group into one of the 16 possible hexadecimal symbols. The most significant group (on the left) is a decimal 12 or a hexadecimal C; the other group of four bits is a decimal 13, which is a hexadecimal D. The hexadecimal representation for this number is thus

CD

Table 1. Numbering Systems

Binary	Octal	Decimal	Hexadecimal
0	0	0	0
1	1	1	1
10	2	2	2
11	3	3	3
100	4	4	4
101	5	5	5
110	6	6	6
111	7	7	7
1000	10	8	8
1001	11	9	9
1010	12	10	A
1011	13	11	B
1100	14	12	C
1101	15	13	D
1110	16	14	E
1111	17	15	F
10000	20	16	10

Because the octal and hexadecimal representations are much more compact, they are easier to express, remember, and work with than a lengthy string of 1s and 0s. Table 1 compares binary, octal, decimal, and hexadecimal representations of the first 16 integers.

Microcomputer Architecture

Referring again to Figure 1, let us discuss in more detail each of the major elements within the microcomputer. We shall begin by describing the structure of the two types of memory found in microcomputers.

Memory Registers

The key element to understanding the organization of a microcomputer's memory is the *memory register.* Memory registers are usually eight bits wide and thus can each store a byte of information. Each register has a unique address so that the microprocessor can select a particular register by placing its address on the address bus. It can then, via the data bus, either set the bits in the selected register by performing a *write operation,* or look at how the bits were previously set by performing a *read operation.* In either case a byte of information is transferred over the data bus. Memory registers occupy most of the available addresses.

From a functional point of view one can picture the memory structure as a vast array of 8-bit registers connected in parallel to the eight lines in the data bus: Bit 0 of each register is connected to data line 0, bit 1 to data line 1, and so forth. Between this array and the address bus is sufficient decoding circuitry such that one and only one register from this array will be activated or enabled for each address that can appear on the address bus. This structure is the same for both RAM and ROM, the only difference being that ROM is designed only to read the microprocessor's registers, not write into them. [It would have been more logical to refer to RAM (random access memory) as read/write memory to distinguish it from read only memory (ROM).] As we see from the structure, the random access to registers within each type of memory is the same. *Random access* means that registers may be accessed (addressed) in any order. The opposite of this is *sequential access* of information, such as that found on magnetic tape storage devices.

The Bus Structure

The microprocessor communicates with the other components of the microcomputer by putting signals on the control, data, and address buses. These buses are parallel conducting paths that electronically connect the outputs and inputs of the microprocessor with the appropriate inputs and outputs on the other electronic components comprising the microcomputer. The bus structure of a typical microcomputer is shown in Figure 1.

The *control bus* carries all of the control signals that the microprocessor must either send or receive as it communicates with other components and parts of the microcomputer. The details of the signal line that make up the control bus differ somewhat from one type of microprocessor to another. There are, however, some control signals that are common to most microprocessors, which we shall briefly describe:

The *reset* signal is an *input* to the microprocessor that sets all of the microprocessor's internal registers to a predetermined state. (We shall say more about this control signal later.)

The *read/write* signal is an *output* from the microprocessor. It tells each component whether the microprocessor is performing a read-from or write-to operation.

The *valid memory address* signal is an *output* from the microprocessor that informs each component when the address code on the address bus is valid.

The *interrupt request* is an *input* signal to the microprocessor from the external components. It is used to signal the microprocessor that some external component requires attention. The microprocessor may be set to ignore this request.

The *nonmaskable interrupt* is also an *input* signal from one or more external components. It is similar in function to the interrupt request signal except that the microprocessor cannot be set to ignore it: It is a demand rather than a request.

The *data bus* for an 8-bit microprocessor has eight lines (Figure 1). This means that data are transmitted or received in bytes (8 bits at a time) between the microprocessor and any of the other components. The data bus is used to communicate with various peripheral devices. The *address bus* usually consists of 16 lines. These are controlled by the microprocessor and used to transmit the address of the register that the microprocessor is reading from or writing to. It does this by placing the binary code for the desired address on these lines.

Let us illustrate this process with a specific example. Assume that the microprocessor is going to store the number 56 in a memory register whose address is 4000. The 16-bit code for this address is 0000 1111 1010 0000 and the 8-bit code for the data is 0011 1000. In order to accomplish the write operation, the microprocessor sets each line in the address and data buses to either 5 volts (representing a binary 1) or 0 volts (binary 0) in accordance with the above code and then sends a signal on the read/write line (one of the lines in the control bus) that causes the code on the data lines to be stored in the selected memory register. Since this microprocessor may address 2^{16} different registers, it has a total address space of 64K addresses, where $K = 1024$ (2^{10}). Although most of this address space will be filled with memory, some of it may be used to address input and output devices.

The Memory and I/O Map

One of the unique characteristics of each kind of commercially available microcomputer is the manner in which the available addresses have been assigned. One point that cannot be overemphasized is that the microprocessor can only communicate by reading and writing to registers. All devices that are part of the microcomputer must appear as a register or a set of registers with a particular address or set of addresses. A large portion of the 64K address space is reserved for RAM or read/write memory. This space consists of 8-bit memory registers that hold the instructions and data that form a user's program. Another sizable portion of the total available address space is utilized by ROM registers, which contain permanent programs and data necessary for the operation of the machine. A portion of the remaining address space is assigned to devices that connect or interface to the microcomputer. Take, for example, the keyboard. This device is interfaced to the microprocessor via a parallel I/O interface like the one shown in Figure 1. To us the keyboard appears much like that of a typewriter, with some 50 or so keys that we can push. To the microcomputer, however, the keyboard appears to be two 8-bit registers that occupy two consecutive addresses. We shall call these the *keyboard data register* and *keyboard status register*. Each contain a

System Memory Map

Page Number: Decimal	Hex	
Ø	$ØØ	RAM (48K)
1	$Ø1	
2	$Ø2	
⋮	⋮	
19Ø	$BE	
191	$BF	
192	$CØ	I/O (2K)
193	$C1	
⋮	⋮	
198	$C6	
199	$C7	
2ØØ	$C8	I/O ROM (2K)
2Ø1	$C9	
⋮	⋮	
2Ø6	$CE	
2Ø7	$CF	
2Ø8	$DØ	ROM (12K)
2Ø9	$D1	
⋮	⋮	
254	$FE	
255	$FF	

Figure 2. The system memory map for the Apple II+ microcomputer. The 64K address space is organized into *pages*. Page numbers are shown in both decimal and hexadecimal representations. There are 256 pages with 256 locations on each page. For example, page C3 contains the 256 locations starting at address C300 and ending at address C3FF. [Reproduced with permission of Apple Computer Inc. from the Apple II Reference Manual, p. 68.]

byte of information. The keyboard data register contains the binary code for the key that was last pressed. The keyboard status register holds information about the status of this device. Likewise, other devices, such as disk drives or a printer, appear to the microcomputer to be registers located at particular addresses that were assigned to these devices when the microcomputer was designed. Figure 2 shows how the address space within an Apple II+ microcomputer is assigned.

Some microprocessors make no distinction between writing data to an I/O device register and writing data to a memory register, but others do. In the former case the I/O device registers occupy part of the memory space (Figure 2); in the latter, the I/O registers have a separate space. This separation is accomplished by the addition of a special control line in the control bus called the *I/O line.* This line is used in conjunction with the read/write line to activate the I/O device register and not the memory register. Using this scheme an I/O device can have the same address as a memory register: There is no conflict because the state of the I/O line is used to determine which one of the two possible registers is selected. Microprocessors that use this I/O line must have additional instructions that differentiate between memory operations and I/O operations. Microprocessors without an I/O control line make no such distinction.

Microprocessor Architecture

The Internal Registers

The microprocessor is the hub of the microcomputer. Its function is to fetch instructions from memory, where the program is stored as a series of bytes, execute each instruction as it is fetched, and then fetch the next instruction. Depending on what the instruction is, the microprocessor may also fetch data from memory in order to operate on it, write data to memory, or send data to a device via an I/O interface as it executes the instruction.

There are several registers internal to the microprocessor that are used to control its operation (Figure 3). All of the functions that the microprocessor can perform can be specified and understood with reference to the contents and changes that take place in these internal registers. Although it is beyond the scope of this chapter to deal with the details of each of these registers, a simplified description of a typical fetch and execute cycle will illustrate the principal of operation of some of them. We shall think in terms of a simplified 8-bit microprocessor that consists of the following components (Figure 3):

1. An *address register,* which is 16 bits wide and controls the 16 address lines on the address bus.
2. A register called the *program counter* (PC), which contains the address of the next instruction or data byte to be fetched.
3. A *data register,* which is 8 bits wide and is connected to the eight data lines on the data bus. It is used to transfer bytes of information to and from the microprocessor unit over the data bus.
4. An *instruction decoder* that determines which logic or arithmetic function the microprocessor will perform.
5. An *arithmetic* and *logic unit,* which actually performs these functions.
6. An *accumulator,* which is an 8-bit register used to store temporarily the results from the arithmetic and logic unit.

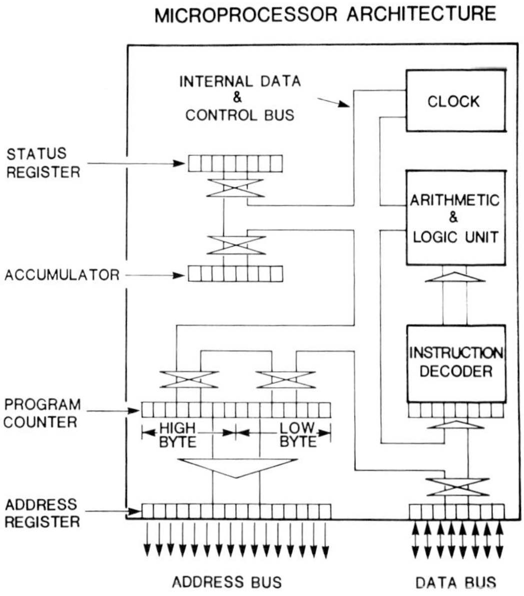

Figure 3. A simplified register model of a typical 8-bit microprocessor. The arrows indicate the direction in which information may flow along the internal data and control buses.

7. A *status register,* which is a collection of bits (often called *flags*) that contain important information concerning the results of the last instruction executed. Many of the instructions that the microprocessor can execute are conditional. They say, in essence: Do this if some "condition" is true, or do that if some "condition" is false. The present status of all the conditions that the microprocessor can test are held in the status register.

The Fetch and Execute Cycle

The fetch and execute cycle is the primary mode of operation of the microprocessor. As long as power is applied to the microprocessor it will continue the cycle of fetching an instruction from memory, executing the instruction, and fetching the next instruction. This cycle begins with the program counter. Its contents are transferred to the address register. This address appears on the address bus, and the program counter is incremented by one. The contents of the memory register pointed to by the address on the address bus are transferred through the data register into the instruction decoder. At this point we enter the execute phase and one of many things can happen depending on which instruction is being executed. Let us suppose that the instruction is to load the accumulator with the byte of data that immediately follows the instruction. The program counter, which was incremented by one during the fetch phase, already contains the address of that byte of data. The contents of the program counter are transferred to the address register, the program counter is incremented, and the data contained at that memory register are transferred through the data register to the accumulator. The instruction has been executed and we are ready to fetch the next instruction. This cycle continues as long as the microprocessor is in its normal operating mode. The timing of the cycle is determined by the period of the clock input and the particular instruction being executed. Simple instructions can be executed in 2 clock cycles, whereas others can take 3, 4, or in some cases 10 cycles. The length of the clock period also varies from one type of microcomputer to another. The slowest machines use a clock period of about 1 microsecond (μsec) and the fastest around 0.1 μsec.

Machine Language Instructions

Microprocessor instructions are 1- or sometimes 2-byte-long codes that select the next logical or arithmetic operation that the microprocessor will carry out. The kind of instructions that a microprocessor can execute are quite elementary. They consist of such things as loading and storing the accumulator or the other internal registers, adding to or subtracting from the accumulator, comparing two registers, and executing a branch to a different part of the stored program if certain conditions occur (for instance, if execution of the last instruction left the accumulator 0). Most microprocessors have about 50 basic instructions. Some instructions have several variations depending on the addressing mode. The *addressing mode* is the manner in which the address of the operand is found. The *operand* is the data that the instruction will act on. For example, there are several instructions that load the accumulator. One of these will load it with the byte of data stored immediately after the instruction. A variation of this instruction uses the two bytes following the instruction to form the 16-bit address where the data to be loaded is stored. A 16-bit quantity, such as an address, can be viewed as two bytes juxtaposed. The byte in the numerically most significant position is called the *high byte;* the other is the *low byte.* There is another variation of this instruction that uses the first byte following the instruction as the low byte of the address where the data is stored, and assumes the high byte is 0.

Again, the nature of machine language instructions and the manner in which the microprocessor fetches and executes these instructions is best illustrated by a simple example. The following short routine is used to create a pause of a certain duration within a running program. Routines of this type are often used for timing purposes and are called *timing loops.*

Mnemonic Machine Language		Address Relative to Start	Machine Code
Label	Instruction		Byte
	LDA, D255	1	1 0 0 0 0 1 1 0
		2	1 1 1 1 1 1 1 1
LOOP	DCA	3	0 1 0 0 1 0 1 0
	BNE, LOOP	4	0 0 1 0 0 1 1 0
		5	1 1 1 1 1 1 0 1
	(Next Instruction)	6	

On the left the program is written in a *mnemonic code.* This is a set of multicharacter symbols that bear a one-to-one correspondence with the machine code on the right. The characters represent an abbreviated version of the operation being performed. This mnemonic language is also called ASSEMBLER. The binary machine code is what is actually stored within the memory at contiguous memory register addresses. The machine code shown here is for the Motorola 6800 microprocessor. The three mnemonics shown in this routine stand for the instructions "*LoaD* The *A*ccumulator," "*DeC*rement the *A*ccumulator," and "*B*ranch if the result of the last instruction executed was *N*ot *E*qual to zero."

The routine begins by loading the accumulator with the number 255. Then the next instruction, which is labeled LOOP, decrements the accumulator by 1. The label is important since it is referred to in the next instruction, which indicates that the program should loop back to that program line if the result of the previous instruction does not leave a 0 in the accumulator. Thus, the decrement accumulator instruction will be executed 255 times. After the 255th cycle, however, the accumulator will be 0, the condition for branching will not be met, and the program will continue on to the next instruction.

Now let us follow the operation of the microprocessor as it fetches and executes these instructions. The routine begins when the program counter contains the address of byte 1. This instruction is fetched and the PC is incremented. The instruction is decoded and "tells" the microprocessor to fetch the next byte, treat it as data, and transfer it to the accumulator. The PC is incremented and now contains the address of byte 3, the next instruction. This byte is fetched and the PC is incremented. The byte is decoded and tells the microprocessor to decrement the accumulator by one. The PC now contains the address of byte 4, the branch instruction. The process continues as this instruction is fetched, the PC incremented, and the instruction decoded. The first step in executing this instruction is to fetch the next byte, which contains the information about where to branch, place it in the data register, and increment the PC. The PC now contains the address of byte 6. The second step in executing this conditional branch instruction is to test the condition. If the condition to branch is met, the contents of the data register are added to the low byte of the PC and 255 will be added to the high byte of the PC.

Here is what happens to the contents of the register:

	High Byte	Low Byte
PC before branch	0 0 0 0 0 0 0 0	0 0 0 0 0 1 1 0
(Contains a 6)	1 1 1 1 1 1 1 1	1 1 1 1 1 1 0 1
PC after branch	0 0 0 0 0 0 0 0	0 0 0 0 0 0 1 1
(Contains a 3)		

Remember, when adding in binary, 1 + 1 = 0 with a 1 carried to the left. We see that in this addition a 1 is carried into the 17th bit of the PC, which means it is discarded. The PC is thus reduced by 3 and now contains the address of the decrement accumulator instruction, which will again be fetched and executed. This looping procedure will continue until the branch condition is not met (namely, until the accumulator has reached 0). When the branch condition is not met the contents of the data register will not be added to the PC and thus byte 6, the next instruction outside the loop, will be fetched and executed.

Read Only Memory

In addition to the read and write memory that we have discussed, a microcomputer will reserve and use some of its address space for read only memory (ROM). Read only memory can only be read by the microprocessor and cannot be overwritten. ROM devices can be written to only once. A variation of this kind of memory device, known as an EPROM (erasable programmable read only memory), may be erased by exposure to ultraviolet light and reprogrammed as needed. ROMs are important components of any microcomputer. Unlike read/write memory (RAM), the information stored within these components does not disappear when the computer is turned off. To emphasize this difference, programs stored in ROMs are called *firmware* rather than software.

If it were not for these devices, the computer, when first turned on, would contain no programs and would not be able to perform any tasks. At the very least, when a microcomputer is powered up it must begin executing an elementary routine that can accept input from the keyboard or other device so that a program (usually the operating system) can be loaded into memory and run. This is accomplished using one of the control lines into the microprocessor known as the *reset line.* The reset button pulls this line to 0 volts when pushed. Microprocessors are designed so that when a reset signal is received the processor responds by placing a predetermined address on the program counter and starting the fetch and execute cycle from that address. What the microcomputer requires is a ROM containing this elementary start-up routine. The first instruction of that routine must be located at the address placed on the PC when the microprocessor receives a reset signal. A short routine of this type that is used to load the operating system from a remote storage device (such as a disk) into memory and begin its execution is called a *bootstrap.* A *monitor* is another type of ROM-based start-up program that allows the contents of registers in memory as well as registers internal to the microprocessor to be inspected and loaded by keyboard entry. In addition, commands that begin the fetch and execute cycle at an address entered from the keyboard may be issued.

The I/O Interface

In order for computers to be very useful, they need to communicate with other devices such as printers, plotters, and CRT monitors. In this section, we shall describe how the microprocessor communicates with any one of the many peripheral devices that are used as part of the microcomputer. We shall consider a peripheral device to be anything that the microprocessor can read from or write to other than memory. Keep in mind that the microprocessor is limited in that it can only communicate to registers that are within the address space. Whatever the device may be, it must appear to the microprocessor to be a register or group of registers located at some unique set of addresses within the address space. It is the function of the I/O interface circuitry to implement this transformation. The interface is a translator. It transforms the bytes of information that are written into its registers into a form or format that the device is designed to accept. Conversely, it also transforms information coming from the external device into bytes and loads it into a register that can be read by the microprocessor.

The Parallel Interface

A parallel interface is used to transmit and/or receive data bits sent simultaneously to or from a device along multiple transmission lines. Parallel interfaces are often used to connect to high-speed data acquisition laboratory equipment.

For a device that is sending or receiving data in byte-size segments, a minimum of eight transmission lines is required. In practice there are more than this because a few control lines between the device and interface are required. These let the device signal the interface that it is either ready or not to communicate data, and likewise let the interface signal the device concerning its readiness. This signaling process between the device and the interface is called *hand-shaking*. Properly employed, the hand-shaking signals allow the device and the interface to communicate at the highest possible speed. They prevent the device or interface from transmitting data before the previous data transmission has been digested, and they allow the device or interface to transmit data as soon as the intended receiver, be it the interface or the device, is ready.

The Serial Interface

The serial interface is used to exchange data one bit at a time between the interface and a device along a single transmission line. Serial interfaces are commonly used to communicate with other computers and peripherals such as printers and plotters. A bit is transmitted by applying one of two possible voltage levels (each possibility representing either a 1 or 0 for that bit) on the transmission line for a specific length of time. This time interval is called the *bit-time*. The reciprocal of the bit-time is the *baud rate* and is the number of bits per second that are being transmitted. Table 2 shows the standard baud rates and bit-times that are used in serial data transmission. A byte of data is transmitted by sending each bit in the byte juxtaposed in time. That is, each bit-time immediately follows the preceding bit-time. In order to separate the bytes in a continuous transmission of data, so that the least significant bit and most significant bit are properly located in the received data, each transmitted byte is surrounded by a set of special bits called the *start and stop bits*. As implied, the start bit, which is always a 0, precedes the first bit of the data byte and the stop bit or bits (there may be more than one depending on the particulars of the transmission protocol) follow the last bit of the data byte.

In order to read a stream of serially transmitted data properly, the receiver, be it either the device or the interface, must "know" the bit-time that the transmitting unit is using. Thus the transmitting and receiving units must be set for the same baud rate. The process works as follows. When the receiver detects the beginning edge of the start bit it waits $1\frac{1}{2}$ bit-times. The end of this wait period comes in the middle of the bit-time for the first data bit. The receiver senses the voltage present on the transmission line at this time and determines whether the first bit is a 1 or a 0. It then waits another bit-time and senses the voltage for the next bit. The process is repeated until all eight bits of data have been "clocked

Table 2. Baud Rates and Their Corresponding Bit-Times

Baud rate	Bit-time (msec)
300	3.333
600	1.667
1200	0.833
9600	0.104
19200	0.052

in" to an internal register. The contents of this register are then transferred to the I/O register to be read by the microprocessor.

Serial and parallel interfaces each have advantages and disadvantages. An 8-bit parallel interface can transmit data about 10 times faster than a fast serial line. This is obvious since the parallel interface transmits eight bits simultaneously rather than sequentially and no start or stop bits are required. On the other hand the serial interface requires only three wires (signal transmit, signal receive, and signal ground) as opposed to the multiple conductor cables needed for parallel transmission. Thus, serial communication is usually used where speed is not critical or where communication must be done over long distances. Parallel communication is used where speed is necessary or distances are short.

The Use of Interrupts by I/O Devices

It is clear that the microprocessor can access a device whenever the program it is executing instructs it to do so, but how does the device access the microprocessor? How does the microprocessor know that a peripheral device is ready to send data? There are two possibilities. First, the microprocessor could poll the device periodically to see if the "ready to send" bit on the interface has been set. Alternatively, the device could make use of one of the control lines on the control bus called the *interrupt request* line (IRQ) to gain the attention of the microprocessor. This line is an input to the microprocessor that functions in some ways like the reset line. When the IRQ line goes low, this signals the microprocessor that some device needs service and wishes to interrupt. If the interrupt flag (one of the bits in the status register) is set to allow interrupts, then the following process occurs.

1. The current contents of the program counter and several other important registers inside the microprocessor are saved in a part of RAM that has been reserved for this function. This area of RAM is called the *stack.*
2. The program counter is loaded with a predetermined address that is part of the microprocessor design and program execution begins at that address.
3. At that address, the first instruction of the routine that will service the device must be stored. The details of the program and what it does depend heavily on the specific device that is being serviced. However there are some features that are common to all interrupt routines. First of all, the contents of any registers that this routine will use should be stored on the stack and then replaced just prior to exiting the interrupt routine. This will enable the microprocessor to continue with the program it was running before it responded to the interrupt. Secondly, the final instruction in the interrupt routine must be a "return from interrupt" command. When the microprocessor executes this command it returns from the stack the original contents of all the internal registers. In particular, the program counter contains the same address it contained just prior to the interrupt request. Thus the program that was running prior to the interrupt continues on as if nothing had happened.

If more than one device can send an interrupt request, we need to include one further section in our interrupt service routine. The first part of the interrupt routine must poll the devices to determine which device interrupted. This puts a constraint on our interface hardware. One bit in the device status register, one of the registers that constitute the interface to that device, must be set when the device interrupts. When the devices are polled, the microprocessor reads these device status registers to determine which device requested the interrupt, and then branches to the interrupt service routine for that device.

Using interrupts as part of an I/O interface is not always necessary but often results in the most efficient use of the microprocessor's time. The situation is similar to, let us say, trying to read a book while expecting visitors. If the doorbell is disconnected, the only thing one can do is to periodically check the door to see if the visitors have arrived. If the doorbell functions, however, checking the door is unnecessary. The bell serves as a request (not a demand) to answer the door. When the bell does ring, one may finish the sentence or paragraph one was reading, place a bookmark on the page, put the book away, and then go to the door. When the visitors leave, one can return to reading exactly where one left off.

Microcomputer Software

We have already seen that a program consists of machine language instructions and the data used by those instructions. These instructions and data are stored in memory as a series of bytes. It is important to realize that this is the only type of program the microprocessor can perform. Most software, however, is not written in machine language but in a high-level language, one that uses English-like as well as algebraic commands and instructions. We therefore need to spend some time describing the process whereby a program written in a high-level language such as BASIC is converted into a runable machine language program.

Interpreted Versus Compiled Code

There are two methods presently used to generate machine language code from a high-level language. One method is to run a program, called an *interpreter,* that "reads" the high-level language program, one line at a time, converts the line into machine code, executes this code, and moves on to the next line of the program.

An alternative to this method is to compile the program into *machine code.* In this process the high-level language program serves as a source file for the compiler, which is itself a machine language program. The compiler reads the program and produces an incomplete machine code version, called *object code.* The machine code produced at this point in the process is incomplete because the compiler has produced machine instructions to branch or jump to subroutines that are not yet part of the program. The subroutines, which among other things are used to calculate mathematical functions, are contained in a special collection of machine coded routines called the *object library.* The second step in the production of a runable machine coded program is to include or "link" those library subroutines with the compiled program to form a complete and runable program. It sounds complicated but it is done quite automatically by the microcomputer using a program called a *linker.* The linker program uses the compiled code from the high-level language program and the library of subroutines as input and produces a runable stand-alone program. We can summarize this process in three stages:

1. Create a high-level language program.
2. Run the compiler program using the program written in step 1 as input. This produces as intermediate object code.

3. Link this object code with the subroutine library using the linker program to produce a runable machine code.

The interpreted code method has both advantages and disadvantages in comparison to the compiled code method. First of all, the interpretive approach is obviously simpler. A program written in BASIC can be "run" immediately. Secondly, a high-level language program occupies much less memory than its machine language counterpart. For example, it does not require much memory space to store the BASIC statement:

10 A = B*SIN(T)

The final machine code, however, that performs this function is quite lengthy. Often a long BASIC program when compiled will not fit into the available memory space of a small microcomputer and must be broken up into segments.

On the other hand, interpreted programs run much slower than the compiled version would. This is because the interpreter requires a significant amount of time to compile a line of the program before it executes it. It then "forgets" the code from that line and compiles the next line. In fact, if the program contains a loop, the same line of the high-level program may be recompiled many times.

Assemblers

An assembler is a program or a collection of programs and routines designed to aid programmers who are writing machine code. Simply described, they allow the programmer to create a machine language program using mnemonics and labels instead of the binary coded instructions. Since any line of the program may be labeled, conditional branches or jumps from one line of the program to another may be specified using the label for that line. The assembler then converts this mnemonic or assembly language into binary coded machine instructions. The process is similar to compiling a high-level language only much simpler because there is a one-to-one correspondence between the mnemonic and the binary coded instructions. Although this type of programming is tedious, routines that have been assembled generally execute in the shortest possible amount of time (see Chapter 2). It is often necessary to use assembly language programming when doing data acquisition. For example, reading data bytes from an A to D converter that is being used in a voltage measurement is often time critical: We may wish to have a series of these measurements made at rather precise time intervals and/ or many measurements taken over a short period of time. One approach to simplify the programming involved is to write a short, fast, and precisely timed machine language routine that takes the raw data bytes from the address corresponding to the A to D converter interface, and stores them somewhere in memory. This machine language routine may be called from inside a BASIC or other high-level language program, which can then read the stored data bytes and process and present this information in its desired form. This approach solves two problems:

1. It allows for very fast acquisition of data.
2. The data processing program, which involves extensive mathematical manipulation, file storage, and printing, may be written in a high-level language.

Programming languages that are useful in data acquisition are discussed in detail in Chapter 2.

Interfacing the Microcomputer in the Laboratory

Two of the most popular microcomputers widely used for data acquisition in the laboratory are the Apple II+ or II E and the IBM PC or XT. One of the reasons for the wide acceptance of these microcomputers in the laboratory is the ease with which they can be interfaced to laboratory measurement equipment using commercially available interface cards that plug directly into one of the slots available for peripherals. It is common these days to find electronic measurement equipment with either a binary coded data (BCD) output port or the so-called IEEE-488 General Purpose Instrument Bus (GPIB) port on the rear panel of the unit. These are intended for interfacing applications, and we shall discuss briefly how each part works and how it can be interfaced to a microcomputer.

The General Purpose Interface Bus

The GPIB was developed by the electrical engineering community to meet the need to standarize methods for interfacing electronic measurement equipment (such as digital voltmeters and frequency meters) or control equipment (such as a programmable power supply) with a computer. The GPIB has a bus structure that is similar in concept to the bus structure within the microcomputer. There is a *controller interface* that controls the action of the GPIB. This is usually the interface between the GPIB and the microcomputer and puts the microcomputer in the position of being the controlling device. All external devices are then connected in parallel to this bus through a *talker/listener interface* that is designed to go between the GPIB and the specific device in question. This interface is generally provided by the equipment manufacturer as an extra cost option. Thus with one GPIB controller interface taking up one slot in the microcomputer, as many as 14 external devices (all connected along the same bus) can transfer data to the microcomputer and/or be controlled from the microcomputer.

The bus itself consists of eight data lines that allow the bidirectional flow of data to and from the controller and eight control lines used for hand-shaking and bus management. Each device on the bus is given its own unique address, which is set with switches located on the device interface. The controller can then communicate with the devices one at a time, which is of course an essential element of any communication scheme using a bus structure. The devices along the bus may be primarily talkers (a frequency meter, for example), listeners (such as a programmable power supply), or talker/listeners. Devices that both talk and listen have both a talk address and a listen address. An example in this last category would be a digital multimeter that we wish to switch from, for example, a voltmeter to an ammeter. To accomplish this the microcomputer, through the GPIB controller, needs to send out the following commands. It would first address all devices to "untalk." There is one address that is reserved on the GPIB for this purpose. It would then address all devices to "unlisten." There is another address reserved for this purpose. It would then address the multimeter to "listen" and would send the proper command bytes to switch the measurement mode of the multimeter. The syntax of the commands that cause the switch-over are not defined by the GPIB protocol; they are specified within the design of the device. The GPIB protocol specifies how communication between devices on the bus is to be established. Once communication is established, the command strings that would be sent to the device and the format of the data transmitted from the device are totally device dependent.

The ways in which data can be transferred along the bus are very flexible. It is even possible for the controller to address one device as a listener and another as a talker and have data transferred from one device to another without passing through the microcomputer. The maximum rate at which data can be transferred over this bus is 1 megabyte per second.

Nonstandardized Binary I/O

Another common type of digital output or input that is often found on laboratory electronic equipment is a multipin connector usually located on the rear panel of the device and given the generic lable *digital I/O* or *binary data input or output.* These are nonstandardized outputs whose details of operation vary greatly from one device to another. However, they usually have the following minimum structure:

1. At least eight data lines, which may or may not be bidirectional.
2. Control lines from the device that indicate whether the device is "ready to be read," "ready to receive," or "not ready," and lines to the device that "request the device to send" or "request the device to receive."

Devices that have this kind of I/O capability may easily be interfaced to a microcomputer through one or more parallel interface ports. There are standard, commercially available interfaces of this type for most microcomputers. The interface plugs into one of the peripheral attachment slots within the microcomputer. An interface of this type will usually appear to the microcomputer as four addresses. Three of these will be 8-bit I/O ports and the fourth will be a control and status register. There will be 24 I/O lines available to the external device. We might use eight of these for the binary data lines and perhaps two more for handshaking signals. This depends on the detailed nature of the device. At this point the device appears as two registers in the address space of the microcomputer, an I/O register and a status register. We must now create the software that will control the device.

A-to-D and D-to-A Conversion

There are many measurement devices found in the laboratory that neither accept nor transmit information in digital form. Their output is instead an analog voltage, which can have any value between some upper and lower voltage limits. The term *analog voltage* refers to a voltage that represents the measured size of some physical quantity. It is another way, a nondigital way, of coding information.

Before we can even think about interfacing the microcomputer to these nondigital devices, we must first be able to convert an analog voltage into a digital code and, conversely, a digital code into an analog voltage. The devices that perform this function are called A-to-D and D-to-A converters (ADCs and DACs, respectively).

The digital side of a D-to-A converter accepts a number coded in bits and the analog side produces a voltage between 0 and some maximum voltage (usually 5 V) depending on the magnitude of the number coded on the digital side. For example, let us consider an 8-bit D-to-A converter. On the digital side the device can accept 8-bit digital data representing some integer between 0 and 255. On the analog side, a voltage proportional to this number is produced. It should be emphasized that an 8-bit converter can produce a maximum of 256 different output voltages, ranging from 0 to 5 V, in equal steps. A 12-bit converter, on the other hand, produces 4,096 voltage steps over the same voltage range, since it accepts a 12-bit input representing numbers between 0 and 4,095.

The analog-to-digital converter works in the opposite fashion. An 8-bit A-to-D converter accepts a voltage (generally between 0 and 5 V) as input and produces an 8-bit code representing a number between 0 and 255. The number produced is proportional to the input voltage with 0 volts producing the number 0 and 5 V producing the maximum number 255. Although the voltage input is continuously variable the digitized output can have only 256 discrete states. Thus an 8-bit A-to-D converter has resolution of $\frac{5}{256}$ V (20 mV). This is the smallest voltage change that could, with certainty, be seen on the digital side of the converter.

A 12-bit converter has 16 times higher resolution and a detectable voltage change as small as $\frac{5}{4{,}096}$ V (1.2 mV).

An important parameter that needs to be considered in a discussion of A-to-D and D-to-A converters is the *conversion time.* This is the time required for an A-to-D converter to sample the signal input voltage and produce the digital code (be it 8, 12, or more bits) that represents that voltage. The higher the resolution of the A-to-D converter, the longer it will take to produce this code. Typical conversion times for 8-bit converters range from 1 to 10 μsec. A 12-bit converter using the same solid-state technology as an 8-bit converter would take about twice as long to complete the conversion. The reciprocal of the conversion time is the maximum sampling rate. There is a clear trade-off. If high resolution is required, lower sampling rates must be accepted.

Conclusions

We have covered a great deal of material in a short space. I would like to conclude this chapter by reviewing the major concepts behind the design and operation of today's microcomputers.

1. A *microprocessor* is used as the *central processing unit* (CPU). It performs all arithmetic and decision-making processes. It also controls the transfer of information between itself and other components within the microcomputer.
2. Communication between microcomputer components is accomplished using a *bus structure.* Components are interconnected through the *address, data,* and *control buses.* A digital code placed on the address bus lines is decoded by circuitry to select a particular register within a particular component. Data are transferred to or from this register over the data bus using command and timing signals sent over the control bus.
3. Other than the microprocessor, the components within a microcomputer may be broadly categorized into three groups: decoders, memory, and I/O interfaces.

a. *Decoders* are used to select a particular register whenever the address for that register appears on the address bus. The selected register is then "active" on the data bus and data can be transferred to or from it.
b. *Memory* refers to those components that collectively form the information storage registers within the microcomputer. There are two basic types of memory storage: RAM and ROM. *RAM* refers to those registers that can be both read to and written from. The contents of a RAM register can be changed. Most of the address space within the microcomputer is occupied by RAM registers. *ROM* contains registers which can only be read. ROMs are important because they contain programs and data vital to the operation of the machine. ROM memory is permanently located at specified addresses within the address space and will not be erased when the power is turned off.
c. An *I/O interface* carries information in the form of electronic signals between any external device and the microcomputer. It acts as a translator. It accepts information coming from an external device, translates it into bytes, and places it into registers to be read by the microprocessor. Conversely, information written into the registers of the I/O interface is translated into signals to operate the device. All of the peripheral components in a microcomputer system communicate with the microprocessor through some type of I/O interface. An important concept to keep in mind is that all devices appear to the microprocessor as a register or a set of registers at some address or set of addresses within the address space.

4. *Software* is the general term used to refer to the information that the computer uses. Information stored permanently in ROMs within the microcomputer is called *firmware* to emphasize its nonvolatile and permanent nature.

The only programs that the microprocessor can execute are machine level programs. The actual program that the microprocessor executes consists of a series of bytes stored in RAM. The microprocessor fetches the byte located at the RAM address dictated by the address register within the microprocessor. If this is the first fetch or the first fetch since the completion of the last instruction, the microprocessor interprets the byte as a code for one of the instructions in its instruction set and executes it. If a program is written in any language other than machine code, it must be converted to machine code before it can be executed. Other programs—interpreters or compilers and linkers—are used to convert programs into runable machine code.

References

Covvey, H.D., and McAlister, N.H. (1980) Computer Consciousness. Reading, MA: Addison-Wesley.

Daley, H.O. (1982) Fundamentals of Microprocessors. New York: Holt, Rinehart and Winston.

Geisow, M.J., and Barrett, A.N., eds. (1983) Computing in Biological Science. New York: Elsevier Biomedical Press.

Microprocessors Book 1 and 2 (1979) Benton Harber, MA: Heath Company.

Newell, S.B. (1982) Introduction to Microcomputing. New York: Harper and Row.

2

Programming Languages for the Laboratory

R.L.A. Sing and E.D. Salin

Introduction

Though initially it may appear that the choice of hardware will dominate the selection of a laboratory computer, the judicious choice of software for the laboratory is as important, if not more important. The selection of a laboratory computer should be made with the full understanding that a symbiotic relationship must exist between the hardware and software. The users will interact with the computer through software languages and, in this respect, familiarity, ease of use, user friendliness, computation power, speed, compactness, and a number of other features of the software and programming languages play major roles.

There exists a myriad of programming languages for microcomputers: ASSEMBLER, BASIC, APL, LISP, COBOL, FORTRAN, C, Z, FORTH, ALGOL, PASCAL, ADA, PL/I, and more. Only some of these languages are generally useful in the laboratory with its particular requirements and constraints. Furthermore, from laboratory to laboratory, the requirements will change, and so a judicious choice of programming language(s) is necessary.

In this chapter, we shall define the features of programming languages and discuss how these features affect the performance of these languages when they are applied to laboratory tasks. Based on these features, we shall examine five microcomputer programming languages most commonly used in laboratories: ASSEMBLER, BASIC, FORTRAN, FORTH, and PASCAL. In the conclusion, we shall present the selection of programming languages that we have used in our laboratory and found to be very effective.

Language Types

A number of the features of programming languages can be attributed to the language type, that is, low level or high level. The latter type can be further divided into interpreted, compiled, and incrementally compiled languages. Before proceeding with the description of the features of programming languages, it would be beneficial to examine briefly how these various language types work.

Low-Level Languages

Low-level programming involves *direct* instruction of the central processing unit (CPU). Instructions take the form of binary operational codes (op codes) and operands that the CPU recognizes. Each instruction is a very precise and limited manipulation of data (bits) by the CPU. The op codes are specific to each CPU (e.g., Z-80, 6502, 8086, 68000). Low-level programs are often referred to as *machine code* because they are machine (CPU) dependent. In this type of programming, every minute detail of each operation must be considered in the program.

The CPU requires that binary coded machine code programs be stored in memory for execution. It will then fetch and execute the instructions in order. The machine code may be entered into the computer memory using toggle switches or, more commonly, using program development software included with the operating system. Since this task of manual code input is quite tedious, humans prefer to use a symbolic representation of the instructions. The symbolic representations are called *mnemonics* and have clearer meaning and are easier to

Table 1. 16-Bit Addition

Machine code	Mnemonics	Comments
11011000	CLD	Clear the decimal flag
00011000	CLC	Clear the carry flag
10100101	LDA 00	Load low byte 1st number
00000000		
01100101	ADC 02	Add low byte 2nd number
00000010		
10000101	STA 04	Store low byte of result
00000100		
10100101	LDA 01	Load high byte 1st number
00000001		
01100101	ADC 03	Add with carry the high
00000101		byte of 2nd number
10000101	STA 05	Store high byte of result
00000101		

remember than the numerical op codes. For example, LDA (load the accumulator) is easier to remember than the machine code 10111011. The conversion of this mnemonic representation to machine code is performed by a program called an *assembler.*

Table 1 shows a side-by-side listing of the machine code and mnemonic representation (often called *assembly code*) of a 16-bit addition routine for the 6502 CPU. The routine performs the addition of two 16-bit values to yield a result that does not exceed 16 bits (i.e., 65535). As can be seen, the machine code is almost meaningless to all except the most dedicated "hackers."* On the other hand, the assembly code, with the appropriate comments, clearly indicates the instructions to be executed. The use of an assembler allows us to use symbolic references to addresses, called *labels,* rather than having to use the actual addresses, which require considerable computation to find.

The writing and execution of an assembly program involves the following steps:

1. The writing of text *(source code)* that contains the mnemonic representation of the instructions we wish the CPU to execute. This step is usually implemented with the help of a text editor.
2. The conversion of the source code to machine code equivalent (called *object code*) performed by the assembler program.
3. The loading of the object code into the computer memory by a linker/loader program.
4. The execution of the program with the help of the operating system Go or Run command.

Generally when we discuss programming at this level we use the term *assembler* or *assembler level* programming. The user should be aware that machine code can be generated without the use of an assembler program.

**Computer Hacker:* A subspecies of the genus Computer Freak with a particular affection for operating system and low-level language programming. Often has difficulty communicating with others.

High-Level Languages

It is very tedious to have to specify to the CPU in full detail all the steps required for even the simplest arithmetic calculation. For this reason, high-level languages exist that allow us to specify the instructions we wish to execute at a higher level, using English and algebraic formulations. As an example, the line

$$A = B + C$$

would in most high-level languages perform the same operation as would the assembler program of Table 1. It can be easily seen that this line performs the addition of two numbers because of the algebraic appearance. What is more, the program of Table 1 performed the addition of 16-bit integer values, whereas the above line in a high-level program could perform the addition of integer *or* real values depending on the language or the declaration of the variables.

With high-level languages, programs are written using instructions or commands that are independent of the CPU but meet the requirements of the particular language. Each of these commands may consist of hundreds of low-level instructions. To execute on a particular computer, the high-level program must be converted to or executed by a program that is appropriate to the particular CPU. There are three ways in which high-level programs are transformed into some executable form: compilation, interpretation, and incremental compilation.

Compiler Languages. For this form of high-level language, the high-level program is converted to its pure machine code equivalent by a program called the *compiler*. The compiler creates a machine code equivalent of the entire high-level program. It is this machine code equivalent that is loaded and executed when the program is run. The steps involved in writing and executing a compiler language program are as follows:

1. Writing a text (source code) containing the high-level commands for the program. This is usually done using a text editor.
2. Conversion of the high-level program to the machine code equivalent by using a compiler program.
3. Linkage of the machine code version of the program with the necessary routines in the language library.
4. Loading and execution of the complete object code module using operating system commands.

Interpreter Languages. In this form of high-level language, the program is not converted to a complete machine code equivalent before being run. Instead, a program called an *interpreter* executes the necessary machine code routines required to perform the high-level language commands. The interpreter will read each line, identify the commands, data, and variables, and execute the machine level routines necessary to perform the operations required. The interpreter will then advance to the next command line. Interpreters allow statements to be executed immediately without being part of a program. This can be a great aid in debugging a program. Most interpreter languages incorporate an EDIT command allowing direct typing of programs without the need of a separate text editor. Since the interpreter operates directly on the source code, only the source code needs to be stored and subsequently loaded into the computer. In comparison, compiler languages usually require that both source and object code be stored.

Incremental Compiler Languages. These languages are a cross between interpreters and compilers. Commands can be used in an interpretive fashion or can be grouped together to form routines that are compiled and added to the existing dictionary of commands. This allows users to write their own language commands, which can be incorporated into the language itself.

Programming Language Features

The features of programming languages for the laboratory that should be considered can be divided into two groups: those that affect the execution of programs and those that affect development and writing. The features that affect execution include speed, regularity of execution, compactness, and device interfacibility and control. The development and writing features include structure, ease of learning, portability, ease of writing, ease of development, modularity, popularity, character handling ability, and other features described below. The version of the programming language should also be considered: Early versions of some languages have been substantially enhanced.

Speed

The speed of execution of a programming language largely depends on its type. The order of execution speed for any given task is generally assembler > compiler > incremental compiler > interpreter. ASSEMBLER executes fastest because it instructs the CPU directly. Compiled programs are essentially machine code programs; however, because they are designed for general use and subroutines are used extensively, they do not run as fast as assembler programs that have been specifically written for a given task. Interpreter languages are substantially slower than compiled programs because a large proportion of the execution time is spent in the interpretive process rather than in productive execution.

Programming language speed in the laboratory is apparent in two major areas: computation and data acquisition. If simultaneous *(real-time)* computation (calculations that must be performed as the acquisition is taking place) and high acquisition rates are necessary, execution speed is of great importance. Since intensive computation with assembly languages should generally be avoided (see section on Ease of Program Writing), compiled languages are preferred for these tasks. Even in cases where real-time calculations are not necessary but extensive calculations are required, slower programming languages can be a great inconvenience because of long waits for results and system tie-up.

For data acquisition, the speed factor plays a more important role. The Nyquist (Malmstadt et al., 1981) criterion specifies that a signal must be sampled at twice its frequency to be accurately recorded. Many laboratory applications can be dealt with adequately at acquisition frequencies below 100 Hz. Any application that can be handled using strip-chart recorders can be accommodated within this frequency limit. This would include most chromatographic applications, polarographic scans, and UV–visible scanning spectrometry. Most interpreter languages can accommodate these rates with no great difficulty, provided that real-time calculations are not necessary. For faster rates, in the 100–15,000-Hz range (Sing et al., 1983) (or for real-time calculations on the acquired data), compiler languages are required. Even with compiler languages, the upper range of rates can often be obtained only when using integer arithmetic. The very high data acquisition rates, above 15,000 Hz, can generally only be achieved with assembly level programming (Sing et al., 1983). In this case, the upper limit of acquisition frequency will eventually be limited by the particular hardware configuration. Further gains in acquisition speed are then only possible by changing CPUs or by using special hardware methods such as direct memory access.

Direct memory access (DMA) involves using a programmable hardware device, usually called a DMA controller, which is designed specifically for high-speed data transfer. These transfers can either be I/O type transfers (e.g., to a fast disk) or from one portion of memory to another. This transfer takes place by effectively turning off the CPU for a period of time while the transfer takes place. Since the transfer is controlled by hardware, rather than software, the rate of transfer can be very high but all of the flexibility of software is lost. Most CPU families have a DMA controller integrated circuit to complement their line of integrated circuits.

Interval Regularity

The rate of data acquisition is not the only timing concern for data acquisition. Equally important is the regularity of the sampling interval, particularly for precise frequency-dependent work or where the programs are relied on for precise elapsed time measurements. The quartz oscillator clocks used to synchronize the hardware of the computer system are very stable and reproducible. Each machine code instruction requires a specific number of machine cycles (clock cycles) for execution, so program execution time is inherently very stable. Problems in interval timing arise because conditional branching (e.g., IF ... GOTO) is almost always necessary in acquisition programs. If the execution times of both paths of a branch are not the same, the acquisition time interval will not be constant. With assembly programming, it is always possible to balance the paths by inserting NOP (null operation, which does nothing but consume time) instructions. This is rarely possible with compilers and interpreters because the languages have been designed to run as fast as possible and branch paths do not necessarily execute with the same duration (number of cycles). In addition, when sections of code are not necessary, as in multiple precision arithmetic, they are often skipped in high-level languages. The user has no control over this phenomenon. The result is often a time interval that oscillates or shifts suddenly. With CPUs that can perform 16-bit integer arithmetic in single precision (i.e., are able to do 16-bit arithmetic in one machine code instruction), it is possible with some high-level languages to obtain regular time intervals, but only if integer variables are used.

When precise timing is necessary and a high-level language is required, hardware timing, rather than software timing, becomes the solution of choice. This involves the use of integrated circuits (e.g., MOS 6522, Intel 8253) designed specifically as timers. These devices often take their clock directly from the very stable crystal clock of the computer, and their outputs can be wired directly to external devices to trigger events such as analog to digital conversions.

Compactness

There are two aspects to the compactness of programming languages: the amount of memory required for execution and the amount of secondary (usually disk) storage they require. The amount of memory consumed by the executing program will ultimately limit the amount of data that can be acquired if the acquisition rate is high enough to prevent the storage of data on a secondary storage medium during acquisition. In addition, programs performing extensive calculations (e.g., Fourier transforms) will run much faster if all the data are resident in main memory. Secondary storage input or output of data can slow program execution enormously.

Assembly programs generally consume the least amount of memory for a given task, since only those routines needed for the task are included in the program. With compiler languages, the object code (run-time module) can be quite large even for a small program. This is due to the inclusion of comprehensive subroutine packages of which only a part is needed. *Subroutines* are sections of code that perform a particular function and are generally "called" (used) from several different parts of the main program. As an example, a multiply-divide-add-subtract-log-exponent package may be linked into the object code module when only the add and subtract functions are needed. An increase in the size of the source code does not proportionally increase the size of the run-time module. Application programs for interpreter type languages are relatively compact. However, the interpreter itself must be present at run time, thereby increasing the size of the total run-time module considerably.

When selecting a secondary storage system for an application it is important to consider the implication of using a particular type of programming language. The type of programming language may be much more important in this regard than the specific program. Compiler languages require a disk-based system for adequate performance. Interpreter-based systems, especially if they are ROM (read only, or permanent, memory) based, may find audio cassette tape storage sufficient. Audio cassettes are usually inconvenient for data storage, but small floppy disk drives are usually adequate. Large assembly programs will also require disk storage for convenient program development.

Device Interfacibility and Control

Many interfacing problems beyond simple analog-to-digital conversion (ADC) and digital-to-analog conversion (DAC) applications may require sophisticated manipulations of bits in memory, either in the CPU registers or peripheral devices. These manipulations may have to be performed very rapidly. Assembly level programming is the only way to ensure rapid absolute control of the computer and its peripherals. However many applications, particularly straightforward ADC and DAC, are adequately handled using high-level languages. This is particularly the case if sophisticated I/O (input/output) ports are available to ensure automatic *hand-shaking* (communication) with conversion or control devices.

With high-level languages, the direct byte and bit manipulation capability is usually version specific rather than language specific. Few of the language standards provide for direct byte and bit manipulations because they were initially designed for large mainframe computers. Adaptation of these languages to the microcomputer environment has resulted in the development of "control" instructions, and these are much better developed in certain versions than others. Some incremental compiler languages (e.g., FORTH) are very well suited to byte and bit manipulations because their core instructions are defined directly in machine code and are accessible to the users.

Structure

Structured programming amounts to a clear, logical way of programming based on using only a small number of allowed programming structures (Figure 1). These structures are characterized by the fact that the flow through them is direct and there is only one entry point and one exit point. Any computing task can be reduced to a combination of these structures and represents the most logical progression through the task.

Structured programming is more a philosophy of programming than a purely language-specific feature. A structured program begins well before any code is written. The task is first defined and organized, the data requirements are established, and the basic structures for accomplishing the task are elaborated. Only then does programming begin with translation of the design into structured code. Some languages naturally reflect structure and, even though they may not be as efficient (fast or compact) as other languages, they are desirable because it is easier to design, upgrade, document, and maintain programs in structured languages. It is possible to write structured programs in a language that is not highly structured; however, this is a more difficult task than with a language that is inherently structured. Unstructured programs are characterized by the frequent appearance of the unconditional (unprofessional) GOTO instruction and a "hopscotch" flow through the program as shown in Figure 2.

Ease of Learning

A number of features can enhance the learning of a programming language. These include simple syntax, English and algebraic (or seemingly algebraic) commands, the availability of documentation, the availability of courses, the presence of people already well versed in the language, and the availability of facilities on which to practice programming. Assembly programming does not have a simple syntax and there are relatively few people who are well versed in its use. Many non-

Figure 1. Programming structures.

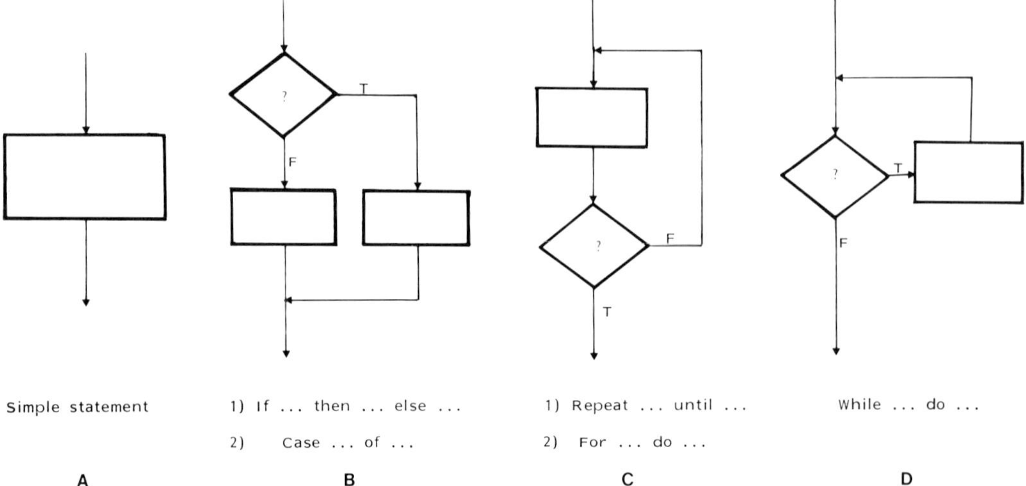

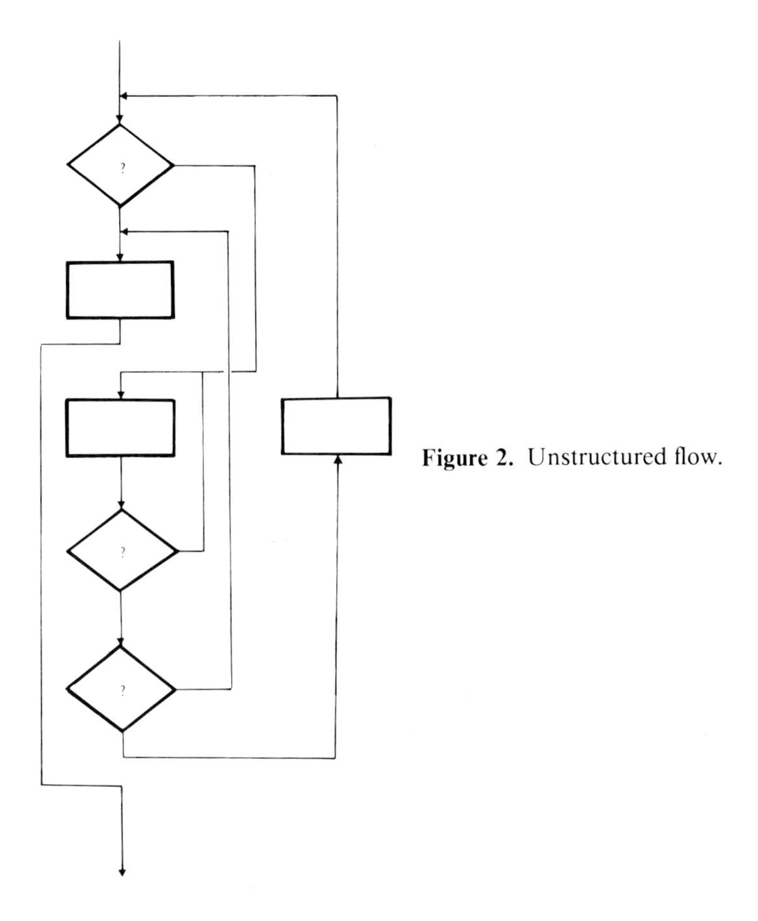

Figure 2. Unstructured flow.

professional programmers find it difficult to dissect the task they wish to perform into the minute detail required for assembly programming. Languages with a simple syntax (e.g., BASIC) are easily learned, as has been demonstrated by many self-taught users.

Portability

As use of the laboratory computer increases, the system will eventually become overloaded and the acquisiton of additional or more powerful facilities will be required. When time, money, and effort have been invested in software for the laboratory, it is very desirable to be able to transfer this software without change from one computer to another. Such *portability* of programs is again greatly dependent on both language and version.

ASSEMBLER programs are portable only if the CPUs of the two computers are the same. If not, programs have to be completely rewritten using the new CPUs specific instruction set. For high-level languages, portability is increased because the languages are not dependent on the CPU but on the definition (instructions and format requirements) of the language. In this respect, standardization of the various versions will greatly enhance portability. Unfortunately, some high-level languages were standardized after many different versions had already been in existence, so the standard was not strongly implemented and many version-specific differences exist.

Often, it is possible to find certain language implementations that are maintained for many different CPUs. Then it is possible to ensure portability by selecting a version that a software manufacturer supports for different CPUs.

Popularity

The popularity of a programming language depends on several factors. If a language is popular, then it is quite possible that a program to solve your problem has already been written. This may result in substantial savings of time and money. If a language is popular, it is also quite possible that laboratory personnel will have had some exposure to it and will feel more comfortable when attacking a programming task. It should also be easier to obtain external help for the design and development of programs if the programming language is in widespread use. A final consideration is the educational aspect. Most computer stores carry a large selection of texts and tutorials on popular languages and programs. These are an invaluable aid in an environment where one must constantly reeducate oneself.

Ease of Program Writing

The ease with which programs can be written in a particular language will depend on the simplicity of the syntax, the power of the commands, the amount of structure, the availability of assistance with that language, and the appropriateness of that language for that task. The ease with which programs can be written will have a substantial effect on the number and quality of programs developed in a laboratory. This factor is especially important in an environment in which new nonprogrammer personnel are required to use the computer. If it is easy to write and develop programs, many experiments or statistical analyses that would be too tedious to perform manually may be attempted and more thorough investigations will result.

Programs to perform extensive calculations or complex data manipulations are sometimes difficult to conceptualize, let alone program at the assembly level. The attention to every minute detail of every operation makes this a formidable task. It is best to tackle these problems with high-level languages. On the other hand, writing programs to maintain absolute control over peripheral devices in a high-level language can be difficult, especially if high speed is required. When high-speed acquisition and extensive calculations are required, interfacing a machine language data acquisition routine to a high-level program is an appropriate solution.

Ease of Program Development

Ease of program development differs from ease of program writing in that development encompasses the typing, compiling or assembling, testing, and debugging of programs, whereas the writing step is limited to the actual composition of the program. Certain programming languages make program development easier than others, and certain versions of the same programming language offer features and utilities that help in the development process.

In general, program development with interpreter languages is enhanced because it is possible to type, execute, edit, and debug programs without ever leaving the interpreter. The program can be conveniently halted to examine the value of variables at various locations in the program. Resumption from that point is allowed. Output (PRINT) statements can be included to verify the progress of the program. With compiler languages, any change to the program requires editing the source code using the text editor and then recompiling and relinking the object code. This process can consume large amounts of time. Program execution cannot be halted and reinitiated at will, and variables cannot be examined at run time unless a *debugger* (a special program to help locate errors) is available for that language. Assembler programs are often quite difficult to develop. A debugger is valuable to check program execution, but one must be well versed in its use to identify errors quickly. Large assembler programs are difficult to write and manage because of the extreme attention to detail required.

Within a given language, certain versions are better equipped for rapid program development with special editors, syntax checkers, variable checkers, reformatters, debuggers, and fast compilers.

Modularity

In a research laboratory, general purpose routines can be written and used in a variety of programs. In order to avoid having to redevelop these routines each time they are needed, it is important to select a language that accepts module development and is designed to use subroutines efficiently. Then a library of routines can be developed that will greatly reduce the task of program development.

Modularity is enhanced in languages that allow parameter passing to subroutines. *Parameter passing* means that subroutines can be written with dummy parameters that are replaced by the actual parameters when the subroutine is called. With other languages, subroutines use variables that are shared with the remainder of the program, and hence subroutines must be carefully used in order not to destroy the contents of global variables. For languages in which line numbers are used, it becomes necessary to check for line number overlapping every time a standard subroutine is added to a program.

Machine Code (Assembler) Interfacibility

The programming of intricate arithmetic operations in machine language should be avoided if possible due to the complexity of writing machine code. Yet where high data acquisition rates are required, assembly programming may be essential. To overcome this dilemma, data acquisition routines should be written in ASSEMBLER and interfaced to high-level language programs for calculations, report generation, and interaction with users. The interfacibility with machine code programs is a version-specific rather than a language-specific feature. Certain versions allow calling machine language programs without allowing parameters to be passed. Others allow the passing of limited parameters within the subroutine call or allow full access to variables. Various versions (usually from different manufacturers) of high-level languages may be different in a number of specific features, even though they are superficially identical. For example, in a recent edition of a popular computing journal it was possible to find six different versions of BASIC and five versions of PASCAL (*Anon,* 1981).

With many versions, the assembly routines must be written using an editor and assembler and then carefully interfaced to the high-level language program. This involves some risk, because the high-level language designers may not provide any information on the language's use of memory, so that the programmer may not be sure of the effects of using certain portions of memory. With other versions, the assembly program can be included *in-line*. This means that it is written directly in the source code of the high-level program. The compiler of these versions ensures assembly and also provides for access to program variables.

Character Handling

When attempting to write user-interactive or -friendly programs, programmers must do extensive character manipulations. For example, many terminals allow the computer to position the cursor or read from the cursor position. This is controlled by character codes. Some languages do not handle characters well and developing user-friendly programs in these languages is very difficult. Other languages are highly character oriented and elegant programs can be developed easily.

Programming Languages for the Laboratory

Having looked at some of the features of programming languages and how they operate in the laboratory, let us now examine the five most common laboratory programming languages for microcomputers: ASSEMBLER, BASIC, FORTRAN, PASCAL, and FORTH.

ASSEMBLER

ASSEMBLER-level programming offers the maximum possible speed of operation, compactness, and device interfacibility and control when done properly. Whenever possible it is best to develop subroutines that are called by a high-level language. The high-level language can be used for calculation and all the input/output to the operator, disks, graphics devices, and printers, whereas the assembler subroutine performs the data acquisition. This minimizes the complexity of the ASSEMBLER routine.

Data acquisition interval regularity can be well controlled at the ASSEMBLER level. However, the very best and easiest acquisition interval control is implemented using a hardware timer. Most CPU systems have a hardware timer, which can be set to output pulses at extremely precise intervals. Although ASSEMBLER-level acquisition software is often very fast, it must invariably wait for a *flag* to be set by an external event or a hardware interrupt. A hardware interrupt can be set by a device such as a timer by transmitting a signal to the CPU. This results in the CPU stopping its regular program execution and almost immediately being transferred to execute a specific subroutine called an *interrupt handling routine*. In either case the program often waits in a very short instruction loop and tests for this flag or waits for an interrupt. There may be an uncertainty of approximately two instructions. This can result in a time window (roughly 5 μsec on a

6502) during which a response to the flag or interrupt may be made. This time interval will be insignificant to some but will certainly be important for others. Under certain circumstances the uncertainty can be eliminated but, if not, the programmer must consider the consequences.

Although ASSEMBLER-level programs have a number of advantages, inherent structure is not one of them. Of all the programming languages, ASSEMBLER is the only one that allows unlimited flexibility. The undesirable consequence is that it is possible to write completely unstructured programs (known in the computer industry as "rat's nests"). These can be essentially impossible for anyone but the program author to unscramble. They may also cause the author considerable trouble during stages of development and maintenance. However, programs can be written in a structured way in any language. With a bit of self-discipline, ASSEMBLER programs can be structured as well.

ASSEMBLER-level programming is moderately difficult to learn, but once the basic skills are acquired, only the specific instruction set must be learned when changing CPUs. Although the instruction sets themselves usually consist of only around 200 instructions, programming can be very tedious and complex due to the necessity of keeping track of *all* of the required functions of the computer. For a human, this would be equivalent to taking conscious control of one's body down to the level of controlling stomach peristalsis, pupil size, and hair growth. Obviously there is a lot to do, and the only reason to go down to this level is if one needs that sort of control. Programming at this level requires a good working knowledge of low-level binary mathematics as well as familiarity with common Boolean algebra functions such as ANDS, ORS, and XOR.

ASSEMBLER-level programs are only portable to computers with the same CPU. If they use input/output devices, then they may also need to be modified for the hardware on the new system. Some translation programs do exist, but they are generally inefficient. This is similar to a human language translation problem. A word-for-word translation often does not convey the true meaning of a text. In the same way, instruction-for-instruction translation of programs may not yield a program that operates exactly the same way as the original, or even necessarily correctly.

Assembly programming was moderately popular in the recent past and consequently a great deal of literature is available for many of the 8-bit (older) systems. It is considerably less popular now and so proportionally fewer texts are written. However, there is at least one text on assembly programming for each of the major CPUs on the market.

Program writing is certainly more difficult in ASSEMBLER than in any other language for the reasons stated above. Program development is also relatively slow at the assembler level. If the program is bulky and is being stored and retrieved from an audio cassette, the entire procedure becomes very cumbersome. Larger programs should either be developed on a disk-based system or in a system with sufficient memory to allow the editor, assembler, source, and object code to exist simultaneously. Such a system is only useful if the assembler can read the code in the editor's memory buffer area.

Because of its flexibility, ASSEMBLER-level programming provides the ultimate in character handling ability. For this reason, and others above, many database management systems are written directly in ASSEMBLER.

BASIC

The popularity and growth of microcomputers in the early days was to a great extent due to their low cost and the ease of learning and using BASIC. BASIC (Beginner's All-purpose Symbolic Instruction Code) is available as an interpreter or as a compiler. It is the interpreter form that is responsible for the widespread use of BASIC with microcomputers. The interpreter can be made compact enough to fit easily into ROM (read only memory), which offers an economical way for computer manufacturers to provide high-level language capabilities on their small computers. The extensive use of BASIC led to the development of many versions, some incorporating special graphics and I/O commands, and eventually to the compiler versions available today.

Table 2 is a comparison of data acquisition rates with various languages. All the data were acquired on the same Z-80 CPU computer running at 4 MHz. The fastest Z-80 systems now run 6 MHz, so the reader can assume that these times could be 50% faster. Execution times would be very similar on 8080 or 8085 systems running at high CPU clock rates. The data clearly indicate that interpreter BASIC is very slow at data acquisition tasks. As a gross generality, the same statement can be made of all interpreter BASIC functions. Clearly, compiler BASIC is a factor of 100 faster than interpreter BASIC and generally compares favorably with other compiler languages. The reader should be aware that relative placement within the compiler languages may well be version-specific. Many of these languages have special features or allow acquisition directly from ASSEMBLER-level subroutines.

Table 2. Acquisition Characteristics of High-Level Programming Languages

Language	Acquisition[a] frequency (Hz)	Size (bytes)	Acquisition interval
ASSEMBLER	51,000	20	regular
FORTRAN-80[b] integer[c]	4,730	6595	regular
MBASIC[b] integer	220	177	regular
MBASIC[b] real[c]	152	175	irregular
BASIC Compiler[b] integer	15,400	6144	regular
BASIC Compiler[b] real	368	6656	irregular
PASCAL MT+[d] integer	3,500	4862	regular

Information extracted from Sing et al. (1983)

[a]All data taken on a Z-80 system with a 4-MHz clock.
[b]BASIC and FORTRAN-80 versions were from Microsoft, 10800 N.W. Eighth, Suite 819, Bellevue, WA 98004.
[c]Integer and real refer to the type of variable used as an index in the acquisition loop.
[d]PASCAL MT+ from Digital Research, P.O. Box 579, Pacific Grove, CA 93950.

Interpreter BASIC exhibits a certain interval irregularity that may be important to some users. The timing uncertainty is usually not more than a few percent of the timing interval. For those wishing to search for a peak or record an average, this should not pose a problem. However, those interested in any sort of waveform analysis may find this precision unsatisfactory. The reader is referred to Sing et al. (1983) on this topic. Compiler BASIC (using integers) was very precise in our evaluation. An examination of the assembler code indicated that the compiler had made intelligent use of the register structure of the Z-80 CPU. The 6502 and certain other processors are not well suited to 16-bit or larger counting in the CPU and consequently will have irregular intervals. The reader should test a BASIC version by observing the hand-shake I/O signals on an oscilloscope while varying the number of interactions in a FOR-NEXT loop.

Interpreter BASIC initially appears to be compact, because only the characters of the program are saved in memory. This is a bit deceptive because the entire BASIC interpreter is also occupying "memory" space, even though it is not necessarily occupying R/W memory on a ROM-based system. ROM interpreter BASICs usually occupy 8 kilobytes (KB) or more of memory. It is quite possible to have a BASIC data acquisition program with a smoothing function and peak search capability in 2 KB of R/W memory on such a system. This explains why the 20-KB R/W memory Commodore VIC 20 and 4-KB R/W memory Rockwell AIM-65 are popular and useful in laboratories. Interpreter programs grow linearly with length.

Compiler BASIC has the same constraints as other compiler languages. For example, additional bulky and inessential subroutines will be linked in.

Most microcomputer BASICs have commands that allow reading and writing directly to memory or I/O devices. These instructions are PEEK, POKE, and also for some computers IN and OUT. The PEEK command reads the value at a user-specified memory address and returns the value to the BASIC program. For example,

```
X = PEEK(40960)
```

would set the variable X equal to the value at address 40960. This could be a reading from an analog-to-digital converter. The POKE command performs the opposite function and allows the user to place a value in a memory address. This could be used to send a value to a digital-to-analog converter. They are not as fast as ASSEMBLER level I/O but generally give the programmer the same level of control. Usually there is also a USR (user) command, which allows the BASIC program to jump directly to an ASSEMBLER-level program. This program could be written by the user or might be a routine provided by a device purchased for the system. Thus, it is possible to get assembler-level performance from a program that is written *mostly* in BASIC.

BASIC is a minimally structured language. Like all languages, programs can (and should) be written in a structured manner. BASIC is certainly the easiest language of those discussed in this chapter. We believe that most scientists and technicians can teach themselves the rudiments of BASIC if provided with a satisfactory teaching manual and a good BASIC on a microcomputer.

Because the standard for BASIC was adopted after many versions existed, BASIC is only moderately portable. Many versions exist and some of these differ greatly from others in the way information is stored on disks, in graphics commands, and in mathematical capabilities, particularly matrix manipulations. Compiler versions also have different requirements than those of interpreter versions. The user will observe distinct changes in capabilities when moving from the rudimentary ROM BASICs available on some small systems up to the sophisticated BASICs available on high-powered minis (e.g., Hewlett-Packard) and mainframe computers. Some of the better BASICs provide translation chapters in their instruction manuals to aid their users in transporting BASIC programs.

The very simple syntax of BASIC usually makes program development straightforward. BASIC has a simple line entry mode that allows the entry of one line at a time. In simple systems, when one makes an error it is necessary to retype the entire line. Many advanced BASICs now have a line editor that allows the modification of a given line of text. This is convenient when one makes mistakes on large lines, and means that short to moderate length BASIC programs can be entered directly in BASIC. Longer BASIC programs, like most other language programs, are best developed using a sophisticated text editor that can provide all the string search and replacement functions that are so convenient during program writing.

The combination of interpreter and compiler BASIC makes both easy program development and rapid execution possible with a single programming language. The interpreter can be used for the development and debugging stage, and when the program is operating correctly it can be compiled for rapid execution.

In BASIC, subroutines are subsections of the main program and usually must share variables with the main program. The development of a library of subroutines in BASIC is hampered by the care that is necessary to prevent unwanted destruction of data due to overlapping variable use and to prevent overlapping line numbers. Furthermore, subroutines must be stored in their source forms and the programmer is responsible for merging the required subroutines with the main program. With some compiler BASICs, it is possible to link subroutines written in other languages; however it is impossible to generate a true library of routines written in BASIC for the reasons stated above.

In summary, BASIC is easy to use but is comparatively slow in running unless compiled. Although it has developed significantly in the past 10 years, it lacks the power of PASCAL and does not require a highly structured approach.

PASCAL

PASCAL is a compiler language developed in 1968 in Zurich by Niklaus Wirth. The language was developed in the spirit of the ALGOL programming language and was designed to "teach programming as a systematic discipline based on certain fundamental concepts clearly and naturally reflected by the language" (Jensen and Wirth, 1974). Because of this, PASCAL is a *highly structured* programming language. It allows the definition and use of complex data types (arrays and records) and supports full subroutines with value and variable parameter passing. Over the past few years, PASCAL has been making large leaps in popularity. In many universities and schools, PASCAL has replaced FORTRAN as the language of choice for introductory courses in computer science. The language has established firm roots in the microcomputer environment and many versions of PASCAL are available for a wide variety of computers. PASCAL is the professional choice for a compiler language to use on microcomputers. A great deal of popular literature is also available for PASCAL.

The speed of PASCAL is roughly comparable to other compiler languages for calculations. Data acquisition rates for some specific versions are compared in Table 2. These speeds will be adequate for many applications. Acquisition intervals are not regular, unless integer arithmetic is used *in the* PASCAL MT/+ *version.* The reader is cautioned that testing should always be done with the version being considered for purchase. The alternative, of course, is to develop an assembler-level subroutine or to use hardware timing. Program sizes tend to be somewhat larger in PASCAL due to the inclusion of the more powerful routines when the program is compiled.

One of the limitations of the original PASCAL is its lack of powerful input/output capabilities. This has resulted in numerous extensions of PASCAL in different versions, each version having been tailored to its specific environment. This has caused certain loss of standardization and portability between versions. Microcomputer versions of PASCAL usually have good machine code interfacibility. ASSEMBLER routines can be developed as subroutines and linked to the program. Some versions of PASCAL allow in-line ASSEMBLER code insertion with direct references to variable storage locations. This last feature facilitates the interface between the PASCAL and ASSEMBLER-level programs. Standard PASCAL itself is not inherently well suited to device control, because it lacks statements comparable to the BASIC PEEK and POKE functions. The ease of ASSEMBLER-level program linkage compensates to a great extent.

PASCAL is popular as a teaching language primarily because of its structured nature. PASCAL is more difficult to learn than BASIC, but its more disciplined programming style results in superior programming skills. Learning the syntax of PASCAL takes roughly as long as learning the syntax of FORTRAN. However, the repeated use of PASCAL develops good programming methodology. Repeated exposure to either FORTRAN or BASIC does not develop or encourage good programming style.

Although forcing the programmer to operate within certain structures, the logical programming style of PASCAL eventually results in easier program writing. The logic requirements will eliminate many errors that could have been made with a less structured programming language. Because it is a compiler language, development is tedious due to the need to repetitively edit, link, and compile during testing. Development is significantly enhanced in certain versions by using special development utilities. These will allow text editing, syntax (spelling) checking, and reformatting prior to compilation and linking.

PASCAL has undergone few revisions since its drafting in 1968. The final report on the language and its definition occurred in 1973. Since then the *Pascal User Manual and Report* by Jensen and Wirth (1974) has been used as the standard for the various implementations of the language. Provided that programs are written using only the standard language specifications, programs are virtually assured of portability. Each version offers extensions on the language and microcomputer versions are specially tailored to their environment.

In PASCAL, subroutines are complete entities with local variables and constants and the ability to pass both value and variable parameters. Because of this, modularity is ensured and subroutine libraries can easily be developed. This enhances program development. Machine language interfacibility is generally good with microcomputer versions because of the subroutine structure adopted by the language. Assembler routines can usually be developed separately and then included in the library or linked directly to the program. Certain versions allow in-line assembly coding with direct references to variable and storage locations.

PASCAL has a built-in character data type so character handling is greatly simplified. User-friendly, highly interactive programs are easily written in PASCAL. Complex data structures can be defined and manipulated, and so PASCAL is the language of choice for complex data manipulations and interaction with data base management systems.

PASCAL is a powerful, highly structured compiler language. It requires more expertise than BASIC but, once learned, allows the programmer to easily develop more sophisticated programs. PASCAL programs are normally much more structured and consequently are generally easier to maintain than programs developed in other languages.

FORTRAN

FORTRAN was the first true high-level programming language and was designed specifically for scientific calculations. This is reflected in the name of the language, FORmula TRANslator. It allowed scientists to formulate their equations algebraically without resorting to ASSEMBLER-level programming. It has not received extensive development for the microcomputer environment due to its weak character-handling capabilities. This limitation makes it difficult to generate interactive programs. It is particularly for this reason that PASCAL has received much more attention and development for microcomputers.

FORTRAN is especially well suited to calculations and formatted column output of data. It is very popular with scientists and engineers who do extensive calculations. Although PASCAL has powerful calculation capabilities, FORTRAN is superior in this aspect.

FORTRAN use is very widespread due to its long lifespan. It has been used as a teaching language, because it was essentially the only suitable language for scientific applications. As a result, many scientists have been exposed to FORTRAN and numerous application programs exist. There are several revisions of the FORTRAN standard, so portability between standards may be compromised. Modern versions of FORTRAN (FORTRAN 77) include a number of programming structure enhancements and character-handling capability. However, FORTRAN still does not approach the sophistication of PASCAL in these respects.

FORTRAN syntax is more complex than that of BASIC. The language is taught at most teaching institutions, and there are many "resource" persons available. Writing of simple programs in FORTRAN is quite straightforward, but the writing of programs requiring extensive data manipulation is hampered by the lack of powerful programming and data structures. Interactive program writing is severely constrained by the lack of effective character input/output. The development stage is characterized by the tedious edit, compile, and link sequence.

FORTRAN is a powerful language for computation and is well suited for tabular output of data. It lacks the inherent structure and text-handling capabilities of PASCAL, but enjoys the speed advantage of compiler languages.

FORTH

FORTH is an incremental compiling language. It uses post-fix syntax as well as stacks. The reader may be familiar with both of these terms from Hewlett-Packard calculators. This format can be somewhat confusing, though it may be very efficient. For example, the equation

$$5*((2 + 3)*(7 - \tfrac{8}{4}))$$

would be

$$5\ 2\ 3 + 7\ 8\ 4 - * *$$

in FORTH. FORTH was originally designed specifically for instrument control and has a number of features that facilitate this function. FORTH allows direct manipulation of memory and input/output devices and permits the programmer to choose the number base to be used. FORTH also allows in-line ASSEMBLER code for sections of program that require especially high speed. FORTH operates very well at the byte level and so is well suited for character handling.

FORTH programs are generally compact and fast. This results from the use of symbols to represent commands. The internal structure of FORTH is very efficient and revolves around a compact core of routines. Programs are generally made up of combinations of these core functions. FORTH allows users to define their own extensions to the language. In this way, users can develop special versions of the language for their applications. This customization capability does not compromise the portability since each of the routines is composed of the central core functions. Essentially, users can have their own custom programming language.

Unfortunately, the flexibility of FORTH is provided at the expense of structure. The post-fix notations and use of the stack can lead to programs that are extremely difficult to read. This seriously endangers the maintenance of FORTH programs. The language is generally difficult to master and not well suited for arithmetic. Many versions of FORTH can do nothing more than add, subtract, multiply, and divide integers. Though the language boasts some very avid followers, it is not very popular and there is little literature available.

Buying a Computer

Buying a computer is really buying a machine that will run the software that you need. All else is secondary. The acquisition choice has been discussed in the literature (Salin, 1982a–d) and starts with an evaluation of the data acquisition requirements. These break into roughly two parts: the amount of data and the rate of acquisition. From Table 2 one can determine what language(s) meet or exceed the requirements for speed. Depending on the language, one can then determine the memory space requirements. This depends on whether the data are stored directly in memory or in the language itself as integers or real numbers.

The next question is one of processing. Is it necessary to process in real time? If so, what are the requirements? There is an incredible difference between a simple peak search and a Fourier transform. If the processing can be delayed and sufficient memory space exists to store the data, then the information can be evaluated after the data are acquired and without storage. This is called *resident delayed processing.* BASIC is often very good for resident delayed processing.

It may also be possible to store the information in a mass storage device for later processing. This is called *delayed processing,* and for this one seldom uses anything but a high-level language. It simply does not make sense for a scientist to develop processing programs in ASSEMBLER unless time is critical.

Conclusions

Our conclusions on languages can be summarized in a few words:

1. Use ASSEMBLER only when necessary.
2. Use BASIC, in one form or another, whenever convenient.
3. Use PASCAL for *major* programming efforts.

Users should temper these to their own environment. In our laboratory we employ a small computer network (Sing and Salin, 1984). We use AIM-65 computers directly for instrument control and data acquisition. It is possible for us to program these in machine code, ASSEMBLER, FORTH, PASCAL, or BASIC. Whenever possible we use BASIC. If speed or finesse are essential, we employ ASSEMBLER-level programming. Whenever we wish to save information or do complex processing, we send the information from the AIM-65 to our "laboratory" computer via an RS-232C standard interface. This is done with a BASIC program. The information is sorted as a standard text file in the laboratory computer. Further processing and graphics are done on this machine, invariably in BASIC. We have seldom found that the speed advantage of compiled BASIC is necessary for our programs, even though we often run in a multiuser environment under the MP/M operating system.

We have developed two programs in PASCAL. One is a modeling program and the other is a data base management package. The modeling program is complex and may be used for a long time by many users. Consequently, it deserved PASCAL. Our initial evaluation of the data base management system was carried out in BASIC, but the limitations of BASIC showed up quite clearly in the evaluation and BASIC was never considered for the final work.

We experimented with FORTH, particularly because it was available in a well developed ROM package for the AIM-65. We concluded that the code was difficult to read, and we could not get the acquisition speeds that we needed. We believe that FORTH would be very valuable in an environment in which complex instrumentation is used. FORTH is not a popular language and is quite different from the familiar BASIC, FORTRAN, and PASCAL due to its intensive use of stacks. Sometimes it seems downright bizarre. For these reasons we believe that it is best used in a relatively stable environment rather than in a laboratory that sees a great turnover of personnel. Also, FORTH coding should be closely supervised and documented to ensure efficient transfer of the code from hand to hand. A number of instrument companies have begun using FORTH because of its excellent control capabilities as well as its portability.

FORTRAN probably belongs only in laboratories that already have an extensive commitment to running software in that language. PASCAL is better suited to long-term maintenance, and BASIC is easier to use. Persons familiar with FORTRAN can learn BASIC independently in less than 2 days. Pascal can also be self-taught by anyone that has endured the rigors of FORTRAN.

References

Anon (1981) Byte 6(4):141 (April).

Jensen, K., and N. Wirth (1974) Pascal User Manual and Report, 2nd Ed. New York: Springer-Verlag.

Malmstadt, H.V., C.G. Enke, and S.R. Crouch. (1981) Electronics Instrumentation for Scientists. Menlo Park, Calif.: Benjamin/Cummings Publishing Co., pp. 406–407.

Salin, E.D. (1982a) Laboratory computer acquisition considerations, Part One: Software considerations. Am. Lab. 2: 80–86.

Salin, E.D. (1982b) Laboratory computer acquisition considerations, Part Two: Language selection. Am. Lab. 3: 35–38.

Salin, E.D. (1982c) Laboratory computer acquisition considerations, Part Three: Components and support hardware. Am. Lab. 4: 105–112.

Salin, E.D. (1982d) Laboratory computer acquisition considerations, Part Four: Computer system peripherals. Am. Lab 5: 129–134.

Sing, R.L.A., S.W. McGeorge, and E.D. Salin (1983) Laboratory data-acquisition capabilities of microcomputer high-level languages. Talanta 30: 805–809.

Sing, R.L.A., and E.D. Salin A versatile low cost laboratory computer network. Talanta (submitted).

3

Selection of Hardware and Software for Laboratory Microcomputers

S. Mark Poler, Steven Akeson, and Dale G. Flaming

Introduction

The objective of this chapter is to describe hardware and software components suitable for data acquisition in cell and neurobiology. The chapter is based upon experience we acquired while developing three different IEEE-696 (S-100) computer systems for laboratory analysis. [The IEEE (Institute of Electrical and Electronic Engineers) is a professional organization that promulgates standards for electrical and electronic products; 696 is the number of the working committee that completed the standards for S-100 computers.] The first system discussed is used to record extra- and intracellular potentials from retinal cells. It can also be used to control experimental stimuli (Flaming, 1982). Its components are listed in Table 1. The second system acquires electron paramagnetic resonance (EPR) data from an IBM EPR device and displays and statistically analyzes the data. It also acquires data and provides nonlinear, programmable feedback control in muscle contractility experiments. The third computer system functions as a multichannel chart recorder and data analysis system in electrolyte transport experiments on the renal proximal tubule (Figure 1). Its components are listed in Table 2.

These systems were assembled during 1980–1983 and their components reflect the state of hardware evolution at that time. The total cost for each of these systems ranged from $4,000 to $13,000, excluding the time invested in software development. Since hardware prices are decreasing while function per card is increasing, comparable systems in 1985 should cost 10%–20% less.

While developing these systems, we gained considerable insight into the advantages and disadvantages of selecting various computer components. In this chapter we shall introduce the reader to the components of a microcomputer and how to assemble them into a customized laboratory system. We shall also cite reasons for our selection of this computer architecture for biological data acquisition. General principles will be emphasized because the evolution of hardware components is rapid. The state-of-the-art component selection available to the reader will undoubtedly differ somewhat from the components described in this text.

We have included a glossary of terms in Appendix 1 to help guide the reader through the maze of computer terms in current use. Terms not defined in the text will be found there. Extensive reference and vendor lists are also provided in the appendixes. These are intended not only to document the text, but also to give

Table 1. $4,000 System Components

Integrand S-100 box
Integrand disk drive box
2 Shugart floppy disk drives
CompuPro Z-80 board
CompuPro 64 KB memory board
CompuPro Disk I floppy disk controller board
Morrow Switchboard I/O board
California Digital 12-bit A-D board
Z-19 Terminal System Components

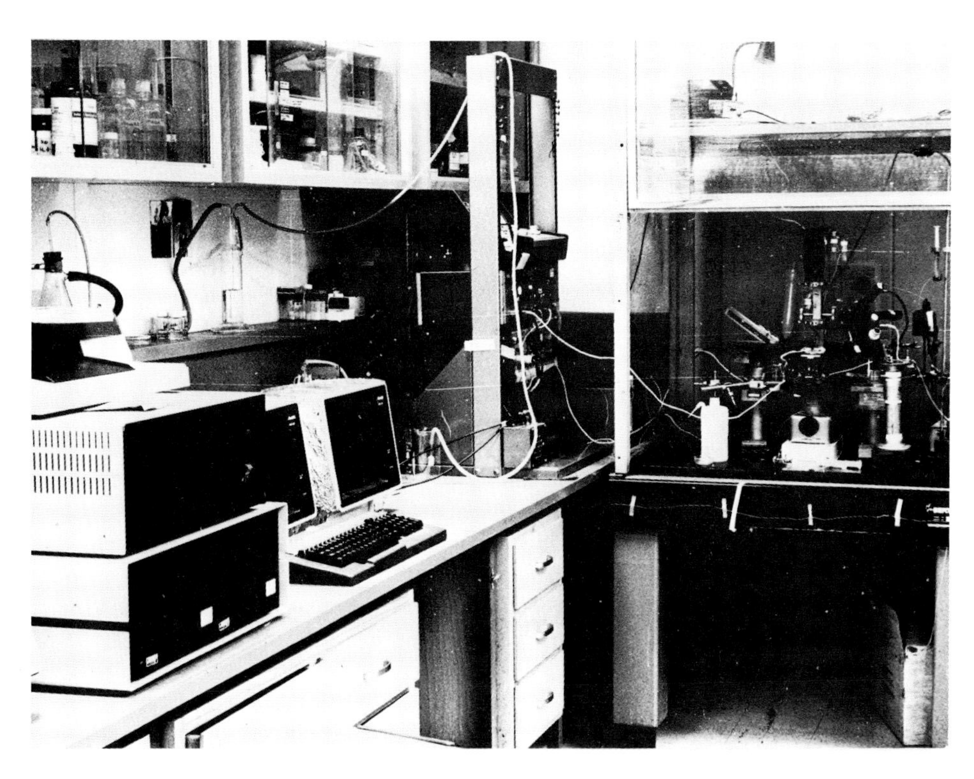

Figure 1. S-100 computer for data acquisition and analysis of experiments in renal tubule epitheliology. The computer enclosure sits on top of the disk drive enclosure. To the right of the computer are two monitors, one for the command console display (driven by the VIO-X), and the other to display graphic data from the Microangelo. The keyboard can be moved freely to a convenient location. The analog instruments monitored by the computer are shown on the right.

Table 2. $13,000 SYSTEM COMPONENTS

CompuPro: 8086/8087 CPU (CSC, 10 MHz)
128KB RAM 21 (CSC, 12 MHz)
System Support 1 (1 RS-232C port, clock-calendar, FPP, interrupt hardware, BOOT ROM)
Interfacer 4 (3 RS-232 serial ports, 1 parallel port)
Enclosure, power supply, and motherboard (12 MHz)
Floppy drive enclosure and power supply, 2 Qume DT-8 8-inch drives
Fulcrum: VIO-X2, terminal board, equivalent to 80,000 baud serial terminal
Scion: Microangelo, 512 × 512 pixel monochrome graphics display
Tecmar: TM-AD212, 12-bit, 2s complement analog to digital convertor, with TM-PGH programmable gain option, 40 KHz multiplexed conversion
IDS: Prism 132 printer, dot matrix
Hewlett-Packard: 7470A plotter, 8½-inch wide bed, 2–4 color.
Amdek: Two 13-inch green phosphor monitors

the interested reader access to a broad spectrum of resources. In the vendors list (Appendix 2), an asterisk indicates those vendors whose products were actually used in our computer systems. In addition, we call your attention to two very useful product directories that have been published (Libes and Libes, 1983b; Libes, Manno and Terry, 1983). We believe this chapter will be useful both to those who wish to develop an S-100 bus component computer system and also to those who desire more general information about computer hardware and software.

Buses and Other Computer Architecture

Overview of the IEEE-696 ("S-100") Bus

The IEEE-696 is the result of an intensive volunteer committee effort by hobbyists, engineers, and manufacturers to establish a standard bus structure that would allow various computer components to operate together. The standard is an outgrowth of the MITS Altair computer introduced in 1974, which was the first widely available hobbyist microcomputer. It used printed circuit boards and edge connectors having 100 contacts because these were readily available as surplus items at that time (Figure 2). Shortly, other manufacturers began to supply the Altair with component boards that provided additional functions for the computer. Because the timing and pinouts of the original Altair bus were used differently by different manufacturers, conflicts developed. Additional complications arose with the expansion of memory capacity in 8-bit systems and the advent of 16-bit microprocessors. It became critically important to standardize the S-100 bus, and the IEEE-696 standard evolved to meet this need.

The S-100 bus was originally intended only for 8-bit microprocessors (starting with the Intel 8080), but with the IEEE-696 standard it has evolved into a bus

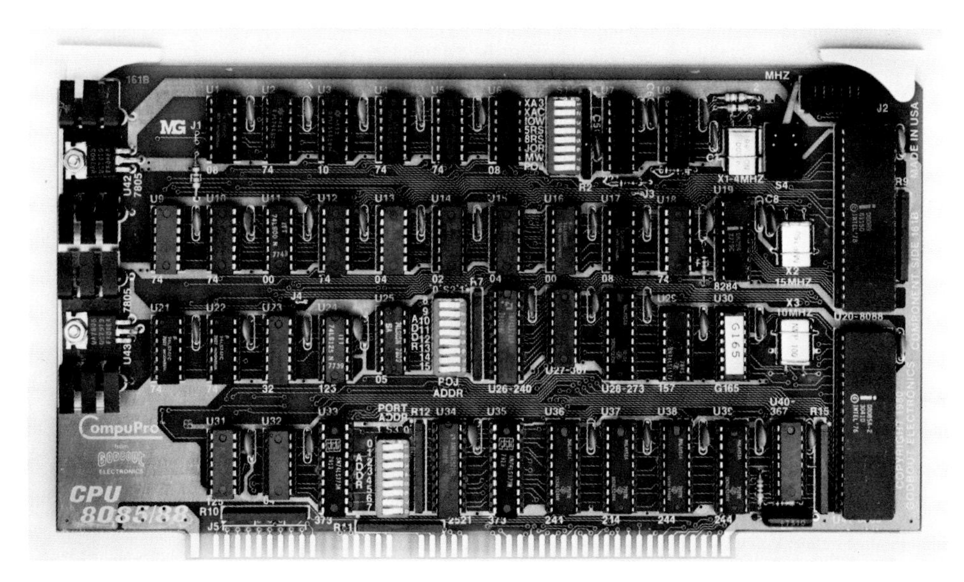

Figure 2. The CompuPro Dual Processor CPU board is an example of an S-100 computer board. [Reproduced by permission of CompuPro, Inc.]

that can accommodate either 8- or 16-bit data transfers and up to 16 megabytes (MB) of memory. An elegant interrupt structure has been specified. Provisions have also been made to accommodate high operational speeds achieved by recent microprocessors and memory and auxilliary devices. The IEEE-696 is the only bus standard for microcomputers that virtually ensures that hardware components from different manufacturers will operate as intended when integrated into a single computer. Operating system modifications required to integrate hardware components into a functionally unified computer system will be discussed below.

The essential details of the IEEE-696 standard are contained in an eminently readable though technical book by Libes and Garetz (1981; see also Libes and Libes, 1983; Garetz, 1983). The authors have been seriously involved in establishing the standard that was ratified in 1983. The book by Libes and Garetz differs from the standard in a few minor details because it was published before the standard was finalized (Libes and Libes, 1983). Of several books that we have read on this subject, it is definitely the most complete and useful. The IEEE-696 standard itself can be obtained from the IEEE (Appendix 2).

Other Possible Computer Configurations

There are a number of other buses available for microcomputers. They include the STD, Multibus, Multibus II, VAX, LSI-11, VME, and IBM PC or XT buses. Some of these buses will accommodate 32-bit data (Libes 1983); all are well-established commercially and are standardized to various degrees. Each has its own strengths and weaknesses. The IBM PC and its descendants deserve special mention because the PC has become an enormously popular microcomputer.

The IBM PC and XT do not match the best attainable performance of IEEE-696 computers because the IBM is a single-board computer running at 5 MHz. Expansion slots are provided but the data path is limited to 8 bits. Scattered published reports suggest that the IBM PC does not match the performance of less expensive 8-bit computers in some common functions. However, the IBM personal computers are receiving very widespread support and many manufacturers now sell second source components for these computers. A wide selection of special function boards, including analog/digital converters (Tecmar), graphics cards, and multipurpose boards are now available for the IBM. If less performance will suffice for your laboratory applications, the IBM may be a lower cost alternative to IEEE-696 computers.

Care must be exercised to ensure hardware and software compatibility when combining "IBM-compatible" products from several manufacturers, especially in the area of graphics. Good support for high-resolution color graphics exists for some of these machines, but unfortunately the graphics interface is the least standard part of the IBM computers. Graphic software incompatibility is thus a common experience among IBM PC-"compatible" computers. [Descriptions of an IBM PC used in neurophysiology have been published (Aitken, 1984; Chapter 20).] Single-board computers, such as the Apple and Kaypro, may be good laboratory acquisition computers for some applications. Special purpose peripherals and analog/digital converters must be added (Place and Bailey, 1983). However, these configurations are rather slow in comparison to our S-100 system because the speed is limited to the transmission rate of the parallel or serial ports used for the interface (Long, 1983; Newrock, 1983). IEEE-488 parallel interfaces are much faster than RS-232 serial interfaces. When applicable, single-board computers often prove to be the least expensive choice for a computer committed to a single task. The Fujitsu Micro 16S is also a single-board computer with a proprietary expansion bus. High-resolution graphics and four 8-bit ADCs are included in the base price, which make it an attractive single unit package for some slow, low-resolution (Table 3) laboratory applications. An excellent compromise between the virtues of single-board and S-100 computers is the Heath/Zenith Z-100, which is an "IBM-compatible" computer with an IEEE-696 expansion bus (Pournelle, 1983a, b). IBM also markets the CS-9000 as a laboratory computer (Clune, 1984).

Customizing a Laboratory Computer

Any mini- or microcomputer used in the laboratory will have "bugs" that must be worked out, and IEEE-696 computers are no exception. Particular difficulties are encountered when constructing a hybrid system that consists of products from several manufacturers. These problems will be minimized if all of the components truly *conform* to the IEEE-696 standard. This is very important since no single supplier makes all the components required for an IEEE-696 laboratory computer system. Because the IEEE-696 standard is so rigorous, you stand a better chance of developing a compatible, working system with this standard than with any other component microcomputer system on the market. Beware of manufacturers who claim compatibility with the standard—this is *not* equivalent to compliance.

The easiest way to create a laboratory computer begins with selection of an integrated IEEE-696 computer system (Bender, 1984a; Darwin, 1984; Hardy and Jackson, 1983; Mau, 1983; Ratoff, 1984). "System A" from CompuPro is such a basic computer (Figure 3) and comes with the operating system preintegrated and software for word processing, electronic spreadsheet calculator, and database management. Appendix 2 lists a few reputable manufacturers of integrated basic IEEE-696 computers. The total cost of a preintegrated system with bundled software is typically much less than the price of the component parts. Additional IEEE-696 compliant ADC, DAC, and graphics boards, and necessary peripherals are then chosen. Thus CPU, disk controller and disk drives, and memory and system support functions are already integrated by the manufacturer. Table 2 illustrates how we used this approach to create our second laboratory computer. The addition of the special function boards requires someone who can integrate the products into one system using the manuals of the basic computer system and of the added components. This ensures selection of appropriate port addresses and setup options for the additional boards. This process is tedious and often

Table 3. Decimal Equivalents of Memory Address Ranges or Digital Resolution

Number of bits	Maximum equivalent decimal integer value	Abbreviation	Designation
8	256 bytes		
10	1024 bytes	KB	kilobyte
16	65536	64KB	
20	1,048,576	1MB	megabyte
24	16,777,216	16MB	
30	approx. 1 billion	1GB	gigabyte

requires a bit of experimentation before a correct configuration is created. Once functional switch settings are set up, the computer is ready for programming. The application program that accesses the additional boards will need to know correct port addresses for various devices, but the operating system itself usually does not need to know how to address these devices.

Although we considered many products, we decided ultimately that an IEEE-696 S-100 bus system was best for our laboratory data acquisition needs. Our reasons for selecting IEEE-696 hardware were as follows:

1. largest capacity/number of functions per card,
2. competitive component prices,
3. concurrent 8- and 16-bit address range and data paths,
4. very large memory capacity,
5. early access to state-of-the-art hardware components,
6. great flexibility in choice of hardware and software, and
7. the high rate of compliance by manufacturers to the IEEE-696 standard.

We discuss some of these strengths below when describing the various components of a bus-based computer system. Understanding the possible component variations and choices will permit you to make sophisticated choices among available alternatives.

Central Processing Units (CPUs)

The CPU (central processing unit) is the primary executor of instructions contained in the operating system and user programs. It performs most of the control, logic, and arithmetic operations of the computer (see Chapter 1). The most common CPUs for microcomputers have either an 8- or 16-bit architecture. This refers to the number of parallel lines available on the data and address buses of the microprocessor: 8-bit CPUs can transfer 8 bits of information in parallel; 16-bit CPUs can transfer 16 bits. Decimal-equivalent magnitudes and address ranges accessible per bit size are summarized in Table 3.

The best known 8-bit processors are the 6502 (found in Apple II computers, among others) and the Zilog Z80 (found in a host of 8-bit microcomputers). The Intel 8086 family of processors are by far the most utilized 16-bit processors, largely because IBM uses the Intel 8088 in the IBM PC. The Motorola MC68000 is another very popular 16-bit chip.

The distinctions among 8-, 16-, and 32-bit microprocessors can be quite unclear. Some microprocessors are considered hybrids because the width of their internal and external datapaths differ (Wong, 1983). Some 16-bit CPUs (the Intel 8088, for example) are hybrid $\frac{8}{16}$ bit machines with a data bus that is only 8 bits wide. Some 16-bit processors have full 16-bit-wide data buses and may have 32-bit internal registers and address range (the Motorola MC 68000 or National Semiconductor 16032, for example).

8 Versus 16 Bits

Speed is the most often touted advantage of the 16-bit microprocessors now available. However, 16-bit processors are not always faster than 8-bit machines. There have been numerous reports in the microcomputer literature of 8-bit software

A

Figure 3. (**A**) A CompuPro IEEE-696 computer enclosure without its cover. All or a few slots can be populated by boards complying with the IEEE-696 standard. The constant-voltage transformer and filtering capacitors occupy the right portion of the enclosure. (**B**) A detailed view of the CompuPro motherboard, showing board edge connectors power connectors, and bus conductor traces. [Reproduced by permission of CompuPro, Inc.]

outperforming its own 16-bit counterpart (Duncan, 1983, 1984; Pournelle, 1983a, b). In part, this reflects the relative immaturity of 16-bit software. In addition, word- and data-processing applications have a natural 8-bit data structure because of the 7-bit word size of the ASCII alphanumeric code. There are now in production 8-bit Z-80 and Z-800 processors with very fast 12–15-MHz clock rates that take full advantage of the large base of mature 8-bit software. These 8-bit processors are faster than many 16-bit processors because of their faster clock rates. Both 8-bit CPUs and 8-bit software are less expensive than their 16-bit equivalents.

There are, however, several major advantages to 16-bit processors. First, they vastly increase the range of memory that can be directly addressed without bank switching (Table 3). They also usually have higher clock rates and thus run faster in operations that require transfer of data in 16-bit units. The limited software available for 16-bit machines is nevertheless a real limiting factor at present. Operating systems and programming languages as well as applications programs are not as readily available for 16-bit machines. As an example, the 8086 and 8088 16-bit microprocessors from Intel have been manufactured since 1977. Yet FORTH and C were the only high-level programming languages available for these CPUs until 1981. PASCAL, FORTRAN, and BASIC have only recently become readily available for the Intel 16-bit line.

B

Dual Processing

The IEEE-696 standard (as well as some other bus structures) makes provision for two classes of bus master: *permanent* and *temporary*. This system architecture allows distributed processing to occur in parallel using two CPUs or a CPU and other "smart" devices that do not conflict in use of the bus. Examples of temporary masters include disk controllers, buffered input/output controllers (which pass data directly to memory locations from peripheral devices), and secondary CPUs (which can carry on auxiliary processes in parallel with the main system CPU). With this architecture, a Z-80 slave processor, for example, can be used to support word processing and database management using mature 8-bit software, while the main data acquisition tasks can be carried out by a faster 8086 or 68000 system master.

Choice of CPU

We believe the choice of microprocessor should be dictated in large part by software needs. You should find software suited to your intended application and then select a computer with a CPU appropriate to execute it. We chose the

CompuPro 8086/8087 microprocessor board (Bender, 1984a) for the following primary reasons:

1. The 16-bit data pathway allowed us to fetch 12-bit digitized data within 16-bit words rather than in two separate 8-bit fetch cycles.
2. The FORTH programming language (from Laboratory Microsystems; see Appendix 2) was available for this processor in an implementation that takes advantage of the file structure of the CP/M operating system. FORTH has a uniform language syntax implemented for several microprocessors, including the Z-80.
3. FORTH extensions were available to take advantage of the 8087 floating-point numeric coprocessor (Redington, 1983).

Because both 8- and 16-bit FORTHS have a common syntax, we were able to develop programs that would operate on both the 8086 and various 8-bit computers. Because word processing and database management programs were not yet available for CP/M-86, we were able to carry out these activities with an 8-bit microprocessor on the S-100 bus. Both types of software, of course, are now available for CP/M-86.

Arithmetic Coprocessors

We chose the 8086/8087 board because it provides a socket for the 8087 arithmetic coprocessor. These devices, sometimes called floating-point processors (FPP), are hardware devices that greatly accelerate computation in floating-point arithmetic. FPPs designed for 16-bit CPUs are often more versatile than those for 8-bit processors because of their enhanced interrupt and multiprocessing capabilities. 8087 or 80287 (Intel) numeric coprocessors are a must for applications demanding extremely rapid or high-precision floating-point arithmetic (Duncan, 1982; Duncan, 1983, 1984; Redington, 1983). The 8087 maintains up to 64 bits in the significand (equivalent to 10^{19}), plus a 15-bit signed exponent, which together occupy 80-bit internal registers. The coprocessors are invaluable for fast Fourier transforms (which retain phase information from input waveforms) and feedback control requiring trigonometric computations. These functions are accelerated 20–1000-fold using hardware arithmetic processors rather than floating-point software (Duncan 1982; Duncan, 1983, 1984). AMD 9511 and Intel 8232 arithmetic coprocessors are 32-bit precision units most often used as peripheral devices with 8-bit CPUs.

Support Cards

Disk Drives and Controllers

A complete S-100 bus system requires disk drives for storage. Disk drives are managed by disk controller cards on the S-100 bus. Many disk controllers are available for S-100 computers. We use the CompuPro Disk I controller for our floppy disk drives (Appendix 2). This controller supports the most widely used 8-inch floppy disk data storage format, the IBM 3740 standard. The Disk I is capable of acting as a temporary bus master device. This means that it can perform temporary master accesses (TMA) on the computer bus in parallel with the permanent bus master or CPU. Thus, the disk controller can access memory locations required by the disk access function without the intervention of the CPU except for the initial command. This parallel processing capability can increase the overall speed of the computer dramatically. Other controller boards support $5\frac{1}{4}$-inch floppy disk drives or hard disks.

Floppy Disks

Floppy disks are flexible storage media that have a thin magnetic disk platter mounted in a protective envelope. Floppy disks come in three common sizes: 8 inch, $5\frac{1}{4}$ inch, and $3\frac{1}{2}$ inch. There are a variety of disk formats. The only widely accepted 8-inch disk format is IBM 3740, a single-sided, single-density disk format with a 256-KB storage capacity. Most 8-inch disk controllers can read and write IBM 3740 disks so that programs and data files in that format can be transferred between otherwise incompatible computers. Other disk formats are largely computer dependent, and thus do not provide software vendors with a reliable and economical medium in which to distribute their software. Unfortunately, the IBM 3740 format does not take advantage of the much larger storage capacity possible with current floppy disk technology. Many 8-inch floppy disk controllers, including the Disk I, have a double-sided, dual-density format that can store up to 1.2 MB of data.

Floppy disk drives must be compatible with the disk formats supported by the user's operating system. You must determine in advance whether your operating system will support single- or double-sided disks, hard- or soft-sectored disks, the number of sectors per track, the number of bytes per sector, the number of tracks per inch, the total number of tracks required, and the track-to-track stepping rate. The simplest solution is to choose the floppy disk drive recommended by the disk controller manufacturer for the chosen operating system.

The formats of $5\frac{1}{4}$-inch diskettes are even less standardized than the 8-inch formats. There are over 100 incompatible $5\frac{1}{4}$-inch disk formats in use today. Thus, particular care must be taken when specifying a $5\frac{1}{4}$-inch drive to ensure compatibility in hard or soft sectoring, the number of tracks per inch, and total number of tracks used. The formats used by the IBM PC are becoming a de facto standard because of their very large share of the market. In general, $5\frac{1}{4}$-inch floppy disks have less than one-half the capacity of their 8-inch counterparts, although there are notable exceptions to this rule.

Hard Disks

Hard (also called "fixed" or *"Winchester"*) disks usually have a recording medium that uses a metal platter permanently sealed inside the drive. The drive is sealed to keep out dirt, which allows the heads to ride much closer to the surface of the disk, which in turn permits higher information densities. The drive also rotates an order of magnitude faster than floppy disk drives, thus permitting much more rapid information transfer. A typical 20-KB program will take 2–3 seconds to load from a floppy, and a few tenths of a second to load from a Winchester.

Storage capacities of $5\frac{1}{4}$-inch hard disks start at 5,000,000 bytes (5 MB) and range as high as 140 MB. Larger-capacity disks are available and can hold 400

MB of data. Costs range from $1000 to $4000 for the 5¼-inch Winchesters. Hard disks and their associated controllers are usually purchased as complete subsystems. Prices start at about $1500 for 5-MB capacity hard disks and range up to $5000 for 80-MB capacity disks.

In summary, hard disks offer several advantages. First, hard disks are actually less expensive per byte of on-line disk storage than are floppy disks. Second, they have 5–10 times faster access speeds than floppy disk drives. Finally, the biggest advantage of the hard disk drive is the very large volume of data it can hold. Your dealer will generally be your most useful consultant in selecting an appropriate hard disk subsystem. For even faster access to data storage, a disk emulator can be used that utilizes RAM to store data. Data emulators are described below. The associated hazard of very large-scale data loss in a disk failure should be minimized by inclusion of a diskette or magnetic tape backup system.

Memory

Main memory, usually known as RAM (random access memory), can be added to S-100 computers in blocks. Currently, blocks of 64 and 128 KB are most common, but memory boards containing 256 KB, 512 KB, and 1 MB of memory per card are now also available. The size of memory required in a computer is a function of both the operating system and the intended application. Table 4 summarizes the memory requirements of some of the operating systems marketed by Digital Research, Inc. Even the smallest CP/M operating systems shown in Table 4 recommend the use of at least 64 KB of memory. Minimum sized systems are rarely useful because so little space remains for application programs after memory has been allocated to the operating system. As prices for memory continue to decline exponentially, they become a minimal part of total system cost, yet large memories often do most to increase the functionality of computer systems.

Memory boards can be constructed from a variety of components. Two major types of memory are commonly available: dynamic and static. When buying memory, the most important considerations should be frequency of memory errors and power consumption. *Dynamic memory* usually has a higher frequency of errors than static memory. To some extent, this deficiency can be counteracted by implementing parity checks to detect and announce memory errors. General purpose RAM constructed from dynamic memory components is most prone to

Table 4. Memory Requirements of Digital Research, Inc. Operating Systems

Operating system	CPU family	Memory size		
		Minimum	Recommended	Maximum
CP/M 2.2	8080,8085,Z80	24KB	64KB	64KB [a]
CP/M 3.0	8080,8085,Z80	32KB	128KB	1MB
CP/M 86	8086	64KB	128KB	1MB
CCP/M [b]	8086,8088	192KB	256KB	1MB
MP/M 86	8086,8088	192KB	256KB	1MB
CP/M 68K	68000	64KB	256KB	16MB

[a] Extensions are available to utilize additional memory with bank switching [see Conn (1983) for public domain references].

[b] Concurrent CP/M.

error and is not recommended, although it is often less expensive. Power consumption depends more on the manufacturing process for component integrated circuits (ICs) than other design characteristics and varies widely from manufacturer to manufacturer. *Static memory* is less error prone because it need not be cyclically refreshed. Overall, we think that the slightly higher cost of static memory is worth the expense in order to ensure data integrity.

Disk Emulators

Another type of special purpose memory is the disk emulator (Weidemann, 1983). These products have a variety of commercial names such as SemiDisk, memory drive, Warp Drive, ramdisk, and ramdrive. In general, these devices appear to the computer as standard disk drives. Since they are actually constructed of memory components, they allow much more rapid storage and retrieval of files than is possible using a disk drive. For example, CompuPro claims that its M-Drive can achieve a 35-fold increase in storage speed compared to the CompuPro floppy disk systems. This is borne out in our experience. Disk operations are 2–10 times faster with an M-Drive/H than with our Winchester hard disk. The main disadvantage is that the contents of the disk emulator must be transferred to permanent disk storage before power shutdown or the data will be lost. Nevertheless, for operations that store large amounts of data in files that are accessed frequently, such as database management and word-processing programs, disk emulators will be more cost effective than similar quantities of system RAM. This speed can also be used to advantage when very little time can be allotted to disk activity, as in writing data quickly to a storage file during data acquisition.

Other System Boards

Multifunction Boards. Our system also includes a System Support 1 board (Appendix 2). The System Support 1 is a multifunction CompuPro board that has one RS-232C serial port, prioritized interrupt handling hardware, and a real-time clock-calendar, which can be set by software commands and is run by a backup battery pack when system power is off. An optional BOOT ROM may be placed in a socket provided on the System Support 1 board. The board also contains a socket for an AMD 9511 or Intel 8232 arithmetic coprocessor, discussed earlier.

Multiplexer Boards. The MPX-1 is a multiplexer board that has enormous potential for enhancement of overall computer performance (Thovson, 1983). The board has a dedicated microprocessor that relieves the main system CPU of time-consuming I/O (input/output) tasks. This multiplexer operates in a manner similar to the TMA disk controller, discussed earlier, by allowing parallel processing of I/O and computation. The MPX-1 is especially useful for systems that have extensive I/O applications, such as data acquisition and display or multiuser/multitasking environments. The board would be of particular value in situations where the system CPU is sufficiently fast to accomplish the intended computational tasks of the computer, but is being slowed down at inopportune moments by slow I/O devices, such as the Microangelo graphics display.

Troubleshooting Tools. Several manufacturers offer boards that are useful for detecting and correcting hardware failures. The "Bus Probe" from Jade and the Mullen TB-4a "Extender Board" (Appendix 2) can be invaluable aides for tracing hardware bugs. The "Bus Probe" has arrays of light-emitting diodes (LEDs) that display the activity of all address, data, and interrupt lines in the S-100 bus. The Mullen board is a less expensive extender board that allows one to remove cards from the bus so that they are easily accessible for repair. Analysis of board function can then be carried out using an oscilloscope without obstruction from adjacent cards. The Mullen board also provides ready access to all lines (pinout) of the S-100 bus for oscilloscope analysis. Oscilloscopes are far more useful than Jade-type boards for detecting mistiming and electrical noise problems.

Input and Output (I/O)

Communication Protocols

A computer communicates with input and output devices, such as terminals and printers, through *ports*. Ports can transmit data serially or in parallel. Serial ports usually adhere to a subset of a standard known as RS-232C (Libes, 1978; Zaks and Lessa, 1979). RS-232C technically supports communication only up to 50 feet, but it can be used over longer distances with shielded cables. RS-422A is a similar serial standard that uses electrical connections and buffering designed to better protect it from electrical noise. RS-422A serial communications are intended for communications at distances up to 4000 feet (Witten, 1983).

Parallel interfaces are less complex and usually faster than serial interfaces, but are usually limited to cables no more than 10 feet in length. Parallel interfaces trigger a strobe in the interface connections to signal appropriate timing for data transmission. There are several different types of parallel interface in widespread use, all of which are incompatible with one another. The Centronics parallel interface is most widely used for printers. The IEEE-488 (GP-IB) parallel interface patented by Hewlett-Packard is widely used for interfacing to digital instruments (Newrock, 1983).

CompuPro Interfacers 2, 3, and 4

We use the Interfacer 4 board from CompuPro on our S-100 systems. This board has three RS-232 serial ports and a Centronics parallel port. The serial ports can operate asynchronously up to 19,200 baud. Two of the serial ports can be reconfigured for synchronous use at rates up to 250,000 baud. The CompuPro Interfacer 3 card is similar but lacks a parallel port. It can accommodate either five or eight RS-232 ports for computers requiring extensive serial communication. The Interfacer 2 board provides three parallel and one serial port. Combinations or multiples of these boards can be used in computer systems requiring additional ports by selecting different addresses for each board in a set. Because the IEEE-696 can support over 65,000 port addresses, there is no limit to the possible interface configurations. Addresses for individual boards are selected by setting toggle switches on each board. Other manufacturers make functionally similar interface boards (Darwin, 1984; Ratoff, 1984; Wilkinson, 1983).

VIO-X and Microangelo Graphics Boards

There are a full range of videoboards available for the S-100 bus at very reasonable prices. Videoboards will either support raster display terminals or full bit-mapped graphics. The boards we use in our data collection systems are the Microangelo monochrome graphics and the VIO-X terminal (Appendix 2). The Microangelo has 80 × 45 dot text addressing and 512 × 512 pixel graphics and uses a long-persistence phosphor display. It costs about $1,000, including the monitor, making it one of the cheapest graphics systems available. Its major drawback is that it uses the eighth, or leftover, bit from every byte used to store a 7-bit ASCII character. Since the eighth bit controls display and cursor-addressing attributes for the Microangelo, unexpected changes in the display can occur when displaying ASCII text files.

The VIO-X is one of the fastest boards we have seen to date. It uses a predefined character generator to construct the characters and, because it takes the data from the S-100 bus directly, there are minimal delays in passing information to the device. In an 8086 system we estimated the actual display rate at 70 to 80 kilobaud. Both of these choices predate the advent of graphics display chips such as the NEC 7220. There are now several commercially available color graphics boards that have higher resolution, faster operation, and very sophisticated hardware implemented graphics algorithms, all for about $1,000. Several vendors of graphics interface cards are listed in Appendix 2.

Analog–Digital Conversion

The core of any laboratory data collection system is the circuitry that translates between the digital logic of the computer and the analog voltages of the laboratory environment. Two complementary devices are employed: analog-to-digital converters (ADCs) and digital-to-analog converters (DACs). Criteria you should consider when selecting ADC and DAC conversion boards include the conversion speed, accuracy, control of conversion rate, format of data output (ones or twos complement, see glossary), and software support. There are a large variety of boards now available (see Appendix 2). If very high speeds are required, the most sophisticated designs have components to read and write data directly in and out of the address space of the central processing unit (direct memory access, DMA).

The format of digital output (binary, ones or twos complement) is an important consideration because negative numbers are normally represented in computers as the twos complement of the positive numbers. This format facilitates direct execution of arithmetic operations. Binary and ones complement numbers must first be converted to twos complement format for arithmetic manipulations. ADCs that do not provide twos complement output data force the CPU to do the conversion under software control, which can result in critical delay problems for real-time feedback applications. *Aliasing* refers to a type of error that occurs in the representation of data-containing frequencies near the limiting conversion frequency of ADC. Both these considerations are too complex for adequate discussion in this context, so the reader should consult engineering references for additional information (Jung, 1982).

The ADC and DAC boards in our S-100 system are made by Tecmar, and have been reviewed by Newrock and Knessel (1983). Those reviewers agree with our assessment that the documentation provided by Tecmar is very poor. This is unfortunate because the hardware is well designed and versatile.

The Tecmar D/A-100 DAC we used is a simple board with 12-bit accuracy (Table 3) and four separate output channels each having a strap selectable range of +/−10, +/−5, +/−2.5, 0 to +10, or 0 to +5 volts. There is no hand-shaking—the user can change the output whenever desired. The transition time for a full-range voltage swing is about 3 Msec, equivalent to about 300 kHz.

The Tecmar A/D-212A ADC is the most complicated board in our systems. It has eight differential analog inputs multiplexed to a single programmable gain and 12-bit (Table 3) twos complement output from an Intek ADC module. Tecmar uses the same printed circuit board with several different ADC options, and thus a small forest of option-selecting jumpers must all be correctly set before the board will function. The board will support practically any type of I/O that can be implemented on the S-100 bus.

Analog-to-digital conversion can be triggered with a pulse generated by software, an external pulse, or the output of on-board timers linked to the external pulse input. The converter can also be run continuously. Unfortunately, there is no good way to obtain different conversion rates and gains for individual input channels.

Using the Tecmar hardware and our own FORTH software, we are able to obtain either ADC or DAC conversions at 5 kHz. We believe that 5 kHz represents the maximum speed attainable with our high-level FORTH software. By translating the software controlling data conversion to FORTH assembly language code, we obtained analog-to-digital conversion at 50 kHz and digital-to-analog conversion at 300 kHz, the limits of the ADC and DAC boards, respectively. Tecmar markets a package of FORTRAN subroutines for use in software drivers of these two boards; however, we have no experience with these (Newrock and Knessel, 1983).

Peripherals

Peripherals are the input/output devices attached to the computer. The trend in microcomputer development is towards smarter, more powerful, and cheaper peripherals. Peripherals are generally least expensive when purchased from direct mail-order houses. It is generally safer to purchase peripherals from these sources than other components because they can be easily integrated into your system without custom interfacing services provided by a dealer.

Peripheral Interfaces

There are two types of interface between a computer and a peripheral. The first is a *direct hardware interface*—the physical connection of data lines and status lines used to transmit information between devices. The second type of connection is in software, where subroutines are needed to control communications between the interface and the operating system or user program. These software routines are known as *drivers*. The addition of drivers to a system is not a trivial programming task and should only be undertaken by programmers experienced in I/O operations. For inexperienced programmers it is usually better to find the software that meets your needs and then purchase hardware to match the software interface provided.

There are two fundamental types of hardware interface: parallel and serial. The names refer to the method of transmission of information. *Serial interfaces* transmit information in series, a bit at a time. *Parallel interfaces* transmit many bits of information (usually 8 or 16) simultaneously. The most common serial interface is the RS-232, described earlier, which provides one signal line in each direction. Information over the lines is transmitted by changes in voltage level, one bit at a time. Transmission of a single character typically consists of a start bit, seven or eight data bits, a parity bit, and a stop bit, or a total of 10 bits. Typical RS-232C transmission rates range from 300 to 19,200 baud (bits per second), corresponding to approximately 30 and 1,920 characters per second.

The RS-232 standard is sometimes known as the RS-232 "nonstandard" because of the wide range of variations found in "standard" serial interfaces (Witten, 1983). For example, either pin 2 or 3 can be used for sending (or receiving) data, depending on the type of device. Transmission formats also vary widely. One can use 7 or 8 bits, 1 or 2 stop bits, and even, odd, or no parity bits. There are also variations in the use of the commonly used status lines designed for *hardware hand-shakes.* The need for hand-shaking between devices stems from their independent processing of data, which leads to situations where one device is feeding information to the other faster than the receiver can process it.

Besides the hardware hand-shake, there is another convention, known as the *software hand-shake,* where each device scans the data for control characters that will tell it to stop or resume transmission. The characters usually used for this protocol are X-ON and X-OFF, generated by Control-S and Control-Q characters on most computer systems. CP/M interfaces are usually designed to scan for these characters, which makes software hand-shaking especially simple on these systems.

There is another, older type of serial interface called a *current loop,* where information is passed based on the presence or absence of current flow on the signal lines (Libes, 1978; Zaks and Lessa, 1979; Witten, 1983). Current loop interfacing is typically restricted to 300 or 1200 baud. Fortunately, most RS-232 interfaces have one port that can be reconfigured to current loop if required.

The second major type of interface is the *parallel interface,* where information is passed through eight or more parallel data lines. There are several types of parallel interface, of which the Centronics is the most widely used. Parallel interfaces are commonly used for printers. The GPIB (general purpose interface bus), also known as IEEE-488, is popular for interfacing digital instruments to computers (Newrock, 1983). Parallel interfaces are cheaper to build, so it is typical to find a $100 price increase on a $350 printer if one needs an RS-232 port rather than parallel. However, serial interfaces are far more popular with computer manufacturers because single circuits can be made to serve several serial connectors, thereby reducing manufacturing costs.

Video Terminals

Most current generation video terminals have refresh rastor displays that use a white, green, or amber phosphor. The green P4 phosphor is considered least tiring to the eye, although the white is also widely used because of the popularity of DEC VT-100 terminals that use this phosphor. Most terminals used with microcomputers have 12-inch screens with an 80 column by 24 line display. Many provide direct cursor addressing and an EEPROM (electrically erasable programmable read-only memory)-defined RS-232 interface. Most video terminals without graphics now cost less than $800. For another $200–$300 one can get monochrome 512 × 512 graphics. For around $1500 one can get all of the above in color. RS-232 at 19.20 kilobaud is standard for the nongraphics terminals, whereas graphics systems typically have their own hardware interface on a dedicated card.

Plotters

The trend toward smarter peripherals has produced a minor revolution in plotters. There are several choices in the $800–$1000 range, most of which have RS-232 interfaces, 10,000 × 10,000 resolution, and full low-level command sets for generation of lines, circles, ellipses, and arcs. Plotters with several colored pens whose selection is software controlled cost several hundred dollars more. We have used two different plotters in our S-100 systems, the Hewlett-Packard 7470A (HP), and the Houston Instruments DMP-40 (HI). The HP cost about $1500 when we bought it, the HI just under $1000. Both of these plotters use ASCII character mnemonics for their commands. This allowed us to install them as system devices under CP/M. It also allowed the printers to be accessed by higher-level graphics software packages such as Graphpak.

Software interfacing of plotters is fairly simple. CP/M includes an auxiliary device (sometimes called a *punch/reader*) in the entry points for system I/O. All we had to do was put the correct addresses into the drivers. The HI has automatic baud rate selection and stop bit matching. The HI has an 11 × 17-inch plotting area, whereas the HP only accepts $8\frac{1}{2}$ × 11-inch paper, although a newer, more expensive version (HP 7475A) accepts 14 × 17-inch paper. The HP is over twice as fast as the HI, has a better command set, and includes two pens instead of one.

The cost of plotters is constantly decreasing as their capabilities increase, which is also the case for terminals and graphics displays. By the time you read this, both plotters should (1) have serious competition and (2) cost less. Several other plotter vendors are listed in Appendix 2.

Printers

In 1982, when we were purchasing these systems, there were two classes of printer: letter-quality and dot matrix. Letter quality printers use a *daisywheel* device or some variation where a wheel with raised characters is rotated into position and then struck. *Dot matrix printers* build characters by selecting a subset of a matrix of pins on a print head, which are impacted against a ribbon in an appropriate pattern to produce an ASCII character. The capabilities of dot matrix printers have steadily improved, and dot matrix printers are now available that produce letter-quality print (at a letter-quality price).

If less-than-letter-quality print is acceptable, the dot matrix printers are vastly superior. Modern dot matrix printers like the Okidata 92 or C. ITOH Prowriter cost less than $500 (for parallel interfaces), print at 120 characters per second bidirectionally, and have rudimentary bit-mapped graphics. These are very intelligent devices with internal buffers, multiple software selectable type fonts, and several alphabets. The Japanese printers (and terminals) have the lion's share of the market, in large part because of superior resolution, which is needed to display the Japanese alphabet.

If letter-quality printing is a must, the current low end of the market has several machines that print at 10–20 characters per second for a price of $400–$700. $1500–$3000 models have added features and faster printing. If higher speeds are desired, one should consider alternative technologies such as laser printers. Ink jet printers show great promise, especially for multicolor printing, but are plagued by low reliability. As always, we must emphasize the rapidly changing marketplace in peripherals, which will undoubtedly provide other cost effective alternatives in the future.

Operating Systems

Introduction to Operating Systems

Software stored on disk cannot be directly executed by the CPU. The software must first be loaded into memory and then passed instruction by instruction to the CPU for execution. A disk operating sytem provides the mechanism to load and run programs stored on disk. The operating system creates an environment in which other programs can run under the overall control of the operating system.

Operating systems perform a number of functions. They usually include two groups of commands. One group can be directly accessed by the user; the other group of commands (utility procedures) can be used by programmers for specific programming applications. Commands that perform basic functions such as listing a directory or moving a file between disks operate in an identical fashion on different computers if the same operating system is used. Likewise the utilities accessible to the programmer have the same protocol on different computers so long as the operating system is the same. The operating system offers a hardware-independent environment in which different computers can run the same programs.

An operating system must be custom installed on a given hardware system. To do this, a systems programmer will create a set of software modules that provide access to the peripheral devices on that computer. These modules must adhere to the protocol specified by the operating system vendor. These modules are collectively called the *BIOS* (basic input and output system) on CP/M or the *kernel* on other operating systems. There are usually BIOS functions to find, read, or write a disk file, maintain a directory of files on disk, and to input and output data to peripherals. Often there are also procedures for allocating and moving data within memory.

As an example of these rather abstract descriptions, consider the events that occur when a user wishes to edit a text stored on disk. First, a request must be issued for the computer to set pointers to the address where the program text will be read into memory, allocate adequate memory, locate the editor program on disk, and load it into memory. The operating system then relinquishes control of the computer to the editor program. A request to edit a particular text file is passed by the editor to utility routines within the operating system; these locate, open, and load the text file into memory. Only after the file is in memory can it be manipulated by the editor. At the end of the editing session, the edited text is placed back on disk using similar calls to the operating system, and the file is closed. The editor then releases control to the operating system, which awaits new instructions from the user. The user might then use operating system commands to obtain a directory listing of the file or to evaluate the amount of disk space remaining on the disk.

The CP/M Operating System

CP/M (Control Program for Microcomputers, Digital Research, Inc.) is the most widely used operating system for microcomputers. Over 1,500,000 licensed copies have been distributed to date, with implementations for virtually all 8-bit microcomputers using Intel 8080 compatible microprocessors (see Table 4). There is thus an enormous selection of commercial and public domain software written for CP/M, nearing 10,000 programs by some estimates. Other vendors have produced derivative CP/M operating systems such as TurboDOS for the Z-80 and CDOS for Cromemco computers.

CP/M version 1.4 was the first version of the operating system to be widely distributed. This version has been largely eclipsed by version 2.2, which is now in widespread use. Version 3.0, also known as CP/M Plus, has recently been introduced. It enhances the capabilities of version 2.2 and overcomes some of the older version's deficiencies. Sixteen-bit implementations for the Intel 8086 family of microprocessors (CP/M-86) and the Motorola 68000 family (CP/M-68K) have also been released. All versions of CP/M have a similar command syntax and compatible file structure. This is a very valuable feature of the CP/M operating system. Although it is generally not possible to run CP/M object (binary machine code) programs prepared for one microprocessor on a different microprocessor, text and data files can be read and manipulated by programs run on any CP/M-based microcomputer.

Other CP/M Configurations

Concurrent CP/M-86 is a single-user version of the operating system that runs on a single unit but allows simultaneous operation of up to four independent tasks. The operator can switch between tasks at will. MP/M is a multiuser enhancement of CP/M that allows shared access to central computer resources such as disks, printers, and plotters from several terminals. We find MP/M most useful for preparing and editing documents. Original drafts can be written at one terminal and then transferred to a central disk so that a secretary at another terminal can edit and print the draft without ever having to handle a floppy disk. Database management and statistical analysis software can also be used quite effectively with the MP/M multiuser operating system. There are many other enhancements provided with MP/M that are not germane to this chapter.

Although multiuser MP/M is a powerful operating system for database management and word processing, it is not suitable for data acquisition tasks for two reasons. First, it cannot adequately maintain prioritized interrupts to ensure that real-time data is acquired on time. Secondly, it uses internal interrupts, which make it particularly vulnerable to power supply aberrations, as discussed below.

Other Operating Systems

MS-DOS (Microsoft Disk Operating System) is currently the dominant operating system in the 16-bit computer market, primarily because of its widespread use of the IBM PC (called PC-DOS by IBM) (Bender, 1984b; Wong, 1984). There are subtle differences between MS-DOS and PC-DOS (Kee, 1984). The speed of disk operations and differences in the protection of file integrity have been hotly debated in the press (See *Byte* and *Dr. Dobb's Journal* during 1984; Duncan, 1983, 1984; Skjellum, 1984). MS-DOS and CP/M each have particular strengths that vary in importance for different users (Kee, 1984). MS-DOS has powerful print spooling capabilities that are lacking on CP/M. The IBM PC uses an 8088 CPU, a member of the Intel 8086 family, so it is also possible to run CP/M-86 on the IBM personal computer. In fact, many CP/M programs are available for the IBM PC, although CP/M-86 does not enjoy the dominance of its 8-bit progenitor. Modifications to current releases of CP/M even allow programs written for MS-DOS to run directly on CP/M-based machines. The competition between the CP/M family of operating systems and MS-DOS has greatly accelerated the production of enhancements for both operating systems.

UNIX has been a popular operating system for minicomputers for some years and has a devoted following of users. It is a very powerful operating system and supports numerous software tools and many programming languages in the minicomputer environment. However, UNIX sacrifices simplicity in order to gain power and flexibility (Thomas and Yates, 1982). The sophisticated functions and extended memory requirements of UNIX overwhelm the capabilities of many 8- and some 16-bit microprocessors. However, the more powerful 16/32-bit microprocessors can easily accommodate UNIX functions. Thus, as more Motorola MC68000 and Intel 80286 16/32-bit computers become available, UNIX will become a more popular microcomputer operating system. Several vendors have introduced user friendly UNIX-like operating systems to help make UNIX a more attractive operating system for microcomputers. UNIX is written in the C programming language and is the native environment in which C compilers work best.

We mention the UCSD Pascal p-System which also has an avid group of users and has a good reputation as a complete, consistent environment for program development. It is primarily intended for programming in PASCAL, but FORTRAN-77 and MODULA II are also available for some versions of the p-System. The Sage computer based on the 68000 microprocessor is the only computer of which we are presently aware that uses the p-System and has these programming languages available. We are not aware of any p-System implementations for S-100 computers.

Programming Languages and Software

Software Selection

One of the more serious errors a computer buyer can make is to purchase either software or hardware that is inadequate for the intended use. It is not safe to assume that software that performs adequately on one computer will do well on another. As a working principle, it is usually best to select your application software first. You should not be misled by the claims of advertisements, a recommendation of a friend, or an impressive demonstration of color graphics by a salesperson. If the computer will not run your application program, it will be a poor buy.

You should always become familiar with available general purpose software packages before buying them. Wordprocessors and electronic spreadsheets can often be tried out on a computer borrowed from a friend or in a dealer's display room. A few companies rent software on lease purchase trial arrangements. Poorly written manuals, awkward commands, sluggish performance, or the inability to perform the intended application often become readily apparent during a personal trial. Perusal of advertisements and articles in computer magazines will illustrate the range of software packages available for your application, but this is no substitute for hands-on testing.

You should select hardware before software only if your application requires special hardware capabilities. If some of your applications are very demanding (i.e., require intensive computation or high-speed data acquisition), a particular type of microprocessor may be required. For example, you may need the 16-MB address range or interrupt processing capabilities of the Motorola 68000 microprocessor (Hardy and Jackson, 1983; Wong, 1983). However, in choosing this hardware, you may severely restrict your utility software, programming languages, and operating system choices. You should be certain that adequate applications software is available for your chosen microprocessor. Otherwise, the computer may not be very useful despite the power of the hardware.

Software Design

Unfortunately, and unlike general purpose software packages, it is rarely possible to obtain software off the shelf to perform laboratory tasks. Custom programming is almost always required. The burden of developing this special applications software will fall either on the investigator, technical laboratory personnel, or hired consultants. The investment of time, effort, and money can be enormous. A programming language should therefore be chosen carefully.

The appropriate language should usually have the following characteristics:

1. It should allow easy transfer of programs to new computers when present hardware becomes obsolete or is upgraded.
2. It should facilitate the development of structured programs. Structured programming insures that a programmer or his successors will be able to understand, maintain, and modify a library of programs. The concept of structured programming is discussed extensively in several FORTH and C language references (Harris, 1981; Huang, 1982; Kernighan and Ritchie, 1978; Kochan, 1983; Linhart, 1983; Pournelle, 1983c).
3. It should have the ability to produce general purpose utility programs, known as *software tools.* These utility programs can perform often repeated procedures without the need to reprogram for each application (Anderson and Tracy, 1984; Cameron, 1983; Harris, 1981).
4. It should have adequate execution speed for the intended application. Standard benchmark programs can help make this judgment for programming lan-

guages (Duncan, 1983, 1984), but are prone to many interpretative pitfalls (Houston, 1984; Roberts, 1984).

Software should also be easy to use. It should help the operator remember necessary procedures and provide easy access to the functions provided by the program, such as display and manipulation of data. The use of menus makes computer programs much friendlier to novice computer users.

Language Selection

Programming languages can be low level, high level, or intermediate (Figure 4). *Low-level languages* use machine coded instructions or their mnemonics (assembly language) to communicate directly with the microprocessor. *High-level languages* use English syntax, which has little direct relationship to the binary machine code used by the microprocessor. *Intermediate languages* lie between these two extremes. They interact indirectly but extensively with the underlying

Figure 4. Symbolic representation of the "software gap." The software gap illustrates the different levels of programming languages available for microcomputers. Solid bars (|---|) represent the standard functions provided by a programming language. Broken bars (|-P-|) symbolize readily available extensions to some languages that facilitate very low- or high-level functions. The hardware must be provided with binary instructions for the CPU, which are unique for each microprocessor family, regardless of which programming language is chosen. The user's interface represents the appearance of the computer's behavior to the user of an application program. Any level of programming language can be used to create the application program. Low-level languages generally favor speed, compactness, and computer efficiency. High-level languages generally facilitate programming efficiency and transportability between families of microprocessors. [Modified after Harris (1981).]

```
                           LANGUAGE LEVEL
   LOW                                                            HIGH
c |binary object                                                  |  u
o |code (machine specific)                                        |  s
m |----|                                                          |  e
p |     assembly language                                         |  r
u |     (machine specific)                                        |  '
t |     |------|                                                  |  s
t |          systems                                              |
e |          languages (C, PL/M)                                  |  i
r |     |- - --------- - - - - -|                                 |  n
  |                      general programming                      |  t
h |                      languages (FORTRAN, Pascal)              |  e
a |                      |----------------------------|           |  r
r |                                      Specific Application|  f
d |                                      languages (PERT,APT)|  a
w |                                      |-------------------|  c
a |                              artificial intelligence          |  e
r |                              languages (LISP, SMALLTALK)      |
e |                              |----------------------------|
  |     FORTH's  FORTH  general purpose    FORTH application      |
  |     assembly        FORTH extensions   specific extensions    |
  |     language                                                  |
  |     |- - - ---------------------------- - - - - - - - -|
  |                                                               |
```

microprocessor and other system hardware. Intermediate languages contain sets of primitive functions and utility routines that can be combined or modified by the programmer to extend the language. This facility is most useful for programming of hardware-dependent tasks such as data acquisition from instruments or creation of special purpose languages. (See Chapter 2 for further discussion of these levels.)

We felt the intermediate language FORTH best fulfilled our data acquisition needs. Most importantly, FORTH has the most rapid program development cycle of languages available. Development is interactive, so incremental stages of development and testing proceed rapidly without delays to compile and link program modules (Harris, 1981). Program functions can be modified on-line whenever needed. FORTH is usually the first higher-level language introduced for new microprocessors. It has excellent self-contained editors and other facilities for program development and good provision for hardware control and collection of instrument-generated data (Moore, 1980). Software can usually be transferred between computers and executed without modification because FORTH implementations on different computers are remarkably similar. In addition, FORTH's interpreted execution speed typically rivals compiled programs and is about 20 times faster than the equivalent interpreted BASIC. FORTH combines features of both high-level languages and assembly-level code. Assembly language routines can be encoded directly into the program when maximum speed is required (code words). In addition, FORTH can create entirely new special function routines and data structures for special purposes (Jesch, 1982; Laxen, 1982a, b, c; Stolowitz, 1982). Other programming languages, C and MODULA II, have many of the same virtues but have a considerably more arduous program development and testing cycle (Kochan, 1983; Linhart, 1983; Pepper, 1983; Pournelle 1983a, b, c; Ward, 1983).

FORTH does have several disadvantages. These include its occasionally cryptic choice of operator symbols, unfamiliar RPN (reverse Polish notation), or postfix, arithmetic notation, and direct use of stacks to manipulate and transfer data. This is the arithmetic format familiar to users of Hewlett-Packard calculators. These are only disadvantages in the sense that novice programmers are not likely to be familiar with their structure (Glass, 1983). RPN notation and stack use can be easily learned and are the basis of many of FORTH's virtues.

In most implementations FORTH has its own self-contained operating system. This operating system has some advantages, but precludes concurrent use of a standard operating system, file structure, and the many valuable word-processing, mathematical, statistical, and database programs available for microcomputers. We chose a FORTH implementation from Laboratory Microsystems that can utilize the CP/M or MSDOS operating systems as well as the native FORTH system (Appendix 2). This provision allows us to create and manipulate standard data files that can be accessed by other software packages. The use of a standard CP/M file format also permits easy transfer to data and program source code between computers.

In summary, we feel that FORTH is an excellent choice for on-line data acquisition laboratory applications. It executes rapidly, can call assembly-level subroutines, allows rapid development of structured programs, and is very flexible. Several good tutorials, textbooks, and reference sources for FORTH are available

(Anderson and Tracy, 1984; Brodie, 1981; Huang, 1982; James, 1980; Ting, 1981).

Motherboards, Enclosures, and Power Supplies

Motherboards

The bus and its card edge connectors are typically called the *motherboard* or *backplane.* The motherboard consists of parallel conductors that provide the various communication and power connections between the component boards of the computer (Figure 3b). There are several important things to consider when choosing a motherboard. First, it should have a large enough power supply to handle the maximum intended number of board components. Secondly, it should have a signal frequency specification that exceeds the intended data transfer rate of the computer. Inadequate motherboard designs have been known to melt power conductors and degrade bus signals at frequencies above 2–4 MHz. In order to obtain maximum performance from current hardware, the buyer should be certain that the motherboard chosen is conservatively rated for signal frequencies higher than those required by the component cards, and that it provides bus termination. CompuPro and Integrand motherboards have adequate power supply conductors and high-speed operation (currently to 12 MHz), which should provide trouble-free operation (Appendix 2).

Number of Backplane Slots

The number of edge connectors provided on the bus is a significant consideration. Minimal S-100 bus systems are usually configured with four slots, which can accommodate a CPU, memory, a disk controller, and an input–output (serial or parallel) interface. On many single-board computers these four functions already reside on a single multifunction board that occupies a single card slot. If integrated multifunction cards of this type are used, three slots will still be available for special-purpose cards for such functions as high-resolution graphics, ADC or DAC conversion, and memory expansion.

Some single-board computers are designed so that S-100 bus slots can be added to the basic system. The Zenith Data Systems (and Heathkit) Z-100 series of computers offer this option (Pournelle, 1983a, b). These Zenith products can operate as single-board computers without any additional cards. High-resolution color graphics and dual processors (8085 8-bit and 8088 16-bit) are standard on these computers. If additional boards are needed, however, four IEEE-696 expansion slots can be added to extend the basic functions of the computer for laboratory use.

Seven-, 10-, and 20-slot motherboards are available from several manufacturers (Appendix 2). As a working rule, we believe you should purchase a motherboard with twice the number of slots actually required for your initial application. The excess capacity is a small fraction of the total cost of an S-100 computer and provides room for future expansion of the system for new functions or memory. The extra space also permits separation of boards, which is helpful during troubleshooting and configuration procedures.

Uninterruptable Power Supplies (UPS)

The power supply should be rated to provide the maximum anticipated power consumption, including power surge demands such as occur during access to a floppy disk drive. We have used CompuPro (Bender, 1984a) and Integrand enclosures, which include integrated power supplies (Appendix 2). These have considerable reserve capacity and provide a good margin of safety for even moderately long interruptions of line power. We have tested the reserve of our CompuPro enclosure (which includes an 8086 CPU, 128 KB of static memory, and a disk controller) by turning off the power at the main switch for 2–10 seconds. This is an extreme test of power supply margin. We have found little or no detectable degradation of memory content in this time interval. However, reserve power is very dependent upon the type and number of boards occupying the bus. Boards such as the Tecmar ADC are especially power hungry and thus diminish the overall system reserve. We do not recommend cutting the main power to a computer, but it is reassuring to learn that there is some demonstrable margin of safety against transient power interruptions using UPSs.

Because many physiological experiments attempt to record phenomena that occur infrequently, or where only short periods of data acquisition are possible, protection against power interruptions is very important. The uninterruptable power supply (UPS) is a device designed for this purpose. It is essentially a battery pack with associated circuitry that can detect sags or failures of the line supply voltage. The output of a UPS is generally either an unfiltered square wave or a sine wave generated by a ferroresonant transformer. The latter type also protects the computer from surges in the power line voltage. The sine wave is the preferred waveform since it contains no high-frequency components that might potentially transverse the power supply, enter the digital logic, and be misinterpreted as data or instructions.

Several companies manufacture relatively inexpensive UPSs (Appendix 2). These units have a power output of about 600 VA (volt–amperes) and cost from $600 to $1000. In the event of power failure, the UPS gives you enough time to store accumulated data on a mass storage device and conduct an orderly shutdown of the computer and any instruments powered by the UPS. The protection against lost data, manuscripts, and programs afforded by UPSs can be a very valuable feature. Consider the number of hours required to recreate some of the types of information you produce and you will have an estimate of the value of purchasing a UPS for your computer. This value is multiplied if you have a multiuser or multitasking operating system.

We are using Saft 600-VA UPSs (Appendix 2) on two of our computer systems. We have been able to save data on several occasions when our laboratory power has failed, which is unfortunately not an uncommon enough experience. The UPSs also reduce the number of mysterious computer failures that occur with no detectable power loss. This reduction is probably because the UPS prevents very short and otherwise undetected power interruptions from ever disturbing the computer.

An alternative means of providing uninterrupted power is to install gel batteries directly in the power supply within the computer enclosure. This solution has somewhat lower reserve power but is also less expensive and should be adequate for most purposes (Hardy, 1984).

Heating and Cooling

All of the power used by a computer has to dissipate as heat. Every chip in the system contributes a little, and some components such as voltage regulators get too hot to touch. Some computer components can tolerate heat, but others are radically affected by it. In our experience, disk controllers are the most heat-sensitive components of a computer. If a system runs fine for several hours and then begins to experience disk errors, it is a fairly good bet that the cause is inadequate cooling.

The simplest method of cooling computers is by *convection*. Convection is achieved by placing holes in the enclosure above and below the components. The components pass heat to the surrounding air, which rises and is replaced by cooler air from below. Convection is used in many small microcomputers, and virtually all peripherals. In larger computer systems, including most S-100 computers, the air is actively moved with fans. A typical design includes two fans housed in the back of the machine that push air into the enclosure, which has strategically placed slots to vent the heated air.

The heat output of a computer can be considerable. Indeed, our laboratory computer systems make effective space heaters. The air flow pattern can be critical. Some designs such as the Integrand enclosure will overheat if the top is left off for an extended period of time. ParaGraphics enclosures are especially well-designed for cooling component boards (Hardy and Jackson, 1984).

Support

The final point to consider is support of the computer system you have purchased. Very few dealers have much experience integrating analog-to-digital (ADC) and digital-to-analog (DAC) converters into the computer systems they sell. Only a few will have adequate experience with special graphics devices such as high-resolution black-and-white or color displays. Optimally you should select a dealer who can provide timely and complete integration of all your hardware components into a working computer and then maintain them as necessary. Fully integrated, customized computer systems are expensive because they require a substantial effort on the part of the dealer. We purchased assembled CompuPro computer systems and then did our own integration of ADC and DAC, printers, and plotters. This approach requires a working knowledge of the hardware and some operating system programming. The reward is a detailed knowledge of our computer system that allows us to troubleshoot most hardware and software failures.

If you cannot locate a highly competent hardware dealer and do not have much computer experience, then you should be certain that you have adequate support services available within your institution. You should also investigate the reputation of all the manufacturers from whom you intend to buy hardware. Their reliability, replacement policy for defective hardware, and technical support are very important. Choice of a manufacturer who is nearby may be an advantage, but selection of a manufacturer with good telephone support, a solid warranty agreement, and a policy of replacing defective parts is more important than physical proximity.

The world of microcomputers has one source of support that is not typically available for mainframe computers: the users' group (Appendix 3). The number of hobbyists and professional and scientific computer users has proliferated as computer costs have plummeted. Many of these individuals are willing to share their expertise. The wide availability of public domain software is also an important source of support for new and sophisticated users.

Conclusions

Computers based on the IEEE-696 bus provide tremendous power and flexibility for laboratory data acquisition and analysis. The ability to add an almost infinite variety of special-purpose boards and peripherals to your computer system is a major advantage of bus-based systems. On the other hand, the hardware integration and software development required to build a component system are not trivial tasks. A component system should probably not be considered unless it provides necessary capabilities that cannot be matched by ready-made, single-board computers or a few simple laboratory instruments.

The use of a microcomputer in laboratory data acquisition can be far less expensive than using traditional analog instruments. In our laboratories, the computer serves the function of chart recorder, tape recorder, physiological stimulator, and word processor. The separate costs of these items would have been much greater than the cost of the present computer system.

The power of the computer as a laboratory tool can not be realized without a substantial commitment on the part of the user. The user will need to invest both time and money in the design, set up, and programming of the system. In addition, the user must be willing to devote hours to learning how to use effectively the resource in which such an investment has been made. Whether a laboratory computer system is used constantly or simply collects dust depends critically upon the motivation and commitment of the user.

Appendix 1 Glossary

ADC: Analog-to-digital convertor. An ADC converts an analog input signal into a digital output that can be used by the computer.

ASCII: American Standard Code for Information Interchange. This is the most widely accepted binary code used to encode alphabetic and numeric data. Each character is represented as a unique 7-bit code of an 8-bit byte. There are 128 unique characters specified by the standard ASCII code. Many manufacturers have expanded the code to 256. Many special ASCII characters perform keyboard operations like a typewriter, such as line feeds and carriage returns. Others perform special functions during digital data transfers.

Aliasing: A type of error found when digitizing analog signals. Aliasing occurs when the frequency of the signal being digitized approaches the sampling rate of the ADC. As a result of aliasing error a signal will appear to have a lower frequency than its true frequency. Aliasing can be avoided by keeping the sampling frequency of the ADC four to ten times higher than the highest frequency measured.

Baud: A term meaning number of bits per second, used in specifying data transfer rates over serial interfaces.

Bit: An abbreviation for binary digit. A bit can only have values of 1 or 0.

Board: *see* Card.

Boot ROM*:* The ROM containing the small program that a computer must run when first starting in order to load the operating system by "its own bootstraps."

Byte: The most common unit of digital data, consisting of eight binary bits. A byte can represent integers up to 255 or any alphabetic or special character.

Bus: A series of electrical connectors to conduct power or information between components of a computer.

C*:* A compiled programming language developed at AT&T Bell Labs, said to be so named because it was a descendant of B.

Card: A component of a computer system to serve some purpose, usually containing a collection of integrated circuits and connecting electrical components. Also known as a *board.*

Character: A single letter, number, symbol, or space.

Chip: A jargon term referring to the small, flat "chips" of silicon from which integrated circuits are made.

Compiler: A programming tool to convert English-based high-level language source programs into executable machine code. The compiled version is saved and becomes the permanent executable program.

CP/M: Control Program for Microcomputers. CP/M is a trademarked single-user operating system created by Digital Research, Inc., and is available for a large number of CPUs.

CPU: Central processing unit. The CPU is the "brain" of the computer and performs most control, data transfer, and arithmetic functions of the computer.

CRT: Cathode ray tube, the picture tube of a television set.

DAC: Digital-to-analog convertor. A DAC converts a digital input to an analog output signal.

Disk drive: A device for high-speed, modifiable permanent storage of data. Once data or programs are stored on a disk drive they are not dependent on electrical power. Data are typically stored as magnetic impulses on the magnetic coating of a rotating platter. *See also* Floppy disk, Hard disk, and RAM disk.

DMA: Direct memory access, referring to devices other than the CPU that are capable of direct read and write operations to memory locations independent of the CPU.

EEPROM: Electrically erasable PROM. These integrated circuits provide semipermanent memory that persists when power is turned off. Unlike most other PROMs, they can be readily reprogrammed.

FIG: Acronym for FORTH Interest Group. FIG is a public volunteer association of FORTH enthusiasts and vendors that has been primarily responsible for popularizing and standardizing FORTH. FIG activities include exchange of application programs and development tools.

FIG-FORTH*:* A dialect of FORTH. FIG-FORTH is a public domain product that includes its own operating system and is available for many CPUs.

Floppy disk: A removable, flexible, magnetic disk platter (contained in an envelope) designed for storage of data using floppy disk drives. Floppy disks are also known as "floppys" and *diskettes*. Floppy disks are less expensive, have a smaller storage capacity, and have slower access than hard disks.

FORTH: An interpretive programming language developed by Charles Moore that utilizes reverse Polish notation and parameter passing on pushdown stacks.

FORTH *block:* A unit of disk allocation in FORTH. A block is usually equivalent to one screen in size.

FORTH *cell:* A 16-bit, or 2 byte, integer that is the basic unit of stack manipulations in FORTH. *See also* FORTH Word.

FORTH *dictionary:* A FORTH structure where data and executable computer instructions are stored in a cross-referenced manner that can be searched by the FORTH interpreter.

FORTH *screen:* A unit of FORTH programming which is usually 64 characters wide and 16 lines long (1KB).

FORTH *word:* A FORTH identifier for an executable command statement. Words are analogous to subroutines in other programming languages. Words are defined by calling upon the compiler to make an addition to the dictionary. Once defined they can be executed directly by the interpreter or referenced from other words.

Hard disk drive: A rigid, usually nonremovable magnetic disk platter sealed in a permanent enclosure. Hard disks are more expensive, have larger storage capacity, and have faster access to data than floppy disks.

IC: Integrated circuit. ICs have many variations, including microprocessors and memory components, which provide the basis for the current computer and electronics revolution.

IEEE: Institute of Electrical and Electronic Engineers, an organization that establishes standardized industry practices, among other activities (Shuford, 1983).

IEEE-488: An electrical industry standard specifying a type of parallel data transmission. The IEEE-488 standard is often referred to as the GPIB (general purpose instrument bus).

IEEE-696: An electrical industry standard specifying the bus connections and timing relationships for the S-100 connector bus and associated component cards. The IEEE-696 standard also specifies interrupt and bus arbitration between devices.

Interpreter: A programming tool that converts English-based high-level language program statements directly into binary machine code one line at a time. Interpreters do not save the machine instructions for execution at a later time. An interpreter allows immediate testing and execution of program statements without the need to compile the program first.

KB: Kilobyte, or 1,024 (2^{10}) bytes. The memory and disk storage capacity of computers are often given in kilobytes.

Mainframe computer: A class of computer typically owned by large institutions for very large scale data processing. They usually have the most efficient architecture and operating systems for multitasking and multiuser environments.

MB: Megabyte, or 1,048,576 (2^{20}) bytes. Hard disk storage is often specified in megabytes.

Microangelo: A high-resolution video graphics system for IEEE-696 computers. Microangelo is a trademark of the Scion Corporation (Appendix 2).

Microcomputer: A class of computer typically designed with 8- or 16-bit CPUs. Until recently, microcomputers were characterized by a limited memory size and slow processing speed. Currently, the distinction between minicomputers and microcomputers is not well delineated.

Minicomputer: A class of computer typically designed with 16- or 32- bit CPUs. Minicomputers usually have multiuser operating systems, large memories, faster processors, and higher purchase prices than microcomputers. Minicomputers were in widespread use for about 10 years before the development of the competing microcomputer technology.

Modem: *Mo*dulator *dem*odulator. This is a device for transmission of digital information over telephone connections.

Monitor: A video display device, usually a CRT, for display of computer output of text and graphics. Monochrome and color versions are available.

Motherboard: A board having edge connectors to accommodate multiple component boards. A motherboard contains bus connections between card edge connectors for data and power supply. Also referred to as a *backplane.*

Multitasking: Term referring to simultaneous execution of several tasks by a computer. Can occur either in single or multiuser operating system environments.

Multiuser: Term referring to single computers and operating systems used simultaneously by more than one person.

Ones complement: The result of replacing binary 1s by 0s and 0s by 1s. Thus the complement of 11101001 is 00010110. The sum of a number and its ones complement is all binary ones.

Operating system: A special purpose control program for computers that performs many utility operations of the computer, such as disk storage access, and data transfer between computer components and peripherals, and execution of application programs.

Parallel interface: A device or communications protocol for transmitting data in parallel (simultaneously), usually 8 or 16 bits at a time.

Peripheral: Any device external to the computer that communicates with the computer. Video display terminals, plotters, and printers are typical peripheral devices.

Pinout: The assignment of unique functions to each pin contact of ICs or the edge contacts of board components. The physical layout is represented by a schematic diagram. Additional specifications on the schematic diagram may specify appropriate electrical behavior.

PROM: Programmable read only memory. Such a device usually contains unchanging programs or data that are not lost when power is off, and that can be read like memory. PROM does not provide modifiable storage.

RAM: Random access memory. RAM is the volatile, working memory of the comptuer where executable programs and data are held while computer power is on. RAM memory contents are lost when the computer is turned off.

RAM disk: A specialized type of RAM memory designed to emulate a disk drive. This device is typically used where extremely fast disk access is required for temporary data storage. Data stored in a RAM disk will be lost when power is turned off, so data must be moved to permanent disk storage to be permanently saved. Also sometimes referred to as *virtual memory.* RAM disks are less expensive than general purpose RAM.

ROM: Read only memory, *see* PROM.

RPN: Reverse Polish notation. When using RPN, all parameters are specified before the operator, in contrast to algebraic notation. Also known as post-fix notation.

RS-232 Interface: A device for transmitting data in series using a serial data communications protocol. The standard supports the rate of data transmission, limited error checking, and connector construction.

S-100: A name given to computers constructed on a bus containing 100 connectors. The S-100 bus was standardized (IEEE-696) in 1975.

Serial interface: A device or communications protocol for transmitting data as a serial sequence of binary bits. The rate of transmission, number of start and stop bits, and bit parity must be specified.

TMA: Temporary master access. This refers to a class of devices defined in the IEEE-696 standard that can temporarily have control of bus activities if granted priority by the permanent master CPU of the computer. Such devices include secondary CPUs, disk controllers, and some peripheral interfacer boards.

Twos complement: The sum of a number's ones complement plus one, used to represent negative numbers. The sum of a number and the twos complement of a second number is the difference between the numbers themselves.

UNIX: An operating system. UNIX was developed at AT&T Bell Labs for use on minicomputers. It is largely written in the C programming language and is designed for use with C.

Word: A 2-byte unit of data, containing 16 bits.

Appendix 2 Vendors

IEEE-696 Boards, Mainframes, Power Supplies and Motherboards

*Dual Systems Control Corp., 2530 San Pablo Ave., Berkeley, CA 94702; 415-549-3854 (Wilkinson, 1983; Darwin, 1984)

*CompuPro, Division of Godbout Electronics, Inc., 3506 Breakwater Court, Hayward, CA 94545; 415-786-0909 (Bender 1984, Hardy and Jackson, 1983)

*Integrand, 8620 Roosevelt Avenue, Visalia, CA 93291; 209-651-1203

Lomas Data Products, Inc., 66 Hopkinton Road, Westboro, MA 01581; 617-366-6434. (Ratoff, 1984)

ParaDynamics Corporation, 7895 East Acoma Drive, Scottsdale, AZ 85260; 602-991-1600

Seattle Computer Products, 1114 Industry Drive, Seattle, WA 98188; 800-426-8936.

Zenith Data Systems, 1000 Milwaukee Avenue, Glenview, Il 60025; 312-391-8865. (Pournelle, 1983a,b)

Video Graphics Displays

IEEE-696 boards

A-1000: Graphics Development Laboratories, 2832 Ninth St., Berkeley, CA 94710; 415-644-3551

*Microangelo: Scion Corporation, 12310 Pinecrest Rd., Reston, VA 22091; 703-476-6100

GB3MP: Illuminated Technologies, Inc., P.O. Box 83348, Oklahoma City, OK 73148; 405-943-8086

Ultra-Res: CSD Incorporated, P.O. Box 253, Sudbury, MA 01776; 617-443-2750

CAT-1600: Digital Graphics Systems, Inc., 935 Industrial Avenue, Palo Alto, CA 94303; 415-856-2500

---

*Denotes vendors whose products were used by the authors.

Peripheral devices

VX128, VX384: VECTRIX Corporation, 1415 Boston Road, Greensboro, NC 27407

Terminal Boards

*V10-X: Fulcrum Computer Products, Distributed by W. W. Component Supply, Inc., 1771 Junction Avenue, San Jose, CA 94112; 408-295-7171

Term85-III: Futronics, P.O. Box AE, Sparks, NV 89431; 702-356-0801

Analog–Digital Conversion Boards

Twos-complement type

*Tecmar, 6225 Cochran Road, Cleveland, OH 44139; 216-349-0600

California Digital, P.O. Box 3087, Torrance, CA 90503; 213-679-9001

CalData, 3475 Old Conejo Road, Suite c-10, Newbury Park, CA 91320; 805-498-3651

Offset-binary type

Automated Control Systems, 1105 Broadway, Somerville, MA 02144; 617-628-5373

Ones-complement type

I/O Technology, P.O. Box 2119, Canyon Country, CA 91351; 805-252-7666

Dual Systems Control Corporation, 720 Channing Way, Berkeley, CA 94710; 415-549-3854

Bus-Probe Boards

TB-4a Extender Board: Mullen Computer Products, Inc., 2306 American Ave., P.O. Box 6214, Hayward CA 94540; 415-783-2866

The Bus Probe: Jade Computer Products, 4901 West Rosecrans Ave., Hawthorne, CA 90250; 800-262-1710 (from CA), 800-421-5500

IEEE-696-1983 Specification

IEEE Computer Society Order Department, P.O. Box 80452, Worldway Postal Center, Los Angeles, CA 90080; $7.50 + sales tax + $2.00 shipping.

Floating Point Processors

AMD 9511: Advanced Micro Devices, 901 Thompson Place, Sunnyvale, CA 94608; 408-732-2400

8232, 8087, 80287: Intel, 3065 Bowers Avenue, Santa Clara, CA 59051; 408-246-7501

Plotters

*HP7470A: Hewlett-Packard, 16399 W. Bernardo Drive, San Diego, CA 92127; 714-487-4100

*DMP-40, DMP-29: Houston Instruments Division, Bausch and Lomb, Inc., 8500 Cameron Road, Austin, TX 78753; 512-835-0900

TEK 4662: Tektronix Inc., P.O. Box 500, D.S.63-336, Beaverton, OR 97077; 503-685-3785, 800-547-1512, 800-452-1877 (within Oregon)

Amdek Corp., 2201 Lively Blvd., Elk Grove Village, IL 60007; 312-364-1180

Printers

*Prism-132: Integral Data Systems, Inc., (IDS), Milford, NH 03055; 603-673-9100

*Prowriter, model 8510: C. Itoh Electronics, Inc., 5301 Beethoven Street, Los Angeles, CA 90066

Uninteruptable Power Supplies

Electronic Protection Devices, Inc., P.O. Box 673, Waltham, MA 02254; 617 891-6602, 800-343-1813

*Saft America, Inc., Portable Battery Division, 931 No. Vandalia Street, St. Paul, MN 55114; 612-645-8531

*Sola, 1717 Busse Rd., Elk Grove Village, IL 60007; 312-439-2800

Sun Research, Inc., Box 210, Old Bay Road, New Durham, NH 03855; 603-859-7110

Topaz, Electronic Division, 9192 Topaz Way, San Diego, CA 92123-1165; 619-279-0831

Single-Board Microcomputers

Micro 16s: Fujitsu Microelectronics, Inc., Professional Microsystems Division, 3320 Scott Blvd., Santa Clara, CA 95051; 408-980-0755

Enhanced Operating System Software Sources

*BIOS-86 with track buffering: Lanier Computer Systems, 3603 23rd Avenue, Shawmut, AL 36876; 205-768-2616

MRS/OS (enhanced CP/M replacement for Z-80 computers): OCCO, Inc., 16 Bowman Lane, Westboro, MA 01581; 617-366-8969

FORTH

Empirical Research Group, Inc., P.O. Box 1176, Milton, WA 98354; 206-631-4855

HS/FORTH (for MS-DOS): Harvard Softworks, P.O. Box 339, Harvard, MA 01451; 617-456-3021

*8086 FORTH, Z-80 FORTH: Laboartory Microsystems Incorporated, 4147 Beethoven Street, Los Angeles, CA 90066; 213-306-7412

Master FORTH and *FORTH TOOLS* (book): Micromotion, 12077 Wilshire Blvd., Suite 506, Los Angeles, CA 90025; 213-821-4340

FIG-FORTH: Mountain View Press, P.O. Box 4656, Mountain View, CA 94040; 415-961-4103

Appendix 3 Users' Groups

CP/M Users' Group, 1651 Third Avenue, New York, NY 10028

SIG/M, P.O. Box 97, Iselin, NJ 08830. Information and catalog $2.00.

PicoNet, P.O. Box 391566, Mountain View, CA 94039-1566. Maintains libraries of CP/MUG, SIG/M, and several other special interest groups, including C and PASCAL.

References

Anon (1983) The C language. Byte 8(8): 46–285 (August).

Anderson, A., and M. Tracy (1984) FORTH Tools. Los Angeles, CA: Micromotion, Inc.

Aitken, P. (1984) Passing the lab test. Tech J. 1(4):74–84.

Bender, A.L. (1984a) The CompuPro System 8/16 Model 86/87 computer. Microsystems 4(8): 64–68.

Bender, A.L. (1984b) MSPRO; MS-DOS on the S-100 bus. Microsystems 5(3): 54–61.

Beser, E.L. (1984) Microsystems reviews: The paraGraphics game board. Microsystems 5(1): 78–87.

Brodie, L. (1981): Starting FORTH. Englewood Cliffs, NJ: Prentice-Hall.

Cameron, A.G.W. (1983) The software tools computing environment. Microsystems 4(9): 58–64.

Clune, T.R. (1984) The IBM CS-9000 lab computer. Byte 9(2): 278–291 (February).

Conn, R. (1983) The ZCPR2 System: An introduction to a Z80 enhanced replacement, in the public domain, for the CP/M CCP. Microsystems 4(6): 90–98.

Darwin, I. (1984) Microsystems reviews: The dual systems S104/DMA serial I/O board. Microsystems 5: 94–98.

Duncan, R. (1982) Numeric data processor: The Intel 8087. Dr. Dobb's J. 7(8): 47–49.

Duncan, R. (1983) 16-bit software toolbox: Language benchmarks. Dr. Dobb's J. 8(9): 120–122.

Duncan, R. (1984) 16-bit software toolbox: Benchmarks. Dr. Dobb's J. 9(3): 92–96.

Flaming, D.G. (1982) A short review of current laboratory microcomputer systems and practice. J. Neurosci. Methods 5: 1–6.

Garetz, M. (1983) The IEEE standard for the S-100 bus. Byte 8(2): 272–298 (February).

Glass, H. (1983) Towards a more writable Forth syntax. Dr. Dobb's J. 8(11): 80–83.

Hardy, D. (1984) The S-100 bus: Simple power failure backup systems for the S-100. Microsystems 4(3): 32–38.

Hardy, D., and K. Jackson (1983) The CompuPro CPU-68K. Microsystems 4(11): 62–63.

Hardy, D., and K. Jackson (1984) Ways to keep your cool: ParaDynamics special ventilation methods increase cooling efficiency on mainframes. Microsystems 5(2): 86–89.

Harris, K. (1981) The FORTH philosophy. Dr. Dobb's J. 6(59): 6–11.

Houston, J. (1984) Don't bench me in. Byte 9(2): 160–164 (February).

Huang, T. (1982) And So FORTH. Mountain View, CA: Mountain View Press.

James, J.S. (1980) What is FORTH? A tutorial introduction. Byte 5(8): 100–126 (August).

Jesch, M. (1982) High-level floating point. FORTH Dimensions 4(1): 23–25.

Jung, W.G. (1982) IC Convertor Cookbook. Annapolis, MD: Howard W. Sams, Inc.

Kee, H. (1984) From the sidelines: Subtle differences among the new releases give more possibilities to choose from. Microsystems 5(3): 16–20.

Kernighan, B.W., and D.M. Ritchie (1978) The C Programming Language. Englewood Cliffs, NJ: Prentice-Hall.

Kochan, S.G. (1983) Programming in C. Rochelle Park, NJ: Hayden.

Laxen, H. (1982a) Defining words I. FORTH Dimensions 4(1): 2–4.

Laxen, H. (1982b) Defining words II. FORTH DImensions 4(2): 20–22.

Laxen, H. (1982c) Defining words III. FORTH Dimensions 4(4): 28–30.

Libes, D., and S. Libes (1983a) IEEE-696/S-100 standard update. Microsystems 4(5): 36–43.

Libes, D., and S. Libes (1983b) S-100 product directory. Microsystems 4(5): 44–63.

Libes, S. (1978) Small Computer Systems Handbook. Rochelle Park, NJ: Hayden.

Libes, S. (1983) News and views: Intel announces 32-bit wide multibus. Microsystems 4(10): 30.

Libes, S., and M. Garetz (1981) Interfacing to S-100/IEEE-696 Microcomputers. Berkeley, CA: Osborne/McGraw-Hill.

Libes, S., J. Manno, and C. Terry (1983) The Microsystems CP/M software directory. Microsystems 4(12): 60–99.

Linhart, J. (1983) Managing software development with C. Byte 8(8): 172–285 (August).

Long, J. W. (1983) Interfacing instruments with laboratory instruments. Microsystems 4(4): 62–69.

Mau, E.E. (1983) Decisions on the decision I. Microsystems 4(4): 98–103.

Moore, C.H. (1980) The evolution of FORTH, an unusual language. Byte 5(8): 76–92 (August).

Newrock, R. (1983) IEEE-488 bus tutorial. Microsystems 4(4): 34–61.

Newrock, R., and W. Knessel (1983) The Tecmar data conversion products: A/D and D/A convertors with associated software. Microsystems 4(7): 94–109.

Pepper, C. (1983) C: The unknown language worth knowing about. Access 2(6): 21–36.

Place, R.L., and K.A. Bailey (1983) The "standard" CP/M-86 hardware system in the lab: Bringing up CP/M-86 on an Intel single-board computer system interfaced to a Summagraphics digitalizer. Microsystems 4(4): 84–86.

Pournelle, J. (1983a) User's column: Zenith Z-100, Epson QX-10, software licensing, and the software piracy problem. Byte 8(6): 411–442 (June).

Pournelle, J. (1983b) User's column: Buddy, can you spare a door latch? Byte 8(12): 59–96 (December).

Pournelle, J. (1983c) The user looks at books: The best and the worst books on CP/M, PASCAL, C, and ADA. Byte 8(12): 519–526 (December).

Ratoff, B. (1984) Microsystems reviews: "The lightning one." Microsystems 5(3): 94–99.

Redington, D. (1983) Stack-oriented co-processors and FORTH. FORTH Dimensions 5(3): 20–22.

Roberts, B. (1984) Benchmarks and performance evaluation. Byte 9(2): 158ff (February).

Shuford, R.S. (1983) Standards: The love/hate relationship. Byte 8(2): 6–10 (February).

Skjellum, A. (1984) The CP/M bus. Microsystems 5(2): 22–30.

Stolowitz, M. (1982) Algebraic expression evaluation in FORTH. FORTH Dimensions 4(6): 14–17.

Ting, C.H. (1981) Systems guide to FIG-FORTH. San Mateo, CA: Offette Enterprises, Inc.

Thomas, R., and J. Yates (1982) A User Guide to the UNIX System. Berkeley, CA: Osborne/McGraw-Hill.

Thovson, D. (1983) Microsystems reviews CompuPro's MPX-1 multiplexer channel. Microsystems 4(5): 106–111.

Ward, T. A. (1983) Annotated C: A bibliography of the C language. Byte 8(8): 268–283 (August).

Weidmann, B. (1983) Will solid state drives replace the hard disk? Microsystems 4(5): 64–69.

Wilkinson, L. (1983) A review of the dual systems 83/20 6800 UNIX system. Microsystems 4(9): 38–40.

Witten, I. H. (1983) Welcome to the standards jungle. Byte 8(2): 146–178 (February).

Wong, W.G. (1983) The new 16-bit super microcomputers: A comparative look at the Intel 80286, Motorola 68000, and National 16032.

Wong, W.G. (1984) MS-DOS: An overview, Part I. Microsystems 5(3): 46–52.

Zaks, R., and A. Lessa (1979) Microprocessor Interfacing Techniques, 3rd ed. Berkeley, CA: Sybex.

MICROCOMPUTER USES IN LIGHT AND ELECTRON MICROSCOPY

4

The Reconstruction, Display, and Analysis of Neuronal Structure Using a Computer

J.J. Capowski

Introduction

Neurons generally consist of a cell body or *soma,* and processes that extend out from the soma, repeatedly branching in a treelike manner as shown in Figure 1. The primary function of the soma is metabolic and keeps the cell alive. The processes that extend out from the soma are of two types: *dendrites*, which gather information from the cell's environment, and *axons,* which transmit information to its environment. The real work of the cell—gathering, modulating, and transmitting neural information—is defined by and performed by its tree structure. Thus a morphologist, interested in understanding how a neuron works and how it performs a task in the nervous system, needs to visualize and analyze the cell's branching pattern.

The branching patterns of a neuron tree are similar to that of a deciduous tree in winter, say, a maple. Indeed, studying a neuron tree structure is similar to studying a maple tree. However, problems exist in viewing a neuron tree that do not exist in viewing a maple.

1. One can look at a maple tree with the naked eye, but one can only look at a neuron tree through a microscope. The microscope forces constraints upon the viewer.
2. The maple tree exists in a transparent medium, air. It is possible to view the entire depth of the tree at once. The neuron tree resides in a semitransparent

Figure 1. A manually reconstructed model of a Golgi-impregnated neuron constructed from cardboard and wire. [Reprinted from Gaunt and Gaunt *Three Dimensional Reconstruction in Biology* (1978) by permission of Pittman Books, Ltd., London.]

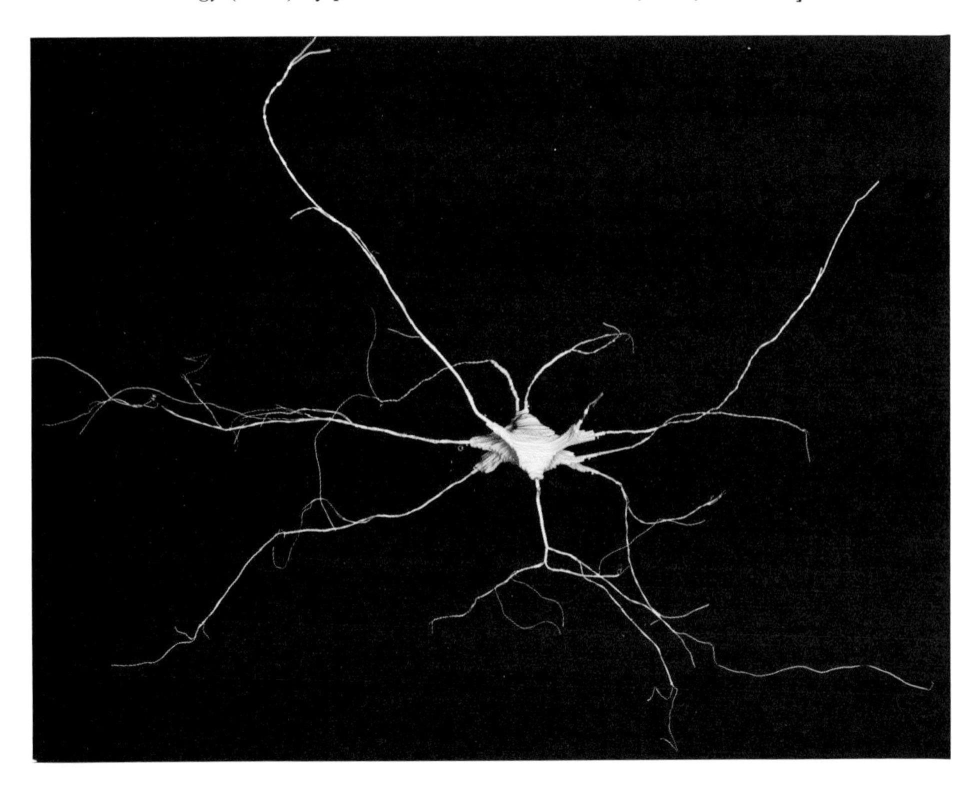

medium, tissue. It is possible only to view a relatively shallow depth of tissue, say 100 microns. This effective viewing depth is almost always less than the total depth of the cell. The cell must therefore be cut into several tissue sections, each containing some part of the tree structure. To reconstruct the intact tree, the individual tree pieces must also be rejoined.

3. The viewer of a maple tree may walk around the tree and view it from any direction. He (or she) may view the global structure of the tree or, with equal ease, study some particular detail of the tree at any focal plane of his eyes. The viewer of the neuron tree is limited to one orientation, defined at the time that the tissue is sectioned. Though he may regard the cell either globally or in regional detail, switching from one view to the other is not a simple transition. Furthermore, since his view is limited to the plane of focus of the microscope's objective lens, he may only look at a shallow two-dimensional slice of the neuron tree at any one time.
4. The maple tree is naturally opaque against a high-contrast background. The neuron is naturally transparent with no contrast relative to its surroundings. Thus the cell must be stained to render it visible. The usual techniques for staining neurons, Golgi impregnation or staining with horseradish peroxidase (HRP), generate reasonable contrast of the cell against its background, but can also stain unwanted tissue elements; even other neurons can be rendered visible by the staining. The cell is also sometimes incompletely stained by these techniques. Questions about completeness of staining and whether the staining changes the cell structure remain a problem.
5. It is possible to view and measure a maple tree today, perturb the tree or its surroundings over a period of time, and then later measure the same tree to record the effects of the perturbations. Unfortunately, viewing the neuron tree destroys it. Thus in any study of the effects of some change on the structure of neurons, one must use large statistical samples to compare one population to another.

The best idea yet devised to overcome all these viewing constraints is to disassemble the neuronal structure and pass it, point-by-point, into a computer. There the neuron, or more properly a mathematical model of it, can be reassembled or reconstructed. Once the mathematical model of a neuron is stored in a computer memory, it is possible, using computer graphics techniques, to generate reasonably realistic views of it. Furthermore, it becomes a reasonable task to generate statistical summaries of the neuron in order to characterize it and to compare one tree or one population of trees to another.

In the rest of the chapter, I will be concerned with the reconstruction, display, and analysis of neuronal structure at the light microscopic level. Studies of neuronal ultrastructure using a computer to reconstruct very small features of a neuron from a series of electron micrographs is a parallel study that is the subject of Chapters 11, 12, and 13 in this book.

History of Stick Reconstruction

Manual Systems

For many years, anatomists have been looking through microscopes at the morphology of neurons and describing their observations qualitatively. Medical illustrators have been used to recreate artistically the three-dimensional images from the microscope. Two famous textbooks by Ramon y Cajal (1909) and Eccles et al. (1967) demonstrate such classical qualitative methods.

The first well-known quantitative descriptions of neuronal structure appeared in a textbook by Sholl (1956). He studied the structure of many cells, classified their types according to general dendritic structure, and counted the number of each type of cell in areas of the cerebral cortex. He also defined and plotted manually a *Sholl diagram,* an analysis that counted the number of intersections that neuron processes make with conceptualized concentric spheres centered at the neuron's soma. Such analysis was the first attempt to quantify the region from which the cell gathers information.

Bok (1959), in his textbook, continued the quantification by tracing dendrites using a camera lucida and measuring the lengths of the dendritic trees. He first made allowances for the three-dimensional structure of dendrites by employing the Pythagorian theorem to the course of a dendrite. He hand-plotted locations of branch points and dendrite endings on polar charts to show the direction of propagation of the dendrites. He also introduced some ingenious optical measuring templates to aid in counting neurons and their parts.

Further manual quantification was done by Aitken and Bridger (1962). More recently, Glenn and Burke (1981) have described a manual method for drawing a three-dimensional representation of a neuron tree. In addition, a textbook (Gaunt and Gaunt, 1978) has been devoted to the subject of manual reconstruction. The reconstruction work of Mannen (1966, 1975) deserves special recognition for the care and effort taken in constructing beautiful physical models of complete cells, an example of which is shown in Figure 1. His interest was in quantifying the volume and surface area of the soma and the length of the dendrites.

Semiautomatic Systems

The experience with manual methods has shown that reconstructing and analyzing neuronal structure is a tedious, error-prone, and lengthy process at best. Many projects were never undertaken because the effort required to perform the analysis was so great. Obviously some sort of automatic data processing capability was needed. Another major flaw in the manual approach was that mathematical summaries were made directly from the cell itself. An investigator who later wished to reanalyze the cells according to some other mathematical parameter had to remeasure the same cells, repeating the tedious procedure. What was needed was some method of transferring the cell structure into some working model from which many different analyses could be done with a modest effort.

In the 1960s, the first affordable, reliable laboratory computers began to appear. If a computer could be attached to a microscope, an operator, given the necessary devices to control the system, could view the neuron in the microscope and trace neuron processes; the computer could accept and manage the lists of coordinates. In such a semiautomatic system, the operator would do what humans do best—pattern recognition—and avoid what humans do worst—bookkeeping. The reverse would hold for the computer, and their marriage would be a fortunate one. Furthermore, at this time early computer graphics techniques were being developed, so that the computer could present to the operator its representation of the neuron as a picture, the form most easily understood by the operator. Finally, with reasonable effort, one could write computer programs to perform quantitative analysis of the neuronal structure now stored in computer memory.

The first semiautomatic neuron tracing system was reported by Glaser and Van der Loos (1965). They attached an analog computer to a light microscope and used the system to measure dendritic lengths. The analog computer, though it was the only affordable computer that could be dedicated to this task, greatly hampered their efforts.

Beginning in the early 1970s, a number of semiautomatic systems using digital laboratory computers were reported. These can be divided into two classes: those in which the operator looks directly into the microscope to view the neurons and those in which the operator views an image, projected optically or with a television camera. Projected-image semiautomatic systems have been well documented by Brown et al. (1979), Hillman (1976), Katz and Levinthal (1972), Levinthal and Ware (1972), Lindsay and Scheibel (1976), Macagno et al. (1973), Paldino (1979), Uylings (1977), and Yelnik et al. (1981). Directly viewed semiautomatic systems have been described by Capowski (1977, 1979), Capowski and Rethelyi (1978), Glaser (1980a), Glaser and Van der Loos (1981), Rethelyi (1979), Rethelyi and Capowski (1977), and Wann et al. (1973).

Because of the relatively immature state of both the laboratory computing field and the semiautomatic neuron reconstruction field, many of the above-mentioned systems were difficult to use and constituted computer science experiments as well as neuronal morphology research. In a report by Foley and Wallace (1974), good techniques for operator control of laboratory computers were formalized, opening this field for subsequent research that continues to the present. Two recently reported semiautomatic directly viewed neuron reconstruction systems, those of Capowski and Sedivec (1981) and Glaser et al. (1983), have attempted to use this human interaction research to produce systems that can be more easily used by biologists.

Some of the problems specific to semiautomatic neuron reconstruction have been addressed in more specialized articles. Problems in the computer graphics display of neuron trees have been addressed by Capowski (1976) and by Haberly and Bower (1982). The use of a Quantimet image analyzing computer in the reconstruction of neurons has been described by Hillman et al. (1977) and by Capowski and Rethelyi (1982). Errors generated during focus axis measurements due to a nonconstant index of refraction of the optical system are described by Overdijk et al. (1978) and Glaser (1982). Different numbering systems for the stored data that describe a neuron tree are described by Glaser (1980b). A comparison of semiautomatic neuron reconstruction systems to purely manual camera lucida techniques is given by DeVoogd et al. (1981). Finally, Buskirk (1978) has described a "semi semiautomatic" system wherein the coordinates of the tree structure are recorded manually and then typed onto punched cards for computer analysis.

The value of semiautomatic neuron reconstruction systems to anatomists will perhaps be best shown when computer companies produce "turnkey systems," complete hardware and software packages for the reconstruction task that require no engineering or computer programming from the anatomist. One system for this purpose is currently available (R and M Biometrics, Nashville), although this is not a recommendation of this system by the author.

Automatic Systems

The next logical step to reduce the tedium of tracing neuron processes is to have the computer take over as much of the tracing process as possible. In this case, the computer rather than the operator would observe the biological image, follow the processes, and capture data points. A decade ago, two attempts at automatic tree tracing were reported by Coleman et al. (1977) and by Reddy et al. (1973). Practical morphological analysis of simple cells has been reported by Coleman in Coleman et al. (1981) and in other articles. Significant operator interaction was required to assist the computer in tracing the cells, however. The automatic tracing systems are, in contrast, excellent practical research in computer pattern recognition of biological images. In the ensuing decade, image digitizing and enhancing techniques were greatly improved, as were the speed and memory capacity of laboratory computers. Taking advantage of these improvements, I have reported some success in the automatic tracing of dendrites (Capowski, 1983a).

Computer recognition of dendrite patterns, however, remains a very difficult problem. In a computer program, one must duplicate to some degree the pattern recognition capabilities of the human visual system. Biological images, including neuron tree structures, are simply far too complex and noisy for the current state of computer pattern recognition.

Conclusion: A Semiautomatic System Is Best

From this brief historical review, I conclude that manual systems are too tedious to be utilized for neuron structure analysis. I conclude also that automatic systems are excellent computer science research today, but are of limited practical value to the morphologist. If one wishes to analyze neuronal structure in the laboratory, one should install a well designed semiautomatic neuron reconstruction system composed of computer science tools that have been proven effective. The project should not develop into a computer science study on how best to handle neuroscience data. The many trade-offs involved in selecting the hardware and programs for such a system are described below.

Semiautomatic Neuron Tracing

General Concepts

The most difficult task involved in using a semiautomatic neuron reconstruction system occurs in entering the tree structure data into the computer. It is therefore the most critical part of the system to design; but it is also the most difficult. In the ensuing paragraphs, the concepts that the system designer must consider when constructing a system for tracing tree structures will be covered in some detail.

Entering a neuron tree into a computer involves the operator tracing the tree structure with a cursor. One can use a *movable cursor* to pass around the tree structure or a *fixed cursor* under which the tree structure may be moved. The computer keeps track of the current x, y, z coordinate of the cursor. The operator tracing the tree structure presses buttons to command the computer to copy the current cursor coordinate into the list of coordinates, already stored in computer memory, that describes or models the tree. Different point types may be indicated with different push buttons. Types of points vary from system to system, but generally, a system should be able to indicate origins of trees, points along the dendrites, branch points, dendritic spine locations, swellings, and different types of process endings. Depending upon the analyses desired, one may enter some indicator of the diameter of the process at each sample point. Several techniques are available to do this; these are explained in the following section.

Specific Design Considerations

When a neurobiologist decides to assemble a neuron reconstruction system, he (or she) should first visit the sites of several existing systems, taking his own cells with him if possible, and do a reconstruction. In this way he can appreciate the capabilities and limitations of these systems applied to his data. He can then intelligently make the necessary decisions about the system he wishes to build. He must decide what image to use, how to measure the microscope coordinates, and whether he need be concerned with dendritic diameter. He must address the problems of tissue origin selection and recovery from errors made while entering data. If his cells extend through multiple sections, he must select some method for rejoining trees cut into pieces by the microtome. Finally, he must decide what interactive device would be best used by him to control the microscope. All of these issues are addressed in the following.

Type of Image. The system designer must decide whether to project the image or whether to have the anatomist observe the image directly in the microscope. Projecting an image, either optically or using television, reduces its spatial resolution and its contrast. However, reconstruction systems using a projected image are faster and somewhat simpler to use than systems using a directly viewed image. If the anatomical study to be performed involves the tracing of very fine processes, complicated tree structures, or poorly stained processes, projecting the image is probably not satisfactory.

Type of Encoders. Another decision is whether to motorize the microscope. If this is done, the computer can both drive the microscope stage to any position and also control the focus axis. Tree structures can be quite complex and, without some assistance from the computer, it would be very easy to overlook branches during tracing. When the operator is tracing a tree and reaches the end of a process, the computer can move the stage to the most recent branch point from which he has not yet traced both branches, thus prompting him not to miss any branches. When the operator is tracing a process and decides to stop tracing temporarily and scan some area in order to understand a complex tree structure better, he may resume by directing the computer to move the stage to the most recently entered point. Indeed he may instruct the computer to move to any point, such as the beginning of a tree, some arbitrary origin, or the center of the soma. The alternative is a passive microscope. In this case, potentiometers are mounted on the microscope stage and focus control knobs, and coordinate values are calculated by the computer from the analog signal of the potentiometers. The operator controls the microscope and the computer reads the current coordinates. Of course, the computer does not have the ability to drive the stage and focus axis. A stepping motor x, y, stage for a research light microscope, interfaced to a common laboratory computer system, is expensive, in the range of $6,000–$8,000. A motorized stage allows the computer to position the stage reliably and repeatedly to any coordinate with a resolution of 0.5 micron. The cost of attaching a driving motor to a direct drive fine focus shaft of the microscope is much less—several hundred dollars. Attaching high-resolution potentiometers, exciting them with a constant voltage, and sending their result to the computer through a set of analog-to-digital converters also costs several hundred dollars.

Entering Process Diameter. Several techniques exist for entering the process diameter at each sample point. The crudest technique is to set some numeric value and assign it to each sampled point. The operator may change this value at his (or her) pleasure. After measuring several fibers with a stage micrometer and obtaining tracing experience, one can achieve reasonable accuracy using this method. Another method, accurate but time consuming, is to sample each point twice, once at each edge of the process, let the computer calculate the distance between the samples, and assign this value as the diameter. A final technique is to use a variable size cursor. Here the operator, by controlling a potentiometer, can vary the size of a circular cursor. He overlays the circle on the process that he is tracing, carefully matching the diameter of the circle with the diameter of the process. The computer keeps track of both the spatial coordinate of the cursor and its calibrated diameter. This is the method used in the UNC system, described later in this chapter.

Origin of Sections. The anatomist must be able to define some arbitrary origin in a section of tissue. It should be a recognizable point to which it is possible to return at any time to verify that there has been no drift in the coordinate system during the tracing. The anatomist must also be able to enter other anatomical landmarks so as to locate the neuron in its environment.

It may not be possible to enter an entire neuron tree in one session. Therefore, provisions must be made to suspend tracing, save the interim results, and resume tracing later. A parallel problem is that a complex tree may be too big to be stored in the computer memory at one time. Provisions must be made in the software to divide complex trees into manageable pieces.

Error Correction. Though a well-designed tracing system will minimize the number of errors that an operator will make while entering trees, they will still be made. Provisions must be made to delete improperly entered points as soon as the error is discovered. One method is to provide a "delete the most recent point" push button. This deletes the coordinate and moves the stage back to the previously entered point, from which the operator can easily continue tracing. Some errors will not be discovered instantly, however. Thus an editing capability must be provided so that a complete tree can be modified after its tracing has been finished. Commands such as "delete a specific point," "change its x coordinate," "change its point type," and "insert a new point" are all helpful in the correction of errors.

Merging Sections. Because neuron trees may continue from tissue section to tissue section, some assistance should be given to locate the continuation of a tree in the next section. This help should be given both at the time of tracing and also at the time of merging. During the actual tracing session, it is helpful if the computer can indicate the point in a new section where it anticipates continuation. Seldom are tissue sections placed on a microscope slide in exactly identical rotational orientation, so some method of rotating the slide to correct this misalignment must be provided. A manually rotatable specimen holder may be purchased for this purpose. Merging after a trace has been completed is discussed in the next section.

Interactive Devices. The concepts of tactile continuity and visual continuity described by Foley and Wallace (1974) should be followed when designing a system for tracing neurons. The operator should not have to take his eyes from the tissue he is tracing in order to give commands to the computer. Nor should he have to move his hands from the device that controls the cursor in order to issue commands. The interactive devices used must be designed to preserve these continuities.

The common interactive devices used to control a computer system are data tablets, potentiometers, joysticks, push buttons, switches, and keyboards. A *data tablet* is a plane surface on which an operator can move a cursor containing cross hairs. Frequently, several push buttons are mounted on the cursor. The computer can sense the current x, y coordinate of the cursor as well as the current status of the push buttons. A data tablet is an excellent tool for controlling a microscope stage in the x and y directions. One can easily write a program to make the microscope stage follow the movements of the cursor. Unfortunately, controlling the focus axis with a data tablet is awkward. One must mount another potentiometer somewhere to control focus movements with the other hand.

A *joystick,* a vertical stick several inches in height that the operator can deflect in the x and y directions, is also an excellent means of controlling the movement of a stage. The computer can sense the current deflection of the joystick and can drive the stage to follow its movements. Furthermore, a potentiometer can be mounted on the end of the stick, so that the operator can twist the stick as well as deflect it. This device, called a *three-dimensional joystick,* can then be used to control the stage and the focus, all with one hand. (The three-dimensional joystick used in the UNC system is shown in Figure 5.)

Push buttons are best used to indicate point types to the computer or to initiate predefined operations, such as "move the stage to the origin." They do not serve well to control the interactive movement of a microscope stage. *Keyboards* are best used to give commands that require numeric values, such as "set the sampling interval to 5.5 microns."

In two contemporary reconstruction systems, those of Glaser et al. (1983) and Capowski and Sedivec (1981), a camera lucida is used to mix the original image of the cell with a CRT image generated by the computer. As a consequence, several new functions can be performed. The generated cursor can be of variable size, allowing easy entry of dendrite diameter. The computer can generate a view of the already reconstructed portion of the cell in its memory, thus providing instant feedback to the operator that the tracing process is proceeding well. Also, a menu of commands can be presented, so that if the interactive device is a data tablet, the operator can enter commands without removing his hand from the data tablet cursor. The operator's view while using such a system is shown in Figure 2.

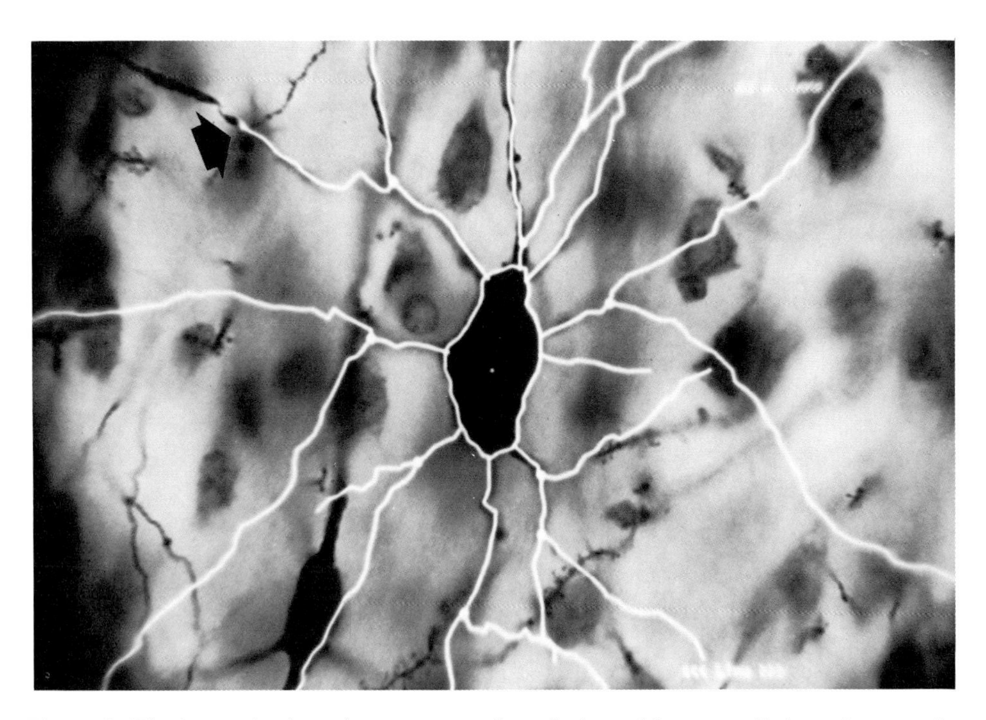

Figure 2. The image in the microscope eyepiece during tablet-controlled semiautomatic tracing. Tracing is currently proceeding at the spot marked by the black arrow. The computer reconstruction to this point is overlaid upon the cell's image. [Modified from Glaser et al. (1983) with permission.]

Four Specific Methods for Neuron Tracing

Capowski and Cruce (1979) described in detail four methods for assembling components into a neuron tracing system and gave advantages and prices for each method. Briefly, the four methods are now reviewed. The first method involves directly viewing the specimen with the microscope. It provides the most accuracy, but requires the most time to trace a cell. It is also the most expensive system. Here the operator views the cell under cross hairs in the microscope binoculars. The stage is moved using either a joystick or a data tablet. The computer drives the motorized stage and focus axis of the microscope, following the operator's commands, at all times keeping track of the current coordinate. When the operator is satisfied that a point of interest along a process is in focus under the cross hairs, he (or she) presses buttons to instruct the computer to sample a point and place it in the point list stored in memory.

The cross hairs may simply comprise a reticle placed in the microscope eyepieces, or may be computer-generated, mixed with the cell image using a camera lucida: In the latter case, it can be replaced by a circle of variable diameter to allow easy measurement of dendritic diameter. Because of the accuracy of the microscope stage and focus motors and because the system is digital, calibration of this system is not necessary, with the possible exception of the variable diameter circle of light.

In a second method the controls of the microscope stage and focus axis are fitted with position encoders that can be read at any time by the computer. The operator moves the stage manually until he observes a point of interest along the process under the cross hairs. He then presses a button to instruct the computer to record a sample. The computer reads the position encoders and stores a scaled value of the readings as a coordinate in the point list.

A disadvantage of this method is that the computer cannot move the stage. It cannot prompt the user to sample missed branches by moving the stage to the most recently entered branch point. Nor can it help the operator restart tracing after a temporary break by moving the stage to the most recently entered point. An advantage of this system is that one does not have to purchase a motorized microscope stage.

In the third method, the microscope image is projected onto a data tablet. The operator traces the cursor over the image, maintaining focus by varying the fine focus knob, following the processes, and pushing a button on the cursor to instruct the computer when to sample a point. The computer reads the tablet x, y coordinates and the z coordinate through an analog-to-digital converter from a potentiometer mounted on the fine focus knob. The anatomist does not have to position the microscope stage for each point along a process, but only to move the stage from frame to frame.

The primary advantages of this system are that tracing can be done quickly and that the cost of the equipment is minimal. The projection of the image reduces its quality, limiting the types of cell that can be traced. Calibration of this system is required because the values that the computer reads from the tablet and from the focus knob potentiometer are relative.

In the fourth method, the microscope image is viewed by a television camera and displayed on a monitor. The operator views the image on the monitor, controls the focus knob of the microscope in order to keep the process in view, and controls data input using a digitizing tablet or some other interactive device. The computer generates a cross on the TV screen proportional to the position of the hand held cursor on the tablet. When a point of interest is in view under the cross, the operator presses a button on the tablet and the computer records the x, y coordinates from the tablet and the z coordinate from a voltage read from a potentiometer mounted on the fine focus knob. The operator must move the microscope stage from one field of view to another; he does not have to move the stage along each process.

The primary advantage of this method is that tracing can be done quite rapidly. Several devices can be used to overlay a cursor on a TV image, ranging in cost from a few hundred dollars to generate a simple cursor, up to $10,000 to digitize and display a high-quality image with any computer-generated information overlaid. This method is quite suitable for a commercially available image analyzer. The television transmission of the image reduces its quality, thus limiting the types of cell that can be reconstructed. The values read by the computer from the tablet and from the focus knob potentiometer are relative, so that calibration of this system is also required.

Intermediate Computation

Focus Axis Corrections

After a block of tissue is cut into sections on a microtome, the sections will shrink and wrinkle to some extent. Additionally, unless one makes focus axis measurements using a tissue mounting system whose refractive index is the same as that of the immersion medium, then the focus axis values read from the microscope will be less than the true values. This is caused by unwanted refraction in the tissue mounting system. Glaser (1982) has described how the problem affects such readings. The two problems, histological shrinkage and unwanted refraction, are additive and yield focus axis readings significantly less than the true values. The error is variable, but may be in the range of 30%. The designer of a study must decide whether this problem is significant. If the neurons to be studied are relatively flat and extend much greater distances in the x and y directions than in the focus (z) direction, such errors may be insignificant. If the cell extends a significant distance in the focus direction, some allowance must be made to solve this problem. UNC's solution to this problem is described in the final section.

Merging Tree Pieces

When tracing a neuron that sends processes into several tissue sections, an operator usually traces all the tree pieces from one section, then all those from the next section, and so on. To complete a reconstruction, all the pieces must be reassembled into complete dendrites or axons. Whenever a branch has been truncated at the edge of a tissue section, the computer must somehow find its continuation in the adjacent section. This may be done in an automatic manner, or in a semiautomatic one. In an automatic merging process, the computer generates a map of all the truncations at the edge of one section and all the potential continuations at the corresponding edge of the adjacent section. Using a mathematical algorithm, the computer attempts to match continuations with truncations, trying different combinations until some error function is minimized. Gentile and Harth (1978) report moderate success with this approach, although the programming effort is significant. Any automatic merging process, however, suffers from the same problem as an automatic neuron tracing process; that is, it attempts, at a great programming effort, to duplicate the pattern-recognition task that a human can perform very easily. For that reason, I strongly favor a semiautomatic approach to the merging problem, as described in the final section of this chapter.

Display of Neuronal Structure

Once a neuron has been completely reconstructed, its image must be presented to the anatomist. This is essential if he is to be certain that the tracing process was done correctly and so that he can study the structure of the cell in a manner which he could not do by simply observing it in the microscope.

Three-Dimensional Display on a Two-Dimensional Screen

The usual devices for presenting a display of a neuron are CRTs and graphics plotters. Both devices are two-dimensional. Thus, to display a three-dimensional structure, one must artificially help the observer understand its depth. The most realistic form of depth cue is smooth rotation of the image. To illustrate this, consider the following. The most realistic way for an anatomist to appreciate the structure of a neuron would be to hold it in his (or her) hands and inspect it as he might do with a basketball. This is, unfortunately, not possible. A reasonable approximation that is possible is to have him control a joystick to manipulate the mathematical model of the neuron. A special-purpose computer called a *graphics display processor* presents a two-dimensional projection of the neuron to the anatomist on a CRT. When the anatomist turns the joystick, the display processor moves the image smoothly to follow his motion. The anatomist will interpret the image as three-dimensional.

Other depth cues, although somewhat less realistic than smooth rotation, are briefly reviewed. One can draw the lines that are further away from the observer in z or depth with less intensity. To a degree, the observer will interpret dimmer lines to be further away. This is called *intensity modulation depth cuing.* Perspective views may also be drawn, utilizing the observer's interpretation that things further away look smaller. The perspective illusion is quite realistic for viewing computer-drawn environments, such as the view that a pilot would see up ahead while training in an aircraft flight simulator. It is somewhat less realistic for computer-drawn objects that do not have such a great depth. Another depth cue is *hidden line removal.* An observer is accustomed to seeing nearer objects block the view of those further away. Such a depth cue is quite valuable in the view of a structure with large visible surfaces; however, for the (almost) stick figures of neuron trees, such a depth cue is less valuable (Figure 3). These depth-cuing techniques must be considered when deciding among the computer graphics hardware described in the following paragraphs. A widely read textbook on computer graphics that explains these techniques in detail is the one by Newman and Sproull (1979).

Six Options for Display Hardware

Six options, ranging in cost from about \$1,000 to \$60,000, are available for computer graphics hardware useful in neuron reconstruction:

1. Felt-tip pen plotter
2. Static TV display
3. Static refreshed line-drawing CRT display
4. Dynamic refreshed line-drawing CRT display with software transformations
5. Dynamic refreshed line-drawing CRT display with limited function hardware transformations
6. Dynamic refreshed line-drawing CRT display with full function hardware transformations

Figure 3. A computer-generated plot of the reconstruction of a neuron. This cell was entered into the semiautomatic system described by Capowski and Sedivec (1981) and the plot generated on a felt-tip pen plotter.

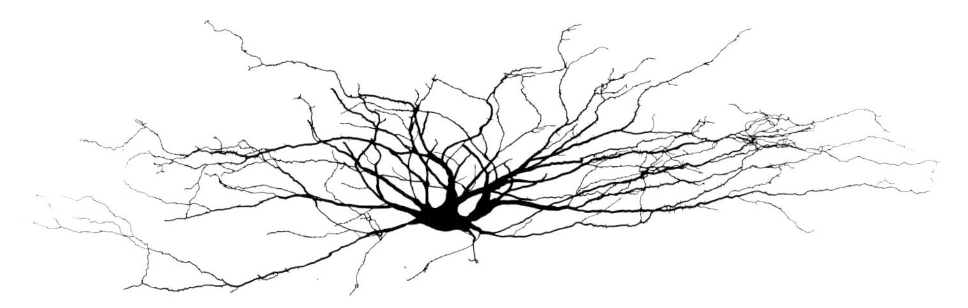

The first option is to purchase a computer-driven felt-tip pen plotter. The computer supplies line end points and the plotter draws lines between the end points on paper. Pen colors and plot sizes can be varied in order to achieve great flexibility in the use of the plotter. This option is attractive to a neuroscientist on a tight budget who is building a reconstruction system one piece at a time because the plotter can be used temporarily as the primary output device and later to make copies of a display when a more complex graphics system is acquired. The cost of a small felt-tip pen plotter with built-in line and character generators is about $1,300.

The second option is to purchase a television monitor and a computer controller card for it. Software is used to change the neuron model into an image composed of a matrix of gray levels, say 512 × 512 picture elements, each having 1 of 64 possible gray levels. This image is stored in a memory on the controller card and displayed on the TV monitor. Depending upon the spatial resolution and the gray-level resolution and whether color or black and white is desired, the cost for the controller card ranges from about $1,000 to $8,000. TV monitors range in cost from about $200 for ordinary resolution black and white up to about $5,000 for high-resolution color.

The first two options generate a *stored static image.* There is no theoretical limit on its complexity, although a complex one might be confusing and might take a long time to generate. The remaining options utilize a *refreshed display system.* Here the image is continually redrawn onto a CRT screen that stores the image for only a few milliseconds, the persistance of the screen's phosphor. For the viewer to think that he is seeing a continuous image, the image must be redrawn or refreshed about 40 times per second. If the image consists of many lines, it may not be possible to draw the entire image so frequently, and it will appear to flicker. For neuron reconstruction work, some flicker is tolerable. If each drawing of the image is slightly displaced from the previous drawing, the image will appear to move smoothly. This technique is used to generate a smooth rotation.

The remaining options require a CRT on which the image is displayed. A normal laboratory oscilloscope can be used. Larger CRTs, up to 21 inches in screen size, can be purchased for approximately $4,500. In the third, fourth, and fifth configurations, the cost of the CRT is not included in the cost of the system. In the sixth configuration, the CRT is included because the manufacturer's electronics are matched to the characteristics of the CRT.

The third option is to purchase a refreshed line-drawing system containing a line generator. The computer supplies line end points and the graphics system draws lines on the CRT between the end points. Because the list of line end points is computed once by the computer and the graphics display controller draws the same end points repeatedly, and because the hardware line generator may be quite fast, say about 5 microseconds per short line, this option may be used to draw a static picture of a complicated neuron composed of 2,000 short lines without appreciable flicker. Several manufacturers make such systems at a cost of about $4,000.

The fourth option is to purchase the same hardware but to use it in a different manner. The computer generates a list of end-point coordinates and the graphics system draws the lines once onto the CRT. The computer then modifies the list of end points slightly, possibly by multiplying each end point by a rotation matrix, and the graphics system draws the slightly rotated image. This process can be used to generate a smoothly rotating neuron tree. Unfortunately, because mathematically modifying the list of end points requires much more time than drawing the image, the speed of the computer limits the number of lines that can be drawn without intolerable flicker. For common laboratory microcomputers, a dynamic image composed of about 200 lines could be drawn in this manner.

The fifth configuration is to purchase a graphics system in which both the matrix multiplication and the line generation is done in special-purpose computing hardware. It is then possible to display a smoothly rotating neuron composed of several thousand lines without significant flicker. Unfortunately, the mathematical transformations required for neuron reconstruction form a small subset of those provided by graphics equipment manufacturers. Also unfortunately, the community of neuroscientists forms too small a market to be profitable for large graphics manufacturers. Therefore, the options available to the neuroanatomist are either to purchase a full function graphics display system (the sixth configuration) and to use only some of its power, or to build a graphics system tailored to his or her real needs. This option is reasonable for investigators who have access to digital electronics expertise. Systems tailored to the neuroscience display problem have been reported by Richardson et al. (1977) and by Capowski (1978a,b, 1983b). These systems may be built for several thousand dollars in parts and several person-months of labor. Recently, however, one dynamic vector system (Eutectic Electronics, Inc., Raleigh, NC) tailored to the needs of neuroscientists has been announced for about $8,000.

The final option is to purchase a full function vector graphics system. Complex neurons can then be displayed with smooth rotation using either an orthographic or a perspective projection. These systems can be purchased from about $23,000 up to about $60,000, depending upon the number of lines that can be drawn without appreciable flicker.

Analysis of Neuronal Structure

An anatomist, looking at a neuron or a population of them, always asks some questions: What are the characteristic features of this cell or population? How does this cell or population differ from that one? Where do the cells grow? In what regions? How do they grow? With many long, skinny processes, or with a few shorter, fatter ones? How does the population change when one changes the animal with age, normal development, abnormal development, drugs, injury, learning, and so on? Once a neuron's trees have been traced into the computer, it is a reasonable task to perform a mathematical analysis to answer these questions. The several categories of analyses that are normally used are described in this section.

Simple Measurements

First and simplest are *counting measurements.* One can write a program to count the number of any feature entered when the cell was traced. Commonly one counts the number of branch points, swellings, spines, and process endings. Standard statistical significance tests are used to determine if the counts of one population are different from the counts in another.

Slightly more sophisticated are *length measurements*. A computer program can calculate the length of all the processes of a cell and compare one cell's process length to another. The lengths can also be used in more sophisticated analyses described below.

Diameter-Based Measurements

If the tracing program stores the diameter of the process at each sample point, diameter-based analyses can be done. The membrane surface area and the volume of each cell can be calculated easily. More complex comparisons are also possible. Figure 4 is a graph of membrane surface area versus radial distance from the cell body for two populations of neurons. Based on this graph, one could state that the two populations are indeed different: One has thicker dendrites at all distances from the soma.

Other Analyses

Other analyses attempt to answer the question, "What is the region of influence of this neuron?" The classical analyses of Bok (1959) contain polar charts or stylized stick-figure views of each cell, indicating in what directions around the compass the dendrites grow. These can easily be generated from computer-stored data. *Histograms* showing the distribution of anatomical features along some anatomical parameter can also be made. Histograms can be three-dimensional, showing, for example, the number of dendritic spines in each element of a matrix of tissue.

Branching patterns can be analyzed. Schematic diagrams of the branching of dendrites can be made in order to describe the branching of a cell's dendrites. A

Figure 4. A graph of membrane surface area versus radial distance from the cell body for two neuron populations. This is representative of many analyses that can be generated to compare one population to another.

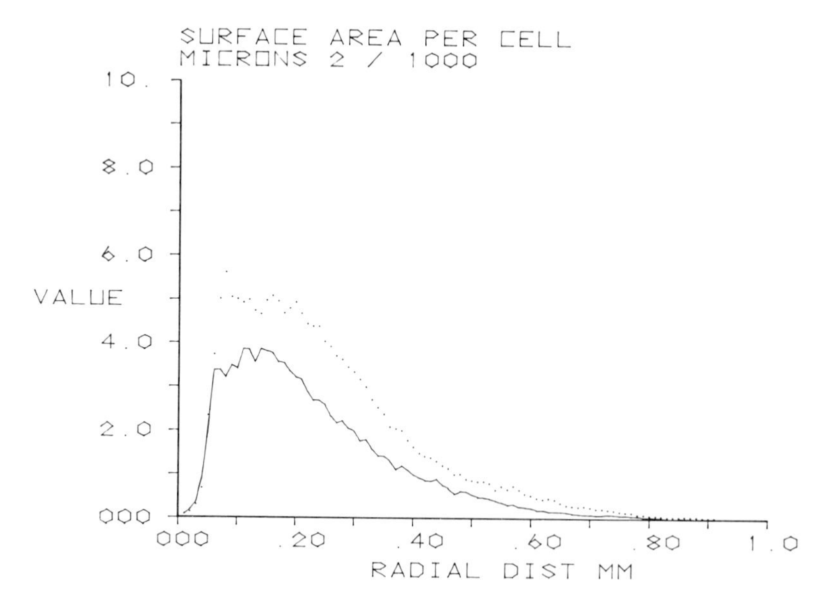

Sholl diagram of a cell can also be drawn. Here a series of imaginary concentric spheres, centered around the cell body, are constructed. Then the number of intersections of dendrites with each sphere is counted and plotted versus the radius of the sphere. This analysis yields a measure of the amount of branching of the cell's dendrites at various distances from the soma. It does not, however, indicate in which directions most of the branching occurs. A good indicator of the direction of branching is a *polar chart* or a calculation of the center of gravity of a cell.

Further branch order analysis can be performed. If one defines the order of a branch point to be the number of branch points between it and its cell body, then one can plot many different parameters such as number of spines or process length versus branch order. This shows whether the features occur distally or proximally to most of the cell's branching. Using this technique one can show the relationship between the particular characteristic and the tree's branching pattern. These types of analysis have produced useful data for quantitatively investigating the changes in structure of neuron trees related to various experimental manipulations.

UNC's Neuron Tracing System

Researchers in the Department of Physiology of the University of North Carolina (UNC) have been using a semiautomatic neuron tracing system since 1974. The latest version of this system is now briefly described. A more detailed description is available in Capowski and Sedivec (1981).

Hardware

Figure 5 shows the most important hardware in the UNC system. A Zeiss Universal microscope has been fitted with stepping motors on its stage and focus axis. These motors can drive the microscope to 0.5-micron resolution along the three

Figure 5. The hardware used in the UNC neuron reconstruction system. Details are given in the text.

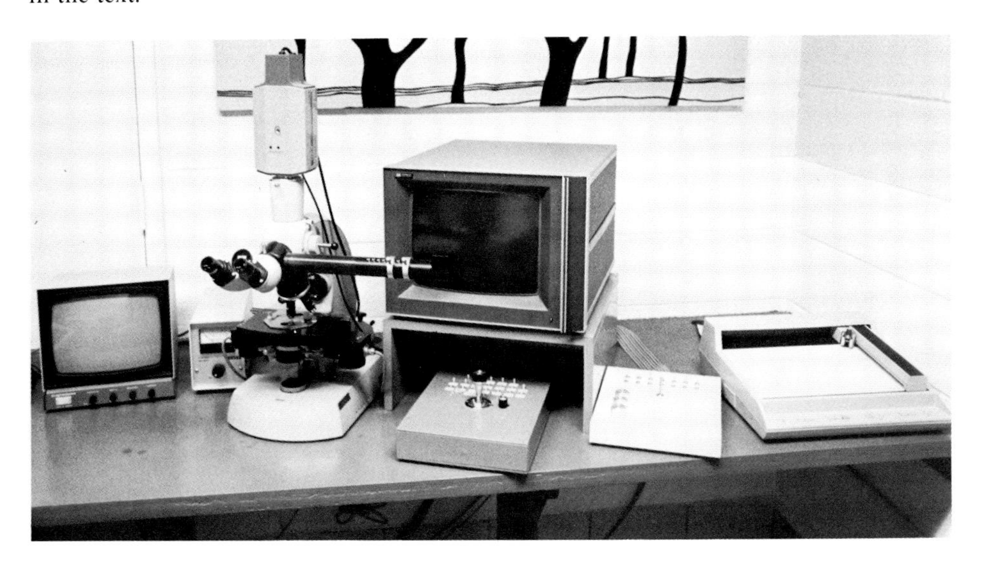

axes. The motors are driven by a Digital Equipment Corporation PDP-11/03 computer system. A rotatable specimen holder can be seen mounted on the microscope stage. A camera lucida drawing tube is also mounted on the microscope and is directed toward the face of a Hewlett-Packard large-screen CRT, model 1304A. When one looks into the microscope, one sees the CRT image overlaid on the biological image. The CRT display is generated by the Neuroscience Display Processor model 2 (NDP2), a dynamic refreshed vector display of our own design, tailored toward neuroscience needs and described by Capowski (1978b). Output from the NDP2 can also be sent to a felt-tip pen plotter, Hewlett-Packard model 7225A. Both the neuron (Figure 3) and the analysis graph (Figure 4) were drawn on this plotter. The operator can control and interact with the reconstruction programs through two devices that are connected only to the computer. A DEC writer is used to give infrequent commands to the programs and receive status messages from them. A box containing a three-dimensional joystick, a potentiometer, and 20 push buttons is used to give frequent commands to the programs.

Operator Control of the Microscope. The microscope control software was designed with the concepts of visual and tactile continuity in mind. The microscope *X, Y* stage is usually moved in a *joystick-tracking mode.* The computer senses the current position of the joystick and drives the stage to that position, always keeping track of the current stage coordinate. The stage travel for full joystick deflection is 150 microns. When the joystick is deflected almost fully in any direction, the tracking logic in the program pauses for 1 second to allow the operator to recenter the joystick without moving the stage. This *clutching* process allows accurate tracing over the entire range of the stage, 25 × 75 mm. A similar scheme is used for the focus control with focus movement of 60 microns for full rotation of the *Z*-axis potentiometer. Clutching logic extends the focus range to 2,000 microns, the range of the fine focus shaft.

A circle, called a *light circle,* is drawn on the NDP2 CRT and superimposed on the center of the biological image. The calibrated diameter of the light circle is variable over 0.3–15.0 microns, according to the setting of the potentiometer to the right of the joystick.

Reconstruction Procedure

The reconstruction, display, and analysis of a neuron that spans several tissue sections are done in five phases. In the first phase, all tree pieces that can be found in a single tissue section are traced into the computer. Then all tree pieces that can be found in the second section are entered. This process continues until all tissue sections through which the neuron projects have been examined. In the second phase, a list of merges is generated by matching all the tree pieces from a section to those of adjacent sections. Thirdly, the tree pieces are scaled to correct for tissue shrinkage, tissue wrinkling, and inaccurate focus axis readings. Fourthly, three-dimensional displays of the neuron are generated on the CRT and plotter so that the operator may appreciate the neuron's structure. Finally, statistical summaries of the cell are generated to show its structure quantitatively. A videotape describing the reconstruction process has been made by Capowski (1979).

Tracing of Individual Trees. In the first phase, all the individual trees from one tissue section are entered. The specimen is mounted on the stage, brought into focus under the light circle, and rotationally aligned so that, for example, one tissue edge is parallel to the stage *y* axis. An easily recognizable point, usually in the center of the cell, is brought into focus and the "slide origin" button is pushed. Subsequently, all coordinates will be measured relative to this origin. A tree is then selected and its origin is brought into focus under the circle. Its origin may be the exit of the tree from the soma or the artificial start of a tree piece at the top or bottom surface of a section. The diameter of the light circle is varied by the operator to match the thickness of the origin of the tree. The "tree origin" button is pushed and the point number, point type, its *x, y, z* coordinates, and its thickness are recorded by the computer. Then the joystick is deflected so that a point some distance along the fiber is moved into focus under the light circle and the "continuation point" button is pushed. This procedure is continued until the entire tree has been traversed. Figure 6 shows the microscope image during the tracing process.

Twelve point types have been defined to enable the tracing of either axonal or dendritic trees; each type can be entered with a distinct push button:

1. Middle tree origin
2. Top tree origin
3. Bottom tree origin

Figure 6. The microscope image during tracing with the UNC system. This shows a portion of a dendritic tree from a complex HRP-stained spinocervical tract neuron from adult cat spinal cord. The microscope power is 630× and the field of view is 103 microns. The light circle is overlaid on the cell's image. Also visible in the eyepieces, but not in the figure, are the number of coordinates already stored in the computer, the current dendritic diameter, and the current *z*-axis coordinate. [Modified from Capowski and Sedivec (1981) with permission.]

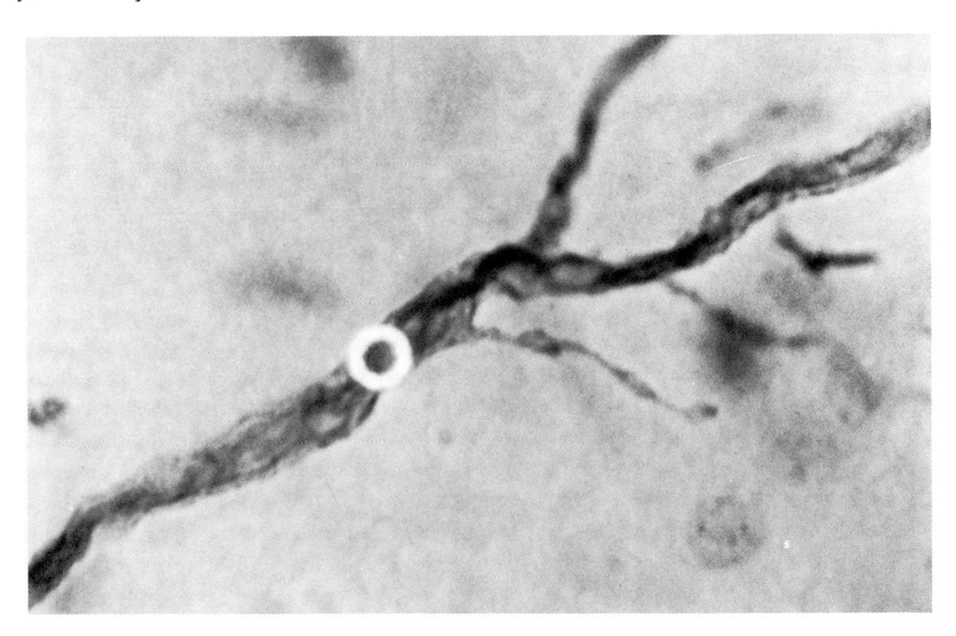

4. Continuation point
5. Fiber swelling
6. Spine base
7. Branch point
8. Natural end
9. End swelling
10. Middle artificial end
11. Top artificial end
12. Bottom artificial end

A *middle tree origin* is used to indicate the start of a tree in the middle of a tissue section, usually emerging from the cell body. *Top* and *bottom tree origins* indicate the start of a tree at a tissue surface. *Continuation points* simply indicate points along a process. A series of them allows the representation of a curved process. A *fiber swelling* indicates a bulb along the course of a fiber, possibly an en-passant enlargement. A *spine base* indicates the location of a spine along a dendrite, but does not indicate the spine's length, direction, or thickness. A *branch point,* obviously, indicates a bifurcation. Five types of ending can be recorded. A *natural end* is used to indicate the tapering of a process to an end. An *end swelling* marks a bulblike growth at the end of a process. *Artificial ends* mark unnatural terminations, processes cut at the tissue edge by the microtome or those that cannot be followed further due to some artifact.

Whenever any process ending is recorded, the computer drives the microscope stage and focus to the most recently entered branch point from which both exiting processes have not yet been traced; this prompts the operator not to miss any of the tree.

Since continuation points comprise about 85% of the points entered during the tracing of a tree, an automatic continuation point entry feature has been devised. Whenever the microscope stage is moved some distance from a previous continuation point, the computer records a continuation point at the current coordinate. The distance is defined by the operator and is usually about 6 microns. The automatic continuation point entry feature speeds up the tracing process considerably.

Merging of Tree Pieces. Whenever a process from a tree piece leaves its tissue section at a top or bottom artificial end, it is necessary to find its continuation in the adjacent section and to attach the continuing piece to the first piece. When all tree pieces have been so connected, the entire intact cell has been reconstructed. An interactive program to aid the operator in performing these merges has been written. Its NDP2 display is shown in Figure 7.

All the tree pieces from two adjacent sections, called the *top* and the *bottom* sections, are displayed simultaneously. A square is drawn around each bottom tree origin and each bottom artificial end in the top section and around each top tree origin and each top artificial end in the bottom section. A circle is drawn around the tree origin or process termination in the top tissue section that is nearest to the center of the picture. The number, type, and thickness of this top circled point is displayed in the upper right corner of the screen. Push buttons on the joystick box can be used to move the circle from one origin or termination to another. A similar circle is drawn around a tree origin or process termination of

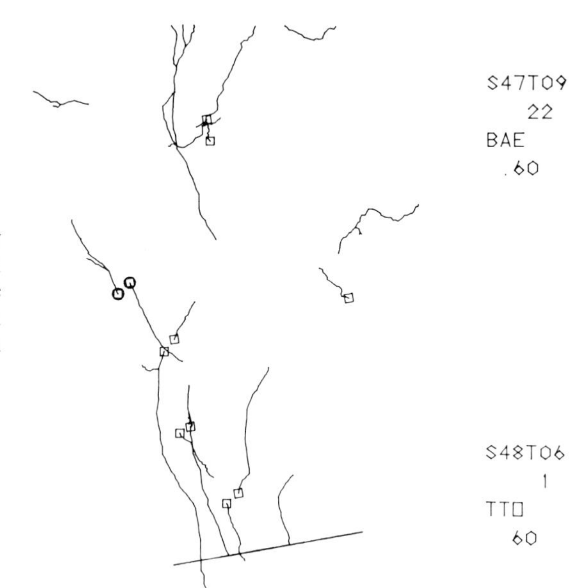

Figure 7. A typical CRT display generated by the merging program of the UNC system. Details are given in the text. [Reprinted from Capowski and Sedivec (1981) with permission.]

the bottom section. Its characteristics are shown in the lower right corner of the screen. With the joystick, the operator can rotate the top section about its z axis relative to the bottom section, translate the top section in x and y relative to the bottom section, and rotate both sections simultaneously about their y axes. Thus he (or she) can move the section displays to inspect their three-dimensional structure and to overlay one on the other by "matching the squares." The operator pushes buttons to move the circles until the top circle indicates a bottom artificial end and the bottom circle indicates its corresponding top tree origin (or vice versa). Once satisfied that these points match in type, location, direction, and thickness, the operator pushes another button to indicate a valid merge. The program checks that the point types are compatible and, if so, generates an entry in a list of merges:

342,TREEB < TREEA

This merge entry indicates that the tree origin of tree A is to attach to artificial end number 342 of tree B. The operator does not have to be concerned with tree names or point numbers; he simply matches the squares and pushes a button.

The trees are not combined when the merging procedure is followed; only a list of merges is generated. Thus it is possible to specify the merges for a very complex cell, all of whose coordinates could not fit into the computer memory at one time.

Focus Axis Corrections. Earlier, the problems of focus axis measurements were described. In order to compensate for these, a scaling program has been written. The operator first indicates the actual section thickness at the time of sectioning, say 80 microns. The program calculates the measured thickness by searching for minimum and maximum focus axis readings in that section, say 50 microns. The program then expands the tree linearly in z to a thickness of 80 microns, thus compensating for shrinkage and improper focus axis readings. In order to compensate for tissue wrinkling, the program first searches through the tree structure for each top tree origin, bottom tree origin, top artificial end, and bottom artificial end. Each, in turn, is called an *edge point*. Then the program searches from the edge point along the process for the first branch point. The portion of the process bounded by the edge point and the branch point is then scaled linearly in z so that the edge point is placed at the tissue edge, the branch point is unchanged, and the points in between are scaled proportionally. The result is a tree whose thickness is that of the microtome setting and whose top and bottom process origins and terminations are exactly at the tissue edges.

Display of Neurons. Two types of neuron display are presented by the UNC system. The first, on the CRT, is a stick figure of the neuron, smoothly rotatable in three dimensions under the operator's control via the joystick. With such a display, the operator can investigate the three-dimensional structure of the cell. Realistic plots of the cells as shown in Figure 3 can be made on the plotter. In these displays, the soma is filled, process thicknesses are given, and swellings are indicated by small circles. Dendritic spines are indicated by short lines of random length drawn in a direction perpendicular to the process at each spine base.

Statistical Summaries. Many statistical summaries of the type described earlier have been generated by programs written for the UNC system. One was shown in Figure 4. These summaries have been and are being presented in many journal articles describing neuroanatomical morphology.

Concluding Remark

In the past few years, the UNC system has had continual heavy use. It is most gratifying to see that all the effort in its design has resulted in an easy-to-use system which is now taken for granted as a normal instrument in anatomical studies.

The author is indebted to Edward R. Perl, Chairman of the Department of Physiology, University of North Carolina School of Medicine, for his financial and moral support over a decade of effort in the field of neuron reconstruction. Many other scientists and engineers have contributed time and ideas to the development of the UNC system. Special thanks for continual help over many years are extended to William L. R. Cruce, Neurobiology Department, Northeastern Ohio Universities College of Medicine, Rootstown, Ohio; to Miklos Rethelyi, Second Department of Anatomy, Semmelweis University Medical School, Budapest; and to C. William Davis, Department of Physiology, University of North Carolina. The most recent funding for the project has come from grant number NS-14899 of the National Institutes of Health.

References

Aitken, J.T., and J.E. Bridger (1962) Neuron size and neuron population density in the lumbosacral region of the cat's spinal cord. J. Anat. 95:38–53.

Bok, S.T. (1959) Histonomy of the Cerebral Cortex. Amsterdam: Elsevier.

Brown, P.B., G.R. Busch, and J. Whittington (1979) Anatomical changes in cat dorsal horn cells after transection of a single dorsal root. Exp. Neurol. 64:453–468.

Buskirk, D.R. (1978) Computer analysis of dendritic morphology. Brain Theory Newsl. 3:184–186.

Capowski, J.J. (1976) Characteristics of neuroscience computer graphics displays and a proposed system to generate those displays. Comput. Graphics 10(2):257–261.

Capowski, J.J (1977) Computer-aided reconstruction of neuron trees from several serial sections. Comput. Biomed. Res. 10:617–629.

Capowski, J.J. (1978a) The neuroscience display processor. Computer 11(11):48–58.

Capowski, J.J. (1978b) The neuroscience display processor model 2. Proc. Digital Equip. Comput. Users Soc. 5:763–766.

Capowski, J.J. (1979) The modeling, display, and analysis of nerve cells. Proc. Digital Equip. Comput. Users Soc. 6:739–742.

Capowski, J.J. (1983a) An automatic neuron reconstruction system. J. Neurosci. Methods 8:353–364.

Capowski, J.J. (1983b) The neuroscience display processor model 3. Proc. Digital Equip. Comput. Users Soc. Fall, 1983:167–170.

Capowski, J.J., and W.L.R. Cruce (1979) How to configure a computer-aided neuron reconstruction and graphics display system. Comput. Biomed. Res. 12:569–587.

Capowski, J.J., and M. Rethelyi (1978) Computer analysis of the distribution of synaptic elements of Golgi-stained axon trees. Brain Theory Newsl. 3:179–183.

Capowski, J.J., and M. Rethelyi (1982) Neuron reconstruction using a Quantimet image analyzing computer system. Acta. Morphol. Acad. Sci. Hung. 30:241–249.

Capowski, J.J., and M.J. Sedivec (1981) Accurate computer reconstruction and graphics display of complex neurons utilizing state-of-the-art interactive techniques. Comp. Biomed. Res. 14:518–532.

Coleman, P.D., D.G. Flood, M.C. Whitehead, and R.C. Emerson (1981) Spatial sampling by dendritic trees in visual cortex. Brain Res. 214:1–21.

Coleman, P.D., C.F. Garvey, J.H. Young, and W. Simon (1977) Semiautomatic tracking of neuronal processes. In R.D. Lindsay (ed.): Computer Analysis of Neuronal Structures. New York: Plenum.

DeVoogd, T.J., F.L.F. Chang, M.K. Floeter, M.J. Jencius, and W.T. Greenough (1981) Distortions induced in neuronal quantification by camera lucida analysis: comparisons using a semiautomated data acquisition system. J. Neurosci. Methods 3:285–294.

Eccles, J.C., M. Ito, and J. Szentagothai (1967) The cerebellum as a neuronal machine. Berlin: Springer-Verlag.

Foley, J.D., and V.L. Wallace (1974) The art of natural graphic man–machine conversation. Proc. IEEE 62:462–471.

Gaunt, W.A., and P.N. Gaunt (1978) Three Dimensional Reconstruction in Biology. Baltimore: University Park Press.

Gentile, A.N., and E. Harth (1978) The alignment of serial sections by spatial filtering. Comput. Biomed. Res. 11:537–551.

Glaser, E.M. (1980a) Computer microscope apparatus and method for superimposing an electronically-produced image from the computer memory upon the image in the microscope's field of view. U.S. Patent 4,202,037.

Glaser, E.M. (1980b) A binary identification system for use in tracing and analyzing dichotomously branching dendrite and axon systems. Comput. Biol. Med. 11:17–19.

Glaser, E.M. (1982) Snell's law: the bane of computer microscopists. J. Neurosci. Methods 5:201–202.

Glaser, E.M., and H. Van der Loos (1965) A semiautomatic computer microscope for the analysis of neuronal morphology. IEEE Trans. Biomed. Eng. 12:22–31.

Glaser, E.M., and H. Van der Loos (1981) Analysis of thick brain sections by obverse–reverse computer microscopy: application of a new, high clarity Golgi–Nissl stain. J. Neurosci. Methods 4:117–125.

Glaser, E.M., M. Tagamets, N.T. McMullen, and H. Van der Loos (1983) The image-combining computer microscope—an interactive instrument for morphometry of the nervous system. J. Neurosci. Methods 8:17–32.

Glenn, L.L., and R.E. Burke (1981) A simple and inexpensive method for 3-dimensional visualization of neurons reconstructed from serial sections. J. Neurosci. Methods 4:127–134.

Haberly, L.B., and J.M. Bower (1982) Graphical methods for three-dimensional rotation of complex axonal arborizations. J. Neurosci. Methods 6:75–84.

Hillman, D.E. (1976) A tridimensional reconstruction computer system for neuroanatomy. Comput. Med. 5(6):1–2.

Hillmam, D.E., R. Llinas, and M. Chujo (1977) Automatic and semiautomatic analysis of nervous system structure. In: R.D. Lindsay (ed.): Computer Analysis of Neuronal Structures. New York: Plenum.

Katz, L., and C. Levinthal (1972) Interactive computer graphics and representation of complex biological structures. Ann. Rev. Biophys. Bioeng. 1:465–504.

Levinthal, C., and R. Ware (1972) Three dimensional reconstruction from serial sections. Nature 236:207–210.

Lindsay, R.D., and A.B. Scheibel (1976) Quantitative analysis of dendritic branching pattern of granular cells from human dentate gyrus. Exp. Neurol. 52:295–310.

Macagno, E.R., V. Lopresti, and C. Levinthal (1973) Structure and development of neuronal connections in isogenic organisms: variations and similarities in the optic system of *daphnia magna.* Proc. Natl. Acad. Sci. USA 70:57–61.

Mannen, H. (1966) A new method for measurement of the volume and surface area of neurons. J. Comp. Neurol. 126:75–90.

Mannen, H. (1975) Reconstruction of axonal trajectory of individual neurons in the spinal cord using Golgi-stained serial sections. J. Comp. Neurol. 159:357–374.

Newman, W.M., and R.F. Sproull (1979) Principles of Interactive Computer Graphics, 2nd Ed. New York: McGraw-Hill.

Overdijk, J., H.B.M. Uylings, K. Kuypers, and A.W. Kamstra (1978) An economical, semiautomatic system for measuring cellular tree structures in three dimensions, with special emphasis on Golgi-impregnated neurons. J. Microsc. 114:271–184.

Paldino, A.M. (1979) A novel version of the computer microscope for the quantitative analysis of biological structures: application to neuronal morphology. Comp. Biomed. Res. 12:413–431.

Ramon y Cajal, S. (1909) Histologie du Système Nerveux de l'Homme et des Vertebres. Paris: Maloine.

Reddy, D.R., W.J. Davis, R.B. Ohlander, and D.J. Bihary (1973) Computer analysis of neuronal structure. In S.B. Kater and C. Nicholoson (eds.): Intracellular Staining in Neurobiology. New York: Springer-Verlag.

Rethelyi, M. (1979) The modular construction of the neuropil in the substantia gelatinosa of the cat's spinal cord. A computer aided analysis of Golgi specimens. Acta Morphol. Acad. Sci. Hung. 29:1–18.

Rethelyi, M., and J.J. Capowski (1977) The terminal arborization pattern of primary afferent fibers in the substantia gelatinosa of the spinal cord in the cat. J. Physiol. Paris 73:269–277.

Richardson, A.D., F.U. Rosenberger, and C.U. Molnar (1977) The coordinate transformer user's guide. Technical Memorandum No. 239, Computer Systems Laboratory, Washington Univ., St. Louis.

Sholl, D.A. (1956) The Organization of the Cerebral Cortex. London: John Wiley.

Uylings, H.B.M. (1977) A Study on Morphometry and Functional Morphology of Branching Structures, with Applications to Dendrites in Visual Cortex of Adult Rats Under Different Environmental Conditions. Amsterdam: Kaal's Printing House.

Wann, D.F., T.A. Woolsey, M.L. Dierker, and W.M. Cowan (1973) An on-line digital-computer system for the semiautomatic analysis of Golgi-impregnated neurons. IEEE Trans. Biomed. Eng. BME-20:233–247.

Yelnik, J., G. Percheron, J. Perbos, and C. Francois (1981) A computer-aided method for the quantitative analysis of dendritic arborizations reconstructed from several serial sections. J. Neurosci. Methods 4:347–364.

5

A Microcomputer Plotter for Use with Light and Electron Microscopes

R. Ranney Mize

Introduction

Cells and synapses are almost always organized into specific patterns in the central nervous system. These patterns or spatial distributions often have functional significance. In the visual cortex, for example, neurons with different physiological and morphological characteristics are distributed within distinct laminae. Corticotectal cells, which project to the superior colliculus, are located predominantly in layer 5. Corticogeniculate cells projecting to the lateral geniculate nucleus are distributed within layer 6. Axons projecting to visual cortex also have selective distributions (Gilbert, 1983).

Sometimes these selective distributions are sufficiently obvious that they can be seen with a low-power microscope if the cells are labeled by histochemical tracers. In other cases, the distribution patterns may be far more subtle either because the specific cell classes are very sparse or because they are not easily visualized with histochemical markers. In the dorsal raphe nucleus, for example, only a few serotonergic neurons project to the superior colliculus and any pattern in their distribution would be difficult or impossible to detect in a single section. A pattern emerges only by combining data from many sections. In other cases, the distribution patterns may be very complex so that it would be desirable to measure the distributions quantitatively and treat the data with statistical tests. At the electron microscope level, detection of distribution patterns invariably requires a measuring device because only a small fragment of the total tissue can be viewed at the high magnifications used with the electron microscope.

To study distributions quantitatively, a mapping device is needed that can measure the positions of cells and other profiles quite accurately. A number of methods have been devised to map the positions of organelles in tissue. One method, now largely abandoned, is to photograph a series of overlapping pictures that together cover a large region of tissue. The pictures are pieced together to form a photographic montage that may be the size of an entire room. Since the individual photographs are taken at high magnification, it is possible to visualize relevant cytological features and also obtain a coarse map of their distribution through the tissue. Needless to say, this procedure is very time-consuming and requires great patience and skill in matching the photographs precisely.

A more modern approach is to attach encoding devices to the stage drives of a microscope so that the positions of cells and processes can be mapped electronically. One system of this type uses linear potentiometers to encode stage position. The potentiometers are used to drive the pen on an analog x–y recorder. The pen moves with the movement of the stage to produce a map of the specimen. The positions of individual profiles can be marked by lowering the pen at the location of each profile. Systems of this type, called *microscope pantographs* or *plotters,* were developed by several laboratories during the 1960s and 1970s (Boivie et al., 1968; Eidelberg and Davis, 1977; Grant and Boivie, 1970; Sterling, 1973, 1975).

These electronic systems also have some drawbacks. Although the x–y plotter produces a convenient visual map of profile distributions, it provides no facility for measuring the distributions for mathematical analysis. To obtain these kinds of measures using a microscope pantograph, it is necessary to count the profiles and measure their distance from the surfaces of the map manually.

This problem can be easily rectified by transmitting the encoded data about stage position to a digital computer. The computer can then perform the mathematical calculations automatically as well as store and display the data. Several laboratories have developed computerized microscope plotters over the past few years (Curcio and Sloan, 1981; Foote et al., 1980; Forbes and Petry, 1979; Johnson and Capowski, 1983; Mize, 1983c; Williams and Elde, 1982). This chapter discusses some relevant principles involved in designing a computer microscope plotter and describes a specific microcomputer-based plotter we have developed for use with both light and electron microscopes.

System Hardware

A computerized system for microscope plotting requires four basic components:

1. a device for encoding the position of the microscope stage and/or histological specimen;
2. a device for converting this positional information to a digital code that can be used by the computer;
3. an appropriate interface to the computer; and
4. the digital computer itself.

Encoding Devices

Stage or specimen position can be encoded by a number of devices. The most commonly used devices are linear potentiometers, motorized stage drives (stepping motors), and shaft encoders. These devices are usually attached directly to the stage drives of the microscope, although slide potentiometers can be attached to the stage itself (see Reed et al., 1980). *Linear potentiometers* generate an analog signal that is proportional to the position of the x and y axes of the stage. The analog signal must be converted to a digital signal for computer input using an analog to digital converter (ADC). Linear potentiometers have been used by Glaser and Van der Loos (1965), Reed et al. (1980), and Sterling (1975). Linear potentiometers are probably the least expensive solution for encoding position. In my experience, however, potentiometers tend to be noisy, are affected by extraneous voltage fluctuations, may not be perfectly linear, and can wear out rapidly.

Stepping motors can also be used to encode stage position. To use them for this purpose, the pulses that drive the motors must be counted. The counts are converted to a measure of distance from a zero reference point. Stepping motors have the advantage of serving two roles: (1) encoding stage position and (2) automatically advancing the stage. Motorized stage drives have been used by Capowski (1977) [see also Chapter 4 of this volume]; DeVoogd et al. (1981); Overdijk et al. (1978); Wann et al. (1973); and Woolsey and Dierker (1978). Stepping motors are both accurate and durable, but commercial motorized stages can be very expensive (Zeiss and Leitz motorized stages cost over $6,000).

Shaft encoders offer an intermediate solution. They generally cost from $350 to $500, substantially less than the cost of motorized stages. Unlike linear potentiometers, shaft encoders are not affected by low-level fluctuations in voltage, are precisely linear, and have a long life span. Because they generate either discrete TTL-level pulses or a binary code, they are directly compatible with digital instruments such as computers.

Shaft encoders are electromechanical devices that generate some type of electrical signal for a given increment in the rotation of a shaft. Because the stages of most light and electron microscopes are manipulated by turning a rotating shaft, shaft encoders are ideal for encoding stage position on microscopes. By appropriate interfacing, shaft encoders can represent the speed of rotation, the direction of rotation, the amount of rotation, and hence the position of the shaft.

Most shaft encoders use an optical device and coded disks to detect shaft movement. Movement of the shaft moves the coded disk across a beam of light, usually generated by a light-emitting diode (LED). A photosensor device generates a pulse each time the beam of light is interrupted by movement of the coded disk. The direction of shaft movement is determined by the phase relationship of the pulses. Shaft encoders are of two types. Encoders that generate pulses are called *incremental* encoders. They require another device that must count the pulses to keep track of position. *Absolute* shaft encoders generate a binary code that uniquely represents a given position of the shaft. The code can be fed directly to the computer. Absolute shaft encoders are usually more expensive than incremental ones.

We use incremental shaft encoders for our plotting system. The shaft encoders attached to our Philips 201 electron microscope are large, heavy-duty industrial encoders (Model VOE-023, Vernitech Corp., Deer Park, NY). They generate 128 pulses per shaft revolution, yielding a resolution of 0.391 microns on the Philips EM. The encoders are attached to the stage drives of the microscope using mechanical gears. The gears are mounted on the microscope and encoder shafts using removable collars (Figure 1A).

A much lighter weight shaft encoder is used to encode stage position on our Zeiss Universal light microscope (Model BIC15-11G4A, Litton Encoder Division, Chatsworth, CA). This encoder has a resolution of 2048 pulses per revolution, which provides a resolution of approximately 12 microns on the Zeiss microscope. The encoders are linked directly to the shafts of the stage drives using a pressure collar (Figure 1B). Both types of shaft encoder require an input power source of 12 V DC and have TTL output logic levels of 0.5 V negative and 5 or 12 V positive.

Counting Devices

If absolute shaft encoders are used, binary input representing position can be fed directly to the computer. If either stepping motors or incremental shaft encoders are used to encode stage position, a counting device is necessary to keep track of the number of pulses generated by the encoder. The count, which can be incremented or decremented, represents the distance the stage has moved from a zero reference point. Counts of pulses can theoretically be handled by the computer itself, but this requires input/output programming routines that can collect data at speeds exceeding the fastest possible generation of pulses by the encoder; otherwise, pulses might be missed and an accurate tabulation of position would be lost. Because it is generally desirable to have the computer free to perform *other* tasks during plotting, it seemed preferable to us to build a separate, dedicated device to count the pulses.

A display/control counting device was built for this purpose by the Biomedical Instrumentation Division of the University of Tennessee Center for the Health Sciences (Figures 2 and 3). In addition to counting the number of pulses output

Figure 1. (**A**) Vernitech optical incremental shaft encoder (SE) attached to one of the drive shafts (SH) of a Philips 201 electron microscope with a gear mechanism (GM). The gears on the shafts are attached by removable collars. (**B**) Litton shaft encoder (SE) attached directly to the drive shaft (SH) of a Zeiss Universal light microscope. The shaft of the encoder is linked directly to the shaft of the microscope stage drive by a pressure collar. The encoder is secured to the stage (S) by a support bracket (SB).

by the shaft encoders, the display/control device converts the counts to 16-bit binary digits for computer input and provides an LED display of the counts for easy readout of the x–y positions of the stage (Figure 2).

A circuit diagram of the system is shown in Figure 3. Each channel has the following components: a pulse shaper to further smooth the encoder pulses to a standard width and voltage level (optional); eight counters to count the encoder pulses; and four display drivers for each four-digit LED display. Four of the counters in each channel are used to output data to the computer in 16-bit parallel binary format. The other four counters output data to the display drivers in binary coded decimal (BCD) format.

The display/control unit is housed in a standard metal cabinet that includes a power on/off switch, a display on/off switch, and a zero reset button. It is 16 in. wide, 8 in. deep, and 6 in. high and can thus be placed unobtrusively near the microscopes (Figure 2).

Computer Interfaces

The signal representing stage position can be transferred to the computer using any of the standard interfaces. Several types of computer interface are commonly available for microcomputers (see Chapters 1 and 3). We decided against using a

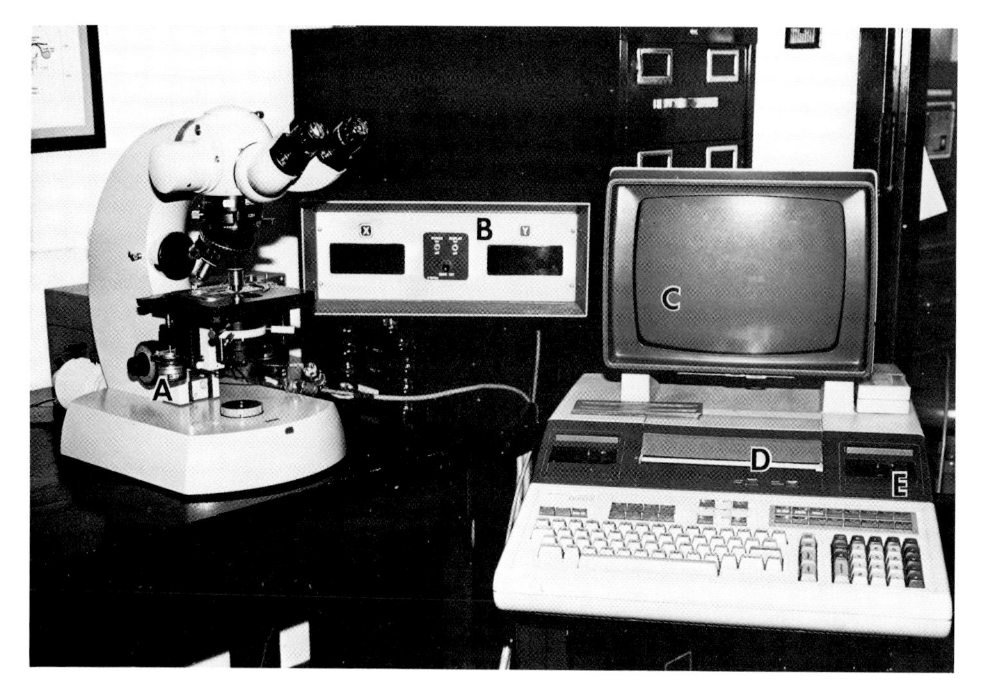

Figure 2. Microscope plotting system attached to the Zeiss Universal light microscope. The system includes two incremental shaft encoders (**A**), a display control counting unit (**B**), two 16-bit parallel interface cards (not shown), and the Hewlett-Packard 9845B microcomputer (**C**). The microcomputer includes a thermal line printer (**D**) and two cartridge tape mass storage drives (**E**).

Figure 3. Circuit diagram of display/control device.

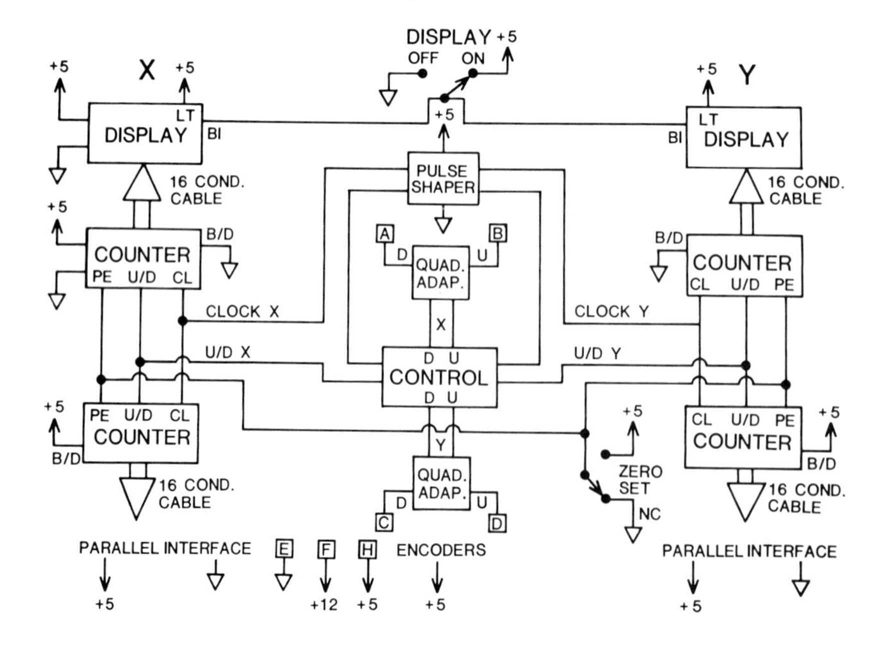

serial interface because it would have slowed the input/output programming routines considerably: Each digit of each count would have to be transferred serially and reconstructed by the computer program. We chose to use a parallel interface because it was somewhat less expensive and could transmit the entire numeric value of the count in parallel. The transfer is thus far more rapid than could be achieved with the serial interface.

We chose 16-bit parallel interface cards because they can transmit 16 bits (± 32,767) of data in parallel to the computer at a very high speed. Counts of 32,000 are well in excess of the number of counts necessary to encode position on either the Philips 201 or Zeiss Universal microscopes.

The Hewlett-Packard 16-bit parallel interfaces (Model 98032A) used with our system are full-duplex cards that have 16 input and output lines as well as handshake, status, and control lines. The flag lines of the interface cards are permanently set so that data are always on the line for computer input. The computer reads the data as a 16-bit word directly from the input data registers of the interface. A READ BINARY BASIC I/O statement is used to read data from the cards. The data acquisition subroutines can store data in core memory at a rate of about 50 Hz using the READ BINARY statement.

Because of this relatively slow data acquisition speed, the subroutines can miss some of the data available on the interface cards. To eliminate this possibility, we designed the display/control device to count at a rate exceeding 50*K* per second. It is physically impossible to rotate the stage drives of either microscope at a speed that exceeds these rates; thus, the count representing stage position is always accurate.

Microcomputer Choices

The computer used to control the microscope plotting system should have several features. It should be portable so that it can be moved from one laboratory site to another. There should be sufficient core memory to hold both the program and data. The computer must be able to accept two interface cards to connect to the shaft encoders. Bit-map graphics should be available to display the plotted maps on the CRT screen. Special function keys are desirable in order to easily access plotter functions. A printer and/or digital plotter are essential to obtain hard copies of the data and plotted maps. There are a number of microcomputers available today that fit these requirements.

We chose a Hewlett-Packard 9845B desktop microcomputer for our microscope plotter because it was one of the few portable, bit-map graphics computers available when we designed our system in 1978 (Figure 2). The 9845B has dual 16-bit NMOS-II microprocessors, 187 KB of RAM memory, a medium-resolution graphics CRT screen, and four interface ports that can accept a number of different interface cards [16-bit parallel, RS-232C serial, BCD, and IEEE-488 (GPIB) instrument interfaces].

The 9845B keyboard includes 16 shiftable special function keys, 17 edit and display function keys, and a numeric keypad in addition to the standard 128 ASCII characters. An 8-in. thermal line printer and two mass storage tape drives for DC-100 magnetic tapes are built into the 9845B cabinet. The graphics system is contained on a graphics ROM and includes 32*K* of resident graphics memory, 40 graphics commands, and a screen resolution of 560 × 455 pixels.

Peripherals for the system include a floppy disk storage device and a digital plotter. The H-P 9895A mass storage device is a two drive system that uses 8-in. dual-density double-sided floppy disks that can store 1.2 MB of data. The H-P 9872A plotter is a four-pen, high-resolution color plotter that accepts up to 12 × 17 in. paper. Although the disk system provides more storage space and higher access speed, the built-in tape drives function quite adequately for storage because the plotter programs do not require frequent mass storage access. The plotter provides very high-resolution hard copies of plotter maps, but lower-resolution maps can also be generated on the thermal line printer using the Dump Graphics feature on the computer.

Many other microcomputer systems with a graphics option would serve the same purpose [see Chapter 3 of this volume]. Hewlett-Packard manufactures a Motorola MC 68000 microprocessor-based computer (H-P Series 200, Model 16) that has most of the capabilities of the 9845, but sells for a base price of under $6,000. Even less expensive microcomputers could be used with our system if a minimum memory of 64*K* were available.

Software

Programming Languages

A number of programming languages could have been chosen for our plotter programs. The 9845B has third-party software packages for FORTRAN and PASCAL. An ASSEMBLER ROM is also available for the computer. We chose the resident BASIC language because of its ease of use, simplicity, and convenient and extensive editing features. The H-P enhanced BASIC has a large I/O command set and a number of advanced graphics commands. It was therefore quite simple to use the language to develop graphics and data acquisition subroutines. The H-P BASIC has other convenient features including six dimensional numeric and string arrays, multicharacter variable names, print using format capabilities, 15 prioritized interrupt levels, variable redimensioning, callable subprogramming, and looping and decision constructs similar to PASCAL (WHILE-DO, REPEAT-UNTIL, LOOP).

Neither FORTRAN nor PASCAL offered any major advantage except speed. Because the H-P enhanced BASIC is an interpretive language and has no compiler, it has a relatively slow execution speed. Speed can be important both for high-speed data acquisition and for complicated reiterative algorithms. Speed of data acquisition was not critical to us because the display/control unit was keeping track of the shaft encoder counts (see above). In fact, data points stored in the trace program were rarely if ever collected at speeds exceeding 50 Hz, which is the approximate sampling rate of the BASIC data acquisition subroutines. Algorithms for calculating profile depths and densities do require substantial amounts of time (approximately 3 seconds/plotted point; 30 minutes to 1 hour for a typical plot). This is not a major problem, however, because we run these routines at night when the computer is not otherwise in use. Thus, execution speed is really not an important factor in microscope plotting as long as data acquisition is handled by an external device.

It should be remembered, however, that other on-line data-acquisition programs designed to collect data from high-speed instruments can depend critically on the execution speed of the language. In these cases, interpretive BASICS can be a poor choice (see Chapter 2). Many BASICs, however, can access assembly language subroutines that will substantially increase execution speed for data acquisition. Compiled BASICs that execute at very high speeds are also available (Chapter 2).

Software Design

There are at least four tasks that should be performed by a computer-based microscope plotter.

1. A computer plotter should be able to generate a digital trace of the tissue. This is desirable both to highlight the features of the specimen and also to serve as a reference for calculating the depth of profiles within the tissue. To produce a digital trace, the computer plotter must be capable of tracing the surface contours of the histological specimen.
2. A microscope plotter must also be able to plot the x–y coordinate positions of profiles within the tissue. It is also helpful if the plotter can annotate the profiles with a code to describe their variable characteristics.
3. The plotter should also be able to reproduce the plot graphically so that the investigator can visualize the spatial relationships in the tissue. This can be done on either a CRT screen or hard-copy plotting device.
4. Finally, the plotter should be able to measure various spatial relationships in the tissue quantitatively.

Our microscope plotting system includes programs for performing each of these tasks. A TRACE program is used to trace around various contours of the tissue specimen. A PLOT program is used to plot the position of profiles within the tissue. An ANALYZE program is used to produce a graph of the plot and to calculate the depth and medial-lateral distance of profiles from the traced contours of the tissue. A DENSITY program computes the density of the plotted profiles within the tissue and calculates the area lying within the traced contours of the tissue. The flow of these programs is diagramed in Figure 4.

Tracing

A digital trace of a histological specimen can be produced by a number of techniques. One approach is to draw the contours of the specimen using a drawing tube (or camera lucida) and a digitizing tablet (Green et al., 1979; Haug, 1979). The specimen is viewed at low magnification through the binoculars of the microscope. The drawing tube is used to superimpose the microscope image and the image of the cursor on the digitizing tablet. The trace is produced by moving the digitizing cursor around the contours of the specimen image viewed through the microscope binoculars. A second approach is to project the image onto a television monitor using a video camera attached to the microscope (Boyle and Whitlock, 1977; Lindsay, 1977; Paldino, 1979; Paldino and Harth, 1977; Uylings et al., 1981; Yelnik et al., 1981). The contours of the specimen image can then be digitized by an electronic cursor that appears on the TV monitor and is manipulated by a joystick, special arrow keys, a digitizing tablet, or other input device. With both of these approaches, the trace is produced by encoding the x–y coor-

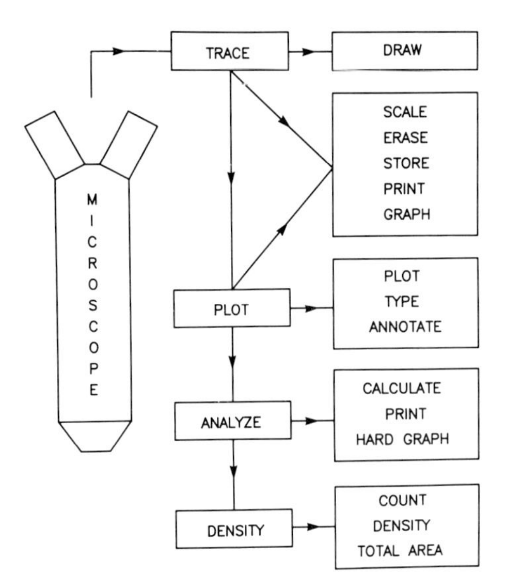

Figure 4. Program flow of the plotter software. The TRACE program is used to draw the outer contours of the tissue (DRAW). This program includes routines for scaling the tissue (SCALE), erasing the tissue (ERASE), storing the tissue (STORE), and printing (PRINT) and plotting (GRAPH) the data. The PLOT program plots the positions of profiles in the tissue (PLOT). This program has routines to code profile types (TYPE) and annotate the data (ANNOTATE). SCALE, ERASE, STORE, PRINT, and GRAPH routines can also be used in the PLOT program. The ANALYZE program calculates depth and medial–lateral distance (CALCULATE), prints these data values (PRINT), and graphs the plot on a digital plotter (HARD GRAPH). The DENSITY program is used to count the number of profiles (COUNT), measure profile densities (DENSITY), and calculate the area of the tissue (TOTAL AREA). [Reprinted from Mize (1983c) with permission of Elsevier Biomedical Press, Amsterdam.]

dinates of the cursor as it is moved around the surface contours of the image. A third method uses interactive video image analyzers to extract the borders of the tissue automatically. A fourth method involves moving the stage drives of the microscope so that the contours of the tissue pass under cross hairs in the binoculars of the microscope (Capowski, 1977; Capowski and Sedivec, 1981; Foote et al., 1980; Forbes and Petry, 1979; Williams and Elde, 1982). The position of the stage as it moves with the tissue is detected by an electronic encoding device. Each approach has advantages and disadvantages, which have been discussed by Capowski and Cruce (1979) [see also Chapter 4].

We chose the fourth approach because it was by far the least expensive alternative for use with an electron microscope. Video reproduction of electron microscope images is expensive and can require extensive modifications of the electron microscope. Image analysis systems are also prohibitively expensive. Zeiss manufactures a camera lucida device for its own electron microscopes, but drawing devices are not generally available for other electron microscopes. By contrast, the shaft encoders used to produce the trace on our system are relatively inexpensive and tracing with the stage drives is a straightforward, easy process using the electron microscope. Tracing is somewhat more difficult using the light microscope, but the technique can also produce an accurate outline of the tissue specimen.

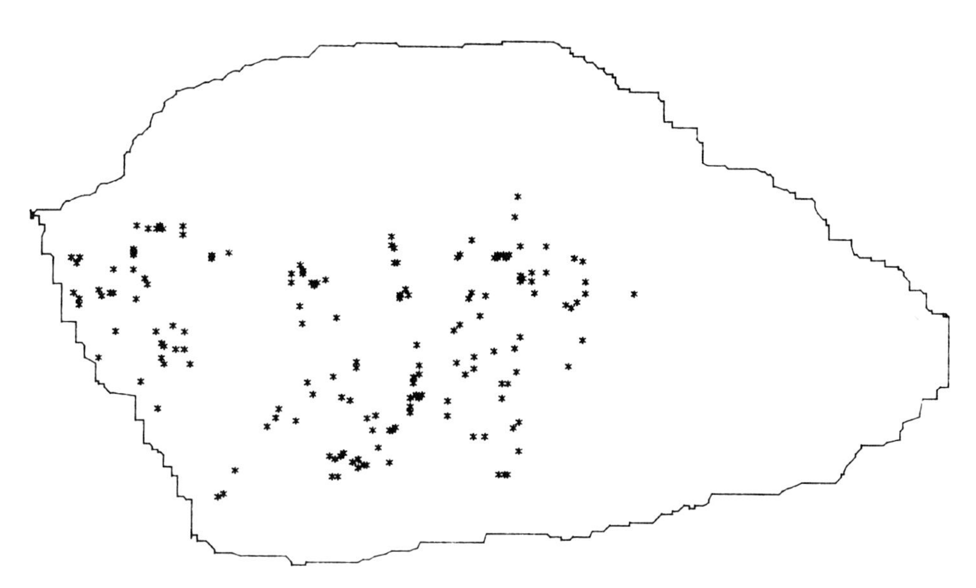

Figure 5. Light microscope plot of the lateral geniculate nucleus, illustrating positions of cells labeled by tritiated L-glutamate.

We use a TRACE program to produce both a map of the contours of the tissue and outlines of various landmarks such as blood vessels and intralaminar boundaries (Figure 5). The TRACE program works as follows. When the program is loaded, instructions appear on the CRT screen for initializing the system. The operator is prompted to turn on the display/control unit, plug the interface cards into the computer and the shaft encoders into the display/control unit, set the system resolution, and assign a file name for storing the data. The operator then enters identifying information about the experiment, including the date, specimen type, microscope conditions, block and section numbers, section thickness, trace and plot magnifications, and the operator's name.

Once the plotter has been initialized, the tracing mode is entered and a digital outline of the tissue can be produced. To accomplish this, the area to be traced must first be scaled. The microscope stage is zeroed by moving to the far upper left corner of the grid or slide, well above the region to be plotted. When this zero reference point has been chosen, the reset button on the display/control unit is pressed, resetting the LED displays and counters to zero (Figure 2). This zero point becomes the reference point for all subsequent positions of the stage. After the stage has been zeroed, a graphics window is defined by moving just above the upper left corner and just below the lower right corner of the tissue. These points are digitized by depressing a special-function scale key. The two points define the window on the graphics device in which the trace will appear. The corners of the tissue section are also digitized using the same procedure. These coordinate positions are used by the ANALYZE program to define the top and sides of the tissue so that distance from these surfaces can be calculated.

After all reference points have been defined and digitized, the operator begins to trace the tissue. To do this, he or she moves to the edge of the tissue, depresses a PEN-DOWN special-function key on the computer keyboard, and begins to move the stage drives of the microscope so that the edges of the tissue pass under the cross hairs located in the microscope's binocular eyepiece. The cross hairs act like a stationary digitizer pen as the tissue passes under it. The stage can be moved without drawing by depressing the PEN-UP special-function key. In addition to digitizing the outline of the tissue, it is helpful to trace a few internal landmarks in order to compare more accurately the trace with the image of the actual histological specimen. The trace is displayed on-line on the computer's graphics CRT so it can be examined for accuracy.

The TRACE program has a number of special functions, which are accessed by special-function keys (Table 1). The trace can be stored at any time under operator control by depressing the STORE special-function key. The trace can be erased at any time using the ERASE key. The trace can be restarted with the RESTART key, or drawn on the CRT screen by depressing the GRAPH key. The actual x–y coordinate data values can be printed on either the CRT or thermal line printer by depressing the PRINT key. The pen can be controlled with the PEN-UP and PEN-DOWN keys. The trace magnification can be changed by rescaling the graphics window (SCALE key). A hard copy of the trace can be produced on the thermal line printer using the DUMP GRAPHICS command.

Table 1. Special Functions Used in Microscope Plotting System

Functions	Key No.	Description
TRACE	1	Enters the tracing mode. Generates an on-line CRT graphics plot of the trace.
PLOT	0	Enters the plot mode. Converts special function keys to 31 numeric codes representing different types of element.
STORE	2	Stores trace or plot data in temporary file. Allows creation of permanent data file when plot is complete.
GRAPH	3	Generates an off-line graphics plot of all trace and plot data. Plot can be reproduced on CRT, thermal printer, or plotter.
PRINT	4	In trace mode, prints x–y data values of trace. In plot mode, prints x–y data values of plotted points, type code and name, and annotation. Prints data on CRT or printer.
ERASE	5	In trace mode, erases entire trace. In plot mode, erases or corrects values of any plotted point.
END	6	Specifies end of plot. Transfers all data from temporary to permanent data file. Ends the program.
RESTART	7	Restarts trace or plot mode after temporary data storage.
PEN-UP	8	Raises the CRT "pen" in trace mode. No data points are collected but cursor moves on screen when stage drives are turned.
PEN-DOWN	9	Lowers the CRT "pen" in trace mode. Data points are collected.
SCALE	10	Used to set the CRT graph window and to set the coordinates for the corners of the tissue.
RE-SCALE	11	Used to reset the CRT graph window or the corners of the tissue.
EXIT	31	Used to exit plot mode. Resets special-function keys to function mode.

Although the tracing procedure sounds awkward, it is mastered quite easily, especially if the tissue contours parallel the stage axes. The tracing is easiest if only one stage drive is moved radically at one time. If the tissue is parallel to one of the axes, the other drive need be moved only slightly to keep the trace on track.

Plotting

Plotting the positions of profiles such as cells and synapses also requires a method for electronically encoding position. Camera lucida-digitizing systems, video display systems, and stage encoders can all be used to plot position. Our system plots x–y coordinate positions of profiles using the shaft encoders.

Our PLOT program allows the operator to store three values:

1. the x–y coordinate values of each plotted point;
2. a numeric code representing the type of profile;
3. a photograph number if a picture of the profile has been taken.

To plot a profile, the tissue is scanned systematically until a profile of interest is located. A special-function key is then depressed, which stores the coordinate values of the profile's position in computer memory. In plot mode, each of the special-function keys on the computer are set to represent a unique code for profile type. It is therefore only necessary to depress a single key to encode any of 32 profile types. The details of the PLOT program follow.

The PLOT program begins by asking the operator whether he or she wants to define some of the special-function keys to represent particular profile types. Twenty of the special function keys (numeric codes 1–20) are predefined to represent common biological organelles (1 = labeled terminal, 2 = glial cell, etc.). The remaining special-function keys can be operator-defined to represent profiles of the operator's choosing. Once the special-function keys have been defined, the operator establishes the width and interval boundaries used for interval scanning (see below).

After these conditions are set, the operator enters the active plot mode and begins scanning the tissue, usually from the top surface to the bottom. Two scanning strategies can be used: block scanning and interval scanning. In *block scanning,* the entire tissue is scanned. In *interval scanning,* selected columns of tissue are scanned (see considerations below). When a profile to be plotted is identified, it is positioned under the cross hairs in the microscope's binocular eyepiece. The special-function key whose numeric code represents that profile type is then depressed. This automatically records the x–y coordinate position of the profile, the code for profile type, and the photograph number if it has been entered from the keyboard. (Any 10 numeric ASCII characters can be used to represent the photograph number.) These data are also printed on the thermal line printer so that an on-line hard-copy printout of the plotted data can be examined at any time.

A number of special functions are also available in the PLOT program, including STORE, GRAPH, PRINT, and ERASE (Table 1). The ERASE key allows the operator either to erase or to correct a plotted data point. When the plotting session is completed, the operator presses the END key, which automatically stores the data on a permanent data file and ends the program.

Graphing and Analyzing Data

Once the plotted data have been collected, they can be graphed and analyzed quantitatively. The ANALYZE program has routines for graphing the completed plot on the digital plotter; for calculating the depth and medial-lateral distance of each plotted point from the top and sides of the trace; and for printing these values on the CRT or thermal line printer (Figure 4).

The graph routine produces a finished copy of the plot on the CRT, printer, or four-color digital plotter using graphics DRAW and PLOT commands. Both the trace and plotted profiles are mapped. The operator can represent different types of profile using any ASCII character as a symbol (Figure 5). Up to four colors can be used on the plotter to represent different profile types. The graph routine also prints an index identifying the plot number, plot type, and date.

The calculation subroutine computes the distance of all plotted profiles from the top and one side of the traced specimen, expressed in microns. To calculate these values, the routine first reads in the trace and plot data from the stored plot file. It next defines which x–y data coordinate pairs of the trace represent the "top" and "one side" of the tissue. This is accomplished by identifying all data values in the sequential trace that were recorded between the two "corners" representing "top" and the two corners representing "right side."

The corners are defined in the following manner (Figure 6). The operator always begins at the lower left edge of the tissue (regardless of its orientation in

Figure 6. Graphic illustration of the method used to compute depth and medial–lateral distance by comparing trace and plot data points. *A* represents the starting point of the trace. *B* and *C* represent corners 1 and 2, which define the "top" of the trace. Points between *C* and *D* represent the "side" of the trace. *E* represents a plotted point. Depth and medial-lateral distance are calculated using an algorithm described in the text (p = plot point; t = trace point; ti = interpolated point; x = x coordinate; y = y coordinate).

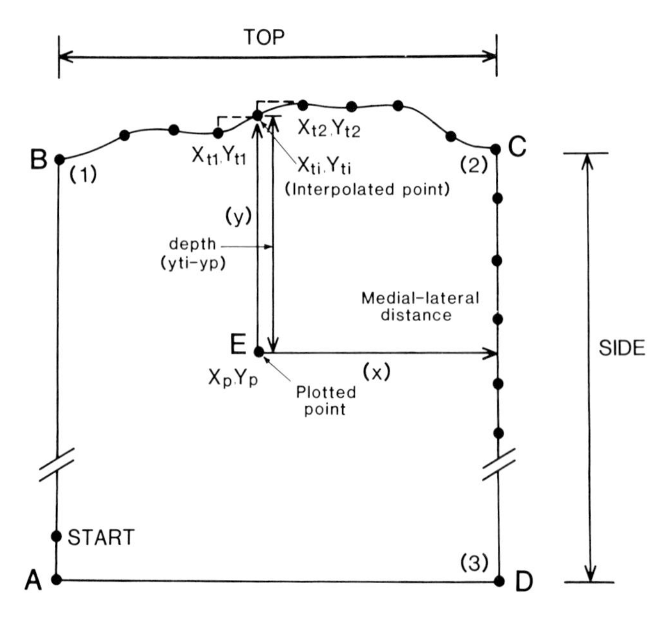

the field of the microscope). Upon reaching the first "corner" of the tissue, he or she presses a special-function key to define the x–y coordinates of that corner. Reaching the second corner, he or she presses another special-function key to define that corner. The third corner is similarly defined. Because the operator always begins the trace at the lower left edge of the tissue, "top" is always defined by the calculation routine as those trace points that lie between corners 1 and 2 (B and C, Figure 6). "Side" is defined as all data points that lie between corners 2 and 3 (C and D, Figure 6).

To calculate depth, a depth subroutine compares every plotted point with every trace point defined as belonging to the "top" of the tissue. The routine identifies the coordinate point in the trace whose value is identical to the x coordinate value of the plotted point. If no trace x value is identical, the algorithm calculates, by interpolation, an imaginary trace point whose x value is identical to the x value of the plot point and whose y value is an interpolated value between the two real y values of the trace points that lie on either side of the plotted point (Figure 6). Once this imaginary trace value has been defined, the depth subroutine compares the y coordinate values of the trace point and the plotted point. The difference between the two points represents the distance (depth) of the plotted point from the top surface. This value is then converted to a unit measure of depth (in microns).

The same process is repeated for each plotted point. A medial–lateral distance subroutine then compares each plotted point with all data values of the "side" trace to obtain a measure of medial–lateral distance (also expressed in microns). When the routines are completed, the ANALYZE program stores the depth and medial–lateral data values on a plot file whose name is defined by the operator.

The print routine of the ANALYZE program produces a hard-copy printout of these data values. The routine prints the number of each plotted point, its x and y coordinate values, the profile type, and picture number as well as the values for depth and medial–lateral distance. Each of these values can be analyzed statistically using commercial statistical packages (see below).

A DENSITY program is used to count the number of plotted profiles, to calculate profile density (number per unit area of tissue), and to calculate the total area of the plot. If the interval scanning mode was used, the DENSITY program also calculates the profile density and area of individual scans. The program begins by computing the number of scans and the x values representing the width of each scan, based upon the scan and interval values entered during the plot routine. The program then counts the number of each type of profile whose x coordinate value falls within the limits of a given scan. These counts are summed across all scans to obtain a total profile count. Profile density (number per unit area) is calculated by dividing the number of each profile type in each scan by the area of each scan.

The scan area is determined using a modified trapezoid function described elsewhere [Chapter 8]. Total scan area is computed by summing the individual scan areas. In cases where the block scanning mode is used, the total area within the trace boundaries is computed, also using the modified trapezoid function. Profile density becomes the total number of each profile type plotted divided by the total area. The program ends by printing the individual and total scan values on the thermal line printer.

Considerations When Plotting

Tissue Preparation

When making quantitative measures of histological materials, it is important to minimize variations in tissue thickness and staining density. This is best accomplished by using standardized procedures for fixing and processing the tissue. Because we plot extensively with the electron microscope (EM), all of our tissue is fixed with aldehydes using standard EM fixation procedures. For normal tissue, we use a 2% glutaraldehyde–2% paraformaldehyde concentration in a 0.1-*M* phosphate buffer. The aldehyde concentrations are sometimes altered for cytochemically treated brains, but we try to use fixatives with identical aldehyde concentrations when sections from different animals will be compared statistically.

Tissue used for light microscope plotting is stained with thionin, mounted conventionally on glass slides, and coverslipped. EM tissue is postfixed in osmium tetroxide, dehydrated with a graded series of alcohols, and embedded in Medcast-Araldite (Ted Pella, Tustin, CA). Our plastic formula is very hard so that it is easy to cut large sections. The plastic-embedded sections are also relatively immune to expansion when exposed to the electron beam.

We cut all of our tissue into consecutive 50–100-μm slices using an Oxford Instruments Vibratome. By saving every slice, we can sample from any region of the specimen. The tissue slices are flat embedded on plastic microscope slides (A.H. Thomas, Philadelphia). To flat embed, the tissue is gently squeezed between the microscope slide and a hardened plastic blank produced from a beam capsule. After the plastic is cured for 48 hours in a 60° C oven, the blanks are popped off the plastic slides. In 95% of the cases, the tissue adheres to the beam capsule. When this fails to occur, the tissue and surrounding plastic are cut from the microscope slide and reglued to the beam capsule. The flat embedding technique prevents the tissue from curling during the curing process and allows us to cut thin sections of large regions of the tissue.

In order to achieve smooth tissue edges, we precision trim mesas using an MT-2B Ultramicrotome (Sorvall, Inc.). The position of the mesa on the block face is determined precisely by using an eyepiece reticule in the stereomicroscope of the Ultramicrotome. We cut large mesas (approximately 800 × 1,600 μm) in order to sample as much of the tissue as possible. Four-mil formvar-coated slot grids are used so that none of the tissue is obscured by a mesh. The thicker grid provides needed support for the formvar and section. The 4-mil slot grids can be purchased from Ted Pella (Tustin, CA).

Choice of Microscope

Shaft encoders can be attached conveniently to most light microscopes. Electron microscopes must have rotating shafts in order for shaft encoders to be used effectively. We use a Philips 201 electron microscope for plotting. This microscope is quite stable and requires little realignment during a plotting session. Gears are used to attach the shaft encoders to the microscope stage drives. Each gear is mounted with removable collars that do not require modification of the shafts (Figure 1A). Other microscopes with rotating shafts to control stage position should also accommodate shaft encoders.

Sampling from Tissue

It is vitally important to adopt a systematic sampling strategy when plotting tissue. We have devised two sampling strategies for plotting. *Block* scanning is used to obtain plots of every profile in the tissue sample. This strategy is effective if the profile types to be plotted are sparsely distributed in the tissue. It is easy to plot profiles without double counting a profile if their distribution is sparse. We use this approach when plotting the distributions of autoradiographically or immunocytochemically labeled neurons.

When profiles are densely packed, it is important not to sample the whole tissue since this would inevitably result in double counting some profiles. Instead, the tissue should be sampled at systematic intervals. We use this *interval* scanning strategy when plotting synapses because their density is very high in nervous tissue.

In block scanning, the entire tissue is scanned. In interval scanning, the operator divides the tissue into a series of columns called *scans*. The columns extend the entire depth of the tissue. Column width is determined by the magnification at which the tissue is viewed. We plot synapses at an indicated scope magnification of 7,000. Viewed through the binocular eyepieces of the microscope, the width of visible tissue is about 6 μm. This distance defines the column width. The amount of visible tissue will be greater or less at higher or lower magnifications. To avoid double counting, an empty interval about 2 μm wide is left between columns.

The plot program can accept column and interval widths of any size. When sampling at low magnification, it is often desirable to have a large column width and a narrow interval width so that a large percentage of the total tissue can be scanned.

Sampling Direction

Tissue can be scanned either vertically, horizontally, or obliquely, depending upon its orientation relative to the axes of the stage. We usually mount our tissue in the scope so that it is parallel to one of the stage axes. It is important that the scanning direction be orthogonal to the direction of any tissue or profile gradients if the gradients are in a single plane of the tissue. The superior colliculus, for instance, is a laminated structure whose layers are arranged in horizontal tiers beneath the tissue surface. We sample this tissue vertically, a direction perpendicular to the laminar gradients; otherwise, one or more laminae might be missed during plotting. When tissue gradients are suspected, it is important to sample perpendicular to the gradient, as Sterling (1975) has emphasized.

Photographing Profiles

We often measure other features of plotted profiles, such as their size, shape, and labeling density. These geometric parameters are measured from photographic prints using a digitizer-based computer morphometry system [see Chapter 8]. When plotting, profile density is often so high that it is impractical to take photographs of every plotted element. In this case, it is important to develop a sampling strategy for choosing which profiles to photograph. The most obvious strategy is to photograph every third, fifth, or tenth profile. This, however, may produce a large sampling error if the plotted profiles are not randomly distributed in the tissue. A preferable strategy is to use a random number table to determine which profiles should be photographed. Other methods for random sampling may also be used (Weibel, 1979).

Section Drift

Once the tissue contours have been digitized, the computer's digital representation of the tissue is fixed. The plotter programs are not designed to compensate automatically for shifts in tissue position during a plotting session. Section drift during plotting must therefore be avoided. Very small movements over many hours probably do not alter calculations very drastically. Nevertheless, the following precautions can be taken to minimize section movement:

1. Sections can be "cooked" in the microscope for about an hour before the plotting session is started by exposing the whole tissue to the beam using a mesh image magnification on the microscope (scan mode).
2. A liquid nitrogen cold trap can be used in order to increase section and stage stability.
3. The beam can be kept away from crossover to avoid localized tissue sublimation.
4. 80–100-kV accelerating voltages can be used, minimizing tissue movement because the beam electrons pass through the tissue faster.
5. At 2–3 hour intervals, the corners of the tissue can be checked to see if they still correspond to the coordinate positions recorded in the computer; if they do not, the x–y coordinate positions can be recentered to correspond to the new tissue positions. These procedures reduce plotting error to a minimum.

Research Applications

We use the microscope plotter to plot the spatial distributions of neurons, synapses, and other organelles in the central visual system (Caldwell and Mize, 1981; Harrell et al., 1982; Mize, 1983a,b; Mize and Horner, 1984). A variety of statistical graphics routines are used to examine these distributions. These include the raw plots of the profiles (Figure 5), histogram summaries of the profile distributions (Figure 7), time plots of profile densities, and two-dimensional scattergrams to illustrate correlations between profile position and other measured variables, such as the cross-sectional areas of cells.

The data are statistically analyzed using comprehensive statistics packages sold by Hewlett-Packard. The packages include summary statistics (means, medians, standard deviation and variance, correlation coefficients, and various order statistics), one-way and multiple analyses of variance, two-sample T-tests, linear and nonlinear regressions, and a battery of nonparametric tests. Distribution analyses are also included with the packages.

These quantitative analyses have been useful for distinguishing populations of synapses and cells in the visual system. For example, we have used the microscope plotter to examine the distributions of synapses in the cat superior collic-

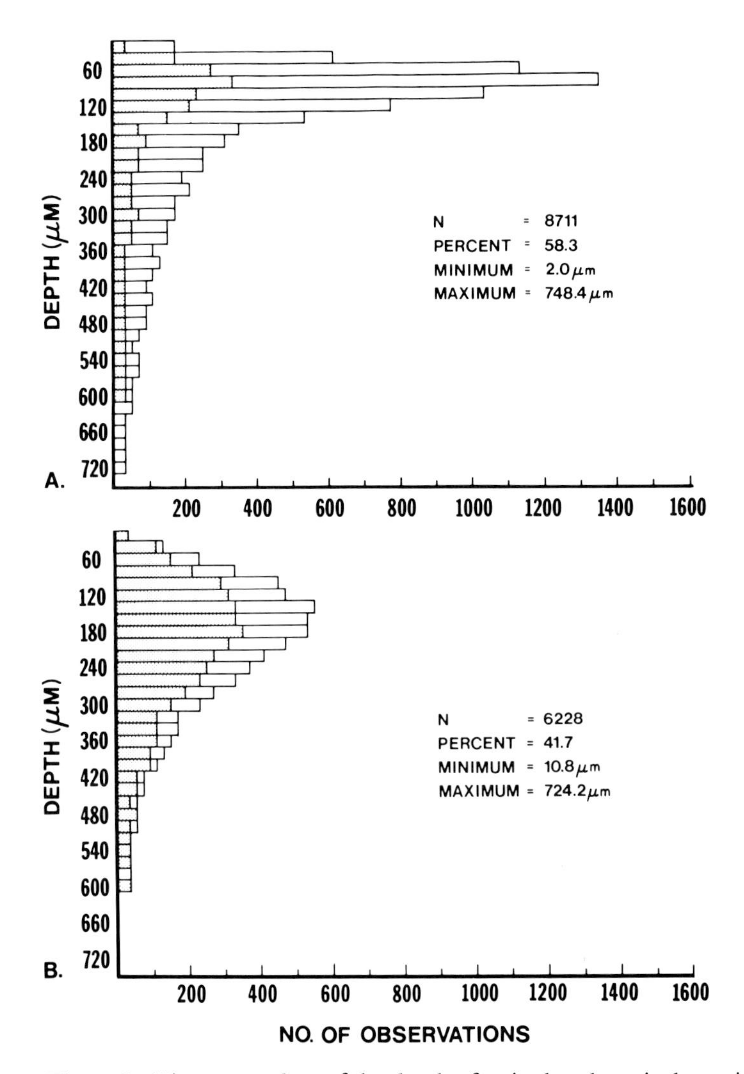

Figure 7. Histogram plots of the depth of retinal and cortical terminals in the cat superior colliculus. Upper plot (**A**) shows that retinal terminals are most heavily concentrated in the superficial 150 μm of the colliculus. Lower plot (**B**) shows the depth of cortical terminals, which are most densely distributed deeper in the superficial gray layer. [Reprinted from Mize (1983b) with permission of Springer-Verlag, New York.]

ulus, (Mize, 1983b). The superior colliculus is a midbrain visual structure involved in orienting behaviors. The superficial layers of the structure are innervated by two major afferent pathways, one from the retina and the other from visual cortex.

A histogram summary of the distribution of these two terminal types illustrates these differences quantitatively (Figure 7). Nearly 60% of all retinal terminals are concentrated in a dense band within the upper 150 μm of the superficial gray layer of the colliculus. The cortical terminals are concentrated in the deep superficial gray layer, below the dense band of retinal terminals. The density of the two types of synapse also varies within the rostral–caudal plane of the colliculus. The retinal terminal density is highest in the most rostral segment of the superior colliculus. The density decreases almost threefold about 1 mm behind the rostral pole, a specialized region of the colliculus representing the area centralis. There is a corresponding increase in cortical terminals in this region (Mize, 1983b).

We have also used the plotter with the light microscope to examine the distributions of neurons labeled with tritiated neurotransmitters (in collaboration with Malinda Fitzgerald). One hour after injection of 25–50 μCi of tritiated L-glutamate into the C laminae of the dorsal lateral geniculate nucleus, heavy grain accumulations were found over some regions of the neuropil and selected cell bodies. Plots of the laminar positions of heavily labeled neurons revealed that the cells were particularly numerous in laminae A_1 and C of the lateral geniculate (Figure 5). To determine if there was a relationship among cell position, cell size, and grain density, we photographed some of the plotted cells. The grains overlying each cell were counted and the cell's size measured with our computer-based morphometry system (Chapter 8). Grain densities ranged from 4.88 to 128 grains/100 μm^2. Significant correlations were found among depth, cell size, and grain density. Cells with average diameters of 15–30 μm had the highest grain accumulations (mean of 45 grains/100 μm^2). Smaller cells had significantly lower grain densities. These results suggest there may be a high affinity uptake of L-glutamate in selected populations of neurons in the dorsal lateral geniculate nucleus. Large cells within laminae A_1 and C appear to be particularly heavily labeled by this putative amino acid transmitter.

The microscope plotter can also be used to study the spatial distributions of more molecular profiles, including mitochondria, synaptic vesicles, and autoradiographic grains in EM tissues. The resolution of the system, however, will need to be increased to accommodate these uses. Shaft encoders with resolutions of 1,250 counts per revolution are sold by both Vernitech and Litton. A 1,250 count encoder would provide a resolution of 0.04 μm on the Philips 201 electron microscope. The 16-bit parallel interface cards could accommodate this resolution (62,500 total counts) only by dividing the counts equally into positive and negative values (because the interface cards can only encode $\pm$32,767 counts). The microscope stage may also produce electrical jitter at this magnification. For this reason, encoder resolutions of 250–500 counts per revolution are probably better suited to electron microscope plotting.

Conclusions

Computer-based microscope plotters have obvious advantages in studies of spatial distributions. Encoding devices can encode the location of organelles with very high resolution. A general purpose graphics microcomputer can handle *all* of the other chores of plotting. The x–y coordinate positions of profiles can be stored automatically in computer memory so that data do not have to be recorded by hand. The computer can keep track of codes for profile type and profile counts. Depth and medial–lateral distance can be calculated from surface landmarks using special algorithms. The maps can be displayed graphically on the computer CRT or a digital plotter. A permanent record of the data can be stored on magnetic tape or floppy disk. Statistical programs can be used to analyze the data mathematically. In short, the microcomputer provides enormous power for collecting, analyzing, and displaying data about distribution patterns found in biological tissues.

Several other computer-based plotters have recently been developed to map biological distributions. Forbes and Petry (1979) describe a plotter that includes a Digital Equipment PDP-12 computer, a Zeiss Universal microscope with motorized scanning stage, and programs written in ASSEMBLER and FORTRAN. The system is used to map degenerating axons in brain tissue, a purpose similar to ours. Williams and Elde (1982) have developed a computer plotter that uses a Hewlett-Packard 9845 microcomputer like ours. The system, however, is used principally to outline and shade regions of tissue that contain different densities of immunocytochemical label. The BASIC programs designed for the system can measure the size of the area that has been labeled, thus producing semiquantitative neurohistochemical maps.

Other recently developed computer microscope plotters have a capability for producing three-dimensional maps of tissue. Foote et al. (1980) describe a system that uses a Zeiss microscope with stage encoders, a DEC PDP-11/34 minicomputer, and an Evans and Sutherland graphics display processor. Sequential sections are plotted with the microscope, similar to the system described in this chapter. The serial sections are then aligned under computer control, using the display processor to rotate and translate the sections until they are in register. With this system it is possible to measure the distributions of labeled neurons within a volume of tissue and compare the distributions in different animals. A similar three-dimensional mapping system is described by Johnson and Capowski (1983) [see also Chapter 11].

To my knowledge, a commercial system for mapping spatial distributions with a microscope is not yet available. There is, however, an intense interest in the neuroscience community to study the distributions of cytochemically labeled cells. One would suspect that a commercial system will be developed in the near future to accommodate this interest.

I thank Joe Laughter of the Division of Biomedical Instrumentation at the University of Tennessee Center for the Health Sciences for designing the display/control unit. Scott Eanes and Bob McCoy of the Hewlett-Packard Corporation helped considerably in setting up the computer. Lee Danley and Linda Horner expertly prepared the figures. Dr. John Harrell helped program the density routines. Mary Gaither typed the manuscript. This project was supported by USPHS Grant EY-02973 from the National Eye Institute, a New Faculty Research Grant from the State of Tennessee, an equipment grant from the College of Medicine, UTCHS, and a donation from the Hewlett-Packard Corporation.

References

Boivie, J., G. Grant, and H. Ulfendahl, (1968) The *X–Y* recorder used for mapping under the microscope. Acta Physiol. Scand. 74:1A–2A.

Boyle, P.J.R., and D.G. Whitlock (1977) A computer-controlled microscope as a device for evaluating autoradiographs. In R.D. Lindsay (ed.): Computer Analysis of Neuronal Structures. New York: Plenum, pp. 133–148.

Caldwell, R.B., and R.R. Mize (1981) Superior colliculus neurons which project to the cat lateral posterior nucleus have varying morphologies. J. Comp. Neurol. 203:53–66.

Capowski, J.J. (1977) Computer aided reconstruction of neuron trees from several serial sections. Comput. Biomed. Res. 10:617–629.

Capowski, J.J., and W.L.R. Cruce (1979) How to configure a computer-aided neuron reconstruction and graphics display system. Comput. Biomed. Res. 12:569–587.

Capowski, J.J., and M.J. Sedivec (1981) Accurate computer reconstruction and graphics display of complex neurons utilizing state-of-the-art interactive techniques. Comput. Biomed. Res. 14:518–532.

Curcio, C.A., and K.R. Sloan (1981) A computer system for combined neuronal mapping and morphometry. J. Neurosci. Methods 4:267–276.

DeVoogd, T.J., F.-L.F. Chang, J.K. Floeter, M.J. Jencius, and W.T. Greenough (1981) Distortions induced in neuronal quantification by camera lucida analysis: Comparisons using a semi-automated data acquisition system. J. Neurosci. Methods 3:285–294.

Eidelberg, E., and F. Davis, (1977) An improved electronic pantograph. J. Histochem. Cytochem. 25:1016–1018.

Foote, S.L., S.E. Loughlin, P.S. Cohen, F.E. Bloom, and R.B. Livingston (1980) Accurate three-dimensional reconstruction of neuronal distributions in brain: Reconstruction of the rat nucleus locus coeruleus. J. Neurosci. Methods 3:159–173.

Forbes, D.J., and R.W. Petry (1979) Computer-assisted mapping with the light microscope. J. Neurosci. Methods 1:77–94.

Gilbert, C.D. (1983) Microcircuitry of the visual cortex. Ann. Rev. Neurosci. 6:217–248.

Glaser, E.M., and H. Van der Loos (1965) A semi-automatic computer microscope for the analysis of neuronal morphology. IEEE Trans. Biomed. Eng. BME 12:22–31.

Grant, G., and J. Boivie (1970) The charting of degenerative changes in nervous tissue with the aid of an electronic pantograph device. Brain Res. 21:439–442.

Green, R.J., W.J. Perkins, E.A. Piper, and B.F. Stenning (1979) The transfer of selected image data to a computer using a conductive tablet. J. Biomed. Eng. 1:240–246.

Harrell, J.V., R.B. Caldwell, and R.R. Mize (1982) The superior colliculus neurons which project to the dorsal and ventral lateral geniculate nuclei in the cat. Exp. Brain Res. 46:234–242.

Haug, H. (1979) The evaluation of cell-densities and of nerve-cell-size distribution by stereological procedures in a layered tissue (cortex cerebri). Microsc. Acta 82:147–161.

Johnson, E.M., and J.J. Capowski (1983) A system for the three-dimensional reconstruction of biological structures. Comp Biomed. Res. 16:79–87.

Lindsay, R.D. (1977) The video computer microscope and A.R.G.O.S. In R.D. Lindsay (ed.): Computer Analysis of Neuronal Structure. New York: Plenum, pp. 1–19.

Mize, R.R. (1983a) Variations in the retinal synapses of the cat superior colliculus revealed using quantitative electron microscope autoradiography. Brain Res. 269:211–221.

Mize, R.R. (1983b) Patterns of convergence and divergence of retinal and cortica synaptic terminals in the cat superior colliculus. Exp. Brain Res. 51:88–96.

Mize, R.R. (1983c) A computer electron microscope plotter for mapping spatial distributions in biological tissues. J. Neurosci. Methods 8:183–195.

Mize, R.R., and L.H. Horner (1984) Retinal synapses of the cat medial interlaminar nucleus and ventral lateral geniculate nucleus differ in size and synaptic organization. J. Comp. Neurol. 224:579–590.

Overdijk, J., H.B.M. Uylings, K. Kuypers, and A.W. Kamstra (1978) An economical, semi-automatic system for measuring cellular tree structures in three dimensions, with special emphasis on Golgi-impregnated neurons. J. Microsc. 114:271–284.

Paldino, A.M. (1979) A novel version of the computer microscope for the quantitative analysis of biological structures: Application to neuronal morphology. Comput. Biomed. Res. 12:413–431.

Paldino, A.M., and E. Harth (1977) A measuring system for analyzing neuronal fiber structure. In R.D. Lindsay (ed.): Computer Analysis of Neuronal Structures. New York: Plenum, pp. 59–72.

Reed, D.J., D.R. Humphrey, and R. Gold (1980) A simple computerized system for plotting the locations of cells of specified sizes in a histological section. Neurosci. Lett. 20:233–236.

Sterling, P. (1973) Quantitative mapping with the electron microscope: Retinal terminals in the superior colliculus. Brain Res. 54:347–354.

Sterling, P. (1975) Quantitative mapping with the electron microscope. In M.A. Hayat (ed.): Principles and Techniques of Electron Microscopy, vol. 5. New York: Van Nostrand Reinhold, pp. 1–18.

Uylings, H.B.M., J.G. Parnavelas, and H.L. Walg (1981) Morphometry of cortical dendrites. In M.A. Galina (ed.): Advances in the Morphology of Cells and Tissues. New York: Alan R. Liss, Inc., pp. 185–192.

Wann, D.F., T.A. Woolsey, M.L. Dierker, and W.M. Cowan (1973) An on-line digital computer system for the semiautomatic analysis of Golgi-impregnated neurons. IEEE Trans. Biomed. Eng. 20:233–247.

Weibel, E.R. (1979) Stereological Methods, vol. 1. Practical Methods for Biological Morphometry. London: Academic.

Williams, F.G., and R. Elde (1982) A microcomputer-aided system for the graphic reproduction of neurohistochemical maps. Comput. Programs Biomed. 15:93–102.

Woolsey, T.A., and M.L. Dierker (1978) Computer-assisted recording of neuroanatomical data. In R.T. Robertson (ed.): Neuroanatomical Research Techniques. New York: Academic, pp. 47–85.

Yelnik, J., G. Percheron, J. Perbox, and C. Francois (1981) A computer-aided method for the quantitative analysis of dendritic arborizations reconstructed from serial sections. J. Neurosci. Methods 4:347–364.

6

Quantitative Analysis and Reconstruction of Retinal Ganglion Cells Using a Color Graphics Computer

Franklin R. Amthor

Introduction

The retina is a laminated, hierarchical, repetitive structure in which neurons communicate primarily by electrotonic spread of slow potentials. The processing of information through the retina is basically serial; at each level (lamina) of the hierarchical structure a different transformation of the visual input may occur. These transformations culminate in neural interactions in the dendritic tree of the retinal ganglion cell, the output neuron of the eye.

Almost all retinal ganglion cells have cell bodies in the ganglion cell layer and axons that leave the retina without bifurcating or synapsing there. The ganglion cell dendrites may occupy one or more sublamina in the inner plexiform layer (IPL), where they receive inputs from bipolar cells and amacrine cells. Bipolar cells have cell bodies in the inner nuclear layer (INL), and conduct information from the photoreceptors and horizontal cells in the outer retina to the IPL. Amacrine cells also have cell bodies in the INL, and subserve mainly lateral interactions within the IPL.

Sublaminar segregation of the inputs to the ganglion cell, and the optical geometry of the eye, impose orthogonal constraints on its dendritic tree structure. The location of a ganglion cell's dendrites in the x–y (tangential) plane determines the specific region of visual space to which the cell responds. This is called the cell's receptive field. The vertical (z) location of a ganglion cell's dendrites within a certain sublamina of the inner plexiform layer (IPL) determines the type of input the dendrite receives from other retinal neurons.

This input scheme is not the only restriction on ganglion cell dendritic geometry. Various excitatory and inhibitory inputs are summed by the ganglion cell dendritic tree and modulate its firing of action potentials, which traverse the optic nerve to the brain. *Feature extraction* or other transformations of the inputs, equivalent to logical operations such as OR or NAND, may involve nonlinear summations of those inputs. This nonlinear summation may depend on aspects of the dendritic geometry, such as the location of spines or the thickness of the dendrites of different branch orders, as well as on the precise arrangement of the inputs.

Knowledge of the dendritic tree structure of retinal ganglion cells is important not only for understanding structure-function relationships, but also for classifying cells and studying their development. For example, it has been estimated in the cat retina that any point in visual space may be simultaneously within the receptive fields of as many as 35 different ganglion cells (Fischer, 1973). Furthermore, it has been shown that the on- and off-center cat alpha cells cover or "tile" the retinal surface independently (Wässle et al., 1983) and completely, with little or no overlap for each type. If such tiling is found to be the case for other ganglion cell morphological classes, then it is very likely that there are as many as 35 distinct types of ganglion cell. These may have different resolutions and be concerned with different visual features, representing, in effect, 35 different and parallel "eyes" examining the visual world. With this many possible cell types, it is essential to analyze quantitatively the ganglion cell dendritic tree structure, in conjunction with other morphological features, to differentiate the various anatomical types. Quantitative dendritic tree analysis will also be necessary to examine what genetic rules or "programs" are followed in dendritic growth, how they are expressed in both normal and abnormal development, and what effects environmental influences might have on this growth.

The problem might be likened to that of trying to understand the branching of a particular tree in a dense forest. Higher branches might get more sun; certain regions of the tree may be shaded parts of the day and experience a different solar orientation. Gravity may dictate that lower-order (closer to the trunk) branches be thicker, while prevailing winds and the location of other trees may create other anisotropies in growth. These environmental factors may determine a growth pattern quite distinct for two trees with nearly identical genetic programs but placed in different areas.

These considerations call for a computer analysis system that cannot only quantitatively compute selected parameters of dendritic morphology, but also can reconstruct the cells completely. Enough detail must be captured to determine (1) the types of interaction that may occur with other retinal neurons and (2) what constraints exist for the processing of these interactions by the dendritic geometry of the retinal ganglion cell itself.

Analysis of the branching pattern of any neuron requires more than just the x, y, and z coordinates of the dendrites. The relation to nearby anatomical landmarks must be considered, and special features of the dendritic tree must also be encoded. These include dendrite thickness, as well as the location of spines, branch points (nodes), and dendritic ends or terminals. Of particular interest is branch order: the relation of a dendrite to other dendrites in the tree, especially the primary and terminal dendrites.

We have developed a computer-based system that fulfills many of these analysis and reconstruction requirements. It relies heavily on the use of color for three-dimensional reconstruction. Although the system uses a DEC PDP-11 minicomputer, there is no inherent requirement for a system of this particular size or performance. Since any of a number of microprocessor-based computers (microcomputers) are available and suitable for this task, the description of this system will be made in as hardware-independent a manner as possible.

System and Components

Histology

Rabbit retinal ganglion cells are stained with horseradish peroxidase (HRP), either extracellularly, following injections into the retina by Hamilton microsyringe (Amthor et al., 1983a), or intracellularly, by iontophoresis during intracellular recording (Amthor et al., 1983b). The tissue is flat mounted, fixed in 2.5% glutaraldehyde (after a short initial period in 1% paraformaldehyde–2% glutaraldehyde) and reacted in diaminobenzidine (DAB) and hydrogen peroxide. After dehydration in alcohols and clearing in xylene, the retinas are counterstained with cresyl violet and coverslipped in histoclad (Clay Adams, Parsippany, NJ).

Computer System

The computer currently being used for cell reconstruction and analysis is not dedicated to only this one task. The PDP 11-34 computer (Digital Equipment Corp., Marlboro, MA) is configured as a time-sharing multiuser system running under RSX-11M and is used by a number of researchers supported by a Vision Science CORE Grant. Essential elements of the overall system for this application are the following:

1. a high-level language such as FORTRAN,
2. a floppy or hard disk memory,
3. a user terminal, and
4. a pen plotter.

In addition to general system elements, there are specific requirements for this system:

1. a graphics digitizing tablet,
2. a (color) graphics terminal, and
3. an extra serial (RS-232) port for optical encoder input.

Graphics Tablet and Graphics Terminal

In our system, a retinal ganglion cell is entered in two stages. The reasons for this two-stage approach will be discussed in more detail later. In the first stage, the cell is drawn by pencil using a microscope camera lucida tube attached to a Zeiss Universal microscope (Zeiss, Thornwood, NY). This drawing is then traced as an x–y plane figure using a Talos digitizing graphics tablet (Talos Systems, Inc., Scottsdale, AZ) by moving a cursor along the drawing of the tree structure of the neuron. This provides a digital stick figure image of the cell, which can be displayed and manipulated by the computer. The Talos digitizing tablet allows data to be entered under either "point" or "increment" modes so that data points can be entered in single or stream fashion. The numeric keypad on the cursor allows a number, such as dendritic order, to be entered with each data point.

The second stage of data entry consists of adding the z dimension to the already present x and y dimensions from the first stage. This is done by displaying the digitized x–y drawing of the cell on a Tektronix 4010 (Tektronix, Inc., Beaverton, OR) graphics monitor. This CRT image is superimposed on the microscope image of the cell through the camera lucida, which is focused on the Tektronix 4010 monitor. The dendritic tree of the cell is stepped through point by point (in the order in which it was entered) by pressing the space bar on the 4010 graphics monitor. The corresponding dendritic feature is kept in focus and an optical encoder attached to the fine focus knob of the microscope is read for each point, giving the z dimension (depth in the retina). This configuration is similar to that of Glaser et al. (1983), although the procedure is somewhat different.

When completely encoded, the ganglion cell is displayed in three views on an AED 512 (Advanced Electronic Design, Inc., Sunnyvale, CA) color graphics terminal. In the main (flat-mount or x–y) view, depth in the retina is encoded by color. This color is maintained for the dendrite in each of two orthogonal views, so that individual dendrites are identifiable in each of the three views. The display uses up to 256 colors in a 512×480 pixel format. Excellent hard copy is obtained with an Image Resources Videoprint 5000 (Image Resource Corp., Westlake Village, CA) automatic photography system. Additional black and white hard copy is obtained with either a printer/plotter (Benson-Varian, Mountain View, CA) or a pen plotter (Houston Instruments, Austin, TX).

RS-232 Port and Shaft Encoder

An extra dedicated RS-232 port is used to interface with the only special hardware for this reconstruction and analysis system. This consists of a Dyneer/Sensor Technology (Chatsworth, CA) optical shaft encoder with 256 counts per revolution. The encoder is attached to the fine focus knob of the Zeiss Universal microscope giving a 0.391-micron/bit resolution. The digital output of the optical encoder is gated to appropriate decoding circuitry (Amthor, 1980; McClelland, 1981), which is connected to an Electronic Systems (San Jose, CA) parallel to serial UART board. The UART transmits standard RS-232 format 8-bit codes to the PDP 11-34.

Program Operation

Rationale for Two-Stage Entry

The first task in the reconstruction of a retinal ganglion cell is the entry of the *x*, *y*, and *z* coordinates of all relevant points of the cell's dendrites. As the word *dendrite* itself suggests (from Greek *dendron*, a tree), these processes form a branched structure proceeding from one or more primary dendrites at the cell soma. Particular points of interest, such as branch points (nodes) and dendritic terminals (branches with end points) must be entered accurately. In addition, points must be picked close enough together (sampled) along the length of each dendrite to represent adequately the curvature or waviness of the dendrite.

The current configuration of the reconstruction system is the result of experience using several different input schemes and display methods. Figure 1 shows the overall procedure. Although entering all three dimensions of the dendritic coordinates at once might seem to be the most efficient, it has several major drawbacks. These are related to the fact that even in cases where staining is very good and there is little or no background, dendritic structures of interest are often at the limit of resolution of the light microscope. Drawing a cell accurately requires the skills of an experienced neuroanatomist who must examine and reexamine apparently ambiguous portions of the dendritic tree such as crossings, imperfectly stained or very thin branches, and material not belonging to the cell. Sorting these things out, while also operating the computer, places an extra burden on the operator. Errors may be difficult to correct without erasing a good deal of previously entered data. The lighting, environment, and the way the microscope is interfaced with the computer may not be optimal when the operator is required to enter all coordinates at once.

After trying such a system and finding these problems, I chose our current method. In this scheme, the cell is first pencil drawn at 1000× under optimum conditions using an oil immersion objective on the Zeiss universal microscope. All decisions regarding the tree structure are resolved at this point. This step really involves no extra work since all cells are accurately hand drawn for reproduction anyway. When the drawing is completed, a photocopy of the drawing is made (reduced if necessary) for entry on the Talos digitizing graphics tablet. The *z* or third dimension is then added to the *x*–*y* data set via the camera lucida CRT device. Although this device bears some resemblance to those of Glaser et al. (1983) and Capowski and Sedivec (1981), the entry system we use is distinctly different because of the two-stage procedure.

Two-Dimensional Cell Digitizing

After the first two steps of drawing the cell and photoreduction (if necessary) are completed, the program TALOS is run (Figure 1) using the digitizing tablet to enter the *x*–*y* coordinates of the ganglion cell. The first data point in every file is

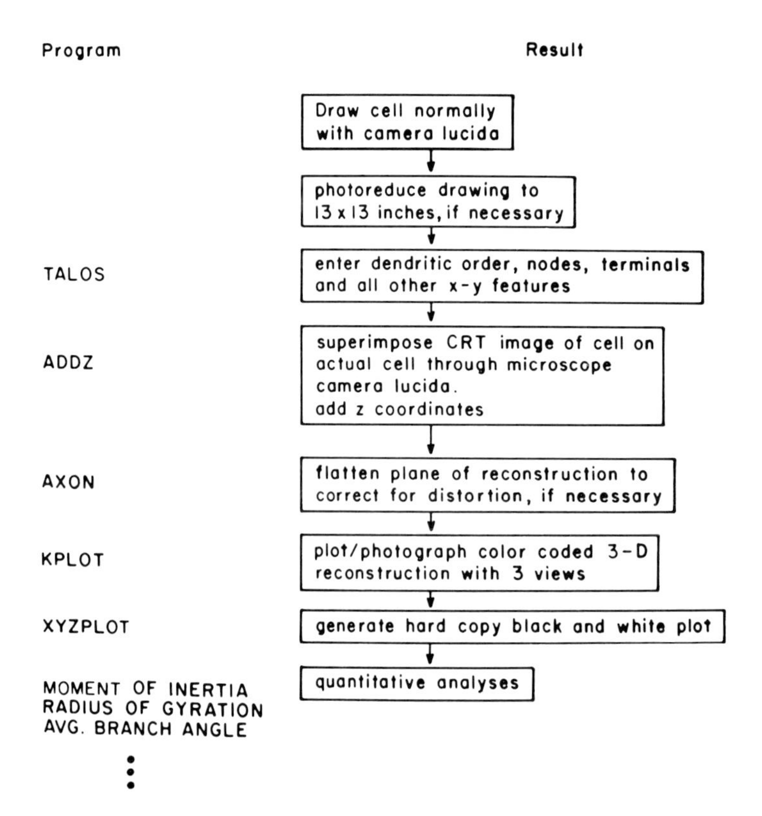

Figure 1. Step-by-step procedure and programs for entering, reconstructing, and analyzing retinal ganglion cells.

the center of the cell soma. This is used for computation of such attributes as radius of gyration and moment of inertia. The second data point is the left-hand edge of a 100-μm scale bar. The third data point is the right-hand edge of this bar. These three points may be used as origins for translations, and for setting up three-dimensional landmarks to be discussed later. The fourth data point entered is the origin of a primary dendrite. Dendrites are then entered by order.

The concept of dendritic order bears upon the relationship between a dendritic branch and other branches in the dendritic tree. The order assigned to a given branch may depend on the ordering scheme used. The scheme, in turn, depends on what aspects of the dendritic tree are of particular interest. In *centrifugal* ordering, the primary dendrites proceeding from the soma are numbered 1, the daughter branches of these 2, and so forth. This scheme expresses the path between a given dendrite and the soma in a particularly straightforward manner. This scheme might be very useful in describing parameters about the growth of the neuron, for example. *Centripetal* ordering, in contrast, numbers the terminal branches (those with end points) 1; branches that give rise to these end-point branches are numbered 2, and so forth. If particular types of inputs were concentrated on outer or terminal dendritic branches, this ordering scheme might be most useful. Variants of centrifugal and centripetal ordering exist, each with its own advantages in particular situations (Uylings et al., 1975).

Since it is much more efficient to enter dendritic order during the two-dimensional digitizing process, an entry scheme was desired that was simple and unambiguous, had simple rules, and expressed the tree connectivity most directly. In addition, it was desirable to associate only one number with the dendrite order and to be able to assign the order in only one pass, which is facilitated by the use of a 12-button hand cursor on the digitizing pad, rather than a pen. These criteria were well met by simple centrifugal ordering, in which primary dendrites are numbered 1, first daughters 2, daughters of daughters 3, and so forth, as in Figure 2.

Once this was decided, it was still not obvious how to deal most efficiently with terminals (end points) and branch points (nodes). After several different schemes were tried, an important entry rule was found that greatly simplified this problem and subsequent reconstruction programs. This rule is as follows: Start digitizing with the lowest-order dendrite, continue through higher orders without lifting the cursor (but changing the order number, as required), until a dendritic end point is reached. Then, when the cursor is lifted and moved to a new (and discontinuous) starting place, a zero is digitized as the first point, and digitizing is then continued as before. The zero indicates a break has occurred, which is always at a node. The use of this digitization scheme is shown in a step-by-step example in Figure 2.

When the dendrites are digitized this way, every node will have one and only one zero point, no matter where in the dendritic tree it is. In addition, all dendritic terminals will always be followed by a zero except for the very last point digitized. All points preceding a zero will be dendritic terminals. With this simple rule, no other constraints of entry are required, yet reconstructing the cell and finding nodes and end points are a simple programming task. As Figure 2 shows, a large number of dendrites may be digitized in relatively few steps and, therefore, very quickly.

The final stage is the entry of the cell soma and axon, which are digitized with an "artwork" code. This is a code for drawing the border of the soma, the axon, and outlining thick dendrites. It is used to render a more complete or pleasing picture of the cell in the reconstruction programs. The analysis programs, which deal only with dendritic features, ignore all points with this code. Artwork-coded points are thus in the file for pictorial reconstruction, but automatically excluded from analysis.

Digitizing Tablet Features

A number of features of the Talos 600 series digitizing tablet have proven to be advantageous for this task. The ability to measure area facilitates measurement of soma area and dendritic field area as quantitative attributes of ganglion cells. The fairly large size of the tablet active region (13 × 13 in.) makes it possible to digitize some cells drawn at 1000× without photoreduction, eliminating a step in Figure 1. The combination of the 12-button hand cursor and increment mode have been particularly useful. The 12-button hand cursor allows the entry of dendritic order without moving the operator's hand from the cursor. Increment mode allows specification of a minimum increment for cursor movement before a point is taken. The cursor button is held down and moved along a dendrite. The result is a uniform and rapid sampling of the dendrite at specified intervals without the need to keep pressing and releasing the button. Cells drawn at 1000× are

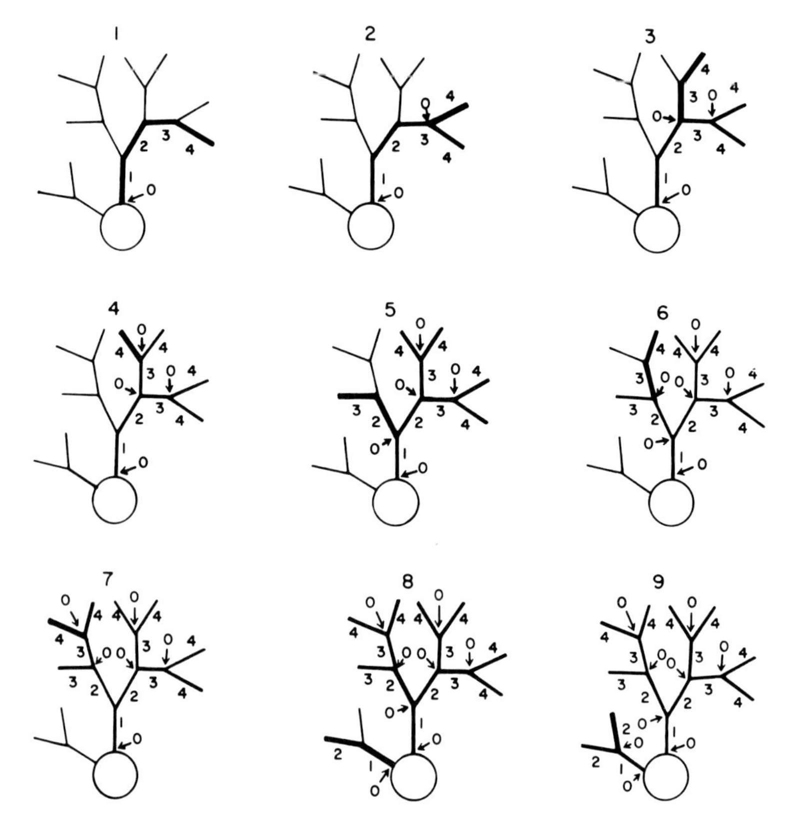

Figure 2. An example of the increment mode "digitization path" for a centrifugal ordering scheme. Before each succeeding frame, the cursor has been moved to a new starting position designated by a 0 (arrow) digitized at that point. In each frame, the very heavy line is the one digitized in that frame; the medium-thickness lines are previously digitized, whereas the lightest lines are not yet digitized.

normally sampled at $\frac{1}{16}$ in., or about every 1.6 μm. With this increment distance on the Talos, files contain on the order of 2000–4000 points. Since a code specifying which button is held down is transmitted with the data points, dendrites of order 1 have a cursor code of 1 and so forth through 9. A provision in the program TALOS allows 10, 20, or 30 to be added to the cursor code to handle higher-order dendrites.

The TALOS program is also interactive by virtue of a time-out feature. When the operator pauses, the dendrites digitized so far are drawn on the Tektronix 4010. Menu-driven options such as "delete vectors since previous time-out," or "add 10 to cursor button" are then selectable.

Third Dimension Addition

What exists at this point (after TALOS) is a two-dimensional stick-figure plan of the retinal ganglion cell. What is needed is to associate a third dimension (*z*, or depth) with each of the sampled dendrite points in the existent file. For this, the program ADDZ is run (Figure1). The CRT screen of the Tektronix 4010 graphics terminal is now imaged through the camera lucida so that the stick-figure computer reconstruction of the cell can be superimposed on the cell appearing in the microscope. The program ADDZ first plots the entire cell on the CRT screen. Then the magnification factor of the CRT display is changed recursively until the cell and the CRT plot are approximately the same scale. Note that it is not necessary to obtain exact scale equivalence since corresponding points on the reconstruction and the cell are easily recognizable even if not exactly superimposed. It may also be necessary to rotate the cell on the CRT about the z axis (in the x–y plane) if a rotary microscope stage is not used, to match the CRT image and the cell.

Once the cell and the CRT image are aligned, it is necessary to reset the optical encoder mounted on the fine focus knob of the microscope. The zero point is carefully set at a point just above the most superficial point of interest so that nearly 100 μm of depth are encodable. After this is done, the program ADDZ begins stepping through all the points on the ganglion cell in the order in which they were digitized. This stepping is done by flashing a very bright cursor point at the location in the display of the dendrite at that point. The corresponding dendritic location on the microscope slide is found, put in focus, and the space bar depressed on the 4010 terminal. This causes the PDP 11-34 to transmit a character, via RS-232, to the box containing the decoding hardware for the optical encoder. This circuitry then transmits to the 11-34 the 8-bit code that corresponds to the position of the encoder. A vector is then written into the new file created by ADDZ which contains the cursor code (for dendritic order), the x and y coordinates (from the Talos digitizing pad), and now, the depth of that point.

The first point (the center of the soma) is now the three-dimensional center of the cell soma. In the reconstruction of retinal ganglion cells, important landmarks are the axon layer, and the border between the inner plexiform layer (where the ganglion cell dendrites are located) and the inner nuclear layer (which contains the cell bodies of the bipolar and amacrine cells). The two points digitized as the ends of the scale bar are arbitrarily given depths of these landmarks as their z dimension.

After these three points, the remaining points (except for the axon and outline "artwork") are all sampled x–y positions along the dendrites. The operator maintains the appropriate dendrite in focus and the cursor point steps along each time the space bar on the terminal is depressed. An important feature of this system is that digitizing is always under the control of the operator, who may go as fast or slow as necessary. The operator may even back up, by pressing the "↑" key, to correct mistakes. It is also noteworthy that the stage is seldom moved, with the exceptions to be described below. This helps to provide greater accuracy because reproducibility of the x and y stage controls is then not a factor in accuracy. Moreover, movement of the stage can distort or lift the slide.

The exception to moving the microscope stage occurs because the cell's total dendritic field is usually greater than the microscope field at the 1000× magnification used for the reconstructions. When the edge of the field is reached, the program ADDZ automatically redraws the cell so that the next point to be digitized is now in the center of the microscope field. When this occurs, the operator translates the stage laterally either left–right or forward–back to bring the cell in superposition with the CRT image, and continues digitizing.

When the entire cell is digitized the new ADDZ ganglion cell file (Figure 1) is closed, and another cell may be entered. This entry method minimizes stage movement to only those occasions when the stage must be moved to an area of the cell not in the present field. The stage then remains fixed until a field border is crossed again. An additional advantage of this method is that no hardware need be added to the microscope stage.

Distortion Correction

There are both linear and nonlinear distortions produced in the processing of tissue for histology. Linear distortion often results from shrinkage during the dehydration process. This overall scaling distortion can be estimated by placing dye marks in the retina, or by measuring the distance between identifiable landmarks before processing. A more difficult type of distortion is wrinkling or lack of flatness in the section. This cannot be corrected by a general formula, or a coarse interpolative algorithm. To solve this problem, we have developed a method of completely reconstructing the plane of the cell and flattening it. This method works well no matter how extensive (in the x–y plane) or random the distortion is. It only fails when the roll or wrinkle is so severe that the section plane is at a significantly nonhorizontal angle.

Conceptually, the plane reconstruction is very simple. It is, in fact, merely an extension of the method by which the z coordinate is associated with the x and y coordinates in the program ADDZ. The program AXON takes as input the file with the cursor, x, y, and z coordinates. Using a procedure identical to that of ADDZ, the operator now keeps a retinal landmark, such as the axon layer, in focus for each dendrite point. Witn this method, the entire *relevant* surface of the retina associated with the ganglion cell is digitized. A simple arithmetic correction is then made on the z coordinates of the original z ganglion cell file to make the axon plane flat. The axon plane is used because it is a fairly discernible landmark in the retina. Other landmarks may be appropriate in other tissues. Alternatively, the surface of the tissue itself might be used. It should be noted that this correction method is easy to implement in a multistage input approach but would be much more difficult using other input approaches. Where this method cannot work well is where the tissue is so distorted that the retina is significantly not parallel to the slide. In this case, the sections are viewed so obliquely that not only z but x and y dimension errors are serious.

Thickness and Other Attributes

The method of superposition used to add the z, or depth, dimension lends itself to entry of other attributes as well as axon plane correction. Dendrite thickness is entered in the program THADD, in a manner analogous to adding dendrite depth (z axis) in the program ADDZ. A bright, blinking line on the CRT is made of variable length and orientation, controllable by the keyboard [see Chapter 4]. This cursor is superimposed on the dendrite and the cell is stepped through, point by point, as before in ADDZ. The length of the cursor line is changed by pressing "<" or ">" keys on the 4010 to make it longer or shorter. The orientation of this line may be changed to conform to the orientation of the dendrite. The resolution of this method is adequate for thick and medium-sized dendrites but is marginal for very small dendrites because of the finite spot size of the 4010. An alternative currently being examined is a high-resolution nonstorage type display that can be superimposed through the camera lucida.

The need to deal with sections across the microscope field (translations) is largely eliminated in our system by photoreducing the drawing before TALOS entry. Combining serial sections (in depth) could be accomplished by separate x–y digitization followed by recursive magnification and rotation matching under computer control [see Chapter 4]. Since I normally deal with retinal whole mounts, however, section combination has not, so far, been necessary.

Cell Reconstruction

Color Display Rationale

At this stage of the reconstruction process, there exists a file with the dendritic order and the x, y, and z dimensions of a retinal ganglion cell, containing up to 4000 points. It is corrected for tissue distortion, if necessary, and may also contain dendrite thickness. The cell is now ready to be displayed. How to display this information is not a trivial question, particularly for its three-dimensional aspects. At first thought, one might suppose this was a minimal problem for retinal ganglion cells, which, after all, are relatively flat. However, one of the very reasons for reconstruction of ganglion cells is knowledge of the *precise* position of the dendrites in three dimensions, so that the location of dendrites within various sublamina of the inner plexiform layer can be determined. This is not just a question of overall accuracy for, in some models of ganglion cell function, the main dendrites might be in one sublamina, whereas higher-order or terminal dendrites might be in another. To investigate such questions, one must know which dendrites are where.

Our first approach was similar to ones tried previously. We presented both a flat-mount (x–y) view of the cell, and one or more side views (x–z or y–z), equivalent to vertical sections. Although these extra views helped, the reconstruction was still inadequate for the precise location of the dendritic tree branches in three dimensions. The reasons for this soon became obvious. Although the retinal ganglion cell stretches over hundreds of microns in the x–y plane, its dendrites may only occupy 5 μm or less of a sublamina in the vertical plane. Looking through the entire cell in side view is like looking at a densely wired circuit board from edge on and trying to figure out what all the connections are.

Fortunately, the technology to solve this problem has become inexpensive enough to be affordable by many neuroscience laboratories. That technology is color graphics. With adequate color resolution, color graphics provides the means to display three dimensions (or possibly more) on a two-dimensional screen, and therefore to eliminate many of the depth ambiguities present in a black and white (monochrome) display.

AED 512

The AED 512 is typical of several new color graphics displays that have recently been introduced. Their price is now within reach of many laboratories. The availability is due not only to advances in solid-state electronics generally but, specifically, to less expensive memory. The limited resolution of the display (512×480 pixels) is due primarily to the use of a standard television type monitor. As configured for this project, the AED can simultaneously display 256 colors (from a possible palette of 2^{24} colors, or three 8-bit image planes). This number of displayable colors is more than adequate. A rough guess would be that 16 or 32 colors would be the lower limit for this application.

A color table was composed to represent depth (z) with a spectrum of 255 distinct colors of about equal brightness, with equal apparent chromatic difference between adjacent colors. Because of the particular requirements for retinal ganglion cells, resolution in the blue-green area of the spectrum was spread out. This region corresponds (in my scheme) to the inner plexiform layer. It is unlikely that there is any color scheme that is best for every reconstruction, because the purpose of the color is to provide, subjectively, a third dimension of information in a two-dimensional image.

The program KPLOT plots a given ganglion cell, in color, with three views. The colorplate at the front of this volume shows a cell reconstructed this way. In the flat-mount (x–y) view (large, lower right panel), the cell processes are color-coded by depth so that those processes closest to the vitreous, such as the axon and soma, are red or orange. Processes closest to the inner nuclear layer (INL) border are blue or violet. These colors are retained for the two side views, which correspond to x–y and y–x vertical sections. The result is that individual processes can be picked out in any of the three views and, because of subtle differences in color, can also be found in the other views even if the processes overlap. Red lines denote retinal landmarks, such as the axon layer and INL–IPL border, and scale bars are shown both for the x–y view and for the two side views. A black and white reconstruction of this cell is shown also in Figure 3. This cell is a uniformity detector that has been stained intracellularly with HRP. This type of cell has a very high maintained firing rate and is inhibited by any change of contrast (lightening or darkening) in its receptive field. There is a regional organization to its dendritic field. Yellow dendrites in the upper left quadrant of the flat-mount view (thick, with heavy arrows in Figure 3) ramify in the inner part of the IPL, closest to the axon layer. Green and blue dendrites (medium width, medium arrows, and fine dotted with smallest arrows, respectively, in Figure 3) seem to ramify in different sublamina of the IPL, and also occupy different regions of the tangential plane.

In cases where even this type of reconstruction is insufficient to determine the three-dimensional dendritic structure, KPLOT allows several other ways to "untangle" things in three dimensions:

1. The cell may be rotated to take advantage of a particularly good viewing angle, and then replotted as before.
2. The cell may be electronically "sliced" as though an actual vertical section of only part of the cell were taken. Since only a limited part of the cell is then replotted, it is easier to locate processes in three dimensions, because overlapping processes are left out of the reconstruction.

The use of color allows other plotting schemes to be used to investigate cell attributes. For example, we were interested in how dendrites originating from different primary dendrites intermingle in the three-dimensional dendritic field. Giving each primary dendrite and all its daughter and subdaughter branches the same color, but different from that of the other primary dendrite systems, allowed us to compare the three-dimensional organization of different primary dendrite systems. This led to an unexpected finding. In certain types of cells, which appeared to have substantial dendritic overlap, the dendrites of different primary dendrite systems never intermingled in both the x–y plane and the same sublam-

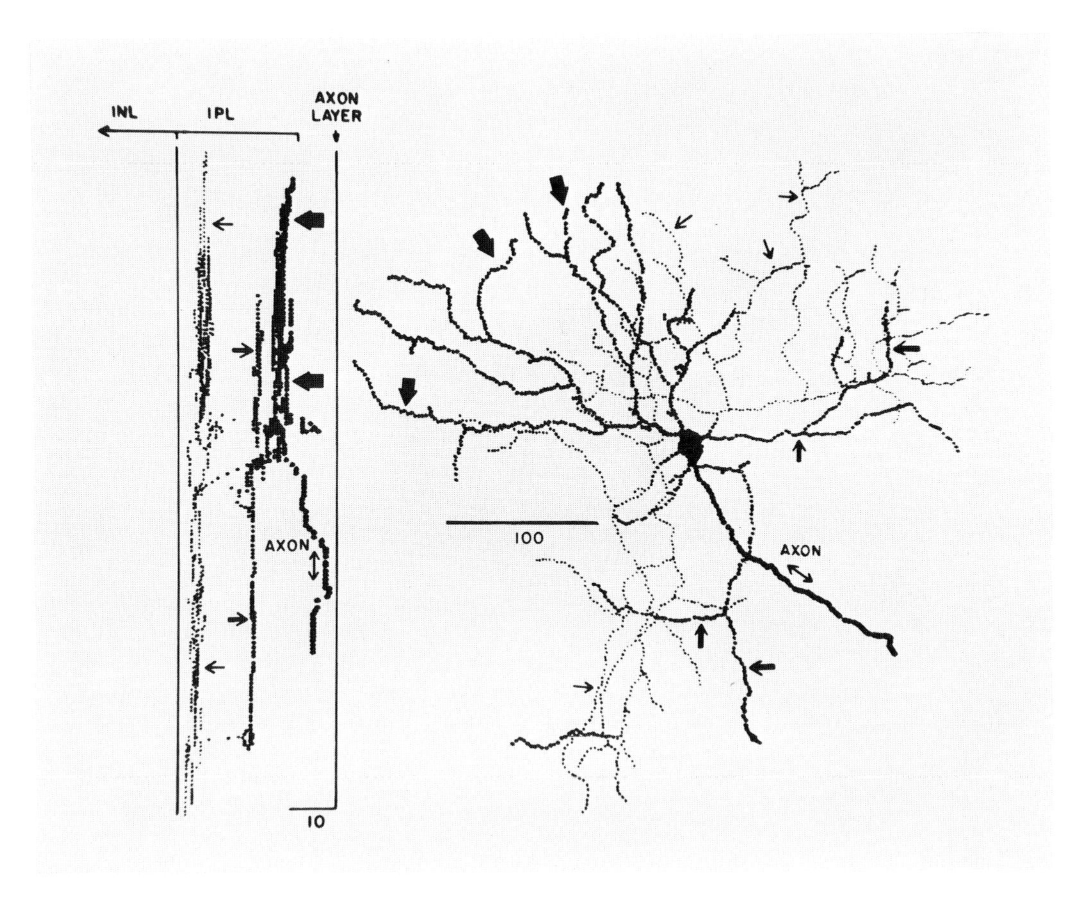

Figure 3. Black and white version of the color-coded three-dimensional computer reconstruction at the front of this volume. Depth (z) is encoded by color in the tangential plane (right lower panel view) in the color reconstruction, by thickness in the black and white figure. Thus, the axon, closest to the vitreal layer, is red–orange (thickest, in black and white), whereas the soma, a little deeper (toward the sclera), is yellow. Note that three levels of dendritic ramification are apparent in the two "sectional" views (top and left). The dendritic ramification closest to the vitreous is primarily yellow (thicker in black and white, with heavy arrows). By noting the color, it can be seen that these dendrites are in the upper left quadrant of the flat-mount reconstruction. Similarly, the green and blue dendritic ramifications can be located by IPL layer and flat-mount region (medium and fine arrows and thickness, in black and white). Red lines correspond to the approximate location of the axon layer in the section views, whereas violet lines denote the INL–IPL border. Note from the 10-μm scale bar in the top section view (left, in black and white) that this dimension (IPL thickness) is expanded relative to the flat-mount view (100-μm scale bar).

ina (z). The color graphics approach revealed a kind of dendritic "exclusion principle." Other questions that can be addressed by the use of color immediately come to mind. Is there a particular distribution or orientation of terminal branches? Do branches in one sublamina tend to be of a particular order? Because the color reconstructions can be done quickly, and the results almost immediately realized, it is possible quickly to generate and test many hypotheses about the dendritic tree structure.

Hard Copy

There are a number of methods for producing a hard copy of the graphical reconstruction. One may, of course, photograph the color monitor directly. The disadvantages of this are a distorted picture (from the screen geometry) and poor color rendition. A better alternative is to use a special-purpose raster scan photographic device such as the Image Resources Video Print 5000 (Image Research Corp., Westlake Village, CA). This unit uses a special flat screen monochromatic CRT and three color filters optimized for this kind of photography.

Color photographic methods are expensive and time consuming and inappropriate for publication in most journals. It is therefore almost mandatory to make black and white pictures of the reconstructions. This can be done using either a pen plotter or a printer/plotter. Programs for pen plotters are in principle very similar to KPLOT, which draws the cell as a series of vectors. An alternative to color is to vary line width in a black and white plot as a function of depth, as in Figure 3. This allows the display of some depth information, but at much less resolution than possible with color. It is, of course, impossible to represent both depth and true dendrite thickness this way. If a printer/plotter raster type printer is used, it is worthwhile in many cases to write a copy program that directly reads from the AED 512 raster format to the printer/plotter. It may also be desirable to expand the scale in one or more views to get a better separation of overlapping dendrites (which are less easily resolved now that they are in black and white). Recently, "color" pen plotters have become available. These are, however, limited to a few colors with no real mixing to produce hues and are therefore suitable only for the simplest types of reconstruction.

Analysis

Rationale

Dendritic tree analysis programs are the final stage shown in Figure 1. They are designed to investigate several different types of characteristic of retinal ganglion cell dendritic trees: (1) features relevant to the electrotonic summation of inputs, (2) features useful for classification, and (3) features related to development. It is, of course, not always obvious beforehand which dendritic parameters will be most relevant to the various issues.

Classification of cell types has been the main issue addressed so far with our reconstruction system. Until recently, most neuronal classification schemes have been more or less subjective. Some of these subjective schemes have proven to be fairly valid in light of later physiological or pharmacological research, whereas others have not. The main point is that subjective, nonquantitative methods of anatomical investigation lack rigorous reproducible measures. Such methods are time consuming, not readily communicated, and not easily reproducible, and they have only really worked well when the number of classes was small, usually less than five. Since it is likely that there are more than five different morphological types of ganglion cell in the mammalian retina, it is desirable to establish the existence of these classes on quantitative grounds. Even more important, however, is the fact that specific hypotheses about structure and its relationship to function and development may well depend on specific numerical values for parameters of the dendritic tree structure.

Development of Methods

Before the computer classification scheme was developed, Dr. Clyde Oyster, Dr. Ellen Takahashi (School of Optometry, University of Alabama in Birmingham), and I attempted a qualitative classification of rabbit retinal ganglion cells. Each of us classified a set of approximately 100 cells independently, and then met and established a consensus classification that contained five classes, several subclasses within each class, and some figure of certainty for both the classes established and whether certain cells were in one class or another. We then began to ask what features, in fact, were really decisive in the classification, and how they might be quantified.

Certain parameters, such as soma area and dendritic field area, were obvious candidates. *Soma area* was obtained by tracing an outline drawing of the soma on the Talos digitizing pad. Area was then computed directly by a routine intrinsic to the Talos. *Dendritic field area* (the area covered by the dendrites of the cell) was measured the same way. This measurement proved not to be so straightforward, however. For some sparsely branched cells, determining perimeter was ambiguous, since the perimeter is normally determined by connecting the dendritic branch end points around the cell. If dendritic end points occur in the interior, inward-shooting wedges from the outside perimeter are created. In a sparsely branched cell, the decision to connect or not to connect to an "inner" dendritic end point may make a very large difference in the area computed. Our solution was to construct a second perimeter by removing all concavities from the perimeter determined by the end point method. This is called a convex hull. Other solutions, such as fitting ellipses or polygons to the dendritic tree, might also have been used.

Unfortunately, neither soma area or dendritic field area, in isolation, proved to be very helpful in distinguishing cells that had obvious dendritic morphological diversity. Other parameters proved much more useful, particularly if expressed in a form normalized for overall dendritic field size. These include total dendritic length, moment of inertia, radius of gyration, and average branch angle. Several of these attributes have their origins in the work of Sholl (1953). Two important differences exist, however, between the attributes used in this system and the pioneering computations of Sholl. The first is that ours are computed automatically. Although this may seem trivial, when hundreds or thousands of neurons are considered, each with as many as several hundred branches, the task of hand computation can be formidable. Second, in contrast to Sholl's concentric rings, attributes can be computed as a continuous function of radius. This allows much more freedom in the choice of parameters, and much greater resolution and accuracy.

Total dendritic length is simply the total length of all the dendrites of the ganglion cell. When divided by the dendritic field area, it gives one estimate of the density of dendritic branching. How this density is radially distributed may in turn be measured by the moment of inertia and radius of gyration attributes. The *moment of inertia* is the sum of the radial distances to all points on the dendrites, divided by the number of points sampled (this normalizes the computation for different sampling intervals). The *radius of gyration* is similar except that the squares of distances to dendritic points are summed. Both of these attributes relate to the relative distribution of dendrites close to or far from the soma. These attributes quantify subjective impressions, such as a large number of dendritic branches at the periphery of the dendritic field giving a strong appearance of a cell filling space densely with a well defined border.

Average branch angle was a crucial measurement that dispersed the population and was obviously related to the subjective appearance of the cells. Average branch angle was measured as the angle between two daughter branches proceeding from each node. The angle was defined as that of the triangle at that node whose other two sides were the lines drawn from that node to either the next node along the daughter branches or the dendritic end point if the daughter branch was a terminal one. This definition, although not always subjectively satisfactory, has the advantage of being well defined. Estimating branch angle at some arbitrary distance from the node may depend critically on the dendritic curvature, and therefore on the sampling interval along the dendrite. Computing the branch angle uses the law of cosines for a general triangle for which all three sides, a, b, and c, are known. To find the angle A at the node, opposite side a (connecting the ends of the two daughter branches),

$$\cos(A) = \frac{b^2 + c^2 - a^2}{2bc}.$$

The branch angles of all nodes in the cell are computed. This attribute distinguishes between cells that subjectively have a very radial appearance, with narrow branch angles, and more intricately or retroflexively branched cells, with much larger average branch angles. These latter have not only larger branch angles but may also have dendrites that actually turn back toward the soma (retroflexive).

Quantification of the retroflexive attribute, and other attributes more directly related to dendritic curvature, was particularly useful. One of these measures is an attribute we call *total radial incremental distance* (TRID). In this attribute, the centrifugal distance of the dendrite going away from the soma is measured. If the dendrite is radially oriented outward from the center of the cell soma, the contribution to TRID of that segment is equal to the length of the segment itself. If the dendritic segment is oriented normal to the radius, it makes no contribution to TRID. If the dendritic segment is actually retroflexive and turns back toward the soma, it makes a negative contribution to TRID. Figure 4 shows cell type histograms for average branch angle and TRID for 82 cells. TRID distinguishes between cells of equal dendritic length but variable nonradial orientation.

Each of these quantitative attributes is measured by a computer program that requires no operator intervention. Although these measurements are automatic, it must be emphasized that the classification using these has never been a totally automatic process. Factors that are not easily quantifiable in the same way as the others, such as nasal or temporal retina, have always also been taken into account. We have looked at the significance of the statistical differences between groups for each single attribute, and attached most importance to those highly significant differences on a case-by-case basis. An open and variable number of programs for quantitative analysis are therefore shown in the overall scheme for encoding and analyzing a given retinal ganglion cell (Figure 1). One compelling result of this quantitative classification effort is that when the computer classification has seriously "disagreed" with our initial subjective one over several attributes, it has

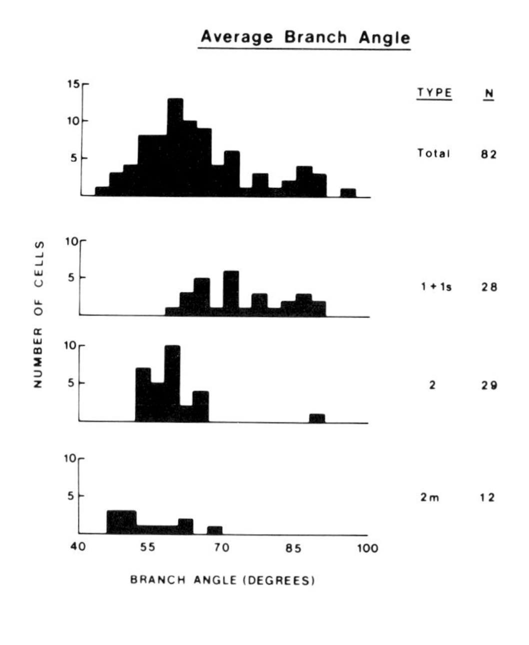

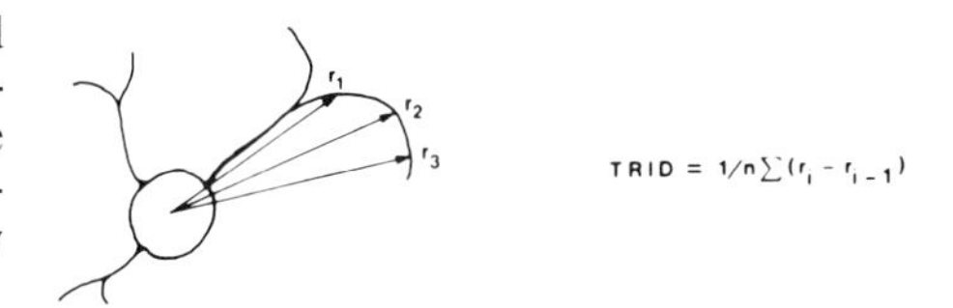

Figure 4. Cell-type histograms plotted as a function of quantitative attributes. Average branch angle is defined for a triangle whose sides extend from each node to either the next node or the dendritic terminal. Total radial incremental distance (TRID) is the sum of differences in *radius* of successive dendritic increments moving centrifugally out the dendrite.

always turned out on reexamination that the first subjective classification was less and less convincing. Usually the ganglion cell has been reclassified in such cases to correspond to the computer classification. In general, at most two or three attributes have been required to distinguish quantitatively the morphological classes identified so far (Amthor et al., 1983a).

We are continuing to investigate these quantitative attributes not only in the context of their ability to distinguish cell groups, but also toward understanding what relation they have to subjectively discernible differences between cells. An even more important question being asked is what the differences in quantitative parameters mean for the function and development of retinal ganglion cells.

Conclusions

This chapter has described a system for the quantitative analysis of the dendritic tree structure of retinal ganglion cells. The system provides for both a three-dimensional color reconstruction of the cell and the computation of a set of quantitative attributes. These attributes include total dendritic length, moment of inertia, radius of gyration, average branch angle, and total radial incremental distance. Novel features of this system are the following:

1. a minimum of special hardware;
2. the use of color to encode depth (z axis);
3. the display of multiple orthogonal views in the three-dimensional reconstruction scheme;
4. a simplified method for automatic computation of dendritic tree attributes; and
5. a multiple-step feature entry system that facilitates precision in coordinate entry and correction for tissue distortion.

Because of the minimum of special hardware, the algorithms and special interfaces needed for the quantitative analysis are discussed in a hardware-independent manner so that the scheme may be implemented on virtually any microcomputer system of adequate memory and graphics capability. A novel aspect of this scheme is the representation of depth by the use of color. The most difficult hardware requirement for this is the need for a high-resolution color display. The AED 512 color graphics terminal used can display 256 simultaneous colors in a 512 × 480 pixel display. Although the number of colors is ample, the pixel resolution is a bare minimum for any realistic display of the ganglion cell dendritic tree. Dendritic thickness, for example, is almost impossible to show at this resolution. I am sure I am not alone in eagerly awaiting continued developments that will bring about lower-priced and higher-resolution color displays.

Five programs are currently used for cell three-dimensional reconstruction. TALOS and ADDZ acquire the x–y and z coordinates, respectively. AXON is used to reconstitute and flatten the axon plane in cases of tissue distortion. KPLOT and XYZPLOT plot multiple views of the cell either on the AED 512 terminal, in color, or on a Houston Instruments digital x–y plotter for hard copy. A variable number of analysis programs are used to acquire quantitative attributes about the dendritic tree structure, such as branch angles, dendritic lengths, and soma areas. Currently, up to 20 such attributes can be computed. These attributes have been used to classify ganglion cells anatomically by quantitative morphological features. They have also been used to investigate structure–function relationships in physiologically identified cells that have been individually dye injected following intracellular recording.

I thank Dr. Clyde Oyster and Dr. Ellen Takahashi for their collaboration in neurophysiological aspects of this project, Richard Sheetz wrote early versions of TALOS, ADDZ, and KPLOT. Jill Gemmill wrote parts of the thickness measuring program, THADD, and generated the black and white plot used in Figure 3. I thank Laura Engstrand and Dave Holder for technical assistance. Dyneer/Sensor Technology donated an optical encoder for the microscope fine focus knob. Special thanks to Caroline Dunn for typing the manuscript. This work was supported by NIH grants EY05070, EY02207, CORE EY03039, and RR05807.

References

Amthor, F.R. (1980) Multiplier increases resolution of standard shaft encoders. Electronics 53(20): 139.

Amthor, F.R., C.W. Oyster, and E.S. Takahashi (1983a) Quantitative morphology of rabbit retinal ganglion cells. Proc. R. Soc. Lond. B217: 341–355.

Amthor, F.R., C.W. Oyster, and E.S. Takahashi (1983b) Do major physiological retinal ganglion cell classes have distinct morphologies? Neurosci. Abstr. 9: 896.

Capowski, J.J. and M.J. Sedivec (1981) Accurate computer reconstruction and graphics display of complex neurons utilizing state-of-the art interactive techniques. Comput. Biomed. Res. 14: 518–532.

Fischer, B. (1973) Overlap of receptive field centers and representation of the visual field in the cat's optic tract. Vision Res. 13: 2113–2120.

Glaser, E.M., M. Tagamets, N.T. McMullen, and H. Van der Loos (1983) The image-combining computer microscope—an interactive instrument for morphometry of the nervous system. J. Neurosci. Methods 8: 17–32.

McClelland, B. (1981) Fewer parts resolve shaft encoder data. Electronics 54(12): 167.

Sholl, D.A. (1953) Dendritic organization in the neurons of the visual and motor cortices of the cat. J. Anat. 87: 387–406.

Uylings, N.B.M., G.J. Smit, and W.A.M. Veltman (1975) Ordering methods in quantitative analysis of branching structures of dendritic trees. In Kreutzberg (ed.): Advances in Neurology, vol. 12. New York: Raven.

Wässle, H., L. Peichl, and B.B. Boycott (1983) A spatial analysis of ON- and OFF-ganglion cells in the cat retina. Vision Res. 23:1151–1160.

7

Microcomputer Control for Electron Microscope Systems

David C. Joy

Introduction

The electron microscope is one of the most complex instruments in common laboratory use, and consequently even routine imaging requires considerable skill and attention on the part of the operator. All but the most experienced microscopist can find problems in effectively running the instrument while simultaneously trying to control the operation of added analytical facilities (such as x-ray or electron spectrometers) and keep track of experimental parameters that will enable the data to be correctly interpreted at a later date. Since data logging and routine control operations are among the things that computers do best, there has been much interest recently in interfacing microscopes with computers (e.g., Joy, 1982; Smith, 1982). This chapter discusses one particular approach to computer control using a personal or "hobby" computer and simple, nonintrusive interfacing. Although the details are specific to a particular instrument, the design concepts are applicable to many comparable systems.

The Scanning Transmission Microscope

The scanning transmission electron microcsope (STEM) used in this laboratory differs in several ways from the conventional transmission electron microscope (CTEM). In a CTEM, one or more lenses magnify the image of the sample onto a fluorescent screen or photographic plate for viewing or recording. In a STEM, the image is formed by rastering a fine focused probe across the sample and using the collected signal to modulate the brightness of a viewing screen scanned in synchrony with the incident beam. In addition to being able to generate images comparable to those available from the CTEM, the STEM has the advantage that the small-diameter electron beam can be used to probe selected areas of the specimen for microanalytical purposes such as energy dispersive x-ray (EDXS) or electron energy loss (ELS) spectroscopy. The fact that the signal is varying in time also makes it possible to process and modify it electronically so as to improve the information content of the image. In particular, by using the electron spectrometer to produce energy-filtered images in both a characteristic inner-shell ionization loss and a suitable background energy loss, a combination image representing the distribution of the element of choice can be generated.

The STEM at Bell Laboratories is the HB501 manufactured by Vacuum Generators Ltd. and is a dedicated unit (i.e., it is not based on a CTEM) operating under ultrahigh vacuum conditions. The electron source is a field emission gun that has an effective diameter of about 50 angstroms (Å). Two condensor lenses and a high-performance objective lens demagnify this to give a final probe of less than 3 Å in diameter at the sample. The instrument is thus capable of atomic-level resolution on suitable specimens. In general, this level of resolution is not required, and the two condensor lenses then provide the necessary flexibility to optimize the beam current and divergence over a wide range of beam diameters. Three postspecimen lenses follow the objective and are used individually or in combination to couple the sample to the annular dark field detector or the electron spectrometer. The microscope is therefore equipped with six lenses, each of which must be precisely set for each mode of operation.

The optics of the microscope allow a great deal of freedom to the operator and the ability to achieve state-of-the-art performance in both imaging and microanalysis. However, this flexibility also makes the operation much less simple than conventional microscopes, as the preset modes and instrumentation typical on such columns are not provided. In order to exploit the full facilities of the STEM it is necessary to provide control and measuring functions that permit the operator to manage the microscope conveniently while maintaining the highest level of performance. The computer system described below was designed to meet this need.

Specifying the System

The first decisions to be made when combining a computer and an electron microscope are the following:

What capabilities are required of the system?
What level of interfacing and control is important in the microscopy laboratory?

The system described here is designed to offer four types of facility:

1. A continuously updated display of the significant operating parameters of the microscope, such as the lens currents, the specimen stage, *x*, *y*, and tilt coordinates, and the capability to monitor items such as the vacuum levels in the electron gun or specimen chamber, which could be crucial for the safe operation of the instrument.
2. A means for the storage and later recall of chosen sets of operating conditions.
3. Allowance for special modes of operation such as those in which control of the beam position and beam blanking is combined with control of some ancillary equipment to give, for example, automated microanalytical operation.
4. A general diagnostic tool for day-by-day checking of the state of the microscope.

Such facilities could be provided by the computer in many different ways. The philosophy of the approach used here was dictated by the feeling that the job of the computer is simply to aid the operator, and not to foster an interest in either computer systems or software development. This consideration led to the formulation of some general rules determining the nature of the computer system and its interface to the microscope.

1. The microscope must remain fully capable of being operated without the computer, if necessary, and the computer interface should in no way inhibit or interfere with the normal operation of the instrument.
2. The interface between the computer and the microscope should require the absolute minimum of wiring modifications to the microscope itself, and any modifications should be of the type that can rapidly be reversed so that normal servicing of the instrument is not affected.
3. All the major items of hardware required for the computer system should be commercially available so that the minimum amount of prototype building is required.
4. The software required to run the system should be easy to write so that extensive development time is avoided; it should be easy to modify so that individual users can customize the software for their own particular experimental

requirements; and it should be easy to document so that operators using the program at a later date can readily determine the capabilities of, and functions available on, the system.

5. The cost in time and money should be commensurate with the benefits to be expected from the use of the system.

These requirements suggest that the most suitable computer for the task would be one of the personal microcomputers now available for home and business use. These units are low in cost and are well supported in both software and hardware by a wide variety of manufacturers. In addition, computers designed for home use tend to be much easier to use, and less intimidating to nonprofessionals, than minicomputers designed specifically for scientific purposes. Of the wide range of units currently available, the Apple II+ was selected because of several important factors:

It has a medium-resolution graphics display (280 × 192 pixels) that is easily accessed from BASIC language programs and that permits both the direct plotting of data and the presentation of low-resolution images.

The machine comes equipped with expansion slots, readily allowing the connection of accessory input and output hardware.

A considerable number of manufacturers sell items, such as analog-to-digital (ADC) and digital-to-analog (DAC) converters, real-time clocks, image acquisition boards, and mass storage devices, that make use of these expansion capabilities. Because these units generally take their power directly from the Apple and are invariably operated as direct memory access (DMA) peripherals, their use simplifies both the interfacing and programming required.

Other systems fulfilling the same general specifications can, of course, be found such as the generic S-100 systems (IEEE 696 specification) [see Chapter 3]. Because these are oriented around a bus structure, the user has available a wide range of processor boards of varying levels of sophistication, a variety of memory options, and a considerable array of special-purpose ancillaries such as color graphics boards. To exploit fully the versatility of such a configuration, however, the user must have advanced technical training. Even with the IEEE specification, the compatibility of various pieces of hardware bought from different suppliers is not always ensured, and little packaged software is available to support the various functions. For most nonspecialist users, therefore, a system built around a particular microcomputer is preferable, because the major design choices have already been made and a high level of compatibility and support for accessories can be assumed.

The Apple System

Figure 1 shows the schematic layout of a computer control system, in the spirit of the principles outlined above, designed to operate with a Vacuum Generators HB501 field emission scanning transmission electron microscope. The Apple II+ was used with 48*K* of memory. This has been found to be adequate even though not all of the memory is available for programming and data storage [about 13*K*

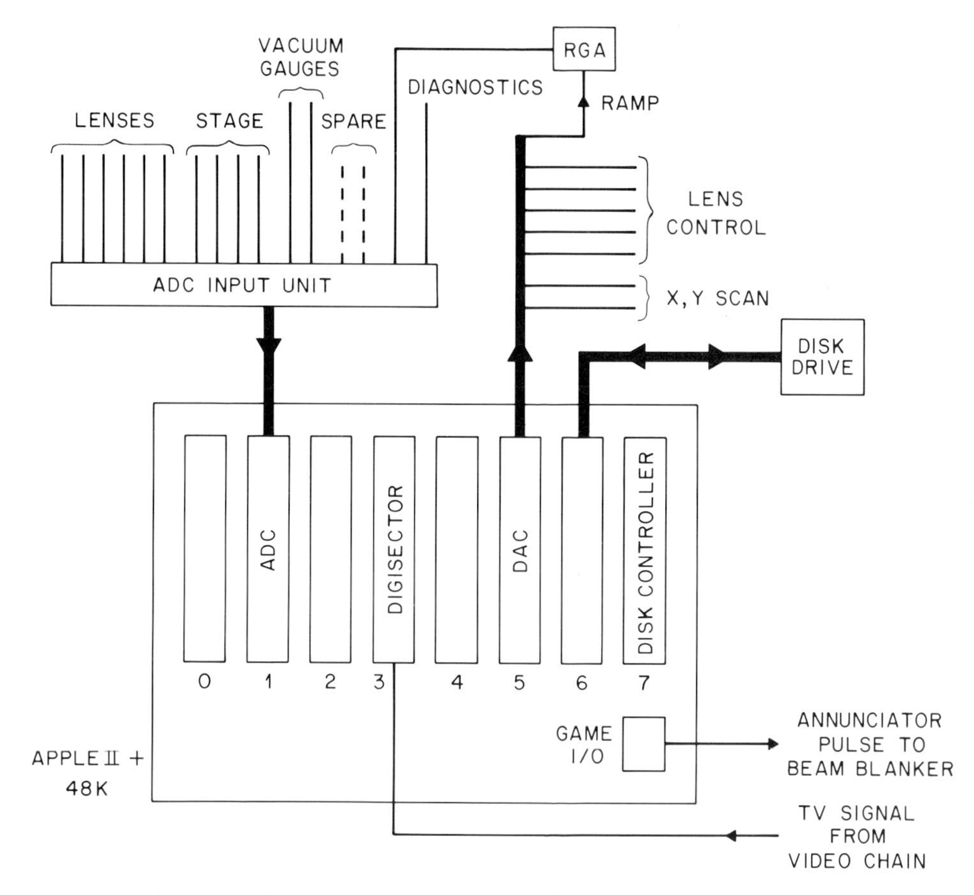

Figure 1. Schematic of computer control system for a Vacuum Generators, HB501 electron microscope.

is reserved for the BASIC interpreter and disk operating system (DOS)]. Four of the eight expansion slots of the Apple are in use. Slot 6 is used for the the disk controller, which drives one or two single-sided $5\frac{1}{4}$-inch floppy disks, each of which provides 114 KB of formatted storage. Although, in principle, the cassette-tape interface of the Apple could be used to give similar storage at a lower cost, the reliability and speed of disk drives is far superior to that available from standard tape units, and cassettes cannot be recommended for this application.

The disks are used to store both the operating programs and data files containing lens conditions or special scan patterns. Typically, any program or file can be accessed and loaded within a few seconds. As hard disk (Winchester) units have fallen in cost they have become increasingly competitive for this type of application, and anyone starting now to build a system of this type might wish to investigate such units because they offer advantages of both speed and storage capability. However, the low unit cost of floppy disks and their ease of duplication for back-up purposes will probably ensure their continued use for a considerable period of time, particularly in situations where users wish to maintain close control over their own programs and data.

Interfaces to the Microscope

Input facilities for the system are provided through an ADC installed in slot 1. This is a standard commercial unit, the AI-13 (made by Interactive Structures Inc.*) which provides 16 channels of analog input, each of 12-bit (i.e., 1 part in 4,096) precision. Any input can be selected under software control, and the sensitivity of each input can be adjusted by sending it a software gain control code (as discussed below) to cover a wide range of applications. The gain change can be accomplished in 20 μsec, whereas the actual analog-to-digital conversion requires 6 μsec. The first six ADC channels (0–5) are used to monitor the lens currents of the STEM (two condensor, one objective, and three postspecimen lenses). In keeping with the philosophy of using the simplest possible interface, the computer system inputs are connected in parallel with the normal digital lens current meters. The signals for these meters are obtained by measuring across the feedback resistor in the lens controller. Since this resistor is 1 Ω in value, and the lens currents are about 4 A maximum, the typical input signal to the ADC is 4 V. Because the input impedance of the ADC units is high (about 1 MΩ), they do not in any way adversely affect the stability or performance of the lens feedback circuitry; but as an additional precaution a 100-kΩ resistor is placed in series with the ADC inputs to avoid upsetting the feedback loop in the event of any malfunction on the ADC board.

Channels 6–9 are used to display the *x, y,* and tilt coordinates of the specimen stage. Since on this particular microscope the stage is remotely controlled through stepping motors, the inputs were taken from the sliders of the position-sensing potentiometers to give signals in the range ±0.1 V. Channels 10 and 11 are used to monitor the vacuum levels in the gun and specimen chamber. The necessary inputs are taken from the analog outputs (0 to −0.1 V) provided on the ion gauge controllers fitted to the instrument. Channel 14 is used to read the analog output of the Vacuum Generators "ANAVAC" Residual Gas Analyzer, which is fitted to monitor both the total pressure and the partial pressure components of the column. Channel 15 is used for system diagnostics and is connected through a switch system on the interface board so that it can be used to check the reading on any other ADC channel and can also be switched to monitor the output of any of the DAC channels. Inputs 12 and 13 are left as spares and to provide for further expansion.

The DACs are in slot 5. The units chosen are the AO-03 from Interactive Structures Inc. and provide eight independent outputs each of 8-bit (i.e., 1 part in 256) precision. Each output is nominally 0 to +10 V but on-board jumpers and potentiometers allow this range to be adjusted in magnitude, or reset to give a −5 to +5 V bipolar swing. The settling time for any channel is about 6 μsec. Although these units are relatively well isolated, some interaction between channels was experienced. An array of simple unity gain buffer amplifiers was therefore built (Figure 2) and placed in the interface unit. These not only improve the isolation but also allow a DC offset to be added to the outputs to permit, for example, exact centering of the scan rasters. The first five DAC channels (0–4) are available for the lens controller. Each output is wired to the summing input of a lens control

*P. O. Box 404, Bala Cynwyd, PA 19004.

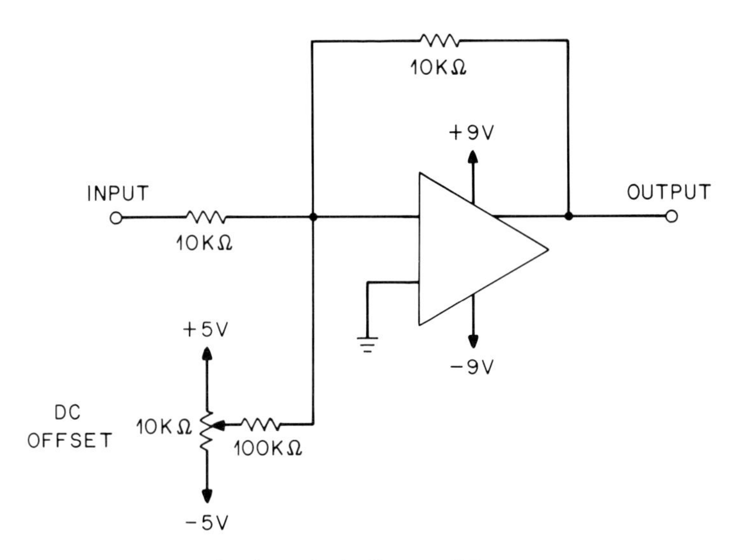

Figure 2. Array of unity gain buffer amplifiers.

amplifier so that, once again, the computer is interfaced in parallel with an existing facility, this time with the normal manual lens control. When the DAC outputs are set to zero the lens controllers are in normal mode; when the manual "medium" control is set to zero then the lenses are under computer operation. Full control of the lenses at the "coarse" level was not attempted because of the limited resolution (8-bit) of the DACs, so the computer is used only to take control once the coarse control for each lens has been set. Channel 5 is used to generate the 0 to + 5 V ramp required to sweep the residual gas analyzer over the 0–80 atomic mass unit range for the measurement of partial pressures. Channels 6 and 7 are set up as bipolar outputs to drive the external scan inputs of the microscope. An additional source of control outputs can be derived from the 1-bit annunciators available at the "Game" control socket of the Apple. Each of the four annunciators is a separate memory location that, on being referenced (e.g., by a POKE call in BASIC), can be set to go high (+4.5 V) or low (+0.5 V) within 3 μsec. One of these outputs is used to control the beam blanking of the HB501, the drive unit for which requires a standard TTL pulse input with which the characteristics of this annunciator signal are compatible. The other outputs are unused, but could be employed to drive relays, warning lights, or even steeper motors.

Image Acquisition and Enhancement

Slot 3 is used to accommodate the Digisector* card, which operates as a simple image acquisition, storage, and enhancement device. The card accepts a standard TV input, provided either directly from the video displays of the STEM or from

*Produced by The Micro Works, P. O. Box 1110, Del Mar, CA 92014.

the TV camera that views the diffraction screen of the microscope, and over a period of about 8 seconds produces a 256 × 256 × 6-bit image that is displayed on the high-resolution graphics screen of the Apple. The input can be interrogated on a pixel-by-pixel basis to give the intensity of individual points or arbitrarily chosen line scans across the image, and images can be accumulated to give some signal-to-noise enhancement. Simple processing of images (such as inversion, contouring, and comparison) is also possible, and images can be stored on the floppy disks (approximately 15 images per disk) and "screen-dumped" onto a dot matrix printer to give hard copy. Although this device was designed for the consumer market rather than for the microscopist, its low cost and versatility make it an excellent accessory to the system.

Electronics

All connections between the Apple (and the ADC, DAC, and Digisector cards that it includes) and the microscope are made through multiway connector cables and 25-pin D-type plugs to the interface board, which is located in the main electronics console of the HB501. Each ADC input and each DAC output is run in a color-coded twisted wire pair (signal and ground return) with a single earth point, connected by $\frac{1}{2}$-inch-thick copper braid, going to the main ground line of the microscope. The buffer amplifiers in the interface board are powered through a floating, solid-state, switching power supply to minimize further any possibility of ground loops. Although the computer is some 10 feet from the interface, no problems have been experienced with pickup or interference because most of the signals are at a relatively high level and are terminated into low impedances. All connections to the microscope terminate in the interface board and are also carried through multiway cables and plugs. Unplugging the Apple leads from the interface board therefore completely isolates the computer from the microscope should this be necessary for service purposes. Despite the proximity of the computer to the column, no problems have been experienced from digital noise ("hash") breaking through onto the image; however, should such problems be encountered they can usually be removed by relocating the computer, screening the leads in a copper sheath, or by placing ferrite beads or rings on individual signal leads.

Software for the System

Much of the attraction of a computer-based system lies in the fact that the same pieces of hardware can be made to perform different tasks simply by changing the program that controls them. Although the program may look very different to the end user, depending on whether its purpose is to store an image or to perform a microanalysis automatically, the subroutines that actually control a specific piece of hardware (such as a DAC output) will usually remain the same. It is therefore convenient to divide a discussion of the software into two parts. In the first, the program routines for driving the hardware are considered. The second covers the way in which these routines can be combined into a program to perform any chosen sequence of operations.

Hardware Subroutines

The first question that naturally arises is that of the language in which programming is to be carried out. As discussed in the introduction it is best to use a high-level language (such as BASIC) because the effort required to write, maintain, and modify programs is much less than that needed for the equivalent programs in machine code (assembly language). Since, in addition, the Apple system runs BASIC whenever it is turned on and booted up, it seems natural to try and use this whenever possible. Whether or not programming in BASIC is satisfactory depends very much on the type of operation to be carried out and the time constraints on that operation. The kinds of points that need to be considered can best be illustrated by taking a specific example—that of driving the DAC outputs to produce a scan raster on the microscope. Because the DACs, and in fact all of the peripherals in the Apple system described here, are direct memory access (DMA) devices, a desired voltage output can be produced in BASIC by the simple statement:

$$\text{POKE(Address),J} \tag{1}$$

where the address (i.e., memory location) of the DAC is given as

$$\text{Address} = -16384 + (256 \times \text{Slot Number}) + \text{Channel Number}. \tag{2}$$

Since the DAC is here in slot 1, channel 3 would have the address

$$-16384 + (256 \times 1) + 3 = -16125. \tag{3}$$

J is a number between 0 and 255. For a DAC set up to give outputs in the range 0 to +10 V, a program that set J = 0 would then give zero output, whereas putting J = 255 would give +10 V. (Note that since BASIC also has an immediate mode of operation, these same commands could also be typed in from the keyboard and executed immediately.) By making an initial assignment of a variable name to each of the channels of interest—for example, calling DAC outputs 0–3 D0 to D3, with addresses

$$\text{D0} = -16128;\ \text{D1} = -16127;\ \text{D2} = -16126;\ \text{D3} = -16125 \tag{4}$$

—a one-line BASIC statement could then set any of the outputs to any value desired. Thus,

$$\text{FOR J = 0 TO 255:POKE D1,J:POKE D3,J:NEXT} \tag{5}$$

would ramp the outputs of channels 1 and 3 over their full range. The straightforward nature of this command sequence makes its use very attractive. However, a penalty is paid for this convenience. Although the DAC hardware can set the output value on any channel within 6 μsec, the actual time required to perform this operation under control from the BASIC program is about 3 msec because of the time required for the BASIC interpreter to read the instruction, translate it to machine code, and perform the desired action. Whether or not this is fast enough will depend very much on the type of operation being performed. If, for example, the DAC is being accessed so as to scan the beam to carry out an x-ray line scan where a few seconds will be spent acquiring data at each point before stepping the beam again, then the 3-msec operation time is of no significance. On the other hand, if we wished to scan the same beam to form an image containing, say, 256 × 256 pixels, then the minimum scan raster time would be nearly 200 seconds because of the same 3-msec limit, and this would clearly not usually be acceptable. Even in cases where a time delay of even tens of milliseconds is acceptable there may still be problems with the simple BASIC command because the execution time can only be changed (for example, by a FOR . . . NEXT loop) in multiples of 3 msec, and these steps may be too coarse.

There are several ways to improve this performance. The simplest is to continue to work in BASIC as before but, after the routine is fully debugged and operational, to compile the program using one of the compilers now available for the Apple. Thus the instructions of (5), which take about 1.6 sec to carry out in standard APPLESOFT BASIC, only require about 0.6 sec when the same statements are first preprocessed with the EINSTEIN* compiler used in our laboratory. Other compilers have been found to offer a similar level of improvement. Provided that copies are kept of the original BASIC source listings, compiling a program introduces no problems of maintenance or documentation, and in some cases the fact that the actual operating program is no longer instantly capable of modification by a user may be a real advantage. It is usually possible to compile subroutines separately and still call them from an uncompiled BASIC program. Thus, critical routines can be sped up while still leaving the rest of the program in an accessible form.

An even more convenient approach, although only so far applicable to these particular DAC and ADC units, is the use of the so-called "amper commands." Each command is of the form & FUNCTION (Address, Value). On encountering the & symbol, the APPLESOFT interpreter jumps to a special memory location, which in turn points to other memory locations in which short machine language programs for each of the allowed operations (e.g., driving a DAC with the function OUT) have previously been loaded from a disk program supplied by the manufacturer. Although the function can thus be called in a simple way from a BASIC program, the operation itself is carried out at the full speed of the machine coded program. Since this arrangement preserves the ease of programming of BASIC while providing the speed of assembly language, it is a very desirable way to proceed where it is available.

If such composite program commands are not available, and high-speed or control flexibility is required, then some programming in assembly language is necessary. Fortunately for devices with direct memory access this is not difficult. The address of channel M of the DAC in slot N is

Address = $CNOM (6)

where the $ signifies that the number is in the hexadecimal base. If a number $J (where $J lies between 00 and FF) is put into this memory location, then the DAC output from channel M will go to the value appropriate to the magnitude of $J. A complete subroutine to drive, for example, channel 3 of the DAC in slot 1 between two given values has the form:

*The EINSTEIN Corp., 11340 W Olympic Blvd., Los Angeles, CA 90064.

```
0304  LDX  $0301   PICK UP START VALUE FROM $0301 (769)
0307  TX   $C103   PLACE IT IN ADDRESS OF DAC
030A  LDA  $0300   PICK UP DELAY VALUE FROM $0300 (768)
030D  JSR  $FCAB   JUMP TO APPLE DELAY ROUTINE
0310  INX          ADD 1 TO X VALUE: COMPARE IT
0311  CPX  $0302   TO END VALUE IN $0302 (770)
0314  BNE  $0307   IF LESS GO TO $0307
```

This routine is loaded in the free memory space from \$300 to \$3FF (768–1023 decimal) of the Apple. It can be run as a subroutine from a BASIC program by the command CALL 772, which forces the program to go to the specified memory location (772 = \$0304). The desired start (S) and finish (F) values for the DAC are stored in memory locations 769 and 770 (769 = \$0301, 770 = \$0302) from the BASIC program by the command POKE 769,S and POKE 770,F where, as usual, S and F are between 0 and 255. When the machine language program is called, the X register of the processor loads itself with the number stored in memory location \$0301 and in turn places it in the memory location \$C103, which represents the DAC and, within about 6 μsec, will go to the proper value. The total time for this whole operation is only about 11 μsec, which for many purposes may be too fast. A delay step is therefore incorporated. Memory location 768 (\$300) has loaded in it a delay value D (where D is between 0 and 255) by the command POKE 768,D. After the DAC is set, the accumulator loads this number and jumps to a built-in subroutine of the Apple at \$FCAB, which generates a time delay proportional to the number stored in the accumulator register. As D is varied from 1 to 255, the delay time changes from about 2 μsec to nearly 0.6 sec. A very wide range of speeds can thus be achieved by simply varying one parameter. X is then incremented by one step, the value is compared to the finish value and, if it is less, the routine loads the new value and runs again. Once the finish value is reached the routine jumps back to the BASIC program that called it.

In practice, machine-coded subroutines have been found to be the most useful because they are adaptable, being equally suitable for both low- and high-speed applications; but in cases where there is no high-speed requirement, the approach of writing the routine in BASIC and subsequently compiling it will probably provide acceptable performance for most purposes.

The example of driving the DACs has been discussed in considerable detail because it also illustrates all of the problems that arise in the routines needed to operate the ADC and the Digisector image card. The ADC is interrogated in BASIC by a command identical to the one used for the DAC:

$$\text{POKE (Address),(Channel)} \tag{7}$$

where the address is calculated as

$$\text{Address} = -16256 + (\text{Slot number} \times 16) \tag{8}$$

and the channel address is

$$\text{Channel} = \text{Channel number} + 16 \times \text{Gain Code} \tag{9}$$

Table 1

Gain code	Input range (V)
0	0 to +5
1	0 to +1
2	0 to +0.5
3	0 to +0.1
4	−5 to +5
5	−1 to +1
6	−0.5 to +0.5
7	−0.1 to +0.1
8	All inputs shorted

where the gain code is as listed in Table 1. The call to the device therefore allows not only the particular input to be selected but its input sensitivity to be changed as required. The ability to change the input sensitivity optimizes the readout for each type of signal examined without the need for changes in hardware. The result of the analog-to-digital conversion is obtained as a 12-bit number, in BASIC, using the statement

$$\text{result} = \text{PEEK(Address} + 1) \times 256 + \text{PEEK(Address)}. \tag{10}$$

The total time required for a gain change and a conversion is only about 26 μsec but, as with the DAC, when the device is driven from a BASIC routine the timing is determined by the speed of the interpreter, and the whole operation takes about 4 μsec. For the majority of applications for which the ADCs are used here, such as reading lens currents or vacuum levels, this is more than adequate. All 16 microscope monitoring inputs can be scanned, their input sensitivity set to the correct value, and the measured result printed out by the Apple in substantially less than 1 sec. If, however, the ADC input is to be used to measure something varying rapidly, such as for signal averaging (Farrow, 1983), a simple machine-coded routine must be written. With such a routine up to 100,000 samples per second can be taken and stored if changes in gain are not needed.

The programming required for the Digisector card is greatly simplified because the hardware is "intelligent." The card contains its own processor and so can carry out its designated functions without any assistance from the Apple processor other than an initial command. For example, with the Digisector in slot 3, the BASIC statements on the left produce the results shown on the right:

PR#3	Turns on the Digisector
PRINT "!"	Clears Apple display screen
PRINT "*"	Acquires image from TV input
POKE 64,X:POKE 65,Y PRINT "." B=PEEK 66	Gives brightness B (0–63) of point (x, y)

Thus, even though the DIGISECTOR is processing data at the rate of many thousands of samples per second, no machine language coding is required for straightforward applications. The processing algorithms required for the production of half-tone images on the Apple display screen by means of dot-density variations, however, are written in assembly language routines supplied by the manufacturer.

Operating Programs

The operating system programs are all written in APPLESOFT BASIC, the extended version of BASIC supported by the Apple. If the time-critical portions of programming (i.e., those portions that control the hardware) have been taken care of by writing efficiently coded general-purpose subroutines, a simple menu-driven BASIC program is sufficient to give the operator access to any of the functions. With the library of subroutines stored on disk, an operating program designed to meet any specific operating requirements can readily be customized with a minimum of programming effort. However, again in the spirit of the concepts discussed in the introductory section, it has been found desirable to incorporate some special constraints into the operating routines.

The most important of these considerations is dictated by the fact that the program is designed for use by a person the majority of whose attention is concentrated on the microscope. Furthermore, the microscope itself is quite likely to be in use in a wholly or partially darkened room. This means that computer operations requiring lengthy, or even accurate, input through the keyboard are going to be subject at best to error and at worst to disuse. These problems can be overcome by writing the program in such a way that hitting any key, or using a key that is particularly easy to find, such as the space bar, provides the necessary response. In the majority of the programs written for use with the hardware described here, a uniform entry format is used such that when a menu of choices is presented to the operator, hitting the space bar moves the cursor from one option to the next. Hitting any other key on the keyboard will select that entry. Only when actual numerical input data is requested is the operator required to be accurate in finding a particular key. This type of input system also eliminates the possibility that an incorrect entry could cause the system to lock up or crash. Because the whole operating program is in core memory, one can recover from a crash simply by hitting RESET and typing RUN to restore operation. This will, however, lose any data previously obtained but not stored on disk. An interesting alternative to this approach might be to use the voice recognition modules now available for the Apple, such as the VIM.* After storing suitable reference files, such a system will respond to a vocabulary of typically 30 or more words from a given operator. A system of this type has already been implemented on one commercial multichannel analyzer (Kahdeman, 1983) and would appear to be the ideal way of freeing the operator from the need to interact with a keyboard.

The other type of input occasionally required is a two-dimensional coordinate, such as that needed to specify a point on an image. This kind of operation is often taken care of by means of joysticks or "game controllers," both of which are read-

*Manufactured by Voice Machine Communications Inc., Santa Ana, CA 92705.

ily available for the Apple. However, experience has shown that the routines used by the Apple to read the positions of such devices are very prone to "jitter," which significantly reduces the accuracy of this kind of input. A simple software joystick has therefore been written that uses the directional keys (<, >, ∧, ∨) to move the cursor around the screen and ignores any other key except "q," which exits from this routine. This is free from any jitter, is much more precise than even the best available controllers, and has the advantage that no additional hardware is required. The obvious disadvantage is that it requires careful attention to the keyboard.

The original and simplest operating program is accessed from a menu (Figure 3) that offers five "pages" of information. On calling for page 1, the excitation currents of the six lenses of the microscope, together with the gun and specimen chamber vacuum levels, are displayed and updated once per second (Figure 4). The lens current values are displayed directly as the 12-bit numbers (i.e., lying between 0 and 4,095) produced by the ADC. The gun and specimen vacuum readings, however, are converted to produce a readout directly in Torr using a correction equation with empirically derived constants. Because the sampling error of each ADC input is $\pm\frac{1}{2}$ least significant bit, some random variation is observed in the last displayed digit of each readout, but this has not proved to be at all troublesome. Touching any key returns control to the menu, which operates as described earlier.

Page 2 makes use of the high-resolution graphics of the Apple to display the *x*, *y* coordinates of the stage relative to the optic axis of the microscope. Using the HPLOT command in APPLESOFT BASIC, the ADC outputs representing the *x*, *y* positions can be plotted directly onto the screen, inside an outer square indicating the safe extremities of motion (Figure 5). This display, together with the tilt coordinates (converted so as to read directly in degrees) and the gun vacuum (in Torr), is updated once per second. Exit to the main menu is again obtained by touching any key.

Figure 3. Simple menu.

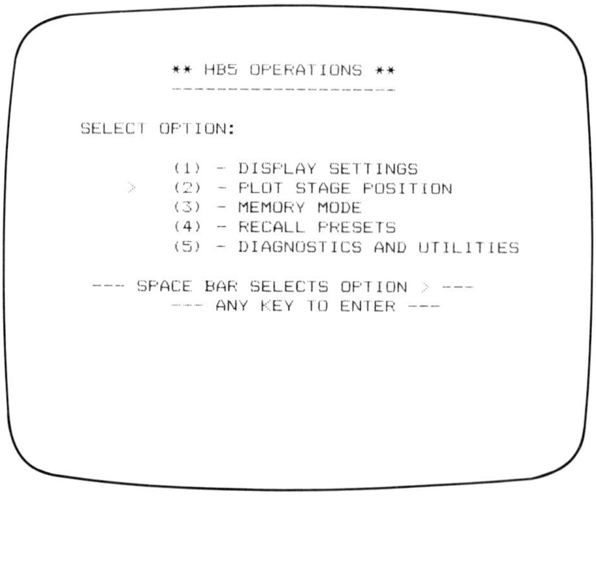

```
              ** HB5 OPERATIONS **
              --------------------

SELECT OPTION:

              (1) - DISPLAY SETTINGS
           >  (2) - PLOT STAGE POSITION
              (3) - MEMORY MODE
              (4) - RECALL PRESETS
              (5) - DIAGNOSTICS AND UTILITIES

  --- SPACE BAR SELECTS OPTION > ---
         --- ANY KEY TO ENTER ---
```

```
-OPERATING CONDITIONS-
----------------------

CONDENSOR 1                 452
CONDENSOR 2                 744
OBJECTIVE                  3177

POST SPECIMEN 1             266
POST SPECIMEN 2             766
POST SPECIMEN 3             000
----------------------------------
GUN VACUUM                 2E-11
COLUMN VACUUM              1E-8

TO RETURN TO MENU - TOUCH ANY KEY
```

Figure 4. Sample readout of current and vacuum level values.

Page 3 of the program allows the current operating conditions of the microscope to be stored on disk under some chosen file name. After displaying warning and prompting messages to ensure that a disk is in the drive and that the drive door is closed, the operator is requested to enter the operating energy (keV) and the operating mode (i.e., imaging, diffraction, microanalysis, etc.) of the microscope, and to choose an identifying name. The ADC signals monitoring the lens currents and the stage position coordinates are then stored on disk together with the other data, after which the program automatically returns to the main menu.

Page 4 of the program allows previously stored sets of conditions to be retrieved. It had originally been the intention that these conditions would then be reset on the microscope by the computer-operated DAC channels interfaced

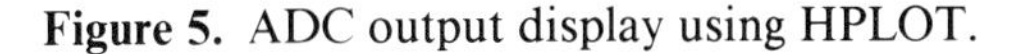

Figure 5. ADC output display using HPLOT.

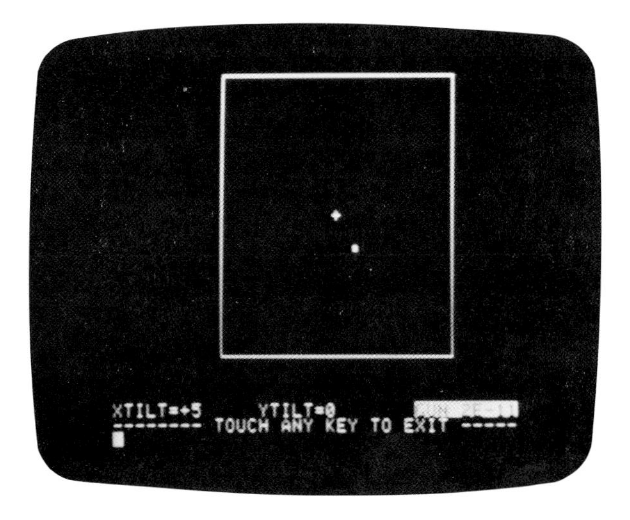

with the lens controllers. It was found, however, that even with carefully optimized assembly language coding, the time required for the iterative process of comparing the measured lens current with the stored value and adjusting the DAC to give a match was considerable, due largely to the relatively slow response of the lens controller and the need to write the routine in such a way that it would not be affected by the sampling jitter of the ADCs. Although the automatic procedure has been made to work, a manual process has been found to be easier and quicker. As now implemented, the recalled preset values are shown with a rapidly updated (10 times per second) display of the corresponding lens currents (Figure 6). The operator then makes the match using the controllers as usual. Although this is less excitingly "automatic" than full computer control, it has actually been found to be rather faster to use under most conditions. Typically all six lenses can be set to the desired values within a few seconds at the required precision.

Page 5 of the program displays a secondary menu that gives the operator access to a variety of programs designed both to monitor the microscope and to check the operation of the computer. These include a continuously updated, unconverted readout of all 16 ADC channels (Figure 7) and a routine for testing, calibrating, and reading any of the DAC outputs. In addition, provision is made to scan and read the ANAVAC residual gas analyzer. The graphics are again used to produce a continuously updated display of the partial pressure (on a logarithmic scale) against atomic mass units (amu) in the range 0–80 amu. Using the assembly language DAC routine discussed earlier, the complete mass range can be scanned in 30 sec. This kind of display has been found to be far more useful and practical for following the state of the vacuum system than a conventional strip-chart record. When hard copy is required this can be produced by a screen dump to the EPSON MX-80 dot matrix printer used with the system. One other option on page 5 allows all the ADC inputs to be shorted to ground, and all DAC outputs to be zeroed, for servicing purposes or in the event of a malfunction in either the microscope or the computer.

The microscope control system discussed above has been in daily use for a

Figure 6. Sample readout of lens currents.

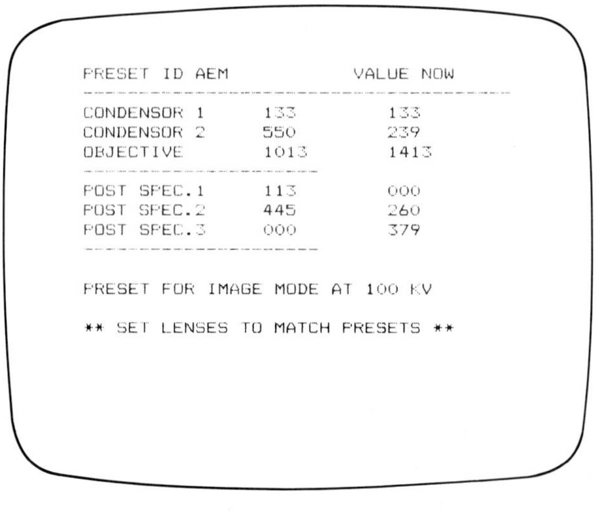

```
PRESET ID AEM                    VALUE NOW
--------------------------------------------
CONDENSOR 1         133          133
CONDENSOR 2         550          239
OBJECTIVE           1013         1413
------------------------
POST SPEC.1         113          000
POST SPEC.2         445          260
POST SPEC.3         000          379
------------------------

PRESET FOR IMAGE MODE AT 100 KV

** SET LENSES TO MATCH PRESETS **
```

ADC DIAGNOSTIC CHECK

CHANNEL #=0	132	OBJ.LENS
CHANNEL #=1	003	COND.2
CHANNEL #=2	319	COND.1
CHANNEL #=3	111	POST.SPEC3
CHANNEL #=4	040	POST.SPEC2
CHANNEL #=5	032	POST.SPEC1
CHANNEL #=6	1998	X-SHIFT
CHANNEL #=7	-190	Y-SHIFT
CHANNEL #=8	003	X-TILT
CHANNEL #=9	-243	Y-TILT
CHANNEL #=10	-1996	COLUMN VAC
CHANNEL #=11	-2109	GUN VACUUM
CHANNEL #=12	000	SPARE
CHANNEL #=13	000	SPARE
CHANNEL #=14	2012	RGA
CHANNEL #=15	000	DAC MONITOR

TO EXIT HIT ANY KEY

Figure 7. Readout of all 16 ADC channels.

period of more than 2 years and has proved to be a very worthwhile addition to the microscope. However, as more experience was gained with the hardware it was clear that even more could be obtained from the system, and a variety of special-purpose program packages using the same operating routines but assembled in a different way have been produced. An example is that designed for semi-automated microanalysis. This again is operated from a menu, the first page of which monitors the microscope functions described earlier. On choosing page 2, the Digisector card is activated and the TV rate image from the STEM is displayed on the graphics screen of the Apple (Figure 8). By means of a "+" cursor, which can be moved around the screen under the control of the software joystick, up to 12 points on the image can be selected. This is accomplished by moving the cursor over each of the points in turn and hitting "q," which stores their coordi-

Figure 8. STEM TV image display.

nates. When page 3 is selected, the beam is immediately blanked under control of a signal from the Apple to prevent any further unnecessary damage to, or contamination of, the sample. After the operator selects the required analysis time, the beam is moved to the first selected point and an audible warning is given. The operator then turns on the energy dispersive analyzer and touches any key on the Apple to unblank the beam. At the end of the chosen time period the beam is again blanked, and a second warning sounded to alert the operator to terminate the acquisition of data on the x-ray system. The DACs now move the beam to the second selected point, and the process described above is repeated. In this way spectra from up to a dozen points can rapidly be obtained with a minimum of operator effort, while limiting the beam exposure of the sample to the lowest possible value. In a variant of this program the Apple itself has been used to drive an electron energy loss spectrometer (using one DAC channel to scan the spectrum) and store the resultant spectra (collected through one ADC channel) automatically onto disk. Although the number of analyzed points is limited, this program package has been very useful, in particular when working with beam-sensitive materials, because it frees the operator from a number of routine tasks, especially that of remembering to turn the beam off when the electrons from it are not being used.

Although the use of this beam scan for lithographic purposes is not relevant in the context of this book, the ability to scan the beam in an arbitrarily controlled way has been found useful for one recurrent problem in microanalysis, that of contamination. When performing either electron energy loss spectroscopy or x-ray analysis using a thin window detector, the problems of hydrocarbon contamination can be severe on many samples, particularly in point analysis mode. An option on the menu allows the operator to initiate a routine that scans the beam to form a hollow square enclosing an area approximately corresponding to the field of view (Figure 9), ideally about 1 μm or less on a side. This scan pattern is repeated three times, a process that takes in all about 1 minute, and produces a

Figure 9. "Window frame" of contamination.

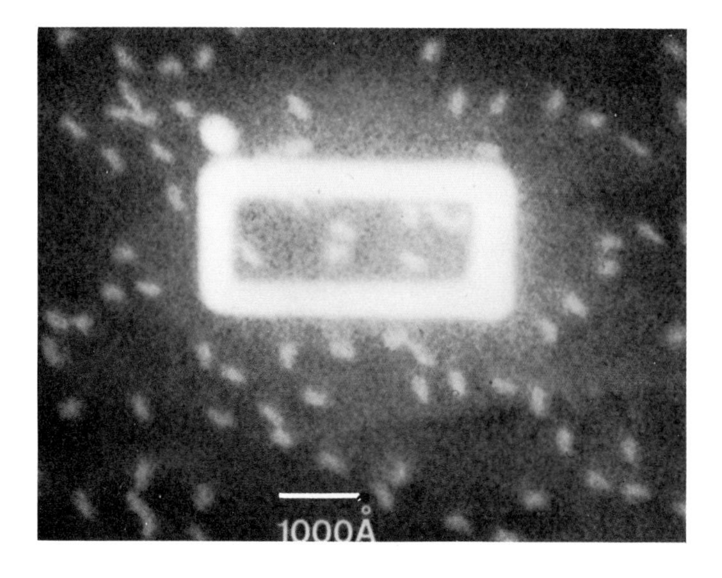

thick "window frame" of contamination around the area of interest. This has two effects. First, it locks up all the mobile surface contamination and effectively removes all such components from within the framed area. Secondly, it prevents such contaminants from moving through the barrier and to the inside of the marked area. After the barrier has been drawn, it has been found that the rate of contamination inside the square is typically only one-fourth to one-fifth of that outside, and so analyses can be carried out for much longer periods without difficulty. As an additional benefit the chosen areas are easy to find again, even when working at low magnification.

Conclusions

The system described here clearly does not represent the state of the art in either hardware or software sophistication. The interface is the simplest possible unit that will allow the required signals to be transferred, and no attempt has been made to allow the computer to interact with the logic systems of the microscope to change the magnification or operating modes. The benefit of these limited objectives has been a system that has been simple to set up and program and that has been exceptionally reliable in operation. The software considerations discussed above have resulted in a family of programs that are efficient to use under laboratory conditions and yet are easily modified to meet individual requirements.

Although by the standards of laboratory minicomputers the Apple is fairly slow for the type of dedicated operation described here, the processing speed is not a limiting factor. On the contrary, the slight limitations in absolute performance are more than offset by the ease of programming and by the ready availability of sophisticated peripherals at competitive prices. Adding the computer to the microscope in the way described does not change or complicate the normal operation of the instrument, but it offers the operator additional freedoms and capabilities.

References

Farrow, R.C. (1983) A model for mini-computer control of the SEM used for basic research. In O. Johari (ed.): Scanning Electron Microscopy 1983. Chicago: SEM Inc. (in press).

Joy, D.C. (1982) Microcomputer control of a STEM. Ultramicroscopy 8: 301–308. [See also other related papers in this special issue of *Ultramicroscopy* devoted to computer–microscope interfacing.]

Kahdeman, J.E. (1983) Application of automatic speech recognition to x-ray microanalysis. Am. Lab. (September): 110–115.

Smith, K.C.A. (1982) On-line digital computer techniques in electron microscopy: A general introduction. J. Microsc. Oxf. 127: 1–3. [This special issue also contains several other papers on related aspects of computer-aided microscopy.]

MICROCOMPUTER USES IN MORPHOMETRY

8

Morphometric Measurement Using a Computerized Digitizing System

R. Ranney Mize

Introduction

It is a long-standing assumption of many anatomists that the shape, size, and other geometric characteristics of cells will reflect, if not actually reveal, their function. It is thus quite common in structural studies to describe *qualitatively* the variety of cell sizes and shapes found in various tissues. These qualitative descriptions often take on an almost anecdotal character. Despite this, anatomists often try to develop classification schemes that categorize cells based on a few qualitative parameters.

A more rigorous approach to cell classification demands a quantitative analysis of a large number of physical parameters. This measurement of the geometries of cells and organelles is called *morphometry.* Morphometry uses systematic mathematical techniques to measure such parameters as size, shape, and volume quantitatively. These types of measurement are clearly important to developing valid cell classifications but, until recently, they have rarely been used in anatomical studies. This is in large part due to the great effort involved in making such measurements.

Manual measurement of such features as area and perimeter is exceedingly tedious and time consuming. As an example, several years ago we wanted to measure the size of cells in a midbrain visual structure, the superior colliculus, in order to compare the accumulation of autoradiographic grains in different sized cells after injection of a tritiated amino acid neurotransmitter, γ-aminobutyric acid (GABA). To accomplish this, we counted the grains overlying every cell, traced the outline of the cell onto high-quality tracing paper, cut out each trace, weighed each cutout, and compared it to the weight of a standard of known cross-sectional area. Using this exhausting procedure, we obtained two significant measures: cell area and grain density (number of grains per unit area). The results revealed a significant (and important) correlation between grain accumulation and cell size (Mize et al., 1981), but the effort involved in obtaining the results was immense.

A number of other manual methods for obtaining geometric measurements have been reported, including the use of planimeters and point counting stereology (Konigsmark, 1970; Weibel, 1979). Many of these techniques are equally time consuming and require separate measurements for each parameter desired (such as area, perimeter, diameter). The recent introduction of inexpensive microcomputers and digitizing tablets, however, has revolutionized morphometry. Using computer-based morphometry systems, it is possible to collect data far more rapidly and accurately than with manual approaches. Mathematical and statistical calculations can be performed by the computer automatically. Several different types of parameter can be measured simultaneously. The computer provides a convenient source of data storage and display. With the advent of the microcomputer, morphometry has become a simple and rapid procedure.

A number of commercial and noncommercial systems are now available for use in morphometry. This chapter describes some basic principles of morphometry, discusses different types of morphometric system, and describes a simple digitizer-based morphometry system we have developed on a microcomputer that is written entirely in BASIC.

Hardware

Computerized morphometry measurements can be made by three types of device: digitizing tablets, video projection digitizers, and automatic image analyzers.

Digitizing Tablets

Digitizing tablets are electronic drawing devices, not unlike "electronic blackboards." They have a tablet surface that is electrically referenced to a pen or cursor. The surface of the tablet is mapped out in Cartesian coordinates. The digitizer generates a signal representing the *x, y* coordinate position of the pen or cursor as it is moved around the tablet. These values are sent to a computer programmed to convert the data to geometric measures of area, perimeter, diameter, shape, and other features. To measure these image parameters, the operator simply moves the cursor or stylus around the profile to be measured. The image can be a photographic print placed on the tablet surface or a histological slide, photographic slide, or movie frame front- or back-projected onto the tablet. The image can also be superimposed on the tablet using a drawing tube attached to a light microscope.

Most digitizing tablets manufactured today use electrical wave sensing devices. The tablets are usually embedded with an electrically active wire grid referenced to the cursor, which contains a pickup coil. Digitizing tablets come in a variety of sizes, ranging from about 12 $in.^2$ to over 4 ft^2. Recently developed digitizing tablets have excellent resolution, ranging from 0.025 mm to 0.25 mm (Table 1). Most commercial tablets have built-in microprocessors to convert the analog signal of the tablet to digital *x, y* coordinate values. They come with a variety of interface options, including RS-232 serial, 8- and 16-bit parallel, BCD, and IEEE-

Table 1. Selected Digitizing Tablet Systems

System	Computer	Resolution	Measurements	Manufacturer/ distributor
Ladd 4000	Rockwell AIM 65	0.10	A,L,P,F,D,S,C, others	Ladd Research Industries, Burlington, VT
Microplan	microprocessor	0.025	A,L,P,F,D,S,C	Laboratory Computer Systems, Cambridge, MA
Zidas	microprocessor	<0.1	A,L,P,F,D,S,C, ang., others	Carl Zeiss, Inc., New York, NY
Numonics 1224EM	microprocessor (8080)	0.25	A,L,P,F,D,S,C, ang., others	Numonics Corporation, Lansdale, PA
Optomax	Apple IIE	0.10	A,L,P,F,D,S,C	Optomax, Inc., Hollis, NH

Note: A = area; L = length; P = perimeter; F = Feret dimensions; D = diameters; S = shape factors; C = center of gravity; ang. = various angles.

488 GPIB instrument interfaces. Code formats include ASCII, binary, and binary-coded decimal.

We use a Hewlett-Packard 9874A digitizer with our morphometry system (Figure 1). The digitizer has a transparent ground glass surface so that images can be back-projected on the platen. The digitizer also has a keypad panel for control of various functions. Ten special-function keys are used to access various digitizing subroutines. A numeric keypad is used to enter data values related to the digitized profile. An LED display prompts the user to enter data. Other function keys are used to set the sampling mode and to align and extend the axis of the tablet. The cursor has a circular glass window with a bull's-eye etched in the glass that is used as the target in tracing. The cursor also has two function keys, one for taking digitized points, the other for activating a vacuum that holds the cursor in place when not in use. The cursor has a resolution of 25 μm (0.0067 in.).

Hewlett-Packard now manufactures a less expensive tablet with an opaque surface and a pen-style cursor that has a resolution of 0.1 mm (HP Model 9111A). Summagraphics, Talos, and GTCo manufacture similar tablets in the price range of $500–$2,000. These are distributed by various companies. The Houston Instruments Hi-Pad is one of the most popular.

Because the less expensive tablets have few or no hard-wired special-function

Figure 1. Microcomputer-based digitizing system used for morphometry. The Hewlett-Packard 9845T desktop computer includes a graphics CRT (A), two magnetic tape mass storage drives (B), and a built-in thermal line printer (C). The computer is interfaced to the HP-9874A Digitizer, which includes a glass platen surface (D), cursor (E), special-function keys (F), and a numeric keypad (G). The computer is also connected to a floppy-disk drive (HP-9895A) and a digital plotter (HP-9872A). [Modified from Street and Mize (1983), by permission of Elsevier Biomedical Press, Amsterdam.]

keys, functions are accessed either by function keys on the cursor or a tablet "softkey" menu. A menu of softkey functions is produced by reserving a block of space on the tablet for this purpose. This space contains software-defined squares representing each function. The computer is programmed so that the function related to that square (softkey) is executed when the cursor is placed over the square and the digitize key pressed. This technique offers considerable programming flexibility and can be easily learned by the user.

Digitizing tablet systems are easy to use, are relatively inexpensive, and provide excellent digitizing accuracy. A number of commercial morphometry systems use digitizing tablets for data entry (Table 1).

Video Screen Projection Digitizers

Video screen projection digitizing systems work just like digitizing tablet systems except the image is projected onto a video monitor rather than on a tablet. The image itself is reproduced by a TV camera. The camera can be attached to a light or electron microscope or can look directly at a photograph on a well-lit working surface. These systems offer certain advantages in cost and convenience, but also introduce additional complications in the procedure.

Video digitizing systems operate by superimposing an electronic cursor over the video image of the profile to be digitized. The operator moves the cursor on the TV screen by manipulating a joystick, "mouse," keypad of arrow keys, tracker ball, or cursor on a digitizing tablet. The cursor on a digitizing tablet is the most commonly used device in commercial systems, probably because the tablet can then also be used to digitize micrographs directly. To digitize a profile on the screen, the operator moves the cursor around the outer contours of the profile. An electronic trace is usually drawn on the screen as the cursor is moved so that the operator can see how accurately the profile has been traced. The measurements made by video systems are under program control and are similar to those performed by digitizing tablet systems. A video digitizing system is described in detail in Chapter 10. Several commercial systems are listed in Table 2.

Table 2. Selected Video Projection Digitizing Systems

System	Computer	Measurements	Manufacturer/ distributor
Bioquant	Apple IIE/IBM PC	A,L,P,F,D,S,C, ang., others	R&M Biometrics, Nashville, TN
Ideas	Apple IIE	A,L,P,D,S, ang.	Computer Instruments, Orrs Island, ME
Optomax	Apple IIE	A,L,P,F,D,S,C, ang., others	Optomax, Inc., Hollis, NH
SMI Unicomp	Apple IIE/IBM PC	A,L,P,D,S, ang.	Southern Micro Instruments, Atlanta, GA
Videoplan	microprocessor (Z80A)	A,L,P,F,D,S,C, ang., others	Carl Zeiss, Inc., New York, NY

Note: A = area; L = length; P = perimeter; F = Feret dimensions; D = diameters; S = shape factors; C = center of gravity; ang. = various angles.

Video systems have two major advantages. First, because the image is reproduced electronically on a monitor, it is not necessary to produce a photographic copy of the image. This can save a lot of money, particularly since photographic supplies have become very expensive. Secondly, a TV image can be digitized electronically so that it can be manipulated by a computer. Some video digitizing systems allow the operator to manipulate gray-scale levels in order to extract particular features of the image. Once extracted, those features can be measured automatically. (See the description of image analyzers below.)

The video projection systems also have several disadvantages. The TV images almost always have a lower resolution than photographic prints or digitizing tablets. Many users also find that remote tracing with a screen cursor is less accurate and more difficult than direct tracing on a digitizing tablet. Some video systems also introduce optical distortions in the image that can lead to inaccurate measurements. Finally, our laboratory found it was somewhat discomforting not to have a permanent record of the data to which one could return for reanalysis and classification.

Image Analyzers

Image analyzers reproduce a digital image of a specimen that can then be manipulated and analyzed quantitatively. The advantage of image analysis systems is that both the extraction process and the measurements are automatic. It is not necessary to outline the objects to be measured manually; rather, the objects are discriminated from background by electronically manipulating the gray-scale levels of the image. Once the appropriate features have been extracted, the analyzer can automatically calculate the area, perimeter, shape, and other geometric features of any or all extracted objects in the field.

An image analyzer can be a very powerful tool for making large numbers of measurements on highly stereotyped profiles. On the other hand, automatic image analysis has not proven very successful for pattern recognition tasks of complex biological specimens. Image analysis has been reviewed by Bradbury (1977, 1979, 1981), Smith (1982), and Sobel et al. (1980), among others. Some of the commercially available image analysis systems are listed in Table 3.

A typical image analyzing system includes an imaging device for reproducing the image, a device for digitizing the image, a device for storing and displaying the image, and a computer and/or other hardware for manipulating the image. Imaging devices include scanning TV cameras and various photometric instruments such as scanning microphotometers and rotating drum microdensitometers [see Chapter 14]. Imaging devices generate an array of picture elements (pixels), each of which contains information about the light intensity present at that point in space. Digitizing devices convert the light intensity at each pixel into a digital code along some photometric scale. The digital values are called *gray-scale values* and range from as few as 16 possible values up to 4,096 in some photometer-based systems. The effective gray-scale resolution can be limited by the scanning device, the digitizing device, the storage device, the computer, or the limited intensity range in the image itself.

An intermediate hardware device is typically employed to store and display the image transiently. A common device for this purpose is called a *video frame store buffer* (Smith, 1982). This device includes blocks of computer memory, each

Table 3. Selected Semiautomatic Image Analysis Systems

System	Computer or microprocessor	Features	Manufacturer/ distributor
IBAS	Z80A-based	extr., meas.	Carl Zeiss, Inc. New York, NY
Joyce-Loebl Magiscan 2	NA	extr., meas.	Nikon, Inc., Garden City, NY
Leitz TAS	LSI 11/23	extr., meas.	Ernest Leitz, Inc., Rochleigh, NJ
Model 2000 Image Analyzer	Apple IIE	extr., meas.	Image Technology, Deer Park, NY
Omnicon 3000	Z80A-based	extr., meas.	Bausch and Lomb, Rochester, NY
Quantimet 900	LSI 11/23	extr., meas.	Cambridge Instruments, San Diego, CA

Note: extr. = extraction; meas. = measurement. NA = not available.

organized to hold a single image frame. Hard-wired or software-based computer algorithms are used to operate on the image in various ways, such as manipulating contrast or arithmetically combining two images. Image processing procedures can also enhance edges by varying contrast and intensity. Once enhanced, the contrast and spatial aspects of the image can be further processed in order to extract objects of given size, shape, or intensity characteristics. When the features have been extracted, it is quite simple to count the objects and measure their size and shape.

Despite this power, automatic image analyzers present two formidable obstacles to the biologist interested in morphometric measurement. The first obstacle is cost. Full-fledged image analyzers with extensive pattern recognition software often cost in excess of $100,000. The second obstacle is the state of development of computerized pattern recognition itself. Pattern recognition of complicated biological materials remains at best a poorly developed art that almost always requires extensive interaction with the investigator. Many investigators find that it is easier to measure each profile manually than to perform a multitude of interpolations that allow one to extract the images from background and artifact. In many cases, successful extraction without significant contamination from artifact is simply not possible.

Of the three approaches to morphometric measurement, I believe the tablet digitizer remains the least expensive and easiest solution for most biological problems. Software to drive tablet digitizing systems is described in the following section.

Software

Programming Language

Morphometric software can be written in machine, ASSEMBLER, or one of many high-level languages. Speed is not an important factor in morphometric digitizing systems so long as the program can read data from the digitizing tablet at an adequate rate. For this reason, many published morphometric software programs have been written in high-level languages. At least three packages have been written in FORTRAN (Albright and Sawler, 1981; Cowan and Wann, 1973; Dunn et al., 1975, 1977). BASIC language programs have been written to run on a variety of microcomputers (Dennino et al., 1978; Mize, 1983c; Pullen, 1982). PASCAL (Curcio and Sloan, 1981) and assembly language (Green et al., 1979; Peachey, 1982) have also been used.

The speed of digitizing depends in part on the speed of the hard-wired programs contained in the ROM of the digitizer itself. These programs convert analog signals to a digital format and perform other housekeeping chores such as skew correction and scaling. Most digitizers can output data at a rate of 30–100 points per second, which is fast enough for most digitizing tasks. This rate can be slowed appreciably, however, by the speed of the input and computation algorithms used in the morphometric computer program. Our programs, for instance, are written in HP's enhanced interpretive BASIC. These programs slow data collection to a rate of about 9 points per second. Methods for increasing this speed are discussed below [page 193].

Software Design

Software requirements for digitizing are fairly straightforward

The operator needs to be able to enter *setup data* about the experiment (date, experiment number, operator name).

A *magnification factor* must be entered so the computer can calculate absolute values for area and perimeter.

The operator should also be able to *enter data values* associated with the digitized profile, such as number of labeling grains, number of synaptic contacts, or number of dendrites.

A subroutine must be developed to *collect data* from the digitizer and *convert* them to decimal values.

Routines for *calculating* area, perimeter, and other geometric measurements are needed.

Finally, the program should have subroutines for *storing and printing the data.*

Our morphometric digitizing program contains separate subprograms for each of these functions, which are described below. A diagram of the program flow is shown in Figure 2. An abbreviated version of the BASIC program is reprinted in the Appendix.

Setup

In order for data to be easily referenced, it is convenient to include index information about the experiment on each data file. Our program has an option for entering the date, block and animal numbers, and operator ID, up to a maximum of 80 characters. This information is stored in a one-dimensional string variable called Title$. The size and dimensions of the variable could easily be expanded to accommodate additional setup data.

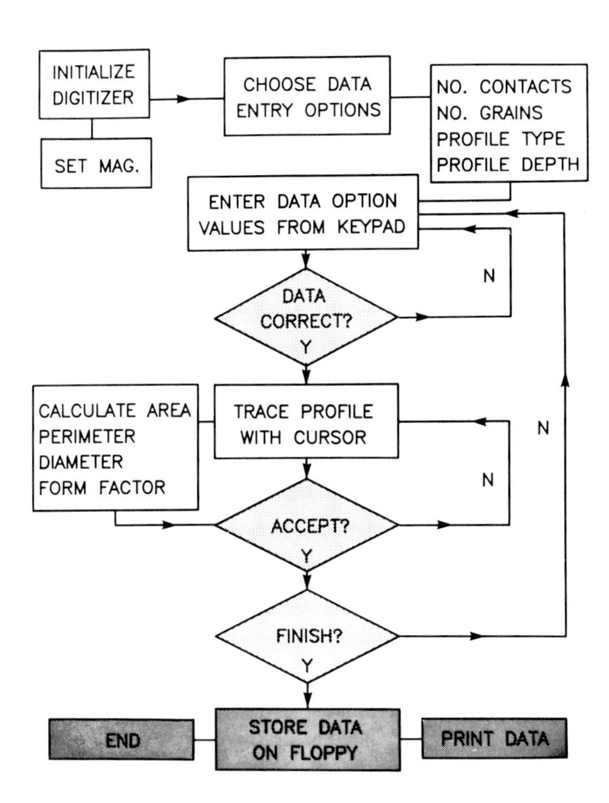

Figure 2. Flow diagram of the morphometry program showing the operating steps involved in collecting data from the digitizer and calculating area, perimeter, average diameter, form factor, and several types of density. [Reprinted from Mize (1984) by permission of Academic Press, New York.]

Magnification Factor

A magnification factor must be entered in order to scale digitizer units to a real unit of measure. Print magnification can be measured manually or under program control. To set the magnification under computer control, a photograph of a carbon grating replica calibration grid or microscope reticule would be placed on the surface of the digitizer. The operator would then digitize the two end points of the grid or reticule and enter the actual (unmagnified) length between the two points. The computer program would then automatically compute the magnification factor. Although the software routine to compute magnification is simple, we find it just as easy to calculate magnification manually and enter the value into the computer from the computer keyboard.

Data Entry Options

In order to associate various features with the digitized profile, it is convenient to be able to enter data values for such things as number of labeling grains, number of dendrites, and a code for the type of profile. If one wants to calculate grain density, for example, one would count the number of grains enclosed within the profile whose area is to be measured. Such counts can be made manually and the value entered into the computer via the numeric keypad. Alternatively, the digitizer can be used to assist in counting. In our system the operator can place the program in *count mode* (Table 4), position the digitizer cursor over each element

Table 4. Morphometry Digitizing Functions

Function	Description
Length mode	Measures length of straight lines or distance between two points. Area and perimeter are not calculated. Operator digitizes two points.
Perimeter mode	Measures length of curved lines. Area and diameters under the curve are calculated. Operator digitizes multiple points.
Area mode	Measures area, perimeter, diameters, and form factor of closed profiles. Operator digitizes multiple points.
Single point mode	Collects a single point each time the digitizer key is pressed.
Continuous point mode	Collects a stream of continuous points for each time or distance increment set by the operator.
Graphics mode	Displays scaled digitized trace on CRT screen so operator can check it for accuracy.
Count mode	Allows operator to take counts of elements inside a profile. Digitize key is pressed once for each count.
Two trial digitizing	Requires operator to digitize each profile twice. Displays percent error between the two trials and averages the trials if error is accepted.
Display features	The LED displays the x, y coordinate position of the cursor. Also displays accumulated length.
Correction features	The LED displays the keypad entries so they can be corrected. Also displays percent error between two trials so operator can accept or reject it.

(i.e., grain) and depress the digitize key to record the count. In order to avoid double counting of grains and other elements that are densely distributed in our tissue, we usually make manual counts using an ink pen to mark grains as we count them. The counts are then entered into the computer from the digitizer's keypad. The automated count mode is used only where elements are sparsely distributed within the profile.

Our program allows the following entries:

1. number of contacts made by the digitized profile (e.g., number of synaptic contact zones of a synapse);
2. number of elements within the digitized profile (e.g., number of autoradiographic grains or number of synaptic vesicles);
3. number of elements extending from a digitized profile (e.g., number of dendrites);
4. numeric code for profile type (e.g., 1 = glial cell; 2 = retinal synapse); and
5. profile depth from the tissue surface.

Three density values are calculated from these entries:

1. the number of contacts per unit of perimeter (contact density);
2. the number of elements per unit of cross-sectional area (element density within the profile); and
3. the number of elements per unit of perimeter of the profile (element density of the profile surface).

The operator is prompted by the digitizer LED display to enter each value using the digitizer's numeric keypad (Figure 1). These values are briefly displayed on the LED so that the operator can correct them if necessary. The values could also be entered from a softkey menu keypad on the digitizer's surface or from the keyboard of the computer. Both of these alternatives are slower, however, because they require the operator either to move the cursor from the digitized profile or to move his or her hand from the cursor to the computer.

Digitizing

The tracing process is similar for all digitizers. The image (a photograph or projected slide) is placed on the platen of the digitizer and the cursor or stylus moved around the outer contour of the profile to be digitized. Our photographs are held in place on the digitizer surface by a thin sheet of transparent plastic held to the digitizer by four magnets (Figure 1). Our morphometric software has three separate modes for digitizing. These include measurements for

1. the length of a straight line or distance between two points;
2. the length of a curved line, which also calculates the area and diameters under the curve; and
3. the area, perimeter, diameters, and form factors for closed figures.

To measure the distance between two points (or the length of a straight line), only two points need be digitized. For curved lines or closed figures, multiple points must be taken. Points can be digitized in one of two modes. In *single point mode* (Table 4) the digitize key is depressed once for each point taken. In *continuous point mode,* points are taken automatically each time the cursor is moved a specified distance or a specified increment in time has occurred. Both distance and time intervals are under program control and can be changed by the operator.

For closed figures, it is important to make a mental note of the position at which a trace is begun; one attempts to return to that point at the end of the trace. A closure routine is used to end the trace automatically when the cursor comes to within 10 digitizer units (250 μm) of the starting point. If this closure window is missed, the operator can end the trace by depressing a special-function key on the digitizer keypad; this also closes the figure.

It is *very* important to close mathematically a figure whose area is being calculated, whether this is done manually or automatically. Unless the figure is closed, the region between the last and first digitized points will be miscalculated in the area calculation (see Figure 3). This can lead to a significant overestimate of the actual area of the digitized profile. The advantage of providing an automatic closure window is obvious. It forces the operator to return as accurately as possible to the starting point of the trace and ensures that the figure is closed. This improves both the accuracy of the calculation and the concentration of the operator.

Our program includes several other features to improve accuracy. The operator can choose to digitize each profile twice (two trial digitizing, Table 4). In this case, the program compares the two traces and displays the percent error on the digitizer LED. If the error exceeds a certain value, the operator can be forced to redigitize one or both traces. If the error rate is acceptable, the program averages the two traces, stores the average values in the appropriate variables, and starts a

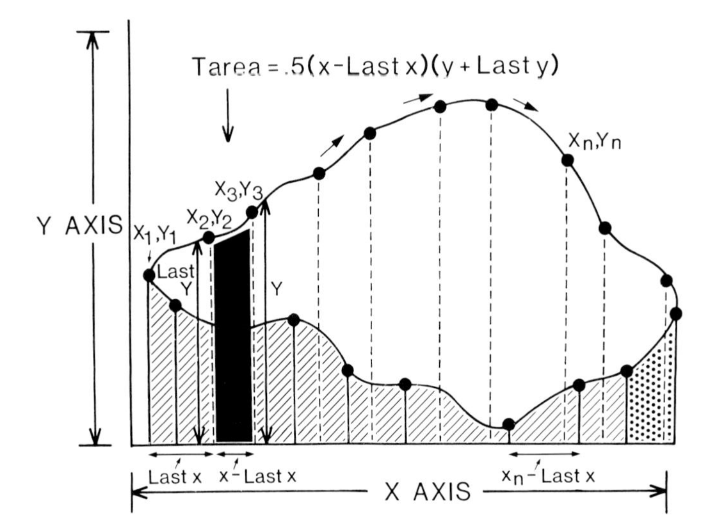

Figure 3. Graphic illustration showing how the morphometry program calculates cross-sectional area using a modified trapezoidal function. Each trapezoid is formed by two sequential coordinate points (for example, X_2, Y_2; X_3, Y_3, shown in black). x − Last x represents the base of the trapezoid and y + Last y represents the summed height of the two parallel sides. Positive trapezoidal areas are computed as the cursor moves to the right (top arrows). Negative trapezoidal areas are computed as the cursor moves to the left (for example, the speckled trapezoid). The negative area is the region lying below the lower border of the digitized profile (cross-hatched region). The total computed cross-sectional area of the profile is the sum of the positive areas minus the sum of the negative areas.

new trial. The operator can also examine the traced outline of the digitized profile on the computer CRT to see how accurately it has been drawn (graphics mode, Table 4). If it is inaccurate, the profile can be retraced.

Geometric Calculations

A number of geometric features can be computed from the digitized trace of a closed figure. Commonly used calculations include cross-sectional area, perimeter, maximum, minimum, and average diameters, Feret dimensions, angles of orientation, the center of gravity, and various form or shape factors. The formulas used to calculate many of these measurements have been published by Bradbury (1977). The calculations we use are reported below.

Area. We use a modified trapezoidal function to compute cross-sectional area. To compute area, the profile being digitized is divided into a large number of small trapezoids (Figure 3). Each trapezoid consists of a rectangle that extends from the base of the cartesian coordinate system, and a triangle at the top of the rectangle, which takes account of the oblique angle at the surface of the profile (black column, Figure 3). The area of a trapezoid is one-half of the base multiplied by the sum of the two parallel sides. The base of each trapezoid is calculated as the distance between the last x coordinate value and the current x coordinate value of the trace (x-Last x, Figure 3). The height of the two parallel sides equals the sum of the last y coordinate value and the current y coordinate value (y +

Last *y*, Figure 3). The area of each trapezoid is one-half of the base multiplied by the heights of the two sides, or

$$A = 0.5*(X-LastX)*(Y+LastY).$$

The program keeps track of the sum of the areas of the trapezoids as the digitizing trial proceeds. The sum is stored in a numeric variable called Sumx. Positive areas are calculated any time the cursor moves to the right (clockwise and upward from point X_1, Y_1 in Figure 3). Negative areas are calculated any time the cursor moves to the left (the lower border of the profile, Figure 3). As the cursor moves to the left, trapezoid areas below the lower border of the profile (cross-hatched area, Figure 3) are subtracted from the positive trapezoid areas calculated as the cursor moved to the right. The speckled column in Figure 3 represents one of the negative areas of a trapezoid below the digitized profile. The total area of the profile becomes the total of the positive trapezoids minus the total of the negative trapezoids that lie below the figure. The last trapezoid area must also be calculated after the figure is closed or a significant error will be introduced. Our program takes the absolute value of the computed area as the real area of the figure. It therefore does not matter whether one digitizes in a clockwise or counterclockwise direction.

Although this method of area calculation is only an approximation, it works accurately for irregularly shaped figures as well as for conventional ovoid and circular profiles. The accuracy for irregularly shaped figures increases with an increase in the number of points taken. A large number of points should therefore be collected for highly irregular shapes so that the figure can be broken into a large number of very small trapezoids. This will minimize the effect of smoothing the irregular outline of the profile. A similar method for calculating the area of irregular figures is Simpson's Rule for Irregular Areas, reported in the *Handbook of Chemistry and Physics.*

Perimeter. Perimeter (or length) is calculated using the Pythagorean theorem for pairs of coordinates, where the perimeter is the square root of the sum of the squares of the current *x* and *y* coordinate values minus the preceding *x* and *y* coordinate values. In the program, this becomes

$$L = SQR((X-LastX)^2 + (Y-LastY)^2).$$

Like area, perimeter is computed on-line while the trace is being digitized. The individual perimeter calculations between each two points are summed and stored in a variable called *L*. This variable keeps track of the total perimeter. In our program, the current summed perimeter can be displayed on the CRT screen and digitizer LED at any time to check the value for accuracy during the tracing process.

Diameters. Our program also calculates three diameter values: the average diameter, the maximum diameter, and the minimum diameter of the digitized profile. These are calculated at the end of each digitizing trial using the *x, y* coordinate data values stored in the data array variables *X* and *Y*. These variables, which are cleared before each trial, can hold a total of 1000 digitized *x* and *y* coordinate values.

Average diameter is calculated from these values using the following steps:

1. The geometric center of the data array is computed by finding the minimum and maximum x and y values of the array and then computing the median of these maximum and minimum x, y points.
2. The data array is centered about the geometric center of the trace by subtracting the center from each data point of the trace.
3. The radius vectors from the center of the new array are computed and summed.
4. The average radius is obtained from the radius sums.

Diameter, of course, is twice the average radius. Maximum and minimum diameters are twice the maximum and minimum radii. These are retrieved from the array holding the radius vectors by searching for the minimum and maximum values in the array. The center of gravity of the figure can also be obtained from the average diameter calculations if the algorithm for average diameter computes the weighted mean center (centroid) of the data array.

The algorithm for calculating diameters works quite accurately if points are collected at equal intervals along the perimeter of the profile. If the distance between points is unequal, the sum of the radius vectors will be skewed. To compensate for this, another algorithm can be used to weight the sum of the radius vectors by the distance between vector points. In this case, each computed radius is multiplied by the distance between the current and last points taken. This weights the value according to its distance from the last point. In practical terms, both approaches yield almost identical results so long as a large number of data points are collected.

Form Factor. For expressing the shape of the digitized profile, we use a formula called a *form factor.* The formula is a measure of the circularity of the profile, where a perfect circle has a value of 1 and a straight line has a value of 0. The formula expresses the ratio of the profile's cross-sectional area to its perimeter. The factor F equals the total area multiplied by 4π, divided by the square of the total perimeter:

$$F = A*4*PI/L^2$$

where A is the total area and L is the total perimeter (length).

Other shape descriptors, such as elongation factor, mean density, mean chord, and orientation, are discussed by Bradbury (1977). We have not found the form factor to be informative in analyzing irregular objects such as synapses, but it can be useful for distinguishing classes of cell where cell shape is thought to be a meaningful parameter.

Printing and Storing Data

During a digitizing session, the data for area, perimeter, and other geometric calculations can be displayed on the CRT and/or printed on-line on a printer. The data can also be stored on a mass storage device after each trial. Because our computer has ample memory (187 KB), our digitizing program holds all data values in RAM memory until the digitizing session is completed. Our program stores a number of variables for each digitized profile, including cell number, section number, magnification, number of contacts, number of internal elements, number of external elements, profile type, perimeter, area, three diameters, and form factor. All values are stored automatically in memory at the end of each digitizing trial. Each value is held in a separate bin of a single matrix array called Area. Storing data in a single array facilitates rapid manipulation of the data and avoids having to merge data files for statistical analysis.

When a digitizing session is completed, the program prompts the operator to store the data on tape or disk. The program creates a data file with a name entered by the user, then stores the data in a format compatible with the Hewlett-Packard statistical packages. A final off-line printout of the data is also produced at the end of the session for use as a permanent data record.

Considerations When Digitizing

Preparation of Tissue

Accurate sampling for quantitative measurement depends upon maintaining as constant experimental conditions as possible. A number of factors can influence measurement accuracy significantly, including (1) variable tissue shrinkage during fixation, (2) variations in staining density, (3) variations in section thickness, and (4) interanimal variability. From my experience, it is important to prepare tissue according to a standardized protocol when quantitative comparisons between animals will be made. This will minimize many of these variations.

To minimize variations in tissue shrinkage, all of our material is prepared for electron microscopy using standard electron microscope procedures. All brains are fixed with an aldehyde fixative (glutaraldehyde–paraformaldehyde). The tissue is cut in 50- or 100-μm sections on an Oxford Instruments Vibratome, postfixed in osmium tetroxide, dehydrated, and embedded in medcast–araldite (Ted Pella, Tustin, CA). To minimize staining variations, we stain light microscope semithin sections with a 2% solution of toluidine blue. Thin sections for electron microscopy are stained with a saturated solution of uranyl acetate for 30 minutes and a 0.1% lead solution for 1 minute. To control for section thickness accurately, we try wherever possible to take measurements from semithin ($\sim$1 μm) light microscope or thin ($\sim$700 Å) electron microscope sections cut on a Sorvall MT2-B ultramicrotome. We estimate the thickness of semithin sections by setting the Sorvall Ultramicrotome to cut at 1 μm. The thickness of thin sections is estimated by the interference color of the section (we use dark silver to pale gold sections). Animal variability can be controlled by collecting data from only one animal or by using a large number of animals and treating the data statistically.

Photography

All of our morphometry measurements are made from light or electron micrographs. Prints of constant magnification are produced on 8 $\times$ 10 in. Kodak Ektamatic paper. At the end of each printing session, we print a negative of a calibration grid (photographed with the electron microscope) or stage reticule (photographed with the light microscope) so that we can calculate the exact magnification of each photograph printed in that session.

Before digitizing, all prints are analyzed by an experienced investigator. We code parameters by marking directly on the print with a fine felt-tip pen. We outline each profile to be measured. We also mark the number of contacts made by the profile, the number of elements within the profile, and the number of elements associated with the profile. We also include a code for the type of profile being digitized and the depth of the profile from the tissue surfaces (obtained from plotted data) [see Chapter 5 and Mize (1983b)].

Figure 4 illustrates this process for a synaptic terminal. The terminal is first examined for synaptic contact zones. These are marked with arrows (3, Figure 4). We then trace around the external membrane of the synapse (2, Figure 4). The number of synaptic vesicles is then counted and the total marked on the print (4, Figure 4). Finally, a code for synapse type and the depth of the profile are marked on the print. The analyzed micrograph is then digitized by trained technical personnel.

Accuracy in Digitizing

The accuracy of measurement depends upon several factors, including:

1. the resolution and sampling speed of the digitizer and computer,
2. the precision of the calculation algorithms, and
3. the skill and reliability of the operator.

Figure 4. Analysis of digitized profiles. The synaptic terminal shown below is measured by tracing its outer surface with a cursor. The morphometry software computes cross-sectional area (1), perimeter (2), average diameter, and form factor from the trace. The number of synaptic contact zones (3) and the number of internal elements (such as synaptic vesicles) (4), are entered from the digitizer's keypad in order to compute contact and element densities. [Reprinted from Mize (1983c) by permission of Elsevier Biomedical Press, Amsterdam.]

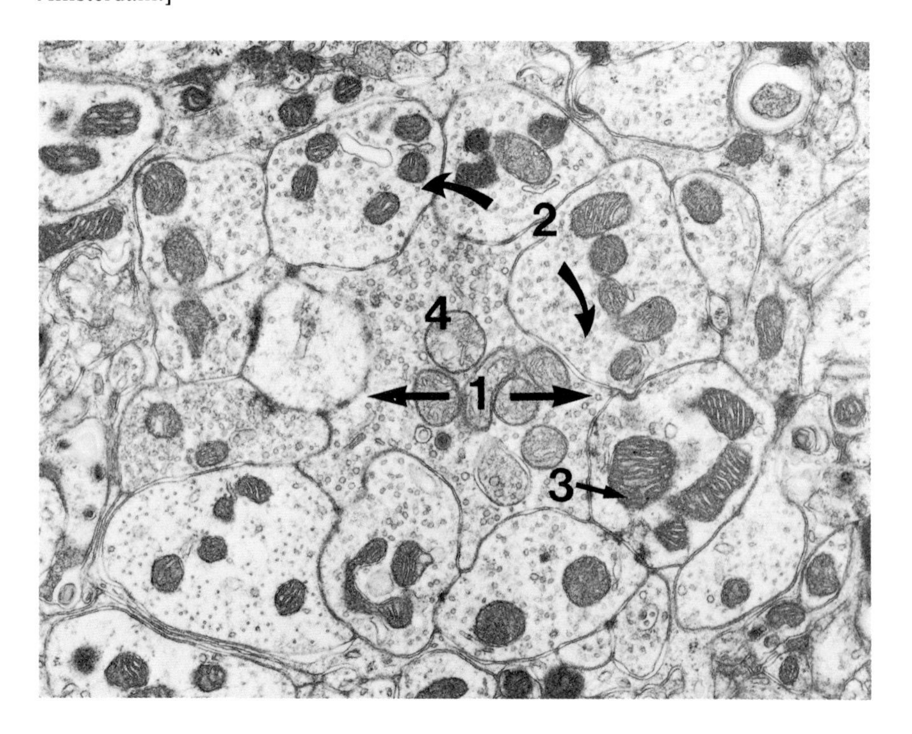

Most digitizers have a resolution of at least 0.1 mm, which is quite adequate if the print magnification of the digitized profiles is great enough. The sampling rate of the digitizer and computer is also important. If the operator digitizes too rapidly, some digitized points will not be captured by the computer even though they have been detected by the digitizer. This can reduce the overall system resolution dramatically. As an example, our digitizer (HP Model 9874A) has a resolution of 0.025 mm and a maximum data output rate of 30 points per second. The effective sampling rate of our BASIC data acquisition algorithms, however, is only 8.70 cycles per second in the continuous sampling mode. Data points are collected by the computer at that rate regardless of the number of data points placed on the bus by the digitizer. Data available on the bus between each data collection cycle are thus lost. Digitized points are collected only at intervals of about 0.5 mm when a technician moves the cursor at a comfortable rate of speed (Mize, 1983c).

There are several solutions to this problem if greater resolution is required:

1. The data acquisition algorithms can be written in machine code to speed data collection.
2. An external data buffer can be interposed between the digitizer and the computer.
3. Data can be collected from the digitizer using direct memory access (DMA) and then operated on once the digitizing trial has been completed [see Chapter 2 for a discussion of this technique].
4. The single point mode can be used. This requires the operator to depress the digitize key each time he or she wishes to digitize a point. The operator can therefore control the distance between data points.

The accuracy of the calculation algorithms depends upon the arithmetic precision of the programming language and the mathematical accuracy of the algorithms. Calculations can be performed in integer, single-precision, or full-precision floating point arithmetic. Although slower, all of our calculation algorithms are performed using full-precision (12-digit) arithmetic in order to optimize calculation accuracy. The algorithms for calculating cross-sectional area and perimeter utilize all collected data points and are thus as accurate as the resolution and sampling rate of the digitizer. The average diameter algorithm uses a subset of the total number of collected points to reduce calculation time. The size of the subset is under program control and can be changed by the operator at any time. The form factor is a derivative of the area and perimeter measurements and thus depends upon their accuracy.

The largest source of error is probably always operator inaccuracy, which can depend both upon skill and fatigue. We have found that an inexperienced technician is often much less accurate than one with extensive experience in digitizing. For this reason, we require that every profile be digitized twice and that the error between the two traces not exceed 1.5%. A skilled technician using the continuous point sampling mode and a natural rate of hand movement can digitize with an error rate averaging about 1% for profiles whose diameter is 5mm (Mize, 1983c). The final print magnification of most profiles will substantially exceed that size.

Research Applications

Our morphometry system has been used to measure the areas and perimeters of neurons and synaptic terminals, the size and shape of synaptic vesicles, the diameter and length of immunocytochemically stained fibers, and the size and density of intramembrane particles in freeze-fractured tissue. The most extensive data have been collected on cell and synaptic terminal size. For example, we have measured the area and depth of neurons in the cat superior colliculus that were retrogradely labeled by injection of horseradish peroxidase into three nuclei: the lateral posterior nucleus, the dorsal lateral geniculate nucleus, and the ventral lateral geniculate nucleus. Most labeled cells projecting to the dorsal and ventral lateral geniculate nuclei were small, averaging about 150 μm^2 in area (Figure 5). By contrast, cells projecting to the lateral posterior nucleus were much larger in size, averaging over 240 μm^2 in area. Our quantitative analysis revealed not only significant differences in cell size in the three populations, but also showed that within each group there was considerable variability in cell size. Differences were also found in the depth of the cells (Caldwell and Mize, 1981; Harrell et al., 1982).

We have also measured the synaptic terminals of retinal ganglion cells that send their axons to different primary projection sites in the brain. In addition to

Figure 5. Histograms from four experimental groups, showing the size distribution of labeled cells after injection of horseradish peroxidase (HRP) into the lateral posterior nucleus (upper left), dorsal lateral geniculate nucleus (lower left), and ventral lateral geniculate nucleus (upper right). A comparison of the least-squares fit of the data for the three groups is shown at the lower right.

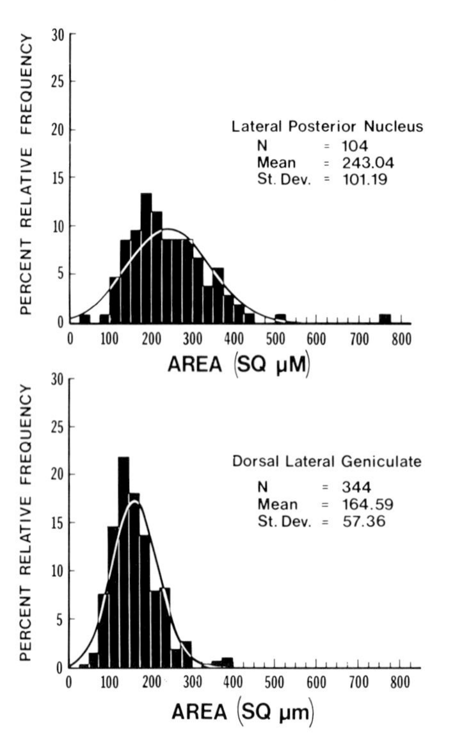

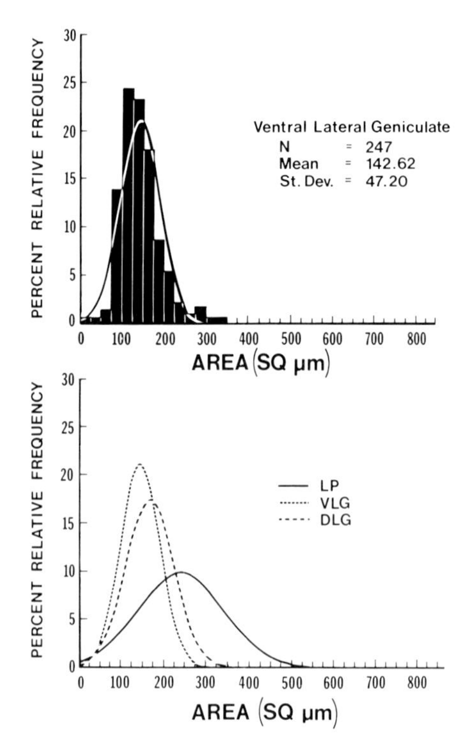

measurements of cross-sectional area, perimeter, and average diameter, we have calculated the density of total synaptic contact zones, the density of contacts with a particular class of dendrite (called *presynaptic dendrites*), and the form factor representing the shape of the terminal. Significant differences were found in synaptic terminal size and organization that were related to the projection sites of the terminals in the visual system. As an example, we found that the retinal terminals in the ventral lateral geniculate nucleus are all small, averaging 0.94 μm^2 in area. By contrast, retinal terminals in the medial interlaminar nucleus have a wide range of sizes, averaging 2.5 μm^2 in area (Mize and Horner, 1984). Differences in terminal size can also be seen in different layers of the cat dorsal lateral geniculate nucleus (Figure 6). These size differences presumably reflect the different physiologies of the retinal ganglion cells that project to different nuclei (Mize, 1983a). Such quantitative estimates of terminal size could not have been made without morphometric analysis.

Having collected these data during the past four years, I believe there are several important constraints that should be observed in collecting data morphometrically. Some of these constraints have been discussed in greater detail by Weibel (1979).

Adopting an appropriate sampling strategy is *essential* for collecting meaningful data. In general, between-animal variability is far greater than within-animal variability. This is due to variability in brain size, brain shrinkage during fixation, the age and sex of the animal, and other factors that vary from animal to animal.

Figure 6. Computer-generated scattergram plot, showing the relationship between the cross-sectional area and depth of retinal synaptic terminals in the cat lateral geniculate nucleus (LGN). Synaptic terminals in the parvocellular C laminae of the LGN (below 1050 μm) are all small. Many terminals in the more dorsal layers are larger.

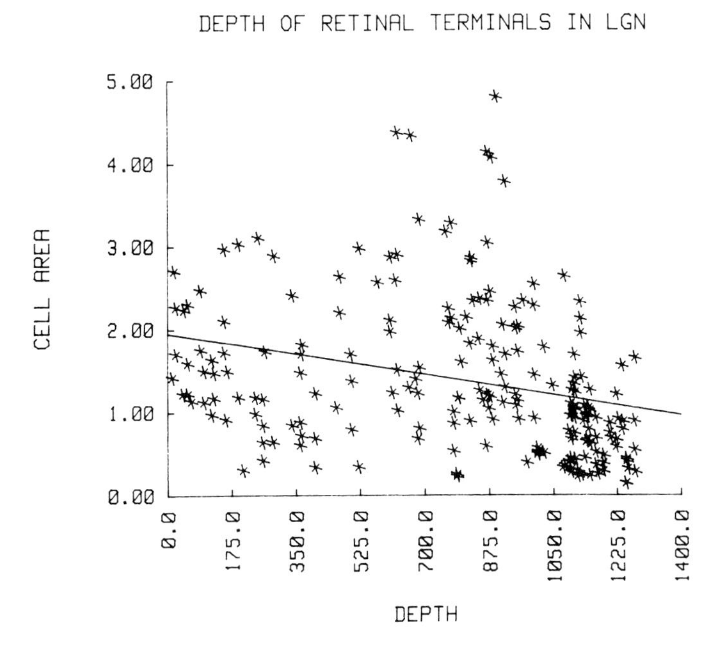

This variability can be dealt with in two ways: (1) by sampling from a large number of animals (30 is a commonly accepted *N*) or (2) by sampling from a single animal where age, sex, and brain shrinkage will be identical. In electron microscopy it is both expensive and time consuming to collect tissue from many animals. For this reason, we believe it is preferable in some cases to make quantitative comparisons from tissue of a single animal. Although the results cannot be generalized to a population of animals, one has a high degree of confidence in the comparisons made between different tissue samples within a single animal.

I also believe it is important to make a large number of observations from a given animal. When measuring synaptic terminals, for instance, we plot the position of virtually every terminal that we encounter in scans made through tissue with the microscope plotter [see Chapter 5]. About 20% of these terminals are photographed for morphometric analysis. Analysis is generally based upon at least 300 observations collected from three to five blocks cut through each nucleus of interest.

We have adopted several sampling strategies in collecting data for morphometric analysis. Photographs of selected synaptic terminals can be taken *randomly* by using a random number table or pseudorandom sampling of every *n*th synapse. Alternatively, tissue can be sampled *sequentially* by photographing every terminal encountered in a scan of the tissue. Our results using either sampling strategy are comparable, suggesting both approaches will yield accurate results if enough observations are collected.

Measurements of parameters are more accurate (i.e., there is less variability) when the profile being measured has a high frequency of occurrence. Where there are few examples of the profile in the tissue, accurate measures are much harder to obtain. Thus, for example, total synaptic density shows less variability from section to section than does presynaptic dendrite density, whose overall occurrence in the tissue is about one-fifth that of the total number of synaptic contacts.

Conclusions

This chapter has reviewed three methods for making geometric measurements of profiles found in biological tissues. It has also described a computer-based morphometry system for quantitative analysis of light and electron micrographs developed by our laboratory. Our system is designed to enter data values from a keypad, trace the contours of profiles with a digitizer, and store and print the data for further statistical analysis. The programs calculate cross-sectional area, perimeter, three diameters, and form factor and compute several types of density.

There are advantages and disadvantages to each of the morphometry methods discussed. *Manual digitizing tablet systems* require extensive interaction with the user. The approach is susceptible to any operator error in determining profile boundaries. The tracing procedure itself is somewhat slow. Photographic preparation of micrographs is time consuming and photographic materials are expensive. On the other hand, properly prepared micrographs have a very high resolution, as do digitizing tablets. The computer can perform all of the calculations automatically. Storing, printing, and statistical analysis of the data can be handled by the microcomputer.

Video projection digitizing systems also require manual tracing of profiles and are thus subject to the same errors as systems that use a digitizing tablet. In fact, error can be greater with the video systems because cursor accuracy on the screen is reduced, the CRT has a lower resolution than do micrographs, and it is more difficult to digitize remotely on the screen. The only significant advantages of the video systems are savings in time and cost because the image can be directly digitized without photographing it.

Image analysis systems potentially provide the most rapid and accurate means for making morphometric measurements. A digital copy of the image can be stored to provide a permanent record. The digitized image can be manipulated and mathematically transformed to extract information that cannot be easily detected by the human eye. The entire procedure can, in theory, proceed automatically without operator intervention. Despite this potential, automatic image analysis systems have proven notoriously inadequate for measurements of complex biological materials. Although significant advances in imaging hardware allow images to be digitized at a high resolution, the development of pattern recognition algorithms for automatic extraction of profiles has not been very successful. Operator interaction is almost always essential when using these systems. It often takes more time to enhance and manipulate the image in order to extract the desired profiles than it does to digitize them manually. The human visual system remains a pattern recognition device of unequaled ability. Computer pattern recognition is far inferior to human capabilities except where detection thresholds exceed the sensitivity of the human eye. It is often not possible to use automated image analyzers to measure the geometries of complex, discrete profiles (Mathieu et al., 1980).

Automatic image analyzers have far greater value in applications requiring accurate densitometric measurements of labeling density. The human visual system is not particularly good at detecting small differences in gray-scale levels (intensity differences) and cannot assign quantitative values to those levels. Thus, analysis of the distribution and density of "analog" labels such as those used in immunocytochemistry requires a photometric measuring device. Computer systems can be of enormous value in these applications.

Although image analyzers may prove to be the technological wave of the future, computer-based digitizing morphometry systems will remain an inexpensive, efficient alternative for measuring discrete organelles. Commercial systems are available for as little as $3500. If you already own a microcomputer with an extended BASIC language, it should be possible to modify the program in the Appendix to run on your computer. A copy of our BASIC program will be supplied on disk or tape or communicated via modem for a modest handling charge.

I am grateful to Joe Laughter of the Division of Biomedical Instrumentation at the University of Tennessee for writing the average diameter algorithms. Some digitizer subroutines were modified from utility routines provided by the Hewlett-Packard Corporation, whose permission to use them is gratefully acknowledged. Lee Danley and Linda Horner prepared the illustrations. Mary Gaither typed the manuscript. This project was supported by USPHS Grant EY-02973 from the National Eye Institute, a State of Tennessee New Faculty Research Grant, and an equipment grant from the College of Medicine UTCHS.

Appendix. Abbreviated Version of BASIC Morphometric Digitizing Program

```
10*************************************************************
20    REM : CLOSED FIGURE AREA PROGRAM FOR CALCULATING AREA, PERIMETER AND
            DIAMETER. PROGRAMMER: R. RANNEY MIZE  REVID 12-15-83. FILE: AREADI
30    REM : COPYRIGHT 1982, UNIVERSITY OF TENNESSEE RESEARCH CORPORATION
            ALL RIGHTS RESERVED  [REG. NO. TXu 104-783]
40*************************************************************


50      OPTION BASE 1
60      OVERLAP                              !Set dual processor overlap
70         COM Area(15,1000)                 !Dimension data array to 1000 points
80         COM Mindiameter1,Maxdiameter1
90         COM Datatitle$[80],Vn$(50)[10],Sn$(20)[10],Nv,No,Ns,Sc(20)
100        COM Feature$(7)[15],Answer(7),Value(7)
110        COM Cellno,Sectno,Sectval,Printdata
120        COM Print,Abortone,Abortboth        !Dimension variables for call
130 !

**************************


140 Readdata:REM                             !Read data into string variables
150      REDIM Vn$(15)
160         MAT READ Vn$
170             DATA CELL NUMB.,MAGNFICATN,SECTION NO,SYNAPSE NO,GRAIN NO.
180             DATA DENDRITES,CELL TYPE,SURF AREA,CELL AREA,DEPTH,ERROR
190             DATA AV DIAMETR,FORM FACTR,MN DIAMETR,MX DIAMETR
200      REDIM Vn$(50)
210         MAT READ Feature$
220             DATA "SECTION NO.    ","X FACTOR       ","NO. SYNAPSES   "
230             DATA "NO. GRAINS     ","NO. DENDRITES  ","CELL TYPE      "
240             DATA "CELL DEPTH     "
250         MAT READ Value
260             DATA 3,2,4,5,6,7,10
270 !

**************************


280 Valuesetup: REM                                 !Initialize variable values
290       Select=706                                !Set digitizer select code.
300       P=25.4                                    !Set user units to metric.
310 Rate: R$="CN32676,1"                            !Set sampling rate of dig.
320          MAT Area=ZER                           !Initialize variable values
330          MAT Answer=ZER
340             Cellno=Sectno=Print=Printdata=0
350             Sectval=Abortone=Abortboth=0
360          OUTPUT Select;"IN;DF;AT;SN"            !Initialize digitizer.
370 !

**************************


380 Functionkeys:  REM                              !Function key definitions.
390            PRINT PAGE
400            BEEP
```

```
410          INPUT "DO YOU WANT A PRINTOUT OF SPECIAL FUNCTION KEYS (Y/N)?",An
s$
420             IF (Ans$<>"Y") AND (Ans$<>"N") THEN 400
430             IF Ans$="N" THEN GOTO 600
440             IF Ans$="Y" THEN PRINTER IS 0
450        PRINT CHR$(132);"DIGITIZER SPECIAL FUNCTION KEY DEFINITIONS";CHR$(12
8);LIN(1)
460        PRINT "Fa -     PRESS TO ACCESS COUNT ROUTINE"
470        PRINT "Fb -     PRESS WHEN FINISHED DIGITIZING A SECTION"
480        PRINT "Fc -     PRESS TO ABORT ONE OR BOTH DIGITIZED TRIALS"
490        PRINT "Fd -     PRESS TO RESPOND WITH 'YES'"
500        PRINT "Fe -     PRESS TO RESPOND WITH 'NO'"
510        PRINT "PreFa - PRESS TO RESTART WITH ANOTHER DATA GROUP"
520        PRINT "PreFb - PRESS TO STORE DATA AND STOP THE AREA ROUTINE"
530        PRINT "PreFc - PRESS TO RESET DIGITIZER"
540        PRINT "PreFd - PRESS TO DRAW DIGITIZED FIGURE ON CRT"
550        PRINT "PreFe - PRESS TO TOGGLE SINGLE POINT / CONTINUOUS POINT MODES
"
560          PRINT LIN(5)
570 !

**************************


580 Specialfeatures: REM                             !Set display and correction
590          PRINTER IS 16                           !special features
600        BEEP
610          INPUT "DO YOU WANT DISPLAY FEATURES ON (Y/N)?",Ans$
620            IF (Ans$<>"Y") AND (Ans$<>"N") THEN 600
630            IF Ans$="Y" THEN Disp=1
640        BEEP
650          INPUT "DO YOU WANT CORRECTION FEATURES ON (Y/N)",Ans$
660            IF (Ans$<>"Y") AND (Ans$<>"N") THEN 640
670            IF Ans$="Y" THEN Cr=1
680        BEEP
690          INPUT "DO YOU WANT TO MEASURE  A REA (CLOSED FIGURES) OR  L ENGTH
 (LINES)  (A/L)",Ans$
700            IF (Ans$<>"A") AND (Ans$<>"L") THEN 680
710            IF Ans$="A" THEN Cl=1
720            IF Ans$="A" THEN 810
730        BEEP
740          INPUT " T WO-POINT (STRAIGHT LINE) OR  M ULTIPLE POINT (CURVED LI
NE) DIGITIZING? (T/M)",Ans$
750            IF (Ans$<>"T") AND (Ans$<>"M") THEN 730
760            IF Ans$="M" THEN Cu=1
770            IF Ans$="T" THEN
780              Pt$="SG"
790              GOTO 910
800            END IF
810        BEEP
820          INPUT "DO YOU WANT  S INGLE OR  C ONTINUOUS POINT DIGITIZING (S/C
)",Ans$
830            IF (Ans$<>"S") AND (Ans$<>"C") THEN 810
840            IF Ans$="C" THEN
850              Pt$=R$
860              Flag=1
870            ELSE
880              Pt$="SG"
890              Flag=-1
900            END IF
910            BEEP
920          INPUT "DO YOU WANT ONE OR TWO TRIAL DIGITIZING (1/2)?",Trial
930            IF (Trial<1) OR (Trial>2) THEN 900
940            IF Trial=1 THEN
950              GOTO Datasetup
960            END IF
```

```
970 Errorrate: REM                                      !Set error tolerance
980        BEEP
990           INPUT "WHAT PERCENT ERROR WILL YOU ACCEPT (0.01-100)",Ans
1000             IF (Ans<.00001) OR (Ans>100) THEN 980
1010             LET Errate=100/Ans
1020 !

***************************



1030 Datasetup: REM                                      !Setup data base values
1040   PRINTER IS 16
1050   PRINT PAGE
1060      A$="9845S ENTRY     "
1070         CALL Display(Select,A$)
1080       PRINT CHR$(132);"DIGITIZER ROUTINE FOR CALCULATING CROSS-SECTIONAL A
REA";CHR$(128);LIN(1)
1090       PRINT "THIS PROGRAM ALLOWS YOU TO RECORD THE FOLLOWING PARAMETERS"
1100       PRINT "OF BIOLOGICAL STRUCTURES:",LIN(1)
1110       PRINT "     1. SURFACE AREA (PERIMETER OR LENGTH)"
1120       PRINT "     2. CROSS-SECTIONAL AREA"
1130       PRINT "     3. NO. OF CONTACTS (I.E. SYNAPSES)"
1140       PRINT "     4. CONTACT DENSITY (NO. CONTACTS/ UNIT SURFACE AREA)"
1150       PRINT "     5. NO. OF INTERNAL ELEMENTS (I.E. GRAINS)"
1160       PRINT "     6. INTERNAL ELEMENT DENSITY (NO. ELEMENTS/UNIT AREA)"
1170       PRINT "     7. NUMBER OF ASSOCIATED ELEMENTS (I.E. DENDRITES)"
1180       PRINT "     8. PROFILE TYPE"
1190       PRINT "     9. PROFILE DEPTH"
1200       PRINT "    10. MINIMUM, MAXIMUM, AND AVERAGE DIAMETERS",LIN(5);CHR$(
27);"1"
1210    BEEP
1220      INPUT "PLEASE ASSIGN A TITLE TO THIS DATA SET (80 CHARACTERS MAX)",Da
tatitle$
1230 !

*****************************************************************
1240 Subdatasetup: REM  ROUTINE TO RESET SECTION NUMBER AND MAGNIFICATION
1250*************************************************************



1260              Print=Printdata=0               !Reset print data exit flags
1270              Cellno=Cellno+1                 !Increment the cell counter
1280    BEEP
1290              Sectval=0                       !Reset the trial counter
1300      INPUT "PLEASE ENTER THE GROUP NUMBER (SECTION NUMBER)",Cellno
1310 !

***************************



1320 Magsetup: REM                                       !Set magnification factor
1330    BEEP
1340       INPUT "ENTER MAGNIFICATION FACTOR FOR THIS DATA SET (0 IF VARIABLE)"
,Mag
1350              Magfactor=0
1360           IF Mag<>0 THEN Answer(2)=0
1370           IF Mag=0 THEN Magfactor=Answer(2)=1
1380           Answer(1)=1
1390 !

*****************************************************************
1400 Feature: REM  ROUTINE TO DEFINE DATA FEATURES TO BE ENTERED FROM KEYPAD
1410*************************************************************
```

```
1420    FOR I=3 TO 7                                  !Set entry features wanted
1430         Ans$="N"
1440         PRINT USING "#,K";CHR$(27)&"&a+24R";CHR$(27);"1"
1450         DISP "DO YOU WANT TO RECORD ";TRIM$(Feature$(I));" (Y/N)?";
1460             BEEP
1470             INPUT "",Ans$
1480                 IF (Ans$<>"Y") AND (Ans$<>"N") THEN GOTO 1440
1490                 IF (Ans$="N") OR (Ans$="") THEN Answer(I)=0
1500                 IF Ans$="Y" THEN Answer(I)=1
1510    NEXT I
1520         PRINT CHR$(27);"m"
1530 !

***************************

1540 Callfeature:REM                                !Call feature subprogram
1550 CALL Feature(Mag,Select,Unit$,P,Length,Cellno,Sectno,Sectval,Printdata,Abc
rtboth,Area1,Length1,Answer(*),Value(*),Feature$(*),Print,Cr,Flag,Pt$,R$)
1560          IF Magfactor THEN Mag=Area(2,Sectno)
1570          Area(2,Sectno)=Mag                    !Enter cell and mag values
1580          Area(1,Sectno)=Cellno
1590             IF Print THEN GOTO Print
1600 !

*************************************************************
1610 Area:  REM  ROUTINE FOR COLLECTING POINTS FOR AREA CALCULATIONS
1620*********************************************************

1630    PRINT PAGE;LIN(20)
1640    DISP
1650       PRINT CHR$(132);"DIGITIZE CROSS-SECTIONAL AREA";CHR$(128);LIN(1)
1660       PRINT "PLEASE TRACE AROUND PROFILE SURFACE."
1670       PRINT "WHEN FINISHED, PRESS FUNCTION KEY Fb.",LIN(1)
1680       PRINT "ACCUMULATING LENGTH IN ";Unit$
1690          FIXED 2
1700 !

***************************

1710 Callarea: REM                                     !Call area subprogram
1720 CALL Area(Mag,Select,Unit$,P,Length,Cellno,Sectno,Sectval,Printdata,Abortb
oth,Area1,Length1,Avdiameter1,Formfactor1,Disp,Trial,Cr,Errate,Pt$,R$,Flag,C1,C
u)
1730                  Abortboth=Abortone=0            !Reset the abort flags
1740                  IF Printdata=1 THEN GOTO Print   !Exit to printdata routin
e
1750                  IF Sectval=0 THEN GOTO Area
1760                  IF Sectval=1 THEN GOTO Area      !Read flags for trial no.
1770                  IF Sectval=2 THEN GOTO Printvalue
1780       GOTO Callarea
1790 !

***************************

1800 Printvalue: REM                                 !Print mean area to printer
1810                  Sectval=0                      !Reset the trial counter
1820                  PRINT PAGE
1830                  PRINTER IS 0
```

```
1840        IMAGE #,5A,X,6Z.D,2X,11A,5Z.2D,X,2A,2X,6A,5Z.2D,X,2A,2X
1850 PRINT USING 1840;"SECT.";Area(3,Sectno);"PERIMETER=";Area(8,Sectno);Unit$;
"AREA=";Area(9,Sectno);Unit$
1860        IMAGE 10A,5Z.2D,X,2A
1870 PRINT USING 1860;"DIAMETER=";Area(12,Sectno);Unit$
1880     PRINTER IS 16
1890     DISP " "
1900     GOTO Callfeature                              !Restart for next cell
1910 SUBEND
1920 !

***********************************************************************
1930 Print: REM  ROUTINES FOR PRINTING DATA TO CRT OR PRINTER
1940*******************************************************************


1950     PRINT PAGE
1960     DISP
1970               Sectno=Sectno-1                   !Decrement the section counte
r                                                     to print proper no. section
s
1980               A$="DATA PRINTOUT   "
1990        CALL Display(Select,A$)
2000           OUTPUT Select;"BP"
2010     BEEP
2020     INPUT "DO YOU WANT A HARDCOPY PRINTOUT OF YOUR DATA (Y/N)?",Ans$
2030               IF (Ans$<>"N") AND (Ans$<>"Y") THEN GOTO 1950
2040               IF Ans$="Y" THEN PRINTER IS 0
2050 !

****************************


2060 Printhead:  REM                                   !Print heading
2070     DISP
2080               Totalsection=Sectno
2090     PRINT TAB(31);"PRINTOUT OF AREA DATA",LIN(1)
2100     IMAGE 20A,5X,K,/
2110             IF Magfactor=1 THEN Mag$="VARIABLE"
2120             IF Magfactor<>1 THEN Mag$=VAL$(Mag)
2130     PRINT USING 2100;"TITLE DATA: ";Datatitle$;"CELL DATA: ";Cellno;"SECTIO
N DATA: ";Sectno;"MAGNIFICATION: ";Mag$;"UNIT OF MEASURE: ";Unit$
2140 !

****************************


2150 Printtable: REM                                   !Print data in table form
2160              T$="_"
2170     PRINT RPT$(T$,80)
2180     PRINT LIN(1)
2190   IMAGE 8A,2X,4A,X,4A,X,4A,X,4A,1X,7A,3X,6A,1X,7A,2X,6A,3X,5A,2X,8A
2200   PRINT USING 2190;"SECT.NO.";"SYN.";"GRS.";"DEN.";"TYP.";"SURF AR";"AREA";
"SYN.DEN";"GR.DEN";"DEPTH";"DIAMETER"
2210     PRINT RPT$(T$,80)
2220     PRINT LIN(1)
2230              I=1
2240        DEFAULT ON
2250        FOR I=1 TO Sectno
2260             Meansurf=Area(8,I)
2270             Meanarea=Area(9,I)
2280             Syndens=Area(4,I)/Meansurf            !Calculate density values
```

```
2290   IF Meanarea>0 THEN
2300                 Grdens=Area(5,I)/Meanarea
2310                 ELSE
2320                   Grdens=0
2330                 END IF
2340             Mag1=Area(2,I)
2350             D=Area(10,I)
2360             Avd=Area(12,I)
2370    IMAGE 5Z.1D,3X,3Z,2X,3Z,2X,3Z,2X,3Z,2X,4Z.DD,1X,5Z.DD,2X,3Z.DD,2X,3Z.DD,
2X,4Z.DD,2X,3Z.DD
2380    PRINT USING 2370;Area(3,I),Area(4,I),Area(5,I),Area(6,I),Area(7,I),Means
urf,Meanarea,Syndens,Grdens,D,Avd
2390         NEXT I
2400         DEFAULT OFF
2410     PRINTER IS 16
2420 !

*************************************************************
2430 Storedata: REM  ROUTINES FOR STORING DATA ON TAPE OR FLOPPY DISC
2440*********************************************************


2450             A$="STORE DATA      "                 !Routine to store data
2460        CALL Display(Select,A$)
2470           OUTPUT Select;"BP"
2480     BEEP
2490             Nv=15
2500             No=Sectno
2510           IF Printdata<>1 THEN GOTO Filedefine
2520     INPUT "DO YOU WANT TO STORE THIS DATA NOW (Y/N)?",Ans$
2530           IF (Ans$<>"N") AND (Ans$<>"Y") THEN GOTO 2480
2540           PRINT PAGE
2550           IF Ans$="N" THEN GOTO Subdatasetup
2560     GOTO Filename
2570 !

***************************


2580 Filedefine: REM                                  !Branch to store or cont
2590     DISP "PRESS CONTINUE TO STORE DATA"
2600     PAUSE
2610 Filename: REM                                    !Set file name to store data
2620     PRINT PAGE
2630     BEEP
2640        LINPUT "PLEASE ENTER THE FILE NAME BELOW (INCLUDE MSFS)",Filename$
2650        DISP "PLEASE WAIT WHILE FILE ";CHR$(132);Filename$;CHR$(128);" IS BE
ING CREATED"
2660           Recordnumb=ABS(Sectno*Nv*8/1000)+2
2670     CREATE Filename$,Recordnumb,1000             !Create file to store data
2680 !

***************************


2690 Store: REM                                       !Store data in format for ne
w                                                       H-P statistical packages
2700     PRINT PAGE
2710     DISP "YOUR DATA IS BEING STORED ON FILE ";Filename$
2720        ASSIGN #1 TO Filename$
2730           READ #1,1
2740              PRINT #1;Datatitle$,No,Nv,Vn$(*),Ns,Sn$(*),Sc(*)
```

```
2750          READ #1,2
2760             I=J=1
2770       FOR I=1 TO Nv
2780          FOR J=1 TO Sectno
2790             PRINT #1;Area(I,J)
2800          NEXT J
2810       NEXT I
2820 !

****************************

2830 End:  REM                                      !End routine for branching
                                                      to continue or stop
2840    BEEP
2850    PRINT PAGE
2860       IF NOT Printdata THEN GOTO 2900
2870    DISP "PRESS CONT WHEN READY TO CONTINUE."
2880    PAUSE
2890       GOTO Subdatasetup
2900             A$="PROG DONE       "
2910       CALL Display(Select,A$)
2920          OUTPUT Select;"BP"
2930    PRINT LIN(22);"YOUR DATA IS STORED AND THE PROGRAM IS FINISHED."
2940    DISP " HAVE A NICE DAY.  "
2950   END
2960 !

*****************************************************************
2970 Subfeature: REM  SUBPROGRAM FOR ENTERING DATA FROM DIGITIZER KEYPAD
2980*************************************************************


2990 SUB Feature(Mag,Select,Unit$,P,Length,Cellno,Sectno,Sectval,Printdata,Abor
tboth,Area1,Length1,Answer(*),Value(*),Feature$(*),Print,Cr,Flag,Pt$,R$)
3000      OPTION BASE 1
3010      COM Area(15,1000)
3020            Sectno=Sectno+1                   !Increment the section counter
.
3030       FOR I=1 TO 7
3040          IF Answer(I)=0 THEN 3060
3050          GOSUB Subfeatureentry
3060       NEXT I
3070    SUBEXIT
3080 !

****************************

3090 Subfeatureentry: REM                           !Routine to enter feature
                                                      values from digitizer keypad
3100        PRINT PAGE
3110        DISP "PLEASE ENTER THE ";TRIM$(Feature$(I))
3120             A$=Feature$(I)
3130          CALL Display(Select,A$)
3140             OUTPUT Select;"BP75,200"
3150             OUTPUT Select;"SK0"
3160             OUTPUT Select;"OS"
3170                ENTER Select;Status
3180                   IF BIT(Status,7) THEN 3220
3190                   IF BIT(Status,0) THEN 3410
3200                GOTO 3160
3210 !
```

```
***************************

3220 Key5:     CALL Input(Select,X,Y,Key,N,Pt$,R$)  !Subexit to print data and
                                                     enter new cell number
3230                  IF Key=1 THEN
3240                     CALL Count(Select,Count,Pt$)
3250                     Area(Value(I),Sectno)=Count
3260                     Count=0
3270                     GOTO 3430
3280                  END IF
3290                  IF Key=32 THEN Printdata=1
3300                  IF (Key=32) OR (Key=64) THEN Print=1
3310                  IF (Key=32) OR (Key=64) THEN SUBEXIT
3320 Toggle1:         IF Key=512 THEN              !Routine to toggle point mod
e
3330                     Flag=Flag*-1              !Single point or continuous
3340                     IF SGN(Flag)=1 THEN Pt$=R$
3350                     IF SGN(Flag)=-1 THEN Pt$="SG"
3360                     OUTPUT Select;"SK0"
3370                     OUTPUT Select;Pt$
3380                     Key=0
3390                  END IF
3400               GOTO 3160
3410            OUTPUT Select;"ON"               !Service enter key
3420               ENTER Select;Area(Value(I),Sectno)
3430     PRINT CHR$(27);"m";PAGE;LIN(23)
3440     PRINT TRIM$(Feature$(I));": ";Area(Value(I),Sectno)
3450        IF NOT Cr THEN RETURN
3460 !

***************************

3470 Correctentry: REM                           !Routine to correct features
3480     DISP "IS THE ";TRIM$(Feature$(I));" CORRECT (SK# Fd=YES;SK# Fe=NO)?"
3490             Value$=VAL$(Area(Value(I),Sectno))
3500             A$="ANS="&TRIM$(Value$[1,7])&" COR"
3510            CALL Display(Select,A$)
3520               OUTPUT Select;"BP"
3530                  Key=0
3540            CALL Input(Select,X,Y,Key,N,Pt$,R$)
3550 Key6:            IF Key=8 THEN RETURN         !Service keys to correct featur
e                                                    values
3560                  IF Key=16 THEN Subfeatureentry
3570        GOTO 3540
3580 Return: RETURN
3590 SUBEND
3600 !

*****************************************************************
3610 Subarea: REM  CALL SUBPROGRAM FOR CALCULATING AREA
3620*************************************************************

3630 SUB Area(Mag,Select,Unit$,P,Length,Cellno,Sectno,Sectval,Printdata,Abortbo
th,Area1,Length1,Avdiameter1,Formfactor1,Disp,Trial,Cr,Errate,Pt$,R$,Flag,Cl,Cu
)
3640     OPTION BASE 1
3650     OVERLAP
3660     COM Area(15,1000)
3670     COM Mindiameter1,Maxdiameter1
3680     SHORT X(1000),Y(1000)
3690         MAT X=ZER
3700         MAT Y=ZER
```

```
3710              DISP
3720                 IF Sectval=0 THEN A$="BEG DIG SURFACE"
                                                          !Display begin dig surface
3730                 IF Sectval=1 THEN A$="REDIG SURFACE  "
3740           CALL Display(Select,A$)                   !Set up digitizer parameters
3750              OUTPUT Select;"SK0"                    !Clear all digitizer keys
3760              OUTPUT Select;"BP90,200"               !Set up tone parameters
3770              OUTPUT Select;Pt$                      !Set sampling rate for dig
3780 !
```

```
***************************
```

```
3790 Datainput:   OUTPUT Select;"OS"                     !Check status for 1st point
3800              ENTER Select;Status
3810                 IF BIT(Status,7) THEN
3820              OUTPUT Select;"OK"
3830              ENTER Select;Key
3840 Key4:              IF Key=4 THEN Abortboth=1  !Service keys for aborting
                                                     both trials
3850                    IF Key=4 THEN GOTO Abort
3860                    IF Key=128 THEN GOTO 3630
3870 Toggle2:           IF Key=512 THEN                  !Routine to toggle point mode
3880                       Flag=Flag*-1                  !Single point or continuous
3890                          IF SGN(Flag)=1 THEN Pt$=R$
3900                          IF SGN(Flag)=-1 THEN Pt$="SG"
3910                       OUTPUT Select;"SK0"
3920                       OUTPUT Select;Pt$
3930                          Key=0
3940                    END IF
3950                    IF (Key<>4) AND (Key<>128) AND (Key<>512) THEN Datainput
3960                 END IF
3970                 IF BIT(Status,2) THEN 3990     !Input points if bit 2 set
3980                          GOTO Datainput
3990                 N=1                                 !Initialize point counter
4000            OUTPUT Select;"OD"                       !Read 1st point
4010               ENTER Select;X,Y                      !Enter coordinate values for
                                                          1st point
4020                 Sumx=0                              !Set area sum to zero
4030                 Xstart=Lastx=X(N)=X                 !Initialize to starting point
4040                 Ystart=Lasty=Y(N)=Y
4050 !
```

```
***************************
```

```
4060 Datacollect:    CALL Input(Select,X,Y,Key,N,Pt$,R$) !Input rest of points.
4070                 GOSUB Sum                           !Add distance to length and
                                                          sum area
4080            IF NOT Collect THEN                      !Set window for auto close
4090                   GOSUB Window
4100                   GOTO 4160
4110            END IF
4120            IF (ABS(X-Xstart)<=ABS(10)) AND (ABS(Y-Ystart)<=ABS(10)) THEN
4130                   OUTPUT Select;"BP200,400"         !Check for auto closepoin
t
4140                   GOSUB Closepoint
4150            END IF
4160               Lastx=X                               !Current point becomes
4170               Lasty=Y                               !last point
4180 !
```

```
***************************
```

```
4190 Disp: REM DISPLAY FEATURES                       !Displays current length on
                                                        CRT
4200    IF Disp THEN                                  !Displays cursor position in
                                                        user units on dig LED
4210       FIXED 4
4220       SELECT Mag                                 !Sets user units
4230          CASE 1 TO 100
4240                Length=L*.984252*P/Mag/1000
4250          CASE 101 TO 35000
4260                Length=L*.984252*P/Mag
4270          CASE >35000
4280                Length=L*.984252*P/Mag*1000
4290        END SELECT
4300          DISP Length
4310          OUTPUT Select USING "K,4D.2D,K,4D.2D";"LB",(X-Xstart)/1016*P," "
,(Y-Ystart)/1016*P
4320    END IF
4330 !

***************************

4340 Key1: REM                                  !Reads keypress to end digitizing
4350 IF Key=4 THEN GOTO Abort                   !Delete current digitized surface.
4360 IF Key=256 THEN                            !Draw section on CRT
4370        CALL Trace(X(*),Y(*),N,Mag,L,Sectval)
4380        Key=0
4390 END IF
4400 Toggle3: IF Key=512 THEN                          !Routine to set point mode
4410              Flag=Flag*-1                         !Single point or continuou
s
4420          IF SGN(Flag)=1 THEN Pt$=R$
4430          IF SGN(Flag)=-1 THEN Pt$="SG"
4440             OUTPUT Select;"SK0"
4450             OUTPUT Select;Pt$
4460          Key=0
4470          END IF
4480 IF Key=2 THEN Closepoint                   !If fb is off, input another point,
                                                  IF fb is on, end digitizing.
4490 IF NOT Cu AND NOT Cl THEN Closepoint
4500 GOTO Datacollect
4510 !

***************************

4520 Window: REM                                !Set window for automatic closure
4530    IF (ABS(X-Xstart))>=ABS(10)) OR (ABS(Y-Ystart))>=ABS(10)) THEN Collect=1
4540 RETURN
4550 !

***************************

4560 Sum:                                             ! Keep running total of length
                                                        and sum area
4570       X(N)=X                                     ! X coordinate points for diam
.
4580       Y(N)=Y                                     ! Y coordinate points for diam
.
4590          H=X-Lastx                               ! Delta x from last point.
4600          J=Y-Lasty                               ! Delta y from last point.
```

```
4610          L=L+SQR(H^2+J^2)                     ! L is total length so far.
4620          Tarea=.5*H*(Y+Lasty)                 ! Tarea is trapezoidal area
                                                     between two points.
4630            Sumx=Sumx+Tarea                    ! Sumx is total area so far.
4640     RETURN
4650 !
```

```
***************************
```

```
4660 Closepoint: REM
4670 !
```

```
***************************
```

```
4680 Calculate:   REM                                       !Calculate final
                                                              cross-sect. area
4690       Tarea=(Xstart-X)*(Ystart+Y)*.5                   !This routine close
s                                                             figure for area
4700       Sumx=Sumx+Tarea
4710       Sumx=ABS(Sumx)
4720   IF C1 THEN                                           !This routine close
s                                                             figure for perime
tr
4730       L=L+SQR((Xstart-X)^2+(Ystart-Y)^2)               !This routine close
s                                                             figure for length
4740   END IF
4750       CALL Diameter(X(*),Y(*),N,Ad,Mind,Maxd)   !Calculate average diamete
r
4760       SELECT Mag
4770          CASE 1 TO 100                                 !Convert to user
                                                              units.
4780                Unit$="MM"                              !Convert to mm
4790                   Length=L*.984252*P/Mag/1000          !Convert length
4800                   Area=Sumx*(.984252/1000)^2*P^2/Mag^2 !Convert area
4810                   Avdiameter=Ad*.984252*P/Mag/1000     !Convert diameter
4820                   Mindiameter=Mind*.984252*P/Mag/1000
4830                   Maxdiameter=Maxd*.984252*P/Mag/1000
4840          CASE 101 TO 35000
4850                Unit$="UM"                              !Convert to microns
4860                   Length=L*.984252*P/Mag
4870                   Area=Sumx*.984252^2*P^2/Mag^2
4880                   Avdiameter=Ad*.984252*P/Mag
4890                   Mindiameter=Mind*.984252*P/Mag
4900                   Maxdiameter=Maxd*.984252*P/Mag
4910          CASE >35000
4920                Unit$="NM"                            !Convert to nanometer
s
4930                   Length=L*.984252*P/Mag*1000
4940                   Area=Sumx*(.984252*1000)^2*P^2/Mag^2
4950                   Avdiameter=Ad*.984252*P/Mag*1000
4960                   Mindiameter=Mind*.984252*P/Mag*1000
4970                   Maxdiameter=Maxd*.984252*P/Mag*1000
4980       END SELECT
4990 !
```

```
***************************
```

```
5000 Formfactor:  REM
5010       Formfactor=Area*4*PI/Length^2           !Calculate form factor
5020            Area(8,Sectno)=Length              !Fill variables with values
```

```
5030             Area(9,Sectno)=Area
5040             Area(12,Sectno)=Avdiameter
5050             Area(13,Sectno)=Formfactor
5060             Area(14,Sectno)=Mindiameter
5070             Area(15,Sectno)=Maxdiameter
5080             Sectval=Sectval+1               !Increment the trial counter
5090          ON Sectval GOSUB Secondtrial,Compare
5100          IF Sectval=0 THEN 5150
5110          IF Trial=1 THEN                    !Exit for one trial only.
5120               Sectval=2
5130               SUBEXIT
5140          END IF
5150       OUTPUT Select;"SK0"                   !Clear all digitize keys
5160  SUBEND
5170 !
```

```
***************************
```

```
5180 Secondtrial: REM                           !Exit routine to prepare for
                                                  second trial
5190         IF NOT Cr THEN 5210                !Skip if not correction feature
5200         GOSUB Printarea
5210            Length1=Length                  !Set 1st length value
5220            Area1=Area                      !Set 1st area value
5230            Avdiameter1=Avdiameter          !Set 1st aver diameter value.
5240            Mindiameter1=Mindiameter
5250            Maxdiameter1=Maxdiameter
5260            Formfactor1=Formfactor          !Set 1st form factor value.
5270     RETURN
5280 !
```

```
***************************
```

```
5290 Printarea: REM                             !Print 1st trial value to CRT
5300     PRINT PAGE;LIN(21)
5310     DISP
5320        PRINT "SECTION NO.";Area(3,Sectno)
5330           IMAGE  24A,1X,5D.2D,X,2A
5340        PRINT USING 5330;"FIRST TRIAL AREA = ";Area(9,Sectno);Unit$
5350        PRINT USING 5330;"FIRST TRIAL DIAMETER = ";Area(12,Sectno);Unit$
5360        PRINT USING 5330;"FIRST TRIAL PERIMETER = ";Area(8,Sectno);Unit$
5370        DISP "(ACCEPT = SK# Fd; REJECT = SK# Fe.)"
5380           GOSUB Lengthdisp
5390        PRINT PAGE
5400        DISP
5410     RETURN
5420 !
```

```
***************************
```

```
5430 Lengthdisp: REM                            !Print first trial value to
                                                  digitizer LED
5440              A$="                 "
5450        CALL Display(Select,A$)
5460                 V$=VAL$(Area(8,Sectno))
5470                 A$="LN="&V$[1,8]&" ACPT"
5480        CALL Display(Select,A$)
5490           OUTPUT Select;"BP200,400"
5500           OUTPUT Select;"OK"
5510 !
```

```
**************************

5520 Key3:       ENTER Select;Key                !Read keys after 1st trial
5530                IF Key=4 THEN Abortboth=1    !Abort both trials: restart
5540                IF Key=4 THEN Abort
5550                IF Key=8 THEN RETURN         !Accept 1st trial
5560                IF Key=16 THEN LET Sectval=0
5570                IF Key=16 THEN RETURN        !Reject 1st trial

**************************

5600 Compare: REM                                !Routine to compare trials
5610    PRINT PAGE                               !Values after 2nd trial
5620    PRINT LIN(22)
5630          Length2=Length
5640          Area2=Area
5650          Avdiameter2=Avdiameter
5660          Formfactor2=Formfactor
5670          Mindiameter2=Mindiameter
5680          Maxdiameter2=Maxdiameter
5690          Abortone=0
5700          Error=ABS(Length1-Length2)
5710             DEFAULT ON
5720                Percent=Error/Length1*100      !Calculates percent error
5730                Area(11,Sectno)=Percent
5740             DEFAULT OFF
5750                Limit=(Length1+Length2)/2/Errate !Sets limit for error
5760             IF Length1<Length2-Limit THEN GOTO Error
5770             IF Length1>Length2+Limit THEN GOTO Error
5780          LET Area(8,Sectno)=(Length1+Length2)/2 !Set final values for two
                                                        two trials
5790          LET Area(9,Sectno)=(Area1+Area2)/2
5800          LET Area(12,Sectno)=(Avdiameter1+Avdiameter2)/2
5810          LET Area(13,Sectno)=(Formfactor1+Formfactor2)/2
5820          LET Area(14,Sectno)=(Mindiameter1+Mindiameter2)/2
5830          LET Area(15,Sectno)=(Maxdiameter1+Maxdiameter2)/2
5840                                             ! Above sets final mean area
                                                    length,diameter,formfactor
5850    RETURN                                   ! if there are two trials
5860 !

**************************

5870 Error: REM                                  !Services error correct
5880    PRINT PAGE;LIN(18)
5890    FIXED 2
5900    IMAGE 10A,2X,5Z.2D,,2A,2X,15A,2X,5Z.2D,,2A,2X,14A,2X,5Z.2D,,2A
5910    PRINT USING 5900;"1ST AREA =";Area1;Unit$;"1ST PERIMETER =";Length1;Uni
t$;"1ST DIAMETER =";Avdiameter1;Unit$
5920    PRINT USING 5900;"2ND AREA =";Area2;Unit$;"2ND PERIMETER =";Length2;Uni
t$;"2ND DIAMETER =";Avdiameter2;Unit$
5930       PRINT LIN(1);"THE PERCENT ERROR IS ";Percent
5940       DISP "(ACCEPT = SK# Fd; REJECT = SK# Fe.)"
5950           A$="              "
5960       CALL Display(Select,A$)
5970           Value$=VAL$(Percent)
5980           A$="ERR="&Value$[1,6]&" ACCPT"
5990           IF Percent>99.99 THEN LET A$="ERR="&VAL$(Percent)
6000       CALL Display(Select,A$)
```

```
6010           OUTPUT Select;"BP"
6020           OUTPUT Select;"OK"
6030              ENTER Select;Key
6040 !

***************************


6050 Key2: REM                                         !Reads keys after 2nd trial
6060              IF Key=4 THEN Abortboth=1            !Abort both trials
6070              IF Key=4 THEN GOTO Abort
6080              IF Key=8 THEN GOTO 5780              !Accept data
6090              IF Key=16 THEN Sectval=1             !Reject 1st trial only
6100              IF Key=16 THEN SUBEXIT
6110         GOTO 6020
6120 !

***************************


6130 Abort: REM                                        !Abort subroutine for deleting
                                                         data from one or both trials
6140             A$="DATA DELETED    "
6150        CALL Display(Select,A$)
6160           OUTPUT Select;"BP85,200"
6170                 IF Abortboth=1 THEN LET Sectval=0
6180                 IF Abortone=1 THEN LET Sectval=1
6190                 PRINT PAGE;LIN(23)
6200                 PRINT "YOUR DATA HAS BEEN DELETED. PLEASE REDIGITIZE, THEN
PRESS SK# Fb."
6210             SUBEXIT
6220 !

*****************************************************************
6230 Input: REM  CALL SUBPROGRAM FOR ACCEPTING DATA INPUT FROM DIGITIZER
6240*************************************************************


6250      SUB Input(Select,X,Y,Key,N,Pt$,R$)
6260         OUTPUT Select USING "K";"OS"            !Check status byte.
6270            ENTER Select;Status
6280              IF BIT(Status,7) THEN 6340         !Find if SFK was pressed
.
6290              IF NOT BIT(Status,2) THEN 6260     !Find if point digitized
.
6300         OUTPUT Select USING "K";"OD"            !Read data point.
6310            ENTER Select;X,Y                     !Input point coordinates
.
6320            N=N+1                                !Increment point counter
.
6330      SUBEXIT
6340         OUTPUT Select USING "K";"OK"
6350         ENTER Select;Key                        !See what key is on.
6360      SUBEXIT
6370 Subend: SUBEND
6380 !

*****************************************************************
6390 Display:  REM  CALL SUBPROGRAM FOR PRINTING TO LED DISPLAY ON DIGITIZER
6400*************************************************************
```

```
6410    SUB Display(Select,A$)                          !LED display subprogram
6420    OPTION BASE 1
6430       FIXED 2
6440       DATA 238,254,156,252,158,142,190,110,96,112,0,28,0,42,252,206,230,10
6450       DATA 182,30,124,0,0,0,118,0
6460       DATA 58,62,26,122,222,142,246,46,32,112,0,96,0,42,58,206,230,10,182
6470       DATA 30,56,0,0,0,118,0,0,96
6480       DATA 218,242,102,182,62,224,254,230,252,1,156,240,130,2,32,68,64,202
6490          INTEGER A(72)
6500          DIM D$[240],Alpha$[72]
6510             READ A(*)                              !Read character data.
6520                D$=""
6530                IF LEN(A$)>15 THEN A$="LINE TOO LONG"
                                                         !Ck line too long.
6540                Alpha$="ABCDEFGHIJKLMNOPQRSTUVWXYZabcdefghijklmnopqrstuvwxy
z 1234567890.[]=-, '?"                                  !List of accepted chars
.
6550                Alpha$[70,70]=CHR$(34)
6560          FOR I=1 TO LEN(A$)                        !For each character,
6570             P=POS(Alpha$,A$[I,I])                  !find position in string
.
6580             IF P=0 THEN P=53                       !If undefined, then
                                                          display blank.
6590             D$=D$&";DD"&VAL$(I)&","&VAL$(A(P))     !Add char to instruction
                                                          string.
6600          NEXT I
6610             OUTPUT Select USING "K";D$             !Print instruction strin
g.
6620    SUBEND
6630  END
6640 !


******************************************************************
6650 Diameter:  REM  CALL SUBPROGRAM FOR CALCULATING AVERAGE DIAMETER
6660******************************************************************



6670 SUB Diameter(SHORT X(*),Y(*),REAL N,Ad,Mind,Maxd)!Laughter program to calc
u                                                        late the average diamet
er
6680        OPTION BASE 1                               !irregular shapes.
6690        SHORT R(1000)
6700        MAT R=ZER
6710        Minx=Maxx=X(1)
6720        Miny=Maxy=Y(1)
6730        REDIM X(N),Y(N)
6740 Step:        Step=N/75                         !Sample selective points to
                                                      speed up algorithm. Reset
6750           Sx=0                                 !denominator to number of
                                                      points you want to sample.
6760           Sy=0
6770              IF Step<=1 THEN Step=1            !Step never less than 1
6780        FOR I=1 TO N STEP Step                  !Loop to calculate the centroi
d                                                     of data array.
6790              No=No+1                           !Number of selected points
6800  REM :        Sx=Sx+X(I)                       !Sum of x.
6810  REM :        Sy=Sy+Y(I)                       !Sum of y.
6820           IF X(I)<Minx THEN Minx=X(I)          !Loop to find minimum and
                                                      maximum coordinates.
6830           IF X(I)>Maxx THEN Maxx=X(I)
6840           IF Y(I)<Miny THEN Miny=Y(I)
6850           IF Y(I)>Maxy THEN Maxy=Y(I)
6860        NEXT I
6870        REDIM R(No)
6880  REM :           Cx=Sx/No                      !X coordinate of centroid
6890  REM :           Cy=Sy/No                      !Y coordinate of centroid
```

```
6900            Avx=(Minx+Maxx)/2                    !X coordinate of geometric
                                                       center
6910            Avy=(Miny+Maxy)/2                    !Y coordinate of geometric
                                                       center
6920         FOR I=1 TO N STEP Step                  !Loop to center data array
                                                       about the origin.
6930            X(I)=X(I)-Avx
6940            Y(I)=Y(I)-Avy
6950         NEXT I
6960        FOR I=1 TO N STEP Step                   !Loop to calculate the radius
                                                       vectors.
6970           J=J+1
6980           R(J)=SQR(X(I)^2+Y(I)^2)
6990        NEXT I
7000              Sr=0
7010        Maxr=Minr=R(1)
7020        FOR I=1 TO N STEP Step                   !Loop to sum the radius vector
s
7030           K=K+1
7040           Sr=Sr+R(K)
7050           IF R(K)<Minr THEN Minr=R(K)           !Loop to find minimum and
                                                       maximum radius.
7060           IF R(K)>Maxr THEN Maxr=R(K)
7070        NEXT I
7080       Mind=Minr*2                               !Calculates minimum diameter
7090       Maxd=Maxr*2                               !Calculates maximum diameter
7100      Ar=Sr/No                                   !Calculate average radius.
7110      Ad=Ar*2                                    !Calculate average diameter.
7120    SUBEND
7130***********************************************************************
7140 Trace:  REM : TRACE CALL SUBPROGRAM TO TRACE FIGURE
7150***********************************************************************
7160   SUB Trace(SHORT X(*),Y(*),REAL N,Mag,L,Sectval) !Program to display figur
e                                                                graphically
7170    OPTION BASE 1
7180    OVERLAP
7190    OUTPUT 706;"BP"
7200    A$="CRT GRAPHICS    "                              !Display on Digitizer LE
D
7210       CALL Display(706,A$)
7220    PLOTTER IS "GRAPHICS"
7230      Minx=Maxx=X(1)
7240      Miny=Maxy=Y(1)
7250      REDIM X(N),Y(N)
7260        FOR I=1 TO N
7270          IF X(I)<Minx THEN Minx=X(I)
7280          IF X(I)>Maxx THEN Maxx=X(I)
7290          IF Y(I)<Miny THEN Miny=Y(I)
7300          IF Y(I)>Maxy THEN Maxy=Y(I)
7310        NEXT I
7320          SHOW Minx-100,Maxx+100,Miny-100,Maxy+100        !Scale screen to fi
g
7330      REDIM X(1000),Y(1000)
7340    GRAPHICS
7350 Input: REM                                          !Input points.
7360       FOR I=1 TO N
7370          IF Flag=0 THEN MOVE X(I),Y(I)          !If first point, don't draw
.
7380          IF Flag<>0 THEN PLOT X(I),Y(I)         !Draw points
7390          Flag=1                                 !Set flag so not first poin
t
7400       NEXT I                                    !Input another point.
7410 Label_trace: REM                                !Label message on CRT
7420     MOVE Minx-100,Miny-100
7430     LABEL USING "K";"PRESS CONTINUE WHEN THROUGH VIEWING"
7440     BEEP
7450     PAUSE
7460 Exit: REM
```

```
7470      GCLEAR                                              !Clear graphics and exit
7480      EXIT GRAPHICS
7490      OUTPUT 706;"SK0"
7500      OUTPUT 706;"BP"
7510      IF Sectval=0 THEN A$="BEG DIG SURFACE"
7520      IF Sectval=1 THEN A$="REDIG SURFACE   "
7530         CALL Display(706,A$)
7540   SUBEXIT
7550 ***********************************************************
7560 Count: REM : CALL SUBPROGRAM TO COUNT PARTICLES
7570 ***********************************************************
7580   SUB Count(Select,Count,Pt$)
7590   OPTION BASE 1
7600   PRINT PAGE
7610   DISP
7620   PRINT LIN(22);"PLEASE PRESS THE CURSOR DIGITIZE KEY FOR EACH COUNT. PRESS
 SFK1 WHEN FINISHED"
7630     A$="COUNT ROUTINE  "
7640     CALL Display(Select,A$)
7650       OUTPUT Select;"SG"
7660       OUTPUT Select;"BP75,200"
7670       OUTPUT Select;"SK0"
7680       OUTPUT Select;"OS"
7690         ENTER Select;Status
7700           IF BIT(Status,7) THEN 7800
7710           IF NOT BIT(Status,2) THEN 7680
7720            OUTPUT Select;"OD;BP75,200"
7730            ENTER Select;X,Y
7740              Count=Count+1
7750              Value$=VAL$(Count)
7760              A$="COUNT="&Value$[1,6]&"    "
7770       CALL Display(Select,A$)
7780       DISP "COUNT =";Count
7790     GOTO 7680
7800        OUTPUT Select USING "K";"OK"
7810        ENTER Select;Key
7820            IF Key=1 THEN
7830               OUTPUT Select;"SK0"
7840               OUTPUT Select;Pt$
7850                  A$="              "
7860                CALL Display(Select,A$)
7870                   A$=VAL$(Count)
7880                CALL Display(Select,A$)
7890            END IF
7900            IF Key<>1 THEN
7910               OUTPUT Select;"SK0"
7920               GOTO 7680
7930            END IF
7940   SUBEND
7950 ***********************************************************
7960 REM : END
7970 ***********************************************************
```

References

Albright, B.C., and B.G. Sawler (1981) A microcomputer system for the data acquisition and quantitative analysis of synaptic relationships in the central nervous system. Biomed. Sci. Instrum. 17:21–26.

Bradbury, S. (1977) Quantitative image analysis. In G.A. Meek and H.Y. Elder (eds.): Analytical and Quantitative Methods in Microscopy. Cambridge: Cambridge University Press, pp. 91–116.

Bradbury, S. (1979) Microscopical image analysis: Problems and approaches. J. Microsc. 115:137–150.

Bradbury, S. (1981) Automatic image analyzers and their use in anatomy. Prog. Clin. Biol. Res. 59B:185–192.

Caldwell, R.B., and R.R. Mize (1981) Superior colliculus neurons which project to the cat lateral posterior nucleus have varying morphologies. J. Comp. Neurol. 203:53–66.

Cowan, W.M., and D.F. Wann (1973) A computer system for the measurement of cell and nuclear sizes. J. Microsc. 99:331–348.

Curcio, C.A., and K.R. Sloan, Jr. (1981) A computer system for combined neuronal mapping and morphometry. J. Neurosci. Methods 4:267–276.

Deninno, M.M., F. Morisi, C. Pallotti, G. Pallotti, and F. Viaggi (1978) Measurements in microscopic images: An inexpensive device for semi-automatic analysis of microscopic images. Med. Progr. Technol. 6:19–22.

Dunn, R.F., D.P. O'Leary, and W.E. Kumley (1975) Quantitative analysis of micrographs by computer graphics. J. Microsc. 105:205–213.

Dunn, R.F., D.P. O'Leary, and W.E. Kumley (1977) On-line computerized analysis of peripheral nerves. In R.D. Lindsay (ed.): Computer Analysis of Neuronal Structures. New York: Plenum, pp. 111–132.

Green, R.J., W.J. Perkins, E.A. Piper, and B.F. Stenning (1979) The transfer of selected image data to a computer using a conductive tablet. J. Biomed. Eng. 1:240–246.

Harrell, J.V., R.B. Caldwell, and R.R. Mize (1982) The superior colliculus neurons which project to the dorsal and ventral lateral geniculate nuclei in the cat. Exp. Brain Res. 46:234–242.

Konigsmark, B.W. (1970) Methods for the counting of neurons. In W. Nauta and S.O.E. Ebbesson (eds.): Contemporary Research Methods in Neuroanatomy. New York: Springer, pp. 315–340.

Mathieu, O., L.M. Cruz-Orive, H. Hoppeler, and E.R. Weibel (1980) Measuring error and sampling variation in stereology: Comparison of the efficiency of various methods of planar image analysis. J. Microsc. 121:75–88.

Mize, R.R. (1983a) Variations in the retinal synapses of the cat superior colliculus revealed using quantitative electron microscope autoradiography. Brain Res. 269:211–221.

Mize, R.R. (1983b) A computer electron microscope plotter for mapping spatial distributions in biological tissues. J. Neurosci. Methods 8:183–195.

Mize, R.R. (1983c) A microcomputer system for measuring neuron properties from digitized images. J. Neurosci. Methods 9:105–113.

Mize, R.R. (1984) Computer applications in cell and neurobiology: A review. Int. Rev. Cytol. 90:83–124.

Mize, R.R., and L.H. Horner (1984) Retinal synapses of the cat medial interlaminar nucleus and ventral lateral geniculate nucleus differ in size and synaptic organization. J. Comp. Neurol. 224:579–590.

Mize, R.R., R.F. Spencer, and P. Sterling (1981) Neurons and glia in cat superior colliculus accumulate (^{3}H) Gamma-Aminobutyric Acid (GABA). J. Comp. Neurol. 202:385–396.

Peachey, L.D. (1982) A simple digital morphometry system for electron microscopy. Ultramicroscopy 8:253–262.

Pullen, A.H. (1982) A structured program in BASIC for the analysis of peripheral nerve morphometry. J. Neurosci. Methods 5:103:120.

Smith, K.C.A. (1982) On-line digital computer techniques in electron microscopy: General introduction. J. Microsc. 127:3–16.

Sobel, I., C. Levinthal, and E.R. Macagno (1980) Special techniques for the automatic computer reconstruction of neuronal structures. Ann. Rev. Biophys. Bioeng. 9:347–362.

Street, C.H., and R.R. Mize (1983) A simple microcomputer-based three-dimensional serial section reconstruction system (MICROS). J. Neurosci. Methods 7:359–375.

Weibel, E.R. (1979) Stereological Methods, vol. 1. Practical Methods for Biological Morphometry. London: Academic Press.

9

A Microcomputer Approach to Point-Counting Stereology

Max C. Poole and W. Daniel Kornegay III

Introduction

Stereology is a mathematical concept that uses geometric principles to determine three-dimensional parameters of a morphological structure from two-dimensional micrographs. Although the number, length, and cross-sectional areas of structures are measurements most often obtained from two-dimensional micrographs, the parameters calculated using stereological methods provide a more comprehensive description of the structure. For example, it is often desirable to determine the number of cells in a reference volume of tissue (i.e., numerical density N_V), rather than just the number of cells in a selection of tissue sections. Likewise it is more desirable to describe the volume of a cell or organelle in a reference volume (i.e., volume density V_V) or the membrane surface area of an organelle in a reference volume (i.e., surface density S_V) than to simply express the cross-sectional area or length per some number of micrographs. Stereological measurements are usually described as a ratio of some structural parameter to a reference volume, expressed per cubic millimeter of tissue, although organelle volumes and membrane surface areas can also be expressed per "average" cell volume.

There are many approaches for quantifying biological structure. These range from antiquated but accurate cut-and-weigh methods to sophisticated and expensive electronic digitizers and image analysis systems. Point-counting stereological techniques are among the most efficient methods and usually involve superimposing a test lattice over an image of a biological structure and counting the number of points falling on the structure and the surrounding tissue. These methods, in addition to being simple, are very accurate, and when quantifying structures that are present throughout the plane of the section (e.g., rough endoplasmic reticulum) they are much faster than the electronic analysis systems. Point-counting methods have been used for a number of tissues including liver (Stäubli et al., 1969), exocrine pancreas (Bolender, 1974), adrenal cortex (Rohr et al., 1976), and anterior pituitary (Poole and Kornegay, 1982) as well as for numerous other tissues. This chapter describes a stereological software package that utilizes point-counting techniques. Both the software and stereological formulas are described in detail.

Mathematics and geometric equations are essential to stereology and therefore the calculation of the stereological parameters are prime candidates for microcomputers. Unfortunately, very few stereological programs have been written. Most likely this is due either to the lack of computer programming experience by some investigators, or to the mathematics, which may overwhelm other investigators. Our point-counting stereology software package is therefore easy to use and is applicable to many problems with just a basic understanding of stereological techniques. A common problem in stereological analysis is the development of an efficient method for accumulating and managing the raw data, which is often voluminous. Our software package also possesses a data base that simplifies the task of data management. The software package consists of nine programs that provide for data management (programs 1, 2, 3), stereological calculations (programs 4, 5), statistical analysis of the data (programs 6, 7), and printing of the raw data and final results (programs 8, 9,).

There are several other good stereological microcomputer programs available from established investigators (Briarity and Fischer, 1981; Bolender et. al., 1982) as well as from commercial sources. Detailed descriptions of the formulas and the point-counting techniques used in this package are given so that the reader can decide if our program is appropriate for his or her application.

Materials and Methods

Computer Hardware

Our point-counting stereology package is written for an Apple II+ microcomputer. The program has been run on systems with the configurations shown in Table 1.

In all cases a two-drive setup with a controller card in slot 6 is used, although the package allows the use of a single drive. Although it is not required, an Apple numeric keypad will expedite data entry. A printer is not required, but does permit more efficient use of the system. The Apple II+ has an Apple Language Card in slot 0 in order to achieve the 64-KB memory.

Tissue Sampling

An adequate sampling protocol is essential for morphometric analysis. For this reason, our software package uses a multilevel sampling approach, consisting of a sampling hierarchy of three levels. Selecting the animals from an experimental or control group comprises the first (and highest) level of sampling. In our studies this is usually a *simple random* process, where all animals have an equal probability of selection. The tissue is removed from the animal and divided into small pieces, which are prepared for light or electron microscopy and are referred to as *blocks.* The tissue blocks constitute the second level of sampling. In our studies, the adenohypophysis is divided into specified regions consisting of multiple blocks, with at least one block being randomly selected for analysis from each region. Our block sampling method is essentially a *systematic random sampling,* as opposed to the simple random sample described at the animal level, since only those blocks within regions systematically defined have an equal probability of being selected.

Once the blocks are selected, the tissue is sectioned for light or electron microscopy, with the sections being collected on slides for light microscopy or grids for electron microscopy. The grid (or slide) level constitutes the third (and lowest) level of sampling. From each slide (or grid) a predetermined number of micrographs are taken. We routinely collect three grids of sections from each block, and from each grid every fifth grid square possessing tissue is selected for a micrograph until eight micrographs are taken from each grid. Once again this involves a systematic random sampling, since we systematically define the grid squares chosen for micrographs without regard for the structures or cells located in that section. A good description of sampling strategies in stereological analysis has

Table 1

Computer	Printer interface card	Printer
Apple II+ 64*K*	Grappler + (slot 1)	Epson MX80FT with Graftrax
Apple IIe 64*K*/80 col.	Grappler + (slot 1)	Okidata 92
Apple IIe 128*K*/80 col.	Wizard-BPO (slot 1) 32*K* buffer	C ITOH 8510A

been given by Weibel (1979) and should be consulted, since the method of defining the sampling at each level will vary depending upon the features of the tissue. Our program will allow data from 10 micrographs to be stored per grid (or slide). The acquisition of raw data from the micrographs is discussed later in this chapter.

A nested analysis of variance (ANOVA) is used to define our sample set by determining the sampling variance at the animal, block, and grid levels. Sampling adjustments at each level are made until the variance at a specific level is reduced. This statistical method has been developed for microcomputers (Kornegay and Poole, 1983) and is included in the software package.

Test Line Lattices for Point and Intersection Counting

The point and intersect counts used in this software package are collected through the use of test line lattices and are manually entered into the computer through the keyboard. The software can accept either square lattices or Weibel multipurpose lattice systems, both of which are described in detail below. The lattices are drawn on transparent acetate film and overlayed onto micrographs. The points and intersections are counted and the morphometric parameters calculated using the software package. In addition to the overlay method, other methods can be used that do not require micrographs. Similar data may be collected using either eyepiece graticules for light microscopy, or projection screens possessing a test lattice onto which negatives of micrographs are superimposed (Weibel, 1979). Regardless of the method of data collection, this package can accommodate the data if the lattices are similar to those described for this program.

Square Lattice Test System. This test system can be a single lattice (Figure 1), or a coherent double-lattice system in which there is a coarse point lattice, with a fine point lattice superimposed. Our program can accommodate data from either lattice if the lattice test line length (in centimeters) and the calibrated micrograph magnification are entered into the program. In our whole cell and organelle analysis programs, the points overlying the cells, nuclei, and organelles are used in calculating the cell and organelle volumes. Furthermore, in the organelle analysis program, the membrane surface area can be calculated by counting the intersections that membrane profiles make with the test lines.

We have found single square lattices with lengths of 1 cm and 0.5 cm to be useful for most analyses. However, a more refined coherent double-lattice system (1:25) is used when measuring organelles with a volume density (V_V) of less than 5% (Poole et al., 1980). The coherent double lattice is designed so that the basic unit of the coarse lattice is a square of 1 cm^2, with a finer lattice of 25 smaller squares superimposed within each coarse square. The details of these lattices are discussed by Weibel (1979).

Weibel Multipurpose Lattice Test System. Square lattices are often difficult to use when counting the intersections that organelle and cell membranes make with the test lines due to the excessive test lines in the lattice. Therefore, Weibel has suggested the use of a multipurpose test system. This test lattice was originally developed by Chalkley et al. (1949) and refined by Weibel et al. (1966). The dimensions for several lattices are clearly described by Weibel (1979). The lattice

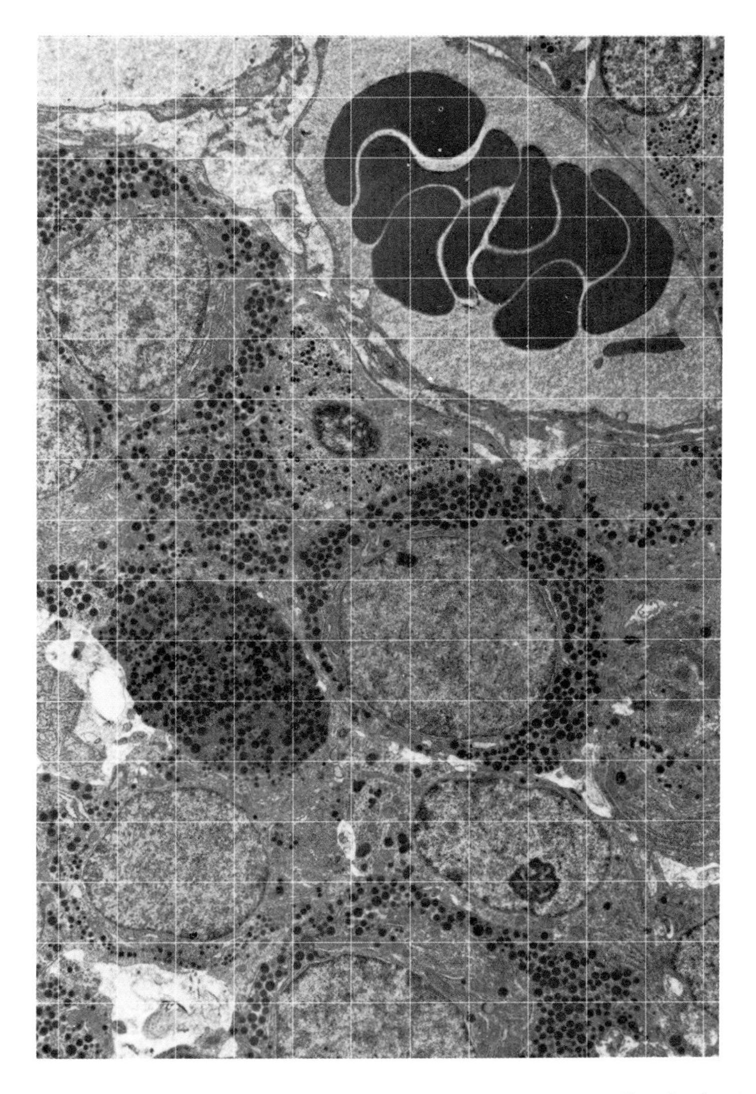

Figure 1. Square sampling lattice. A 1.0-cm square sampling lattice is superimposed over a low magnification (5,561×) micrograph of the anterior pituitary. The points (as denoted by the line intersects) overlying specified cell profiles, the nuclei within the cells, and the total points on the micrograph are summed and entered into program 4. The same lattice can be used on higher-magnification micrographs of cell profiles for an organelle analysis. In that case, not only the test points but also any intersections that the organelle membranes make with the test lines are counted and entered into program 5.

we use most often consists of 300 1-cm lines with tabs at both ends, which produces 600 test points (Figure 2). Each point is spaced exactly 1 cm from its adjacent points, and the lines and points are arranged so that the basic unit is an equilateral rhombus, with angles of 60°. This lattice reduces the work required in counting the organelle points and the membrane intersects when compared to a square lattice, and we have found it to be advantageous in calculating the organelle volumes and surface areas.

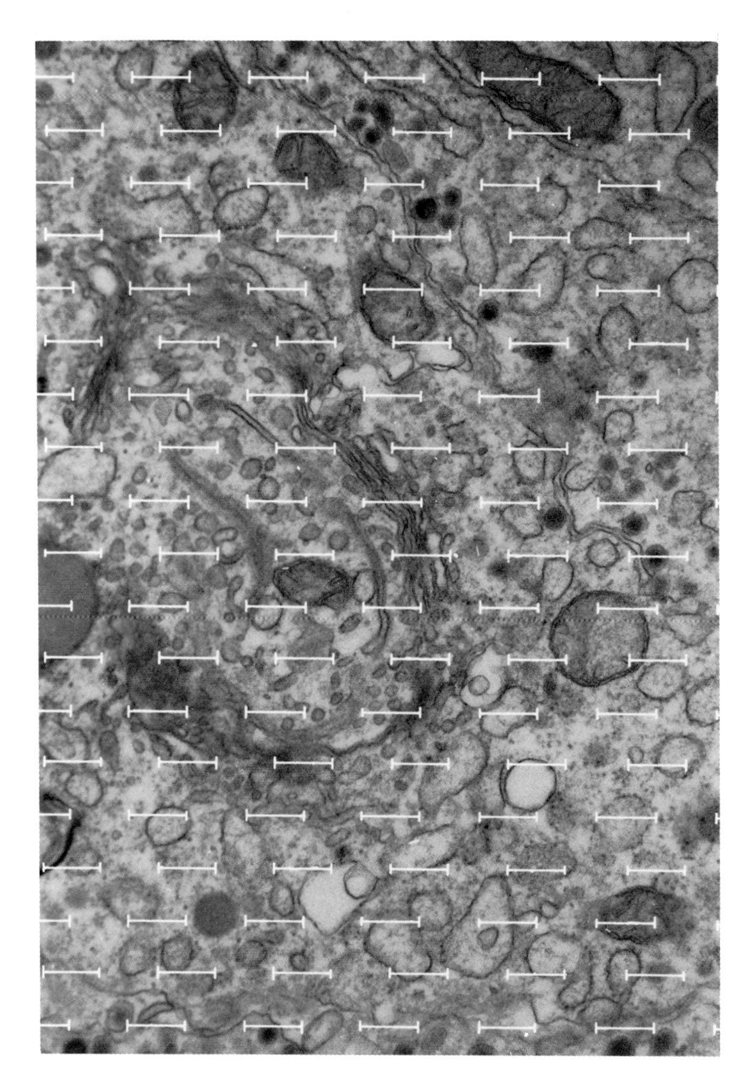

Figure 2. Weibel multipurpose sampling lattice. A 1.0-cm multipurpose lattice is superimposed over a high-magnification micrograph (20,160×) of a portion of a cell profile. In this example the test points overlying the organelle as well as the intersections that the organelle membrane makes with the test lines are counted and entered into program 5.

Program Operation

The point-counting stereology package is written in APPLE SOFT BASIC and operates under DOS (disk operating system) 3.3. Although this BASIC may lack some of the benefits of other microcomputer languages (such as PASCAL), it is a very easy language to learn and it is adaptable to most microcomputers with little revision.

Software Design

The program package begins with a HELLO program, which gives basic instructions and displays a menu of program options. The package includes nine programs (Figure 3):

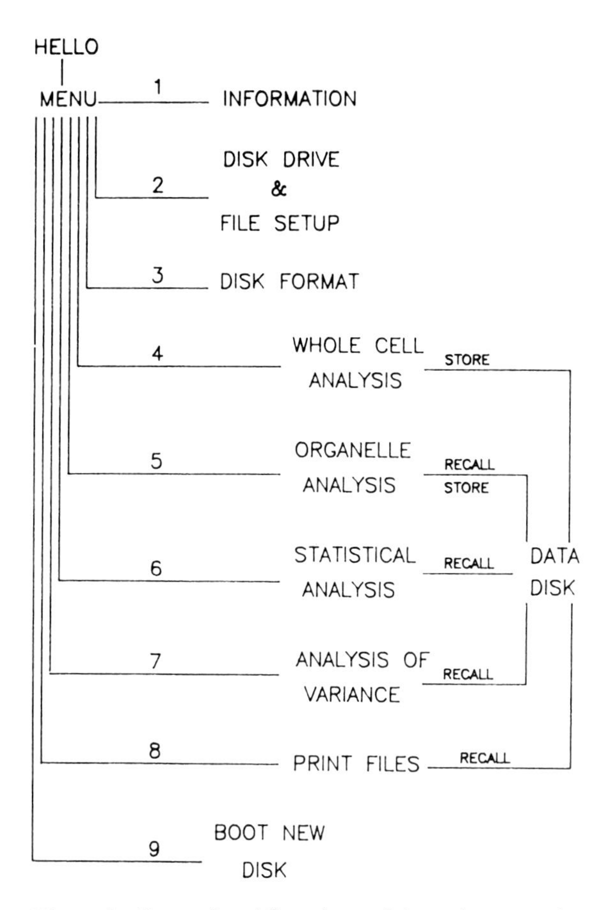

Figure 3. Operational flowchart of the point-counting stereology package. In this flowchart the nine programs in the package are illustrated. In addition, the interaction of the programs with the data file disk are shown. Only programs 4 and 5 store data on disk, and programs 5–8 recall data in order to execute their functions.

1. INFORMATION provides essential instructions for the remaining programs (occupies 1,154 bytes of memory).
2. DISK DRIVE AND FILE SETUP allows one to document the data files by entering project titles and whether the analysis is a whole cell or organelle analysis (occupies 1,399 bytes of memory).
3. DATA DISK FORMAT initializes the data disk (occupies 699 bytes of memory).
4. WHOLE CELL ANALYSIS allows one to enter the raw data and calculate the numerical densities and cell volumes of the respective cell type being analyzed (occupies 7,351 bytes of memory).
5. ORGANELLE ANALYSIS allows one to enter the raw data and calculate the volumes and surface areas of the organelles within a cell type (occupies 8,879 bytes of memory).
6. STATISTICAL ANALYSIS allows one to calculate a mean (± its standard error) of the morphometric parameters calculated in programs 4 and 5 at the grid level, the block level, and the animal level (occupies 8,350 bytes of memory).

7. NESTED ANOVA performs a nested analysis of variance on the morphometric parameters calculated in programs 4 and 5, and allows one to monitor the efficiency of the sample set (occupies 8,144 bytes of memory).
8. PRINT PROJECT FILE is a utility program that gives a hard-copy printout of the raw data and the calculated morphometric parameters contained within a file (occupies 6,589 bytes of memory).
9. QUIT terminates the package (occupies 208 bytes of memory).

Program Steps

1. INFORMATION: This program provides information explaining the functions of the remaining 8 programs in the MENU. It is a simple help feature for our software.

2. DISK DRIVE AND FILE SETUP: Prior to entering data for a morphometric analysis, the data files must be created or opened. This program requires the investigator to denote whether two disk drives are being used and in which I/O slot the drives are interfaced (in our system two drives are used in slot 6). The program also requires a project title of up to 25 characters. If the data being stored are from an organelle analysis, then the organelles must also be identified. An option of four organelles including mitochondria, rough endoplasmic reticulum, golgi apparatus, and secretory granules are given in a menu since these are the ones most often used in our analysis. In addition, four other organelle files can be specified by the user. Data being entered for a cell volume and numerical density analysis do not have to be specified since a file is established for these parameters regardless of the organelles being analyzed.

3. DATA DISK DRIVE: This program is used to format a new data disk. New disks must be initialized if the files specified in program 2 are to be stored.

4. WHOLE CELL ANALYSIS: This program is the core of the morphometric package and performs three primary functions. First, the data for a whole cell analysis are entered. Secondly, the "average" cell volume and the numerical density are calculated. Finally, the raw data, as well as the calculated volumes and numerical densities, are stored for future reference. This program also allows for the amending and expansion of the raw data sets.

Specifically, the program begins with instructions explaining how the data are to be entered. The program then requests the project title, and asks whether the data files already exist. The investigator then enters the animal number (i.e., identification), the block number, and the grid number of the data being entered. If the data files already exist, they may now be changed or expanded.

The input data required for this analysis include the total number of points on a micrograph, the number of points falling on the cells being quantified, the number of points on the nuclei of the cells being quantified, and the number of nuclei of the cells being analyzed. These values are entered for each micrograph. The program can accommodate 10 micrographs per grid (or slide). Following the data entry from the micrographs, the operator presses the "G" key if additional data are to be entered for another grid (or slide) from the same block. "B" is pressed if the next set of data is for a grid from another block. "A" is pressed if the data being entered are for another animal. Following the entry of all data for the various animals, blocks, and grids, the letter "S" is entered to instruct the program to print the raw data per micrograph for each grid, block, and animal.

Following the raw data entry, the program requests the sampling lattice overlay size (only a square lattice is used in this program) and the micrograph magnification. The calculations are then performed, yielding the average cell volume, the numerical density (i.e., number of cells/mm^3 of tissue), and the area of the micrographs being sampled. These parameters are given for each grid. An option allows the investigator to store them for use in the organelle analysis (program 5). The flowchart for this program is given in Figure 4.

5. ORGANELLE ANALYSIS: The organelle volumes and the surface area of any organelle specified in program 2 can be calculated using this program. It is absolutely necessary that the "average" whole cell volume be previously calculated and stored using program 4 for each grid being analyzed at the organelle level, since the calculated cell volumes will be used in their respective organelle calculations.

Essentially the organelle raw data are entered in the same format as the whole cell data in program 4. Data are entered per micrograph for each grid, with up to 10 micrographs per grid permitted. If more than 10 micrographs are required, the

Figure 4. Flowchart of program 4. In this program the whole cell data are entered and the stereological parameters calculated. The raw data and the stereological parameters are stored for later use, and a data change routine allows for the continuous updating of the data set.

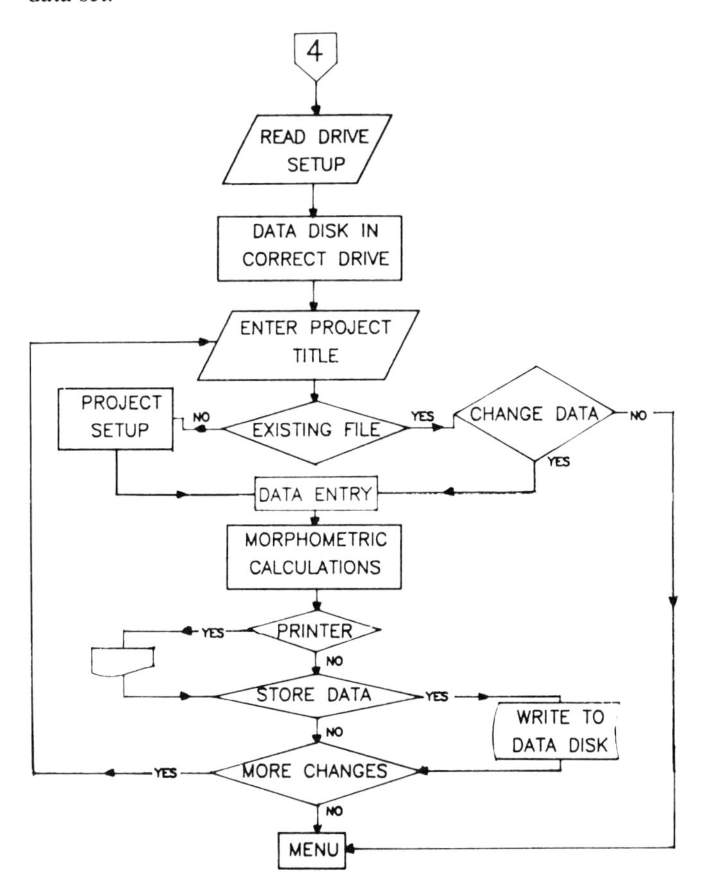

investigator can sum the data from the remaining micrographs and enter the sums as the 10th micrograph. This will not alter the calculations. The organelle volume calculations require that the following raw data be entered: the total number of points on the micrograph, the number of points on a specified cell type, and the number of points on the organelle within the cell. Furthermore, the organelle surface area requires entry of the number of intersects that the membranes of the organelle make with the test lines of the sampling lattice. The organelle volumes and membrane surface areas can be calculated using data acquired from either the square lattice or the Weibel multipurpose sampling lattice. The sampling lattice type and size must be specified, as well as the magnification of the micrographs. The organelle volumes and membrane surface areas are stored as well as the raw data, and the data sets can be amended or expanded at either the grid, block, or animal levels. The flowchart for program 5 is similar to that of program 4 (Figure 4).

6. STATISTICAL ANALYSIS: In this program, the mean and the standard error of the mean are calculated for all of the grids (or slides) per block, for all of the blocks per animal, and finally, for all of the animals per sample set.

7. NESTED ANALYSIS OF VARIANCE: The investigator has the option of analyzing the calculated parameters in programs 4 and 5 with a nested analysis of variance. The program requires no additional data, but the calculated values must have been previously stored. The program will construct an ANOVA table and give the sum of squares, the degrees of freedom, and the mean squares for each level of sampling. Furthermore, the F ratio is given for both the animal and block levels. The variance table provides an index as to the levels most responsible for the variance, and this gives the investigator some indication as to those levels where additional sampling is required. Following the ANOVA analysis the investigator can reenter programs 4 and 5 and amend the existing data sets until the sampling variance is reduced. The flowchart and the calculations for this program have been previously described by Kornegay and Poole (1983).

8. PRINT FILES: This program allows a printout of the entire data file as specified in program 2.

9. QUIT: This program provides for an orderly termination to the package.

Stereological Formulas

In order to use our software package most efficiently, the investigator should understand the morphometric formulas incorporated in the programs. The following paragraphs describe these formulas in detail.

Whole Cell Calculations (Program 4). Weibel and Gomez (1962) have determined that the numerical density of a cell (N_V), and ultimately the volume, can be derived from the following relationship if the size distribution K and shape (β) of a cell reference structure can be determined.

$$N_V = \frac{K}{\beta} \times \frac{N_A^{3/2}}{V_V^{1/2}} = \frac{1}{1.38} \times \frac{(2)^{3/2}}{(4)^{1/2}}, \tag{1}$$

where N_A is the number of nuclear profiles per unit area, and numbers in parentheses refer to forthcoming equations. Our program uses the nuclei as a reference structure. This approach should be applicable to most cell types if some assumptions about the size distribution and shape of the nuclei are made. The size distribution coefficient K is related to the coefficient of variation of the spread of the size of the nuclei. Weibel et al. (1969) have determined that the average sphere diameter D for hepatocyte nuclei is 7.95 $\pm$ 1.12 (S.D.) μm or 14% variation, which yields a K value of 1.027 when read from a K distribution graph (Weibel, 1979). We have performed a similar study on 234 profiles of mammotrope nuclei from the anterior pituitary and found that the average sphere diameter D for the nuclei is 6.75 $\pm$ 1.15 (S.D.) μm or 17% variation, which yields a K value of approximately 1.05. Weibel (1979) has noted that in most practical applications K will range from 1.02 to 1.10 and will have little effect on the final calculations. Therefore, in our program, it is assumed that the nuclei size distribution will fall within this range and K is given the value of 1.

A second assumption concerns the shape coefficient β of the nuclei. In our studies the axial ratios of the nuclear profiles are measured and related to their shape coefficient using the nomogram as given by Weibel (1979). For example, the nuclear axial ratios of mammotropes (major axis/minor axis) varied from a minimum of 1.13 to a maximum of 1.21. Panesse et al. (1972) have suggested that a nucleus with an axial ratio of 1.5 or less should be considered spherical. Therefore, the nuclei of the cells are considered spherical, and a β coefficient of 1.38 is used in the program. This stereology package applies to many cell types with "spherical" nuclei. However, if this is not true for the nuclei, or if the nucleus is not selected as the reference structure, then the calculated K and β coefficients within the program can be changed with little effort.

The number of nuclear profiles per unit area (N_A) is calculated as

$$N_A = N/A = N/(3), \tag{2}$$

where N is the sum of nuclear profiles for a specific cell type and the sample area A is defined as

$$A = P_T \times (Z \times 10{,}000/M)^2, \tag{3}$$

where P_T is the total number of points sampled, Z is the grid lattice size (in cm), M is the magnification, and the centimeters are converted to micrometers by multiplying the centimeters by 10,000 μm/cm.

The volume density V_V is expressed as

$$V_V = P_N/P_T, \tag{4}$$

where P_N is the sum of points per total number of nuclei for the cells being analyzed. The final equation for the volume of an "average cell" ($\overline{V}_c$) is defined by Weibel and Bolender (1973) as

$$\overline{V}_c = V_{Vc}/N_V = (6)/(1), \tag{5}$$

in which the volume density of the cell V_{Vc} is

$$V_{Vc} = P_c/P_T, \tag{6}$$

where P_c is the sum of the points per total number of cells being analyzed.

Organelle Analysis (Program 5). This program can accommodate data from the square grid lattice or the Weibel multipurpose lattice. The volume density V_{Vo} of the organelles is defined as

$$V_{Vo} = P_o/P_c, \tag{7}$$

where P_o is the sum of points falling on specified organelles within the cells and P_c is the sum of points falling on the cells. The volume of the specified organelle $\overline{V}_o$ is determined by multiplying the cell volume $\overline{V}_c$, as calculated in program 4, by the volume density of the organelle:

$$\overline{V}_o = V_{Vo} \times \overline{V}_c = (7) \times (5). \tag{8}$$

The surface density S_{Vo} of the organelle can also be calculated using the intersect data from either the square lattice or the Weibel multipurpose lattice. The surface density is calculated using the cell as the sample reference (i.e., $S_{Vo,c}$). When calculating the surface density $S_{Vo,c}$ using a square lattice the following relationship is used:

$$S_{Vo,c} = 2 \times I_o/L_T = 2 \times I_o/(10), \tag{9}$$

where the sum of the intersects made with the organelle membrane (I_o) is divided by the test line length L_T. The total test line length per sample reference (i.e., cell) is calculated as

$$L_T = P_c \times 2 \times (Z \times 10{,}000/M), \tag{10}$$

where the sum of the points within the cell profiles P_c is multiplied by the length Z of one test line divided by the magnification M. The surface density $S_{Vo,c}$ using a Weibel multipurpose lattice is described in the following relationship:

$$S_{Vo,c} = 4 \times I_o/[P_c \times (Z \times 10{,}000/M)]. \tag{11}$$

The surface area $S_{o,c}$ of the organelle membrane in the average cell is obtained from the following equation:

$$S_{o,c} = S_{Vo,c} \times \overline{V}_c = (11) \text{ or } (9) \times (5). \tag{12}$$

Research Applications

The described programs have been used extensively in the stereological analysis of secretory cells in the adenohypophysis (Poole et al., 1981; Poole and Kornegay, 1982). However, they can be used for many tissues if the tissue is isotropic (i.e., if all structural orientations in space are of equal probability). For those tissues that are anisotropic and show some preferred orientation of membranes or boundaries (e.g., myofibrils in striated muscle), the sampling lattices used in this program package are not appropriate since the lattices also possess a degree of anisotropy. In tissues where anisotropy is present, the investigator must use different sampling systems (Weibel, 1979), which cannot be used with this package.

A sample run of data using program 4 is presented in Figure 5. The project was entitled MP-4 TYPE 1 and was used to calculate the morphometric parameters of a gonadotrope cell type. The raw data were calculated from 17 low-magnification micrographs (5,782×) from grid 1 of block 1 from animal 1. The prints were analyzed using a square lattice overlay of 0.5 cm. The number of micrographs exceeded the 10 allowed in the program. To compensate for this, the sum of two micrographs were entered for eight of the entries, thus yielding the large number of counts (3,666). Also, in the ninth data entry, there were no data points for the cell type being analyzed. The numerical density calculations require that the data from all micrographs be entered, even if the cell type of interest is not found on

MP-4 TYPE 1

ANIMAL NUMBER = 1

BLOCK NUMBER = 1

GRID NUMBER = 1

TOTAL NUMBER OF POINTS	TOTAL NUMBER OF POINTS ON CELL	NUMBER OF POINTS ON NUCLEI	NUMBER OF NUCLEI
3666	252	87	2
3666	481	166	4
3666	211	7Ø	2
3666	49Ø	154	4
3666	327	99	2
3666	614	189	4
3666	5ØØ	85	2
3666	53	12	1
1833	Ø	Ø	Ø

MAGNIFICATION = 5782
COUNTING OVERLAY SIZE = .5

NUMERICAL DENSITY = 117871.57 CELLS/MM3

VOLUME = 797.17 CUBIC MICROMETERS

AREA = 233Ø2.11 SQUARE MICROMETERS

Figure 5. Data printout from program 4. The raw data are entered for grid 1 from block 1 of animal 1. The average cell volume (μm^3/cell) and numerical density (cells/mm^3 of tissue) are calculated for this specific grid. The area refers to the cumulative test area of the micrographs. A similar analysis and printout is made for each grid of each block and animal.

that micrograph. The numerical density (cells/mm^3 tissue), cell volume (μm^3/average cell), and the total sampling area of the micrographs per grid are shown in the printout. Although this analysis is for one grid, the software program gives a similar analysis for each grid, and program 6 (STATISTICAL ANALYSIS) can calculate the means and standard errors for the volumes and numerical densities of the collective sampling set.

A similar analysis is performed at the organelle level using program 5 (Figure 6). The volume and surface area of the rough endoplasmic reticulum (RER) are calculated using higher-magnification micrographs (18,700×) of gonadotrope cell profiles taken from the same grid, block, and animal analyzed in Figure 5. The gonadotrope cell volume, which was earlier calculated and stored using program 4, is recalled by program 5 and used in calculating the RER parameters. The sampling system is a Weibel multipurpose lattice of 1.0 cm. In this analysis both the number of points and intersections of the RER membrane are counted, and both the RER membrane surface area (μm^2/cell) and volume (μm^3/cell) are calculated. Similar analyses can be completed for any organelle using micrographs of higher magnifications.

We have found the cell and organelle stereologic parameters to be most helpful in describing the intercellular and intracellular "dynamics" of secretory cells in the adenohypophysis. For example, we have determined the numerical densities of the six cell types within the pituitary and described not only the relative abundance of the cells but also the regional variations in the cell distributions (Poole and Kornegay, 1982). In addition, by determining the numerical density of the

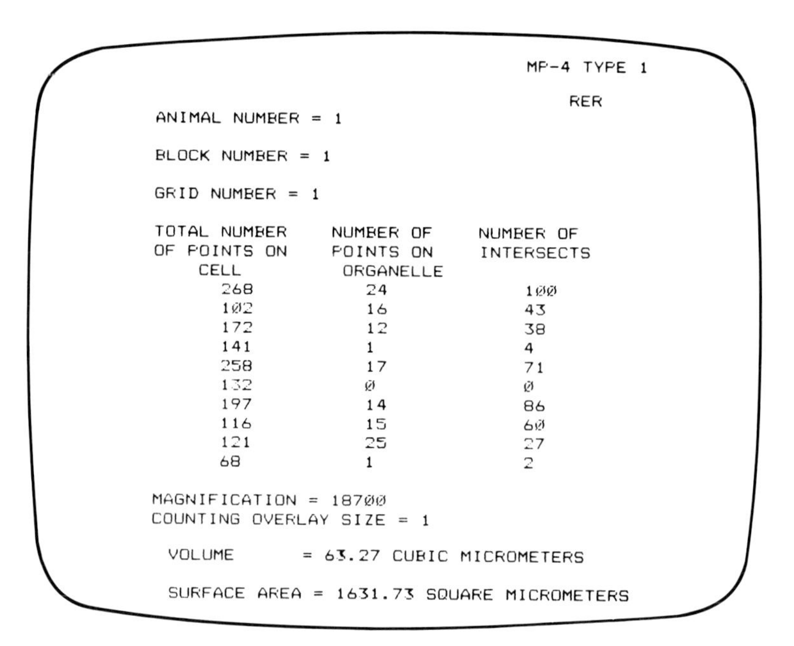

MP-4 TYPE 1

RER

ANIMAL NUMBER = 1

BLOCK NUMBER = 1

GRID NUMBER = 1

TOTAL NUMBER OF POINTS ON CELL	NUMBER OF POINTS ON ORGANELLE	NUMBER OF INTERSECTS
268	24	100
102	16	43
172	12	38
141	1	4
258	17	71
132	0	0
197	14	86
116	15	60
121	25	27
68	1	2

MAGNIFICATION = 18700
COUNTING OVERLAY SIZE = 1

VOLUME = 63.27 CUBIC MICROMETERS

SURFACE AREA = 1631.73 SQUARE MICROMETERS

Figure 6. Data printout from program 5. The raw data are acquired from high-magnification micrographs (18,700×) of cell profiles from the same grid used in Figure 5. In this figure the point and intersect data for the rough endoplasmic reticulum (RER) are entered and the organelle volume (μm^3/cell) and membrane surface area (μm^2/cell) are calculated.

mammotropes per cubic millimeter of gland and assaying for the pituitary prolactin content we were able to determine the hormone content per cell (Poole et al., 1980). Currently, we are monitoring populations of gonadotrope cell types in response to hormone treatment in order to discern the transformations between the cell types.

The volumes and membrane surface areas are also important in describing the cell and organelle changes following stimulation or suppression. For example, we have described the volumes of autophagic and crinophagic lysosomes in mammotropes following periods of increased secretory activity (Poole et al., 1981). Furthermore, we described in detail the organelle changes in mammotropes throughout the rat estrous cycle (Poole et al., 1980). As is apparent, the most exciting and imaginative use of stereology comes in describing the cell and organelle changes following physiological stimuli or suppression. Furthermore, stereological techniques provide the investigator with a powerful tool in correlating morphology with physiological and biochemical functions.

Conclusions

The point-counting stereology programs presented in this chapter comprise a comprehensive package that calculates the cell numerical densities, cell and organelle volumes, and the membrane surface areas using various sampling lattices at multiple levels of sampling. The package also calculates basic statistics and a nested analysis of variance at each sampling level. The ANOVA program allows the investigator to monitor the sample set and decide at which levels the sampling efficiency can be improved. Unlike other stereology programs in which the raw data are tallied using special-function keys, our package only allows the entry of data per micrograph after they have been counted and summed. Therefore, the microcomputer is not committed during the hours of data acquisition and can be used for other functions.

The programs in this package have been developed over the past six years (Poole and Costoff, 1979; Kornegay and Poole, 1983). The present point-counting stereology package is a substantial improvement over the earlier programs since this package is centered around a data base. The data base permits the storage of current data as well as the addition of data collected at a later time. Because of this, the morphometric parameters can be continually updated and the sampling efficiency monitored using the ANOVA program.

The principal programs were originally developed as utility programs for a Hewlett-Packard 9845T microcomputer. However, the point-counting stereology package presented in this chapter was written for Apple II+ and Apple IIe microcomputers primarily because these microcomputers are within the budget of many laboratories.

The authors are grateful to Alma Haddock for her word-processing services.

This work was supported by USPHS grant HD 15259 from the National Institute of Child Health and Human Development, National Institutes of Health, U.S. Public Health Service (Bethesda, MD).

References

Bolender, R.P. (1974) Stereological analysis of the guinea pig pancreas. I. Analytical model and quantitative description of non-stimulated pancreatic exocrine cells. J. Cell Biol. 61:269–287.

Bolender, R.P., E.A. Pederson, and M.P. Larsen (1982) PCS-I-A point counting stereology program for cell biology. Comput. Programs Biomed. 15:175–186.

Briarity, L.G., and P.F. Fischer (1981) A general purpose microcomputer program for stereological data collection and processing. J. Microsc. 124:219–223.

Chalkley, H.W., J. Cornfield, and H. Pork (1949) A method of estimating volume–surface ratios. Science 10:295–297.

Kornegay, W.D., and M.C. Poole (1983) A computer program using a nested analysis of variance in morphometric sampling. Comput. Programs Biomed. 16:155–160.

Pannese, E., R. Bianchi, B. Calligaris, R. Venturs, and E.R. Weibel (1972) Quantitative relationships between nerve and satellite cells in the spinal ganglia. An electron microscopical study. 1. Mammals. Brain Res. 46:215–234.

Poole, M.C., and A. Costoff (1979) A computer program for the morphometric analysis of cell profiles. Comput. Programs Biomed. 10:143–150.

Poole, M.C., and W.D. Kornegay (1982) Cellular distribution within the rat adenohypophysis: A morphometric study. Anat. Rec. 204:45–53.

Poole, M.C., V.B. Mahesh, and A. Costoff (1980) Intracellular dynamics in pituitary mammotropes throughout the rat estrous cycle. I. Morphometric methodology and hormonal correlations with cellular and nuclear volumes. Am. J. Anat. 158:3–13.

Poole, M.C., V.B. Mahesh, and A. Costoff (1981) Morphometric analysis of the autophagic and crinophagic lysosomal systems in mammotropes throughout the estrous cycle of the rat. Cell Tissue Res. 220:131–137.

Rohr, H.P., G. Bartsch, P. Eichenberger, Y. Rasser, Ch. Kaiser, and M. Keller (1976) Ultrastructural morphometric analysis of the unstimulated adrenal cortex of rats. J. Ultrastruct. Res. 54:11–21.

Stäubli, W., R. Hess, and E.R. Weibel (1969) Correlated morphometric and biochemical studies on the liver cell. II. Effects of phenobarbital on rat hepatocytes. J. Cell Biol. 42:92–112.

Weibel, E.R. (1979) Stereological Methods, vol. 1. New York: Academic Press.

Weibel, E.R., and R.P. Bolender (1973) Stereological techniques for electron microscopy. In M.A. Hyatt (ed.): Principles and Techniques of Electron Microscopy, vol. 3. New York: Van Nostrand Reinhold, pp. 239–296.

Weibel, E.R., and D.M. Gomez (1962) A principle for counting tissue structures on random sections. J. Appl. Physiol. 17:343–348.

Weibel, E.R., G.S. Kistler, and W. R. Scherle (1966) Practical stereological methods for morphometric cytology. J. Cell Biol. 30:23–38.

Weibel, E.R., W. Stäubli, H.R. Gnagi, and F.A. Hess (1969) Correlated morphometric and biochemical studies on the liver cell. 1. Morphometric model, stereologic methods and normal morphometric data for rat liver. J. Cell Biol. 42:68–91.

10

A Computer-Based Video Microscope for Cell Measurement

Daniel DuVarney and Raymond C. DuVarney

Introduction

In this chapter we describe in detail a microcomputer-controlled measurement system that greatly automates the process of measuring microscopic objects.

The measurements are made on the magnified image of the object as it is displayed on a video monitor. The measurement system may also be used on photographs, but here we concentrate on describing the video measurement system. The system we describe operates on either the Apple II+ or the IBM PC and is commercially available. Readers interested in obtaining this general-purpose measurement system may contact one of the authors (RCD) or Southern Micro Instruments Inc.*

Overview

Figures 1 and 2 show a photograph and a block diagram of the video microscope system, respectively. The system consists of a closed circuit television (CCTV) camera that mounts on the microscope, a monitor, a microcomputer, a graphics tablet, and an electronic interface that interconnects the microcomputer to the camera, the graphics tablet, and the monitor. The interface allows the CCTV camera picture to be overlayed with the microcomputer video output, so that one can see the microscope image and the computer output displayed simultaneously on the same monitor. Measurements are made directly on the video image by employing graphics software to outline the object of interest. The measurement routine displays small blinking cross hairs on the monitor. The position of the cross hairs is controlled from the graphics tablet. The cross hairs on the monitor move in proportion to the movement of a hand-held cursor on the tablet. A button on the cursor is pushed to indicate that the current x–y coordinate should be used in the measurement routine. In addition, a command strip on the graphics tablet allows system commands to be issued with the cursor.

The system software consists of calibration, measurement, and file-keeping routines. In addition to these there are utility programs for statistical data analysis and histogram plotting. All of the software is menu driven. This means that at each point in the program where a choice must be made, the operator is presented with a menu screen that defines all of the allowed choices and the associated letter command to execute each choice. A typical session with the system would consist of the following steps:

1. *Calibration:* Before any measurements are made, a calibration routine is used to calibrate the field of view for each microscope objective to be used. While viewing a standard calibration slide, the blinking cross hairs are positioned successively over two points on the calibration slide and the cursor button pushed to enter the x–y coordinates into the computer. The computer is then told the actual distance between these points and a calibration factor is derived. The computer also saves several previous calibrations on disk so that the user does not have to recalibrate each time the data acquisition program is run.

2. *Data entry:* Once calibrated, geometrical measurements are obtained directly from the video screen. For example, the area, perimeter, and form factor

---

*120 Interstate North Parkway East, Suite 308, Atlanta, GA 30339.

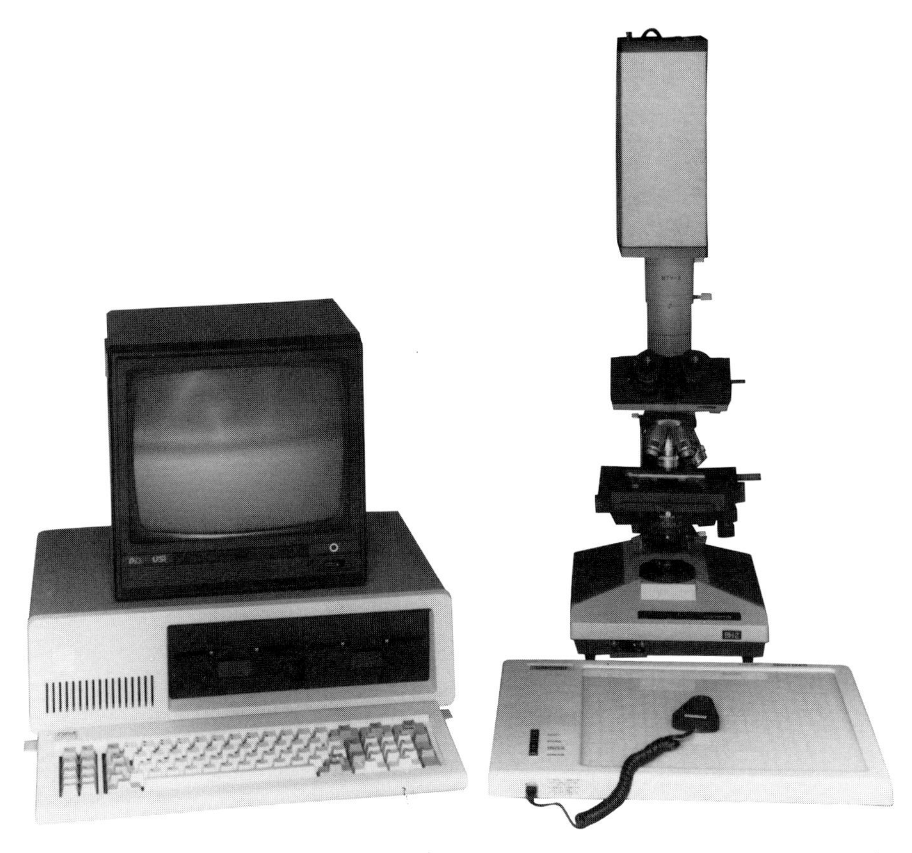

Figure 1. A photograph of the computer-based video microscope system showing the Hipad graphics tablet, an Olympus microscope with an MTI-Dage CCTV camera, a monitor, and an IBM PC microcomputer.

Figure 2. A block diagram showing the system components and their interconnections.

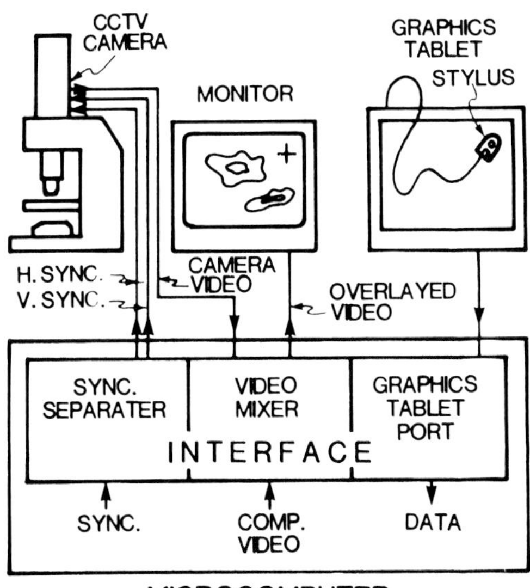

for an object such as a cell may be easily determined. To do this, one simply traces around the perimeter of the object with the cross hairs. Each data point is connected with the one preceding it whenever the button is pushed (or held down), and this produces a digital trace that approximates the shape of the object as a series of straight-line segments. When the object has been satisfactorily traced, the user tells the computer that he or she is finished by positioning the cursor over the box marked "C" (continue) on the command strip and pressing the button.

3. *Measurement*: Once the image appearing on the screen has been reduced to digital representation as described above, the area, perimeter, and form factor are instantaneously displayed on the screen. In addition to these measurements, particle counts, spatial frequency, density, and angles may be computed. The results of all measurements appear at the bottom of the screen and can either be ignored, sent to the printer, or saved on disk.

The accuracy with which one can perform a measurement on the video microscope system is partially limited by the resolution of the graphics screen. This is the number of individual dots that comprise a horizontal or vertical line. The Apple graphics unit has a resolution of 280 dots horizontally across the screen and 192 dots vertically. The IBM graphics unit has 640 dots horizontally across and 200 vertically. In practice, repeated measurements on the same object are reproducible to about 1%. This depends to some extent on the complexity of the object being traced and the experience of the operator.

4. *Analysis:* The software includes a basic statistics package and histogram plotter. These utility programs facilitate the analysis of data files that were created during the measurement phase of operation. The statistics package computes the sum, mean, standard deviation, and standard error, and includes routines to perform a Student's t test and to pool sets of data files. The histogram plotter allows the user to select the x and y ranges, plots logarithmic, linear, or percentage histograms, and lets the user easily modify the labeling of each axis. Since many people will probably want to perform some special analysis, the data is written on the disk in an easily readable ASCII format.

System Hardware

Video Display

The principle of the video measurement system involves overlaying the video from the microscope camera with the video output from the computer. In order to understand how this process works the details of the video display process must be described. The monitor produces an image on the face of a cathode ray tube by sweeping an electron beam rapidly across the inside face of the tube. Phosphors on the inside surface of the tube glow when they are struck by the electron beam at an intensity proportional to the intensity of the electron beam. In order to reproduce an image, a monitor must receive three signals: a vertical synchronization signal, a horizontal synchronization signal, and the brightness signal. The *vertical synchronization signal* is used to trigger the vertical sweep. Each time it reaches the bottom of the screen, the beam returns to the top and waits until the next vertical synchronization signal is received before it starts the next downward sweep.

While it is sweeping downward at approximately 60 Hz, it is also sweeping horizontally at a much higher frequency. The beginning of each horizontal sweep is triggered by another input signal called the *horizontal synchronization signal.* Each time the beam is swept from left to right it returns to the left-hand side of the screen and waits until the next horizontal synchronization pulse has been received. If you look closely at a monitor, you can see the hundreds of horizontal lines that fill the screen. This pattern is called a *raster.*

The third signal contains the picture *brightness* information. It is used to modulate the intensity of the electron beam, thereby generating the shades of gray that render the picture on the screen. These three signals—the vertical synchronization, the horizontal synchronization, and the brightness signal—can be combined in a special fashion to form a standard composite video signal. This type of composite signal is generated by most video cameras and is accepted by most monitors.

Any good quality monochrome monitor with a composite video input may be used with our system. It should have at least a 10-MHz input bandwidth to display fully the finest details of the computer-generated graphics and the CCTV picture.

CCTV Camera

The closed circuit television camera that transmits the microscope image to the monitor produces the three signal voltages that the monitor requires. These are combined in the form of a composite video signal that contains the horizontal and vertical synchronization pulses and the brightness signal. The optical microscope image is brought to focus on a light-sensitive surface within the camera. An electron beam scans the surface and produces a voltage signal that is proportional to the image brightness at each point within the scan. The way in which the camera's light-sensitive surface is scanned is identical to the way the monitor raster is produced. The scan begins in the upper left-hand corner when the vertical synchronization pulse occurs. The beam sweeps horizontally across the top of the surface, producing a voltage signal that is proportional to the brightness of the image at each point along that line. The beam then returns to the left-hand edge and waits for the next horizontal synchronization pulse to occur. When this pulse occurs the next line in the raster will be scanned. It takes about $\frac{1}{60}$ sec to scan completely all the lines in the raster. The electron scan beam then returns to the upper left-hand corner and waits for the next vertical synchronization pulse to occur before the next "frame" is scanned. The two synchronization signals that control the timing of the picture transmission may be either generated within the camera or supplied to the camera from an outside source. In this system,the synchronization signals to the cameras are derived from the computer video output. Thus, when the composite video signals from the camera and the computer are mixed and displayed simultaneously on the monitor, the two images will overlay one another in a stable fashion. The vertical and horizontal synchronization pulses produced by the camera and the video output section of the computer are synchronized with one another.

Our system employs an MTI-Dage CCTV camera. This is one of the few commercially available cameras with less than 0.5% nonlinearity in both the vertical and horizontal display directions. This means that an object measured in different sectors of the field of view of the camera will measure the same wtihin 0.5%. This is a crucial requirement for any camera used in a video measurement system. CCTV cameras not designed for measurement applications may vary by as much as 15% in horizontal linearity. Obviously, these would not be acceptable.

Microcomputer

We have implemented this measurement system on both the Apple II+ or IIe and the IBM PC or XT microcomputers. The minimum configuration is 48*K* of RAM and two floppy disk drives for the Apple, and 128*K* of RAM and one floppy disk drive for the IBM PC.

The Apple system has an Epson MX-80 dot matrix printer with Graphtrax, a firmware enhancement that enables the printer to reproduce hard copies of whatever appears on the graphics screen. The IBM PC system has a similar printer with similar graphics hard-copy capabilities. It is interfaced to the PC through the parallel port on a multifunction card that also functions as a Clock/Calendar, Serial Port, and Expansion Memory Card.

The Apple Graphics is part of the system architecture and is included on the motherboard. The IBM PC requires the Color Graphics display card. On both systems our software operates the graphics in the high-resolution black and white mode. The Apple graphics system can generate a total of 53,760 monochrome dots and the IBM PC with the IBM Graphics Card a total of 128,000 monochrome dots. The system will also operate on the IBM XT system as well as many of the IBM and Apple compatibles.

From the hardware point of view the computer performs four external tasks.

1. It provides the horizontal and vertical synchronization signals to an interface that, in turn, drives the camera.
2. It controls (through an interface) the mixer, which mixes the microscope video with the computer video.
3. It reads (through an interface) the x–y coordinate pairs that are transmitted from the graphics tablet.
4. It provides computer video signals in the form of computer graphics and text, which overlay with the microscope image on the monitor.

Graphics Tablet

The graphics tablet we have employed is the Houston Hi-Pad. This device consists of a rectangular translucent tablet and a movable hand-held cursor. When the cursor is in close proximity to the surface of the tablet, its position on the tablet is sensed, converted into an x–y coordinate pair, and sent to the computer. The coordinates are transmitted as two 14-bit binary numbers. The tablet can digitize and transmit about 100 coordinate pairs per second. Each coordinate pair is transmitted as a series of five bytes. The first byte transmitted is a control byte. It carries information concerning the status of the tablet. It is easily identified since it is the only byte in the five-byte series whose eighth bit is 1. The next two bytes contain the x coordinate. The eighth or most significant bit of these bytes is always 0. The remaining seven bits of each byte must be combined to give the

14-bit result. In a similar format, the last two bytes in the five-byte series contain the *y* coordinate.

The tablet may be operated in either point mode or stream mode. In *point mode* a coordinate pair is transmitted to the computer only when the button on the cursor is pushed. In *stream mode,* however, coordinate pairs are transmitted continuously. This is the mode used in our system. Our measurement software detects the cursor button being pushed by checking the control byte as described above.

Interface

This is the electronic circuitry that ties the computer to the camera, graphics tablet, and monitor. Specifically, it does the following things:

1. It has a parallel input port that receives the data transmitted from the graphics tablet.
2. It separates the horizontal and vertical synchronization signals from the computer composite video output and sends them to the camera. These signals synchronize the camera's raster scan to the computer's raster scan.
3. It takes the composite video from the camera, mixes it with the composite video from the computer, and sends this mixed signal to the monitor. Since the camera and computer are synchronized, the monitor displays a perfect overlay of computer text and/or graphics with the TV camera picture.

The Apple interface consists of two circuit boards which plug into the expansion slots on the Apple motherboard. The interface to the graphics tablet is supplied by the tablet manufacturer. The CCTV camera controller is our own design.

The IBM PC interface, also of our own design, is constructed on one plug-in circuit board. It incorporates the graphics tablet interface, the CCTV camera controller, and a 1-of-8-level mixer that allows the user to select, under program control, one of eight possible picture/graphic mixing ratios ranging from all picture to all graphics.

System Software

Software Goals

The basic philosophy behind the software design was to make it easy for a person who knew little or nothing about computers to be able to take advantage of the superior speed and accuracy with which a computer can read and analyze data. The programs used by the computer were designed to be user friendly so that the person running the computer has to deal with a minimum of commands.

First of all, the programs make up what is called a *turnkey system.* This means that in order to use the computer, one needs only to insert the floppy diskette containing the software into the main disk drive, and turn the power on. The diskette is configured so that after the computer's operating system is loaded, the computer automatically executes an initialization routine on the diskette that will leave the user in a main menu program (Figure 3). This program displays all the possible options the user can select (followed by an associative letter) and asks the user to select one. No commands need to be listed; nothing needs to be mem-

```
Microruler-- version 5.1
Copyright (C) 1982 1983 1984
by Dan Du Varney
All Rights Reserved

Today's Date: 02-20-1984
Enter next option
<A> Align tablet
<D> Acquire Data
<F> Manipulate files
<S> Statistical Analysis
<H> Draw Histograms
<E> Change system environment
<Q> Quit

Option=?
```

Figure 3. The main program menu as implemented on the IBM PC. Each option, with the exception of QUIT, has its own menu. Some of these option menus are depicted in later figures.

orized except how to insert a floppy diskette properly and how to turn on the machine.

The second important aspect of the programs is that they are all menu driven—the user is never confronted with a star and blinking cursor while the computer waits for him or her to enter a command. At each step in all of the programs, the user has all options clearly labeled and explained; this reduces the amount of memorization by the user to a minimum. The major menus used by the system are shown in Figures 4–6.

Figure 4. The ACQUIRE DATA option menu. Note that the user is reminded that he or she is using calibration (1) and that distance units are in microns. Each menu has the option to return to the main menu.

```
The data disk is C:
The current units are MICRONS (1)


          Enter next option
           0 - Calibrate
           1 - Compute geometries
           2 - Compute density
           3 - Compute volume
           4 - Compute angles
           5 - Point count
           6 - Spatial Frequency
           7 - Open new output file
           8 - Adjust camera intensity
           9 - Return to main menu

          Option=?
```

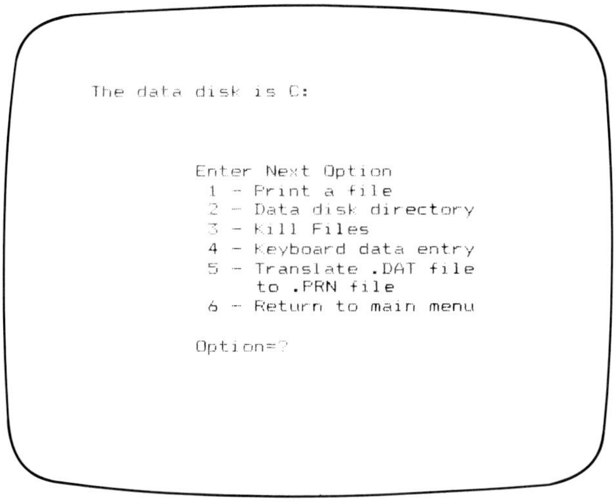

```
The data disk is C:

Enter Next Option
 1 - Print a file
 2 - Data disk directory
 3 - Kill Files
 4 - Keyboard data entry
 5 - Translate .DAT file
     to .PRN file
 6 - Return to main menu

Option=?
```

Figure 5. The MANIPULATE FILES option menu. Option 5 is used to translate our data file format into Lotus 1-2-3-compatible format.

Software Organization

As can be seen from the chart presented in Figure 7, the software is organized around the main menu. This main menu is used to call the other system routines, properly referred to as *overlays.* Our system software has been broken up into a series of relatively independent overlays. Specifically, these are the data acquisition program, the file manipulation program, the histogram program, the statistics program, the environment program, and the tablet alignment program. These overlays are fetched from disk to memory as needed by the main menu. A small section of memory is used to maintain global variables so that data may be passed among overlays as needed.

Figure 6. The STATISTICAL ANALYSIS option menu.

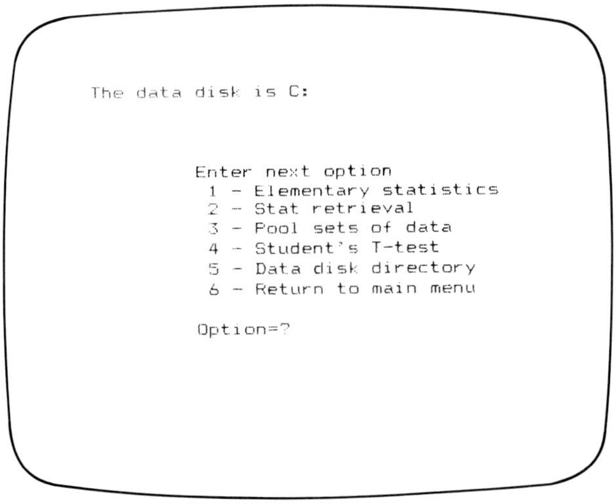

```
The data disk is C:

Enter next option
 1 - Elementary statistics
 2 - Stat retrieval
 3 - Pool sets of data
 4 - Student's T-test
 5 - Data disk directory
 6 - Return to main menu

Option=?
```

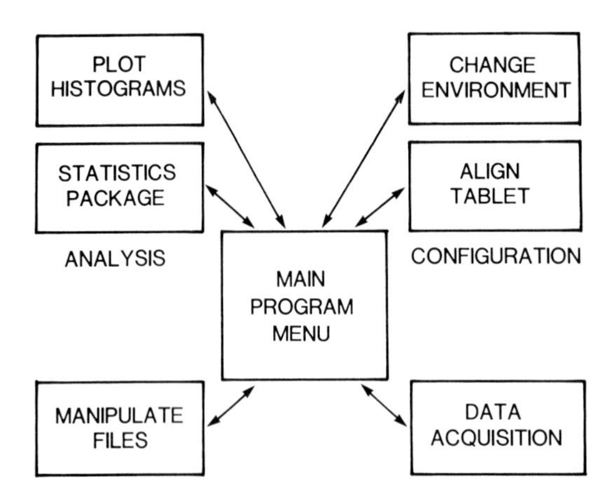

Figure 7. An organizational diagram of the entire software system. Note that access to all parts of the system is through the main menu.

Data Acquisition

The central feature of the software system is the data acquisition program, which one can select from the main menu by typing a "D" followed by a return. The computer's screen clears and, after a moment of disk access, another menu appears (Figure 4) showing all the possible types of data measurement, along with a few other helpful miscellaneous options.

The first option one normally selects is to calibrate the computer's input. The digitizing tablet works by returning the x–y coordinate of the cursor's position on the tablet. The coordinates use an arbitrary scale that ranges from approximately 0 to 20,000 on both axes. It is the computer's job to take this information and convert it into something more meaningful. There are 10 possible calibrations that the user can save and choose from, each consisting of a name and a ratio of real-world units to digitizing tablet units. Thus, for example, to convert a distance measurement to microns the computer would take the distance in tablet coordinates (say 12,084) and multiply it by a scaling factor (perhaps 0.01) to get a real distance (120.84 μm).

Setting a new calibration factor is relatively simple: The operator takes an object of known length, places it under the microscope, and measures its length in tablet coordinates. A light microscope reticle with a 1-mm scale divided into 100 divisions is ideal for this purpose (Graticules, Ltd., Tornbridge, Kent, England). The operator then tells the computer the real length of the object and the unit this length was measured in. The computer calculates a scaling factor and saves this new unit on the disk so that it may be selected from a menu in the future. Once a unit is calibrated, one must not change the magnification of the microscope or the data will be incorrectly scaled. It is wise to make several calibrations of the same unit at different magnifications, and select the correct calibration from the saved list (Microns 50×, Microns 100×, etc.).

The acquisition software is designed to measure distances, areas, perimeters, form factors, angles, volumes, particle densities, and perform particle counts. The first four of these quantities are calculated simultaneously under the general heading "compute geometries." The other calculations are performed separately. The equations used to calculate the distance D, perimeter P, area A, and form factor F are as follow:

$$D = \sum_{i=1}^{n} [(x_i - x_{i+1})^2 + (y_i - y_{i+1})^2]^{1/2}; \tag{1}$$

$$P = D + [(x_n - x_1)^2 + (y_n - y_1)^2]^{1/2}; \tag{2}$$

$$A = \left| \left[\sum_{i=1}^{n} \frac{1}{2} (x_i + x_{i+1})(y_i - y_{i+1}) \right] + \frac{1}{2} (x_n + x_1)(y_n - y_1) \right|; \tag{3}$$

$$F = 4\pi A/P^2; \tag{4}$$

where x_i and y_i are the calibrated coordinate pairs. The distance formula (1) is a sum of the distances between successive coordinate pairs, and is a straightforward application of the Pythagorean theorem. The perimeter (2) is the distance as measured along the traced path plus the length of the segment from the final point (x_n, y_n) to the starting point (x_1, y_1), which closes the figure. The area (3) is the absolute value of the sum of the products of the average x coordinate and the change in the y coordinate between successive coordinate pairs including the closure segment. The sign of the individual terms in the sum depend on the change in the y coordinate. Thus, for any closed figure the net sum over these positive and negative terms gives a value whose magnitude equals the enclosed area. The form factor (4) is proportional to the ratio of the area enclosed by a closed figure to its perimeter squared. This ratio is at maximum for a circle, where the multiplicative constant (4π) sets this maximum value at 1. Any shape other than circular yields a form factor of less than 1.

The measurement algorithms used by the program execute rapidly because the calculations are interleaved with data input. Intermediate calculations occur during the time the operator is tracing the object. Once the user specifies that he or she is through, the computer simply performs some minor calculations to get the perimeter of the closed polygon (created by adding a segment connecting the first and last points), calculates the form factor, scales the data, and prints results to the screen. This method is much more efficient than the more obvious method, which reads data into an array and waits until measurement is completed to begin "number crunching." This latter method uses large amounts of memory, depending on the number of coordinates typically read, and on most machines will produce a noticeable pause before any results are displayed.

The communication between the user and the computer while drawing on the screen is another example of the user-friendly ethic that inspired the program. The leftmost edge of the tablet and the graphics screen are broken up into a vertical row of boxes called a *command strip.* These boxes contain the numbers 1–10 and the letters C, D, P, and E. They serve as an uncomplicated and friendly method for the user to interface with the computer. A picture of the command strip appears in Figure 8.

The numbers stand for files: The user may hit a file number, and the last data measured will be sent to a file corresponding to that number. This allows one to manipulate large numbers of files simultaneously. Previously created files may be opened, or new files created, via an option on the main program menu. Each file specification consists of the name of the file, the quantity that the file contains, and the units in which that quantity is measured. Different files are used for each measurement (perimeter, area, form factor, etc.) and the software rigorously

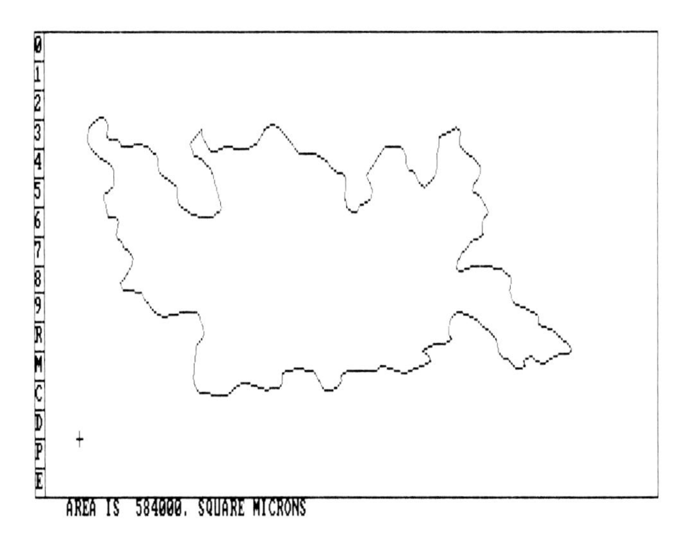

Figure 8. A copy of the graphics screen as it appears after the area of an enclosed figure has been calculated. The command strip appears on the extreme left of the display. Commands to the data acquisition program are issued by placing the cross hairs in the appropriate box and pushing the cursor button.

forces all data in the same file to be measured in the same unit. This prevents the files from containing mixed data types, which is important because most of the statistics generated by the analysis routines would produce absurd results if performed upon different quantities or quantities not measured in the same units.

Data can be stored in files written using the ASCII format or binary code. If stored in floating-point binary, with one series of bits representing the mantissa and another series of bits the exponent, fairly large numbers can be stored efficiently in groups of five bytes. Our data, however, is stored in ASCII format, where each byte represents a single decimal digit, letter, or some special symbol. Although this format takes more memory and is slower because the CPU must convert numbers from their internal format to ASCII, it pays off in the long run because the files created may be edited by standard text editors, and software written in all languages will be able to read the data files. Speed and size are being traded for greater portability.

There are four basic commands available to the user while drawing. They are CONTINUE, DELETE, PRINT, and ERASE. As mentioned previously, the first letter of each of these commands appears in one of the boxes on the screen. To issue a command, one simply positions the cursor over the box containing the appropriate letter and pushes the button. A standard measurement consists of the user drawing an outline around some object to be measured, then hitting "C" to tell the computer that the drawing is complete. The computer then responds with the desired data about the geometry of the object. The user can then hit a "C" to continue, a "P" to get a hard copy of the data, or a number to send it to a file. The software will not allow the user to trace a new outline until at least one CONTINUE is entered; thus one could calculate the area of an object, print the area, send the area to files 1, 4, and 5, and then calculate a new area. "E" and "D" can be hit at any time. "D" tells the computer to delete the last point entered (during point counts or density calculations only), and "E" tells the computer to erase the screen in preparation for another drawing. On the IBM system, function key 1 sends an interrupt and returns the user to the acquire data menu (Figure 4).

Data Analysis

There are several types of analysis from which the user can choose. The data analysis programs were designed to manipulate and analyze previously created files. There is no provision to analyze data as they are read into the computer. This makes sense for several reasons. It allows the analysis and acquisition routines to be separate programs, and any program that can create a file that conforms to the input format of the statistics programs can be used to generate test data. This means that the user could type in data using a text editor and use the statistical package to analyze it. It also means that old files can be reanalyzed at a future date.

The supplied analysis programs consist of a statistics package, a histogram plotter, and a file manipulation program, which allows the user to print data files in an attractive format and enter sample data from the keyboard. Although these programs may not meet the needs of all users, it is relatively simple to find commercial software on the market or write a program to perform some specialized analysis on the data files.

Configuration Software

There are two other programs in the system that are of significance. These programs fall into the category of *configuration software;* that is, they are used to tell the other programs in the system the unique properties about their environment.

The first program is labeled CHANGE SYSTEM ENVIRONMENT on the main menu, and is used to change the hardware input/output configuration of the computer. It maintains a file containing the locations of all disk drives as well as the use of that drive, the location of the printer, and the location of the digitizing tablet interface board. When an "E" is pressed from within the main menu, this program displays all the environmental data and allows the user to make changes. This prevents the computer from being locked into one specific configuration, and makes this program compatible with any hardware arrangement. The configuration data are kept both on disk and in a common block of memory, so that the computer has a permanent record and does not need to reread the disk each time a new program is loaded.

The other configuration program is designed to align the digitizing tablet, and is selected by an "A" from the main menu. The program calculates a coordinate offset and scaling parameter after reading three coordinate pairs from the tablet, specifically, the upper left corner, the lower right corner, and the lower right corner of the command strip. This program is important because it allows the user to change how the tablet is read. One could, for example, reduce the area of the tablet that is mapped to the computer's screen in order to enlarge a small photograph placed upon the tablet. The alignment software also allows for the variances that occur between different tablets. For example, over the same x, y grid some tablets may transmit values between -12 and $+19{,}988$ whereas others would give values between 0 and 20,000. The alignment program ensures that all tablets will be read correctly. Thus the alignment program serves the same basic purpose as the environment program. It lets the user configure the hardware with flexibility.

Conclusion

We have described one of many possible approaches to implementing a microcomputer-controlled video microscope measurement system, and we conclude by reviewing the main features of our system.

We employ special hardware to overlay a CCTV picture of a microscope image with the graphics display generated by the microcomputer. Graphics input is taken from a digitizing tablet. Our particular design emphasizes the use of the microcomputer as a data collection, storage, and analysis tool. The pattern recognition process and the decisions regarding which objects in the field of view will be outlined for measurement or which objects will or will not be counted is left entirely to the judgment of the operator.

The software is a user-friendly turnkey system organized around a main menu. Each option within the main menu selects a submenu that specifies all available options and commands.

The algorithms that are implemented in our measurement software perform general-purpose geometrical measurements. The calculations are interleaved with data taking so that instantaneous results are available at the completion of the data acquisition process.

Results of measurements may be stored in files and/or printed. In the IBM version, the files are in standard ASCII format. The contents of these files may be displayed on the monitor, edited with standard text editors, or used as input to other application programs.

Other system programs perform file-keeping tasks, statistical analysis, data plotting, and any necessary input/output reconfigurations.

The main menu structure linking the measurement, file-keeping, analysis, and configuration subsystems makes the software simple to use, easy to understand, and self-contained.

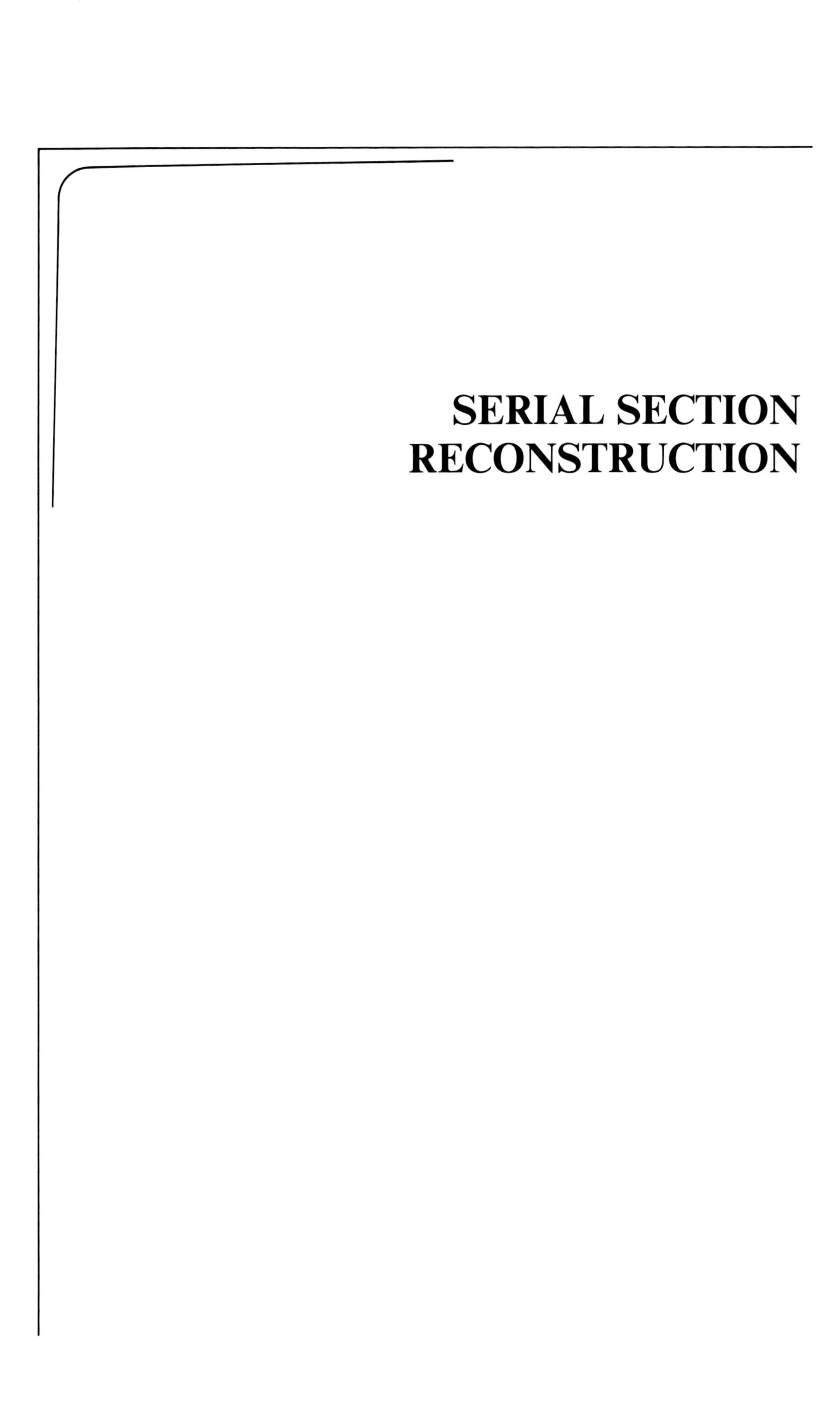

SERIAL SECTION RECONSTRUCTION

11

Principles of Reconstruction and Three-Dimensional Display of Serial Sections Using a Computer

Ellen M. Johnson and J. J. Capowski

Introduction

The Research Problem and Some Early Solutions

Three-dimensional reconstruction helps to solve one of the oldest problems of research in cell and neurobiology. Perhaps the earliest morphological method was dissection. The first biologist could not have been content to study only the outside of an organism. However, once the organism has been cut open to reveal its inside, much of its form and structure have been destroyed. Immediately, the need arises for some means of reconstructing and presenting the information revealed by dissection. The earliest method for accomplishing this was artistic interpretation. Using perspective drawing and shading, the artist could record what the scientist revealed. The anatomical drawings of Leonardo are examples of artistic interpretation where the curiosity of the scientist was combined with the skill of the artist in one person.

The techniques of serial sectioning add some systemization to the essentially destructive process of dissection. In the late 19th-century the development of precise microtomes and the wax embedding of tissue led to quantitative methods of reconstruction. Once it was possible to produce consecutive serial sections relatively free of imperfections such as wrinkles and folds, graphical methods could be used with accuracy. One of the earliest of these graphical methods was developed a century ago by the German anatomist His (1880). This method involves the parallel projection of points from the outlines of features in the sections onto a series of straight lines separated by a distance proportional to the thickness of the sections. The result is a drawing of the original structure viewed perpendicular to the direction of sectioning.

Kastschenko (1886) introduced the more obvious graphical method of superimposing outlines that have been traced from serial sections onto transparent sheets. The separation of the transparent sheets represents depth within the structure. Although this method gives an accurate perspective of depth, these reconstructions can be viewed only in the direction of sectioning. This method is still used with some success in a number of laboratories (Stevens et al., 1980). Additional methods for graphical reconstruction were developed by Odhner (1911) and others that allowed reconstructions produced from superimposed outlines to be pictured from any angle with respect to the plane of sectioning.

Solid models (in clay or wax) of biological structures had been made before the advent of tissue-fixing methods, but a great deal of artistic talent was required to produce accurate freehand models. Once solid reconstruction could be based on unblemished tissue sections, its accuracy was greatly improved. The sectioned tissue components could be represented by sheets, cut out of a material such as wax or wood, whose thickness corresponded to the separation of the sections. The sheets could then be assembled in the proper order to create a solid model. Born, also working in the 1880s, was one of the pioneers in the building of solid models from wax plates (Born, 1883). The model by Špaček and Lieberman (1974) shown in Figure 1 is a beautiful example of some recent work in solid modeling. They have used metal foil separated by pillars to space the sheets by a distance proportional to the separation of the sections. The spaces in the model are filled with putty.

Serial section cinematography is a reconstruction technique that was used as

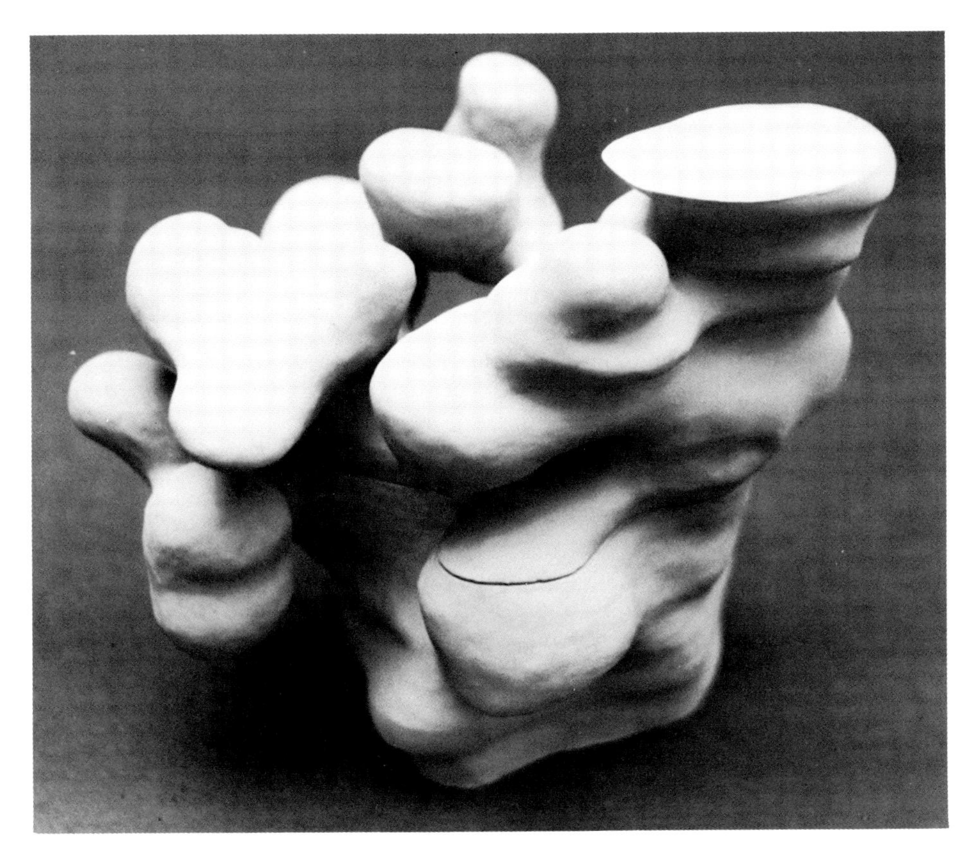

Figure 1. Model of a dendrite and the cluster of excrescences generated at the site of a synaptic glomerulus. [Reprinted from Špaček and Lieberman (1974) with the permission of Lékařské fakulty KU v Hradci Králové.]

early as 1907 by Reicher and is still being developed (Livingston et al., 1976). This method employs time rather than space as the third dimension in the reconstruction process. Each section is recorded in the correct sequence on a frame of a movie film strip. When the film strip is projected, the viewer is given the impression of traveling through the structure from which the sections were taken.

The above is intended as a brief outline of the various methods of reconstruction. Ware and LoPresti (1975) discuss the history of various reconstruction techniques more extensively. Another detailed description of the techniques can be found in Gaunt and Gaunt (1978). Both of these sources append large bibliographies. Gaunt and Gaunt also present stereomicrophotography and holography as techniques for recording and displaying data used in reconstruction.

The Computer as a Research Tool in Reconstruction

The digital computer was introduced as a tool for reconstruction in the late 1960s. It was programmed to assist in the task of manually tracing the outlines of features in serial sections. The outlines were represented in computer memory by

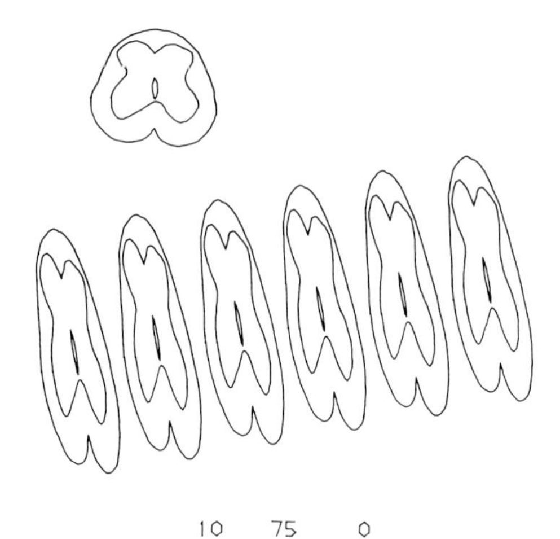

Figure 2. Three-dimensional model of a block of spinal cord tissue. The sections are stacked by depth value. A tracing of an individual section appears at the upper left. [Reprinted from Johnson and Capowski (1983) with the permission of Academic Press, Inc.]

the coordinates of the vertices of polygons approximating them. Once the outlines from each section of a structure were entered into the computer, along with a value representing the depth of the section, the computer could stack the outlines together to form a three-dimensional model (Figure 2). Of course, some sort of device is necessary for the display of such a model and, with a reasonable amount of computer programming, that device should allow the viewing of the model from any direction.

Some of the earlier computer reconstruction systems attempted to be automatic; that is, the computer digitized images of the serial sections and tried to extract data from them with pattern recognition algorithms (Garvey et al., 1972; Wann et al., 1973). This method requires the use of large amounts of computer memory for storing the digitized image and for the pattern-recognition algorithms. However, computer image-processing techniques could not, and still cannot, duplicate the pattern recognition and classification capabilities of the human brain. Primarily for these reasons, the computerized systems developed during the 1970s are semiautomatic: The user traces the outlines with a device such as a data tablet and aligns them under visual control. The computer then automatically displays, rotates, and measures the reconstructed objects. There are several systems of this type, the best known being that of Levinthal and Macagno (Macagno et al., 1979) [see Chapter 12]. Continued improvements in image processing hardware and software, however, are motivating new work on automatic systems (Sobel et al., 1980; Bradbury, 1981).

Most of the early computer systems were used in the study of neurons. Indeed, until recently, computer-aided reconstruction has been a tool used primarily at the microbiological level. However, the development of computerized tomagraphy and nuclear magnetic resonance imaging is rapidly changing this. These techniques produce images of serial sections of the human body that provide input to reconstruction systems. Such systems have become indispensible diagnostic tools. For instance, they allow tumors to be precisely located and measured. Siddon and Kijewski (1982) describe one such system and Cook et al. (1983) discuss three-dimensional display for diagnostic imaging in general.

Designing a Computer System for Three-Dimensional Reconstruction

In this section we consider the design of a semiautomatic reconstruction system for a microcomputer. We discuss the necessary hardware and software components of a system that may be used for the modeling of many sorts of biological structures varying from tiny cells to large tissues. Chapter 4 of this book deals with the reconstruction of neuron tree structures. The reader who is interested in neuron modeling may benefit from reading this chapter to get an overview of system design.

Hardware

System design starts with a microcomputer and a terminal to control its operation. The size of the programs in the system require at least 64 KB of random access memory (RAM). The peripherals that are essential to the reconstruction process are devices for data entry, data storage, and three-dimensional display.

Data Entry. Data entry consists of tracing the outlines of the features to be modeled in each section. As the tracing is done, the x–y coordinates of points along the outline are stored in computer memory. A data tablet is most often used to record points (Moens and Moens, 1981; Perkins and Green, 1982; Street and Mize, 1983; Glaser et al., 1977). Photographs or drawings of sections can be placed on the tablet, or microscope slides containing the sections or transparencies of the sections can be projected onto the tablet. As the data tablet cursor is moved along an outline, the tablet coordinates are sent to the computer for storage. Because the coordinates are in an arbitrary system defined by the manufacturer, calibration is necessary.

An alternative approach is to connect a microscope to the computer. This provides a method for tracing outlines directly from a microscope slide. Use of the microscope eliminates the loss of spatial resolution and contrast that may result from photography or projection. To digitize outlines, the computer must either sense or control the movement of the microscope stage. A photo-sensing device can be used to track the movement of the stage as described by Foote et al. (1980), or the computer can receive signals from potentiometers mounted on the microscope stage control knobs or the stage itself. Photo-sensing or optical shaft encoders offer more resolution than potentiometers but are significantly more expensive. In either case, the operator moves the microscope and instructs the computer when a point is to be stored. If the stage is motorized, the computer can control its movement as directed by the operator through some interactive device. Push buttons may be used, but a joystick allows a smoother and more natural control. The movement of a cursor on a data tablet can also be followed by the computer and used to direct the stage movement. Push buttons and joysticks must be connected to the computer through analog-to-digital converters (ADCs).

Tracing outlines with a microscope is convenient, since no photography and no projection devices are necessary, but the technique is somewhat slower than tracing with a data tablet. Motors designed to move microscope stages normally have a step size of 0.5 μm and sensing devices can detect movements of about 0.7 μm. Both provide more resolution than a data tablet (0.01 in.), which makes the tracing of the details of a structure more accurate. However, tracing long smooth outlines with a microscope is much slower than it would be with a data tablet cursor. Microscope tracing is also more expensive. A data tablet can be purchased for around $1,000, whereas an accurate stepping motor for moving a microscope stage costs $6,000–$8,000. The cost of devices for sensing the movement of the microscope stage can vary from $400 for a pair of potentiometers to several times that much for optical shaft encoders of high accuracy interfaced to a computer.

Data Storage. Data entry is usually not completed in a single session with the computer. In addition, the total number of points that are entered for a reconstruction will usually exceed the capacity of computer memory. Thus, some form of external storage is needed. Floppy disks, magnetic tape, or a hard disk can be used. A hard disk will provide the most versatile and rapid data access, but floppy disks are portable and less expensive.

Display. The purpose of a three-dimensional reconstruction system is to display the two-dimensional components of a structure so that they appear as a three-dimensional object. Thus the choice of a display device is very important. The options fall into two general categories: (1) plotters and graphics printers and (2) cathode ray tubes. Cathode ray tube (CRT) devices can be line-drawing displays or raster displays. In a *line-drawing display* (sometimes called a *vector display*) the image is made up of continuous lines and curves, whereas in a *raster display* it consists of separate picture elements *(pixels).*

To produce a three-dimensional effect, perspective or stereoscopic views of the object may be drawn on either a plotter or a CRT. Hidden lines may be removed before drawing to enhance the effect. Lines that are closer to the viewer can be made darker on a plotter or brighter on a CRT. Such depth cues are adequate, but if the object can be rotated smoothly while it is displayed on a CRT, it can be viewed while it is in motion to create a realistic three-dimensional illusion. The user can interactively manipulate a complicated reconstruction to extract the most visual information about relationships among substructures. A joystick is most often used in combination with push buttons or switches to direct the rotation.

The section on the display of neuronal structure in Chapter 4 describes six options for display hardware. The choices range from a plotter to a line-drawing CRT display with hardware for rotation and translation. The issues discussed there apply also to the design of a more general system. However, unlike the scientist who is studying neuron trees, the user of a more general reconstruction system may wish to model the surfaces of structures. A line-drawing display can only show the outlines of structures. A raster display is needed to provide the texture for realistic surface reconstruction. Raster displays allow one to shade or color each pixel of the surface of a reconstructed object. A color raster display would be ideal for surface modeling, at a cost of $4,000–$30,000, depending on the number of colors and the spatial resolution provided. Such displays are in use in diagnostic reconstruction systems where their cost is small when compared to the cost of the scanning devices that produce the images of the serial sections.

It is possible to develop a system that uses an inexpensive *x–y* plotter as the only display device, but because smooth rotation is such a powerful tool, a dynamic line-drawing system is probably the best choice for a reconstruction system for general use in research in cell and neurobiology. Dynamic raster graphics systems are available but they cost hundreds of thousands of dollars. As discussed in Chapter 4, the cost of a line-drawing system will vary from $4,000 for a line generator with no rotation and translation hardware to $23,000–$60,000 for a system with complete rotation and translation hardware.

Even if a dynamic line-drawing system is available, an *x–y* plotter or a graphics printer is still an essential part of a reconstruction system if hard copies are to be produced. Plotters can be used to enhance the line drawings with color and the filling in of specified areas. Contrasting colors of ink can be used to draw different features of a model to aid in the visualization of its structure.

Software

Software for a computer reconstruction system usually consists of two major programs, one for data entry and storage and one for three-dimensional display. For maximum portability and ease of programming and maintenance, these should be written in a high-level language such as FORTRAN, BASIC, or PASCAL. It is usually necessary to write small subroutines in assembler language to control the digitizing and display devices. In some cases it may be possible to avoid assembler language by using subroutines for examining (peeking into) and putting information into (poking) memory locations. These subroutines allow the program to test and set bits in device control registers and to transmit data to and from device buffer areas. Such subroutines can usually be found in the libraries provided for most high-level languages.

In any case, the details of these interfacing subroutines will depend on the devices chosen. For example, if a data tablet is used for tracing outlines, a subroutine that furnishes the *x* and *y* coordinates from the data tablet output buffer must be available. This subroutine should also return the status of any special function keys or buttons on the digitizer.

Data Entry and Storage. Assuming that there exists a subroutine for reading the data tablet, coordinates from the data tablet can be accepted continuously or point by point. For continuous entry, the computer can be programmed to record the coordinates of a point only when it is a specified distance from the most recent previously recorded point. This distance can be varied to accommodate the sort of outline being traced. If it is jagged or sharply curved, we want the distance from point to point to be shorter than it would be for a smoother outline. Similarly, a time-dependent recording rate can be used. If the operator moves the tablet cursor slowly, as for a complex outline, more points would be recorded. For point-by-point entry a data tablet coordinate is recorded only when the operator presses a button on the tablet cursor. This makes for slower data entry, but it is convenient when the operator wants to enter a particular point on an outline. For the most flexibility, the operator should be given the option of using either point-by-point or continuous entry. Some reconstruction systems (e.g., Moens and Moens, 1981) accept a minimum number of coordinates with point-by-point entry and then use a smoothing algorithm to generate more points.

Before tracing with a data tablet can begin, the tablet must be calibrated. This is usually done by prompting the operator to enter two individual points and the distance, in microns, by which they are separated. A scaling factor for converting from data tablet units to the operator's units can then be computed.

For tracing with a microscope, a subroutine is needed to direct or to sense the movement of the stage. The main program must keep track of the stage movement so that the coordinates of the points that are entered may be stored. If the computer is directing the movement of the stage, then there must be an additional subroutine to detect the movement of a joystick or the depression of push buttons. Such a subroutine reads the output buffer of an analog-to-digital converter. As in the case of tracing with a data tablet, the entry of points can be continuous or discrete.

As an outline is traced, either with a microscope or a data tablet, it should be displayed on a CRT or other graphics device. Thus the operator can detect errors as they are made. The computer can be instructed to delete incorrect points and the tracing can be restarted at the most recently entered correct point.

Each outline that is traced is represented in the computer by the coordinates of points along the outline. One way to store these coordinates in a file is to make each outline a single record in the file. In addition to the x–y coordinates, a z coordinate must be stored to indicate the depth into the structure of the section from which the outline was traced. Although the z coordinate is usually the same for each point along an outline, it is sometimes necessary to represent linear structures passing through several sections. Then it becomes necessary to store a z coordinate with every point.

When an outline is drawn on the display device, there must be some distinction between the first point of the outline and the remaining points. This is so that the plotter pen or the electron beam of the CRT can be directed to move to that point without drawing a line. To accommodate this, each point accepted and stored by the data entry program can be tagged as either a move or a draw point. However, if each file record consists of only one outline, it can be assumed that only the first point in the record is a move point. It may also be desirable to tag and store the coordinates of dot points as part of a reconstruction. These are unconnected points that appear as dots on the display. They may be used to outline or represent certain features in a section.

Often the object to be reconstructed consists of various substructures. The data entry program should provide a means of distinguishing these substructures. This can be done by allowing the operator to enter an identifier to be stored in the file record with the coordinates of the outline of a substructure. This is useful during three-dimensional display when one may wish to select a particular substructure for display. Perkins and Green (1982) have a useful scheme for categorizing substructures.

A file structure must be chosen for the records containing the coordinates of the outlines. A *sequential file structure* allows an unspecified number of variable length records, whereas a *direct access file* requires that the number of records in a file and the maximum record length be specified in advance. Sequential files are difficult to alter and may require sorting by z coordinate before they can be used as input to hidden-line-removal algorithms. A random access linked-list file structure, such as that used by Street and Mize (1983), provides easy insertion and deletion of records as well as the space efficiency of variable length records. Whatever file structure is chosen, the data entry program or a separate program should contain routines for editing records in the file. These routines should provide a facility for displaying the outlines, deleting incorrect or unwanted outlines, and inserting corrected or additional outlines.

An alignment procedure is also essential to the reconstruction process to ensure that the sections are stacked in perfect registration. For proper alignment, the position of at least two points in each section or the orientation of one line and the position of one point must be known. Artificial landmarks such as pinholes can be used, but it is usually easier to use anatomical landmarks. Artificial landmarks have the advantage of being in the same position in every section whereas anatomical landmarks may change position and shape from section to section; nevertheless Foote et al. (1980) have found that the use of artificial landmarks in their work resulted in a section-to-section drift of more than 30 μm. They present a method for obtaining a reference template based on the outline of an anatomical landmark that remains at a constant orientation from section to section and a point whose position changes but may be tracked from section to section. When a data tablet is being used for tracing, a simple alignment method is to place a transparent sheet onto which the landmark has been traced in a fixed position over each section to be traced. Before each section is traced, the landmark on the section is matched with the landmark on the transparent overlay. Using this procedure, the alignment is left solely to the judgment of the operator.

If the computer is to be used in the alignment procedure, the alignment software is usually incorporated into the data entry program. The systems of Street and Mize (1983), Perkins and Green (1982) and Chawla et al. (1982) are examples of systems that employ interactive computer alignment. In these systems each section is rotated and translated to match a reference section. The operator decides when a section is properly aligned with respect to outlines in the reference section.

For some applications it may be desirable to compute the perimeters of the outlines and the areas that they enclose. Once the outlines have been entered into the computer, such computations are easy to incorporate into a data entry program. Once all the outlines representing a structure have been entered, volumes and surface areas may also be computed for the structure.

Three-Dimensional Display. Software for the three-dimensional display and rotation of the reconstruction is an important component of a reconstruction system. It is possible to combine the software used for the input of data from the serial sections with the software used for assembling those data into a three-dimensional model. However, the resulting program most likely will be too large to provide any more than a simple static display of the model. This might be useful for monitoring the progress of the reconstruction process, but a separate program for displaying, manipulating, and measuring the final reconstruction is usually necessary.

The input to the display program consists of the coordinates of the points representing the outlines traced from serial sections. The primary function of the display program is to draw the lines connecting these points onto a two-dimensional plotter or CRT. It is not very useful simply to draw the outlines one after the other to produce a view from an angle perpendicular to the planes of the sections. The user should be able to view the model from any direction. This requires rotation of the points representing the structure about any axis of the three-dimensional coordinate system.

Thus the contents of the display program will depend largely on the requirements of the display device. Many line-drawing CRT display devices have a controller or display processor that directs the movement of the electron beam that draws the image. The more sophisticated versions of these processors are preprogrammed to execute the rotation and translation algorithms discussed below. In other words, these devices contain a combination of hardware and software for rotation and translation. If such a device is not available, these functions must be incorporated in the reconstruction system software. Algorithms for the rotation and translation of points about the axes of a three-dimensional coordinate system involve multiplication of a column vector representing the coordinate triple x, y, z for each point by a matrix representing the rotation and translation transformations desired. For an orthographic projection, the z coordinate is used in the calculation of the transformed x and y coordinates but is not transformed itself. For a perspective projection, the transformed z coordinate is divided into the transformed x and y coordinates. Angell (1981) gives an excellent and concise introduction to the concepts of three-dimensional geometry and presents some transformation algorithms written in FORTRAN.

Ideally, the model should rotate smoothly in response to the user's manipulation of an input device. Whether or not this is possible depends on the display hardware and software, the number of lines in the display, and the speed of the computer. Obviously, the best that can be done on a plotter is to draw a series of static views from different angles. If a plotter is the only device available, drawing stereoscopic pairs may be the most effective display strategy. To produce stereoscopic pairs, a view of the structure is drawn from one viewing angle. Then the structure is rotated by about 7° about the y axis and drawn again. This creates an image for each eye. A stereo viewer can then be used to direct the image on the left to one eye and the image on the right to the other eye to obtain a three-dimensional effect. For examples of the use of stereoscopic pairs see Moens and Moens (1981) and Stevens et al. (1982). Hidden line removal adds to the effectiveness of static views drawn on the plotter, but hidden-line-removal algorithms, discussed below, are difficult to program and debug.

The lines making up the three-dimensional display are the edges of polygons lying in parallel planes. For some applications it may be desirable to draw lines connecting the polygons in one plane to the polygons in the two adjacent planes. This is the first step in representing an object's surface. A number of algorithms have been developed that use various polygons to approximate the surfaces between the outlines that are traced from parallel sections. If a raster display (which allows the intensity of each pixel in an image to be varied) is being used, algorithms for shading the surface representation can be applied. Cook et al. (1983) discuss surface representation and shading techniques that produce realistic models. However, they are difficult to program and take up a considerable amount of computer memory. Furthermore, such models can be smoothly rotated only if they are displayed on a very expensive dynamic raster system. Fortunately, for most applications in cell and neurobiology, models consisting only of the outlines drawn in parallel planes are sufficient. Rather than totally realistic models, what is needed are models that reveal fundamental structural features. As we stated in the section on display hardware, a refreshed line-drawing display is the best choice for accomplishing this goal.

For complex biological structures, hidden line removal is often essential for producing unambiguous displays. Once the structure is rotated to a pleasing and potentially revealing view, the operator instructs the display program to produce a static display with the hidden lines removed. To produce such a display, an algorithm must be applied to each polygon comprising the structure to determine which parts of the polygon are hidden from view. The design of the algorithm is simplified if connections between polygons in adjacent planes are not present: There are fewer hidden surfaces to deal with and the algorithm can operate directly on polygons rather than having to separate into polygons the polyhedra that are produced by connecting the planes.

There are some well-known techniques for determining which parts of a polygon are hidden by polygons in the planes between it and the viewer. First of all the polygons must be accessed in the order of their depth value (the z coordinate). Starting with the polygon in the plane closest to the viewer and the polygon in the next closest plane, a minimax test as described by Sutherland et al. (1974) can be applied to each pair of polygons to determine if they overlap one another.

If the polygons overlap, there must be further processing to determine which points of the polygon in the plane furthest from the viewer are hidden by the polygon in the closer plane. The *surrounder test,* also described in Sutherland et al. (1974), can be applied to determine whether a point of one polygon is hidden by the surface of another one. Once the hidden points are known, it is still necessary to determine which parts of the lines joining those points are hidden. The line joining a visible point to a hidden point must be partially hidden because it will intersect an edge of the closer polygon. If only the visible portion of that line is to be drawn, the coordinates of the intersection point must be calculated.

The calculation and processing of intersections can cause hidden line removal from a complicated structure to consume hours of computer time. If the partially visible portions of lines could be omitted from the display, the calculations of intersections would be unnecessary and the computer time consumed would be reduced to minutes. Fortunately, because the displays are composed of short line segments, the deletion of partially visible segments often produces a display that is quite satisfactory (Figure 3).

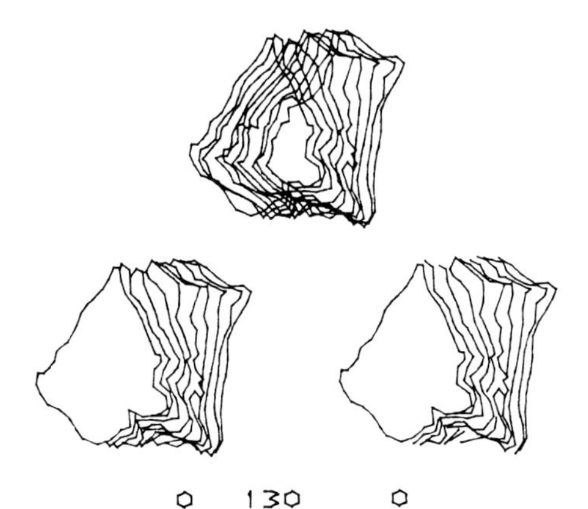

Figure 3. Three-dimensional reconstruction with and without the display of partially hidden lines. The reconstruction before hidden line removal is shown at the top. At the bottom left, hidden lines have been removed and the partially hidden portions of lines have been drawn. At the bottom right, the partially hidden portions have been omitted. The structure being modeled is a portion of a dendrite.

If partially visible lines are included in the display, the problem of a line that intersects the lines of more than one other polygon must be addressed. Here the concept of quantitative invisibility introduced by Appel (1967) is useful.

Research Applications

In our department, a three-dimensional reconstruction system has been used for a number of years. The earliest versions of the system, reported by Capowski (1973) and Johnson and Capowski (1983), were written for a Digital Equipment Corporation PDP 11/45, a moderately expensive minicomputer. It is, however, possible to operate the programs on any computer system with an LSI-11 microprocessor as the central processing unit

The software for the system consists primarily of three programs written in FORTRAN. There are also several small assembler-language subroutines that control the data entry and graphics display devices. There are two data entry programs, one for tracing with a data tablet and the other for tracing with a microscope. The third program is for three-dimensional display.

The Neuroscience Display Processor, Model 2 (NDP2) is used for both on-line display of outlines as they are traced and three-dimensional display of the completed models. The NDP2 is a line-drawing system that can be driven by any PDP-11. It can store up to 4,096 x, y, z coordinate triples to define the end points of the lines to be drawn on a CRT. Because the triples are multiplied by a rotation and translation matrix that can be dynamically updated, the image can be rotated and translated smoothly using a joystick and switches. The NDP2 uses an orthographic projection to map the three-dimensional model onto the two-dimensional display surface. This device is described in detail in Capowski (1978). Its successor, the NDP3, is described in Capowski (1983) and is now commercially available (Eutectic Electronics, Inc., Raleigh, NC).

Hidden line removal is used to enhance the three-dimensional effect provided by smooth rotation and translation. The amount of time needed for producing a static view with hidden lines removed depends on the number and complexity of the polygons in the model. The user can choose to generate a view in which the visible portions of partially hidden lines are not displayed. This shortens processing time and produces views that are quite satisfactory when the model consists of many short lines (see Figure 3).

An example of the use of this reconstruction system is shown in Figure 4. The large outlines represent the boundaries of thalamic regions of a cat. They were traced with a data tablet and assembled into a three-dimensional model, which was rotated and translated. Hidden lines were removed and the resulting image was plotted. The interior outlines and dots, also entered with a data tablet, represent the locations of labeled thalamic neurons. They were displayed in the same orientation as the larger outlines and plotted on the same piece of paper. Thus the distribution and organization of these neurons in relation to the tissue boundaries is revealed.

This reconstruction was used in a study of the projections of the thalamic ventral posterolateral nucleus (VPL) onto cerebral cortical area SI (Spreafico et al., 1983). In Figure 4 the interior outlines represent the boundaries of a "core" zone of neurons and the interior dots represent the boundaries of a surrounding or

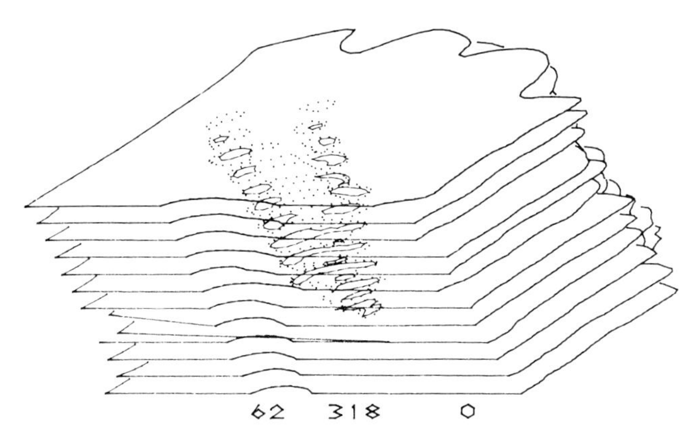

Figure 4. Three-dimensional model of the ventrobasal nucleus of the cat thalamus. [Reprinted from Johnson and Capowski (1983) with the permission of Academic Press, Inc.]

"shell" zone of neurons in VPL. The three-dimensional shape and extent of these zones as well as their relationship to each other would have been very difficult to visualize without computer reconstruction. The surface areas and the volumes enclosed by these structures can easily be calculated by the computer.

There are many examples of how three-dimensional reconstruction can be used to obtain much more information than can be obtained from the study of individual sections. Ware and LoPresti (1975) provide an extensive review of the applications of reconstruction methods. They demonstrate the value of three-dimensional modeling for revealing internal pathways and connections and for allowing the researcher to study spatial relationships among biological structures.

Conclusion

Manual methods of reconstruction are time consuming and tedious. Computer-aided reconstruction systems have relieved much of the tedium and they are certainly less time consuming. The advent of the microcomputer and reasonably priced graphics hardware has made it possible for many researchers in cell and neurobiology to have the power of a computer graphics system at their disposal. Until image processing and pattern recognition hardware and software are developed to the extent that fully automatic systems are feasible, such systems will be interactive. That is, the human operator of the system will make the decisions about what data are to be extracted from the serial sections and the computer will do what it does best—store, manipulate, measure, and display the data.

The authors wish to thank Dr. J. Špaček for providing Figure 1 and Dr. B. L. Whitsel for providing Figure 4. We also wish to thank Dr. E. R. Perl for his support and Dr. C. S. Johnson, Jr. for his comments on the manuscript. The work on the reconstruction systems designed in this department was supported by the National Institutes of Health under grants NS 11132 and NS 14899.

References

Angell, I.O. (1981) A Practical Introduction to Computer Graphics. New York: John Wiley and Sons.

Appel, A. (1967) The notion of quantitative invisibility and the machine rendering of solids. Proc. ACM Natl. Conf.: 387–393.

Born, G. (1883) Die Plattenmodellirmethode. Arch. Mikrosk. Anat. 22:584–599.

Bradbury, S. (1981) Automatic image analyzers and their use in anatomy. Progr. Clin. Biol. Res. 59B:185–192.

Capowski, J.J. (1973) A general purpose three-dimensional modeling program. Comput. Graphics 7 (3):24–28.

Capowski, J.J. (1978) The neuroscience display processor model 2. Proc. DECUS 5:763.

Capowski, J.J. (1983) The neuroscience display processor model 3. Proceedings of the Digital Equipment User's Society, Fall, 1983. 167–170.

Chawla, S.D., L. Glass, S. Freiwald, and J.W. Proctor (1982) An interactive computer graphic system for 3-D stereoscopic reconstruction from serial sections: Analysis of metastatic growth. Comput. Biol. Med. 12:223–232.

Cook, L.T., S.J. Dwyer III, S.L. Batnitzky, and K.R. Lee (1983) A three-dimensional display system for diagnostic imaging applications. IEEE Comput. Graphics Appl. 3 (5):13–19.

Foote, S.L., S.E. Loughlin, P.S. Cohen, F.E. Bloom, and R.B. Livingston (1980) Accurate three-dimensional reconstruction of neuronal distributions in brain: Reconstruction of the rat nucleus locus coeruleus. J. Neurosci. Methods 3:159–173.

Garvey, C., J. Young, W. Simon, and P.D. Coleman (1972) Semiautomatic dendrite tracking and focusing by computer. Anat. Rec. 172:314.

Gaunt, W.A., and P.N. Gaunt (1978) Three-Dimensional Reconstruction in Biology. Baltimore: University Park Press.

Glaser, S., J. Miller, N.G. Xuong, and A. Selverston (1977) Computer reconstruction of invertebrate nerve cells. In R.D. Lindsay (ed.): Computer Analysis of Neuronal Structures. New York: Plenum.

His, W. (1880) Anatomie menschlicher Embryonen. Leipzig: Vogel.

Johnson, E.M., and J.J. Capowski (1983) A system for the three-dimensional reconstruction of biological structures. Comput. Biomed. Res. 16:79–87.

Kastschenko, N. (1886) Methode zur genauen Rekonstruktion kleinerer makroskopischer Gegenstande. Arch. Anat. Physiol. Abt. 388–394.

Livingston, R.B., K.R. Wilson, B. Atkinson, G.L. Tribble III, D.M. Rempel, J.S. MacGregor III, R.E. Mills, T.C. Ege and S.D. Pakan (1976) Quantitative cinemorphology of the human brain displayed by means of mobile computer graphics. Trans. Amer. Neurol. Ass., 101:99–101.

Macagno, E.R., C. Levinthal, and I. Sobel (1979) Three-dimensional computer reconstruction of neurons and neuronal assemblies. Ann. Rev. Biophys. Bioeng. 8:323–351.

Moens, P.B., and T. Moens (1981) Computer measurements and graphics of three-dimensional cellular ultrastructure. J. Ultrastruct. Res. 75:131–141.

Odhner, N. (1911) Eine neue graphische Methode zur Rekonstruktion von Schnittserien in schrager Stellung. Anat. Anz. 39:273–281.

Perkins, W.J., and R.J. Green (1982) Three-dimensional reconstruction of biological sections. J. Biomed. Eng. 4:37–43.

Reicher, K. (1907) Die Kinematographie in der Neurologie. Neurol. Zentrabl. 26:496.

Siddon, R.L., and P. Kijewski (1982) Perspective display of patient external and internal contours. Comput. Biol. Med. 12:217–221.

Sobel, I., C. Levinthal, and E.R. Macagno (1980) Special techniques for the automatic computer reconstruction of neuronal structures. Ann. Rev. Biophys. Bioeng. 9:347–362.

Špaček, J., and A.R. Lieberman (1974) Three-dimensional reconstruction in electron microscopy of central nervous system. Sbornik Věd Praci Hradrec Králové. 17:203–222.

Spreafico, R., B.L. Whitsel, A. Rustioni, and T.M. McKenna (1983) The organization of nucleus ventralis posterolateralis (VPL) of the cat and its relationship to the forelimb representation in cerebral cortical area SI. In G. Macchi, A. Rustioni, and R. Spreafico (eds): Somatosensory Integration in the Thalamus. Amsterdam: Elsevier, pp. 287–307.

Stevens, J.K., T.L. Davis, N. Friedman, and P. Sterling (1980) A systematic approach to reconstructing microcircuitry by electron microscopy of serial sections. Brain Res. Rev. 2:265–293.

Street, C.H., and R.R. Mize (1983) A simple microcomputer-based three-dimensional serial section reconstruction system (MICROS). J. Neurosci. Methods 7:359–375.

Sutherland, I.E., R.F. Sproull, and R.A. Shumacker (1974) A characterization of ten hidden-surface algorithms. Comput. Surv. 6:1–55.

Wann, D.F.,T.A. Woolsey, M.L. Dierker, and W.M. Cowan (1973) An on-line digital computer system for the semiautomatic analysis of Golgi impregnated neurons. IEEE Trans. Biomed. Eng. 20:233–247.

Ware, R.W., and V. LoPresti (1975) Three-dimensional reconstruction from serial sections. Int. Rev. Cytol. 40:325–440.

12

Serial Section Reconstruction Using CARTOS

Noel Kropf, Irwin Sobel, and Cyrus Levinthal

Introduction

The CARTOS (Computer Aided Reconstruction by Tracing of Sections) system provides computational capabilities needed for the study of the three-dimensional structure of organelles, cells, and aggregates of cells. In order to visualize and analyze such structures it is often necessary to section the tissue in which they are embedded. This is especially true when several overlapping or intertwined structures are involved. The CARTOS system allows structural and geometric information to be transferred from serial section images into a microcomputer. It then reconstitutes these data in three dimensions and provides for their display and analysis in a flexible manner. Typically the outlines of objects are collected from many sections, after which they can be displayed and manipulated in a variety of ways. Although designed primarily for analyzing light and electron microscope images of serially sectioned neural tissue, the CARTOS methodology is applicable to three-dimensional reconstruction of arbitrary structures from any series of planar cross-sectional images regardless of whether they are obtained from microscopes or from CT, PET, ultrasonic, or NMR scans. For high-resolution images such as those produced by NMR, techniques similar to those described here seem to be particularly important.

Reconstruction is often used to obtain "qualitative" descriptions of the structures being studied, but it is inherently quantitative in that the coordinates of feature positions are digitized and structures are individually identified in each section. Quantitation, in the sense of delineating cell profiles and stacking them with correct alignment in three dimensions, is necessary even to determine such basic information as synaptic connectivity among a large number of cells. From the resulting three-dimensional numerical model, volumes, surface areas, and distances along a surface can be calculated or measured directly. With the addition of fluid dynamics or electrical transmission models, the reconstruction can help predict flow in a blood vessel or impulse propagation in a nerve fiber. Three-dimensional reconstructions are used regularly by hospitals to plan dose distribution in radiation treatment for cancer, and are being tested for use in reconstructive and plastic surgery (Bloch and Udupa, 1983). There are many other potential applications in clinical medicine, teaching, and biology in general.

Computer-Based Approaches to the Problem

One class of solutions to the problem of visualizing three-dimensional structures sidesteps the reconstruction process completely. In this approach, the original section images (from CT scans, etc.) are digitized in bulk and the resulting three-dimensional array of image densities is manipulated by thresholding, whereby a subset of density values possessed solely by the structures of interest are selected. Array elements (voxels) whose values lie within the threshold set are then displayed. Farrell (1983) illustrates how this technique can be applied with color enhancement of depth to provide excellent views of certain objects on two-dimensional display devices. Fuchs et al. (1982) and others have gone one step further and generate three-dimensional virtual images in space using a vibrating varifocal mirror in front of a standard CRT to change the apparent image plane at high speed. These methods are very storage-intensive when the input images have high resolution. Moreover, neither the surface, volume, nor structure of the objects is calculated.

There are several systems that form the surface of objects defined as a connected set of 3-D volume elements. The algorithms described by Liu, Artzy, Udupa, and Herf (Liu, 1977; Herman and Liu, 1978; Udupa, Srihari and Herman, 1979; Herman, 1979; Artzy, Frieder, and Herman, 1980) have been successfully used for 3-D boundary following in CT scan images. Herman's implementation operates on a voxel array and tracks connected faces that bound the object. The highest resolution currently attainable with CT scanners is about 256 $\times$ 256 per section, with at most 64 sections recorded for a single individual.

Existing computer methods for entering anatomical detail from serial section micrographs can be divided into three broad categories. In the first, a human operator recognizes structural features on a video image, a film, or print and manually traces over them using a stylus whose position is known to the computer. Secondly, manual tracing can be performed over digitized images stored in the computer. Finally the computer can be programmed to recognize features in digitized images "automatically" and record their coordinates.

One distinguishing aspect of CARTOS is that it produces high-resolution reconstructions from high-resolution images on low-cost microcomputers with limited memory. The CARTOS-ACE (CARTOS with Automatic Contour Extraction) system represents a step toward automated feature recognition and effectively reduces the amount of tedious manual labor involved in obtaining the reconstructions, providing the input images meet criteria of reasonably high-contrast cell profiles and low noise levels. For an overview of techniques for computer-assisted reconstruction in neuroanatomy see Macagno et al. (1979) and Sobel et al. (1980). Examples of systems similar to CARTOS can be found in Stevens et al. (1980) and Schlusselberg et al. (1982).

Brief History of CARTOS

Computers have been used in our laboratory for the three-dimensional reconstruction of neuronal material for over 12 years. The first version of CARTOS was written for and implemented on an Adage Graphics Computer (AGT-50, Adage Inc., Billerica, MA) during 1968–1971 (Levinthal and Ware, 1972). A second version on the Adage was much more convenient for users than the first and was used for a variety of scientific studies in the general area of developmental neurobiology. About 1975 small computers with adequate displays became available and CARTOS was implemented on a DEC GT40, which was based on a DEC PDP-11/05 computer (Digital Equip. Corp., Maynard, MA). This system continued to be used for research purposes until it could no longer be maintained, at which point it was replaced by a system based on a DEC LSI-11/02.

One of the LSI-11 based systems was developed and used as a "loaner." All of the CARTOS projects have been supported by the Division of Research Resources of NIH and we have used the loaner so that investigators who are not near our facility in New York can carry out the time-consuming work of tracing data into the computer in their own laboratories. The image combining and studying of the output on high-performance displays are still done at Columbia. The new Codata system programmed by N. Kropf will replace the older loaner by the summer of 1984.

The Reconstruction Process

Tissue Preparation for Reconstruction

The preparation of serial section micrographs for 3-D reconstruction requires careful attention to staining the tissue and particular care in the embedding and cutting of the sections. Even for light microscopy we have found it necessary to use material embedded in plastic to avoid differential distortion from section to section. Many of the investigations in which CARTOS has been used required that more time be spent on developing the methods for fixation and staining than for the actual reconstruction process. This has been especially true for studies of developing neural tissue where penetration of the fixative and visualization of unmyelinated nerve fibers present serious difficulties. In following individual optic fibers within the developing optic nerve, for instance, it is very helpful if the cut is as nearly as possible perpendicular to the nerve. In order to do this it is frequently useful to carry out the reconstruction in two stages: the first with light microscopy to determine the path taken by the nerve, and the second with electron microscopy to visualize the individual fibers in cross section.

Efforts to automate the process of contour extraction described below involve a continual interplay between the investigators preparing the tissue sections for microscopy and those who are developing the image analysis programs. The methods used for staining and microscopy can frequently determine whether a series of micrographs are suitable for work with CARTOS-ACE. It seems clear that the efficacy and usefulness of the automatic procedures will increase significantly in the near future as methods of staining and microscopy improve and a richer set of pattern recognition algorithms is used.

Image Alignment

In order to reconstruct from micrographs, it is important to line them up so that the spatial arrangement of the original tissue is preserved. Large numbers of electron micrographs are needed to record the data on even a small region of tissue. Typically an experiment will require as many as 500–1,000 electron microscope sections to record the anatomical detail in 50–100 microns of tissue.

Until recently, the systems we have developed have used 35-mm film as the recording medium for aligning photographs of serial section micrographs. The aligned film strips have been made on a device we call an *image combiner.* Figure 1 is a photograph of the image combiner apparatus. This device uses a video image of a light or electron micrograph that is viewed on a TV monitor and stored in a video frame buffer (FOR.A FM-60, FOR-A Corporation of America, West Newton, MA). The next micrograph in the series is then mounted in front of the TV camera and viewed on the monitor in rapid alternation with the previously stored image. This is accomplished with a video switch (Panasonic WJ-521, Panasonic Industrial Co., Seacaucus, NJ). When alternated rapidly, the current image, say the nth section, can be aligned manually with that of the $(n - 1)$th section by manual rotation and translation of the negative, which is being viewed live by the camera. After each micrograph is aligned with the previous one in the series, it is photographed on a pin-registered 35-mm instrumentation camera (Automax Industries, Woodland Hills, CA) using a high-resolution copy lens. Thus one obtains a series of aligned photographs, or "movie," on a 35-mm film

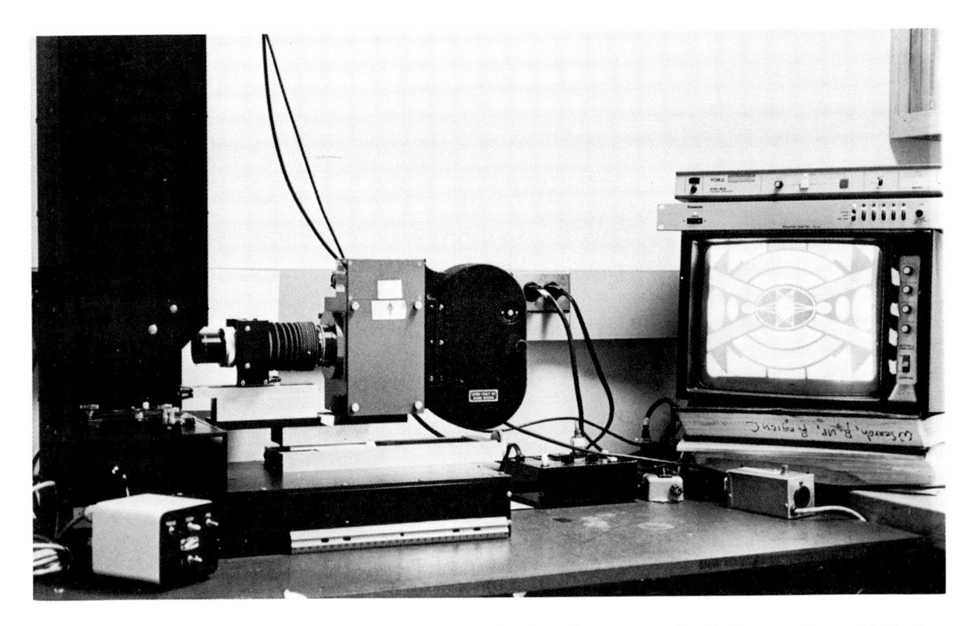

Figure 1. The video image combiner. The black column on the left contains a light box and movable stage, and the video camera above them. Focus, zoom, and iris settings for this camera are controlled by the white box in the left foreground. The column is part of the optical bench that extends underneath the 35-mm pin-registered film camera in the center of the picture. On the right, from top to bottom, are the video frame store, the video sequence switcher, and the TV monitor.

strip. These movies have been the basis for most of our reconstruction work to date.

The decision as to what constitutes good alignment between two sections deserves careful attention. Reference marks or planes are structures embedded in or affixed to the original tissue, which "maintain a constant relationship to, and appear on, every section throughout the entire series" (Gaunt and Gaunt, 1978). Orientation marks, on the other hand, are the cross-sectional profiles of integral structures whose shape is well known (e.g., the outline of a symmetrical ganglion). When two or more distinct fiduciary or orientation marks are present in the series their profiles can be lined up while compensating for any known curvature. Otherwise alignment is achieved when the alternating images produce the least apparent jitter over some part of the video field. Note that unless accounted for or controlled this latter procedure can distort the reconstruction (e.g., a structure that is actually curved could be straightened). Electron micrographs present a special challenge for alignment. To compensate for differential tissue shrinkage or warping due to beam heating or other causes, we find that several areas in a large field of view may require separate alignments. Stevens's system (Stevens et al., 1980) maintains a separate alignment vector for each nerve fiber. At present, we handle this problem using digital images of subfields of the filmstrip that are realigned when they are scanned and digitized. The resulting "microaligned" movie is stored on computer disk. This alignment procedure is similar to the image combining step described above in that digitized images alternate on a video screen (Grinnell GMR270, Grinnell Systems, San Jose, CA). Translation but not rotation is varied under control of the operator. We have found that this compensates adequately for distortions introduced by the beam in an electron microscope.

Some types of biological reconstruction problems can be solved without cutting the tissue into many sections. For example, single neurons can often be stained and the surrounding tissue cleared such that when viewed with suitable optics in a light microscope, their processes can be traced directly in three dimensions. In such cases there is no problem of alignment. In some situations where a small number of sections suffice, aligned photographic prints may be used instead of a filmstrip, or alignment can be postponed until after the tracing is complete.

The Tracing Process

Once the sections of a series are aligned on 35-mm film, we trace the outlines of individual profiles for computer input. The manual tracing system we have been using for many years works well because when the movie is projected at a rapid rate, it is simple to observe continuity through many sections of any single structure, such as a nerve fiber. Using a digitizing tablet, the observer can outline the structure's profile in each section and create a stack of contours to delineate the structure. It is more rapid and easier for the investigator to follow one structure at a time through many sections than to outline many structures in one section before going on to the next. While concentrating on one 3-D structure at a time, the observer can also record the 3-D position of synapses, nuclei, or any other relevant features. The data recorded for an individual structure becomes a *nerve file,* and the connectivity between nerve files can be established by searching for synapses that appear at the same location in two nerve files. In this way "brain files" are created that logically relate one nerve file to another.

Computer Display

Most of the useful information presented to the investigator is in the form of displays of two or more nerve files in three dimensions using either stereo pairs, orthogonal views, hidden line removal, shaded surface display, or dynamically moving display. Interactively controlled smooth rotation of a wire frame model is the ideal method for examining and exploring the large amount of information in a complex reconstruction. For publication, one or more static views are typically sufficient, but animated 16-mm films (Engel, 1981) or videotapes are often preferable.

Computer Hardware and Software

Computers

CARTOS has been implemented in our laboratory in several different ways using diverse hardware. Four separate systems are currently in use for tracing and reconstruction and a fifth system is being built. CARTOS runs on DEC PDP-11 and VAX minicomputers, and on Codata 3300 (Codata Systems Corp., Sunnyvale, CA), ONYX C8002 (ONYX Systems, Cupertino, CA), and Tektronix 4051 (Tektronix Corp., Beavertown, OR) microcomputers. The microprocessors in the

Codata and ONYX systems are Motorola MC68000 and Zilog Z8000 microprocessors, respectively. The Codata system utilizes an IEEE 796 interface bus ("multibus").

The most recent and up-to-date microcomputer-based version of CARTOS runs on the Codata system and, except where noted, that is what we describe below. All of the peripheral input and output devices that we use can be connected to this system using four serial RS-232 (EIA) or RS-422 ports. No parallel ports are required for the basic manual system. We are in the process of implementing the automatic CARTOS-ACE on the Codata; the hardware and software required for this system are described below.

Graphical Displays

Graphical displays are the primary form of output from the CARTOS system. Appropriately, several different types of display devices can be connected and are supported by the software. Line figures (stroke or vector representations—see Figure 2) can be created on pen plotters (IBM XY-750, IBM Instruments Corp.;

Figure 2. Three-dimensional reconstructions of neurons from the visual system of the crustacean *Daphnia magna.* Seven neurons that branch in one-half of the laminar neuropil of the bilaterally symmetric optic ganglion; dorsal and medial views. One inch equals approximately 10 microns. Reconstructed using CARTOS from serial section electron micrographs, displayed in views 90° apart on Evans & Sutherland PS2. A, anterior; P, posterior; D, dorsal; V, ventral; L, lateral; M, medial. [S. Sims and E. Macagno.]

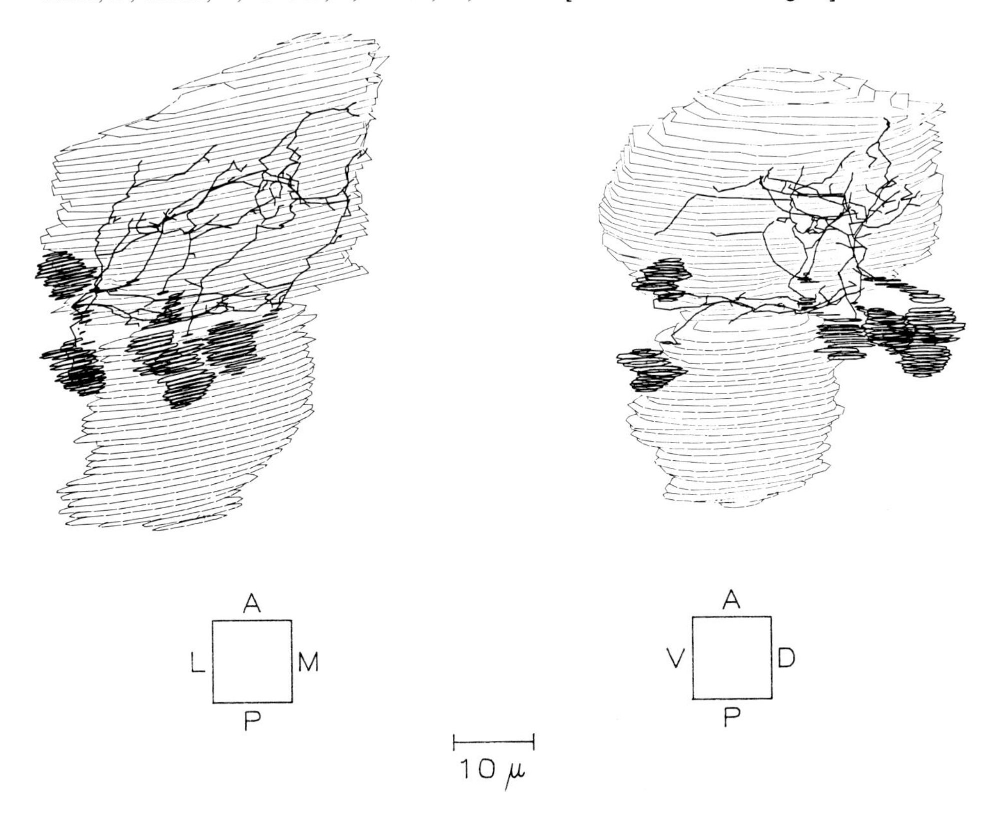

HP7470a, Hewlett-Packard, San Diego, CA), dot-addressable printer/plotters (Versatec 1200A, Versatec Inc., Santa Clara, CA), or on Tektronix 4014 and similar CRT displays. For color line output we are currently using a Grinnell 270 color terminal on the VAX system and a Chromatics CT4300 eight-color display terminal (Chromatics, Tucker, GA) with the Codata. The CT4300 has a resolution of 1024 × 1024 pixels (picture elements or points), and therefore produces fewer jagged lines than most 512 × 512 raster displays. One of the most effective ways for conveying complex 3-D structures using a 2-D display medium is through the use of "motion cues." We use an Evans and Sutherland PS2 (Evans & Sutherland, Salt Lake City, UT) controlled by custom-interfaced potentiometers and a DEC PDP-11/34 host computer to generate changing views of a wire frame model at a rate of 30 views per second, sufficient to give the viewer a perception of smooth motion. The animated models can then be recorded on 16-mm movie film using a standard camera (Arriflex Corp. of America, Woodside, NY) triggered by a computer-generated signal.

CARTOS-ACE Hardware

Our current implementation of CARTOS-ACE runs on a DEC VAX-11/780 computer, and uses a custom-built laser flying spot film scanner (built by Altman Associates, Stamford, CT) called the ANT, to obtain high-resolution (1800 × 1200 pixels × 10 bits/pixel) digitized information from 35-mm filmstrips. The film is mounted in a pin-registered transport (Vanguard Instruments, Melville, NY) under computer control. The ANT deflects a 20-μm diameter laser beam across the film frame with two separate systems: front-surface mirrors mounted on galvanometers for large-scale positioning and acoustooptical deflectors for performing a 32 × 32 raster scan at high speed. Two optical detectors measure the ratio of transmitted to incident laser light through a 20-μm area of film. The computer can sample density from any 32 × 32 pixel "tile" within the "current" film frame in about 60 msec. A frame buffer (Grinnell 270 with Mitsubishi color monitor, Grinnell Systems, San Jose, CA) displays 512 × 512 portions of the digitized images in black and white or pseudocolor with color graphic overlays. Viewing parts of images on the monitor actually reveals more detail than is obtained by back-projecting the 35-mm image onto a ground glass screen. High resolution combined with its random access nature make the ANT particularly well-suited for use with CARTOS-ACE. Unfortunately the digitized images obtained from it are not always uniform because of residual uncompensated temporal and spatial "shading" in the laser optics.

We are in the process of integrating a video frame buffer to provide the Codata system with image acquisition, digitization, storage, and display capability. This frame buffer (IP-512, Imaging Technology, Woburn, MA) provides 512 × 480 pixels × 8 bits/pixel for digitizing and display of input images. It can also be used to display shaded surface representations in 256 levels of gray with a standard black and white TV monitor. With optional color capability, the IP-512 consists of two to five circuit cards that plug directly into the multibus motherboard. We plan to add computer-controlled zoom and pan of a video camera to allow digitizing portions of images to produce a mosaic with higher spatial resolution.

On the present Codata system, long-term data storage is provided by a 20-MB Winchester disk. A 1-MB floppy disk drive provides data archiving and transport capability. This will be expanded as needed to store digitized images.

Peripherals and Lab Equipment

For manual tracing from 35-mm filmstrips we use a Vanguard pin-registered film transport and projection stand to back-project the serial section images onto a ground glass screen. The computer controls the movement of the filmstrip forward and backwards, and it can read out the frame number to ascertain which section is being projected and traced from. The only custom made hardware in the Codata tracing system is the interface between the film transport and a serial RS232 port, which was designed and built in our laboratory.

Several types of *x–y* digitizers are used for manual tracing. The tracing can be done from prints, digitized image displays, projected images, and "image mixing" devices such as a light microscope with drawing tube. We use a spark pen digitizer (SAC GP-8, Science Accessories Corp., Southport, CT) and an optical *x–y* digitizing frame (Numonics 1220, Numonics Corp., North Wales, PA) to enter data points on the back-projected screen for 35-mm movies. Figure 3 is a photograph of the Codata-based manual tracing apparatus. We are using these same positional digitizing devices to interact with the image display in the Codata-based CARTOS-ACE system.

We have also built a CARTOS work station with direct coupling to a light microscope using a drawing tube attachment. This arrangement permits tracing while looking through the eyepiece, and works very well for reconstructing dye-

Figure 3. Photograph of manual CARTOS tracing station. From right to left are the Codata microcomputer system, the color display terminal, and the projection stand with ground glass screen and digitizing frame. The film transport is atop the projection stand. Interface electronics can be seen on the top shelf.

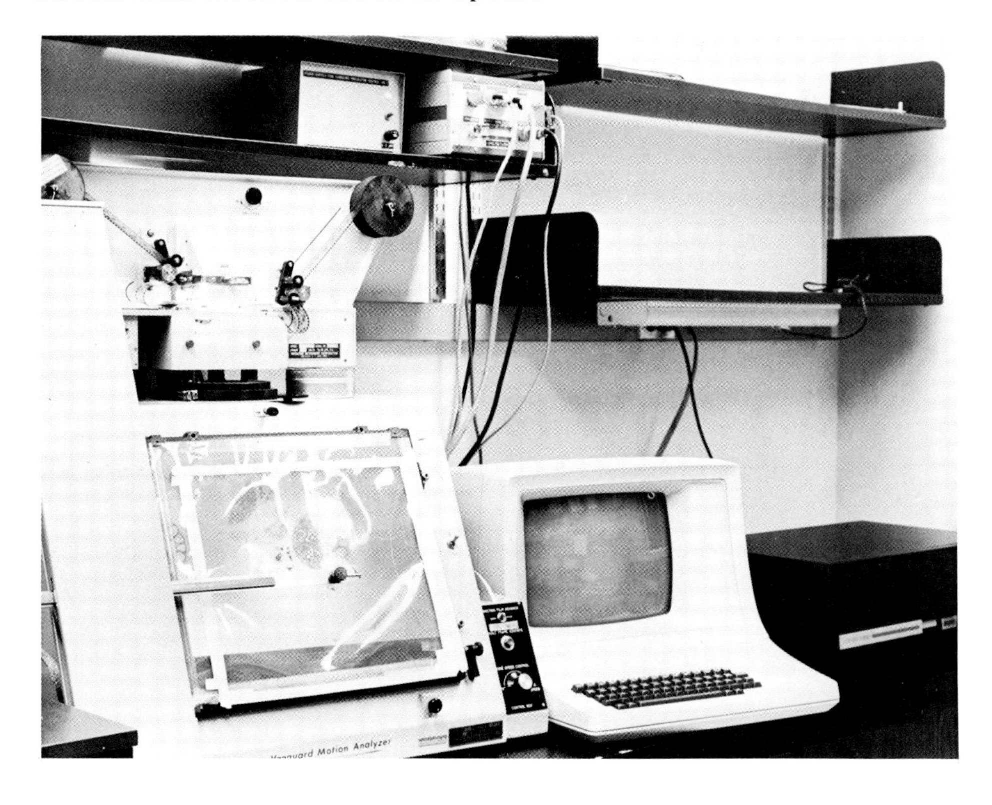

injected or Golgi-stained cells or counting cells in thick slabs of cleared tissue. In this system analog-to-digital converters provide the computer with the microscope stage position and the position of the fine focus knob. Custom-built circuits interface a shaft encoder (fine focus) and linear potentiometers (detecting stage motions) to the computer. This system presently runs on a Tektronix 4051 microcomputer using an IEEE 488 bus. We plan to construct a serial interface for a microscope-coupled system and add it to the Codata system.

The Software Environment

The programs for CARTOS are written in the C language and run under several different versions of the UNIX* operating system. The largest program in the manual system requires about 100 KB of main (RAM) memory. We believe that manual CARTOS could be transported with few program modifications to virtually any microcomputer with adequate memory that runs UNIX. The automatic system requires about 1 MB of main memory.

UNIX and C were chosen as the development base in part because they provide a complete set of application and system development tools, and have proven outstanding for enhancing programmer productivity. We also felt that UNIX and C improved the opportunities for collaboration with other laboratories in the image processing and computer graphics research communities. Equally important, properly written C programs can be easily transported from one computer to another. Flexible data structure definitions in C allow data in files to be stored conveniently and accessed in variable length records. Other qualities of C programs include compact and efficient code, good readability, and easy maintainability.

UNIX provides the C programmer with a powerful set of standard library functions for controlling input and output functions. For example the CARTOS input software uses a supplied function called IOCTL to ascertain how many characters are available to be read at any given time from the digitizing tablet. Graphics capabilities are not built into the C language. CARTOS makes use of several programs and subroutine libraries that were developed in our laboratory, and in one case a commercial program. Graphics output from CARTOS is handled by the following software:

GPIC: a system of 3-D interactive vector graphics display functions similar to the GSPC Core System (GSPC, 1979);

SHADE: a realistic surface rendering program based on Whitted's Raster Test Bed (Whitted and Weimer, 1981);

HL: hidden line removal for "stacked pancake" style display of contours. This program is a commercial product (Graphic Information Systems Technology, New York, NY).

The right-hand column in Figure 4 shows how these graphics modules are connected to the CARTOS software. GPIC (or some GORE-like graphics package) is normally needed to operate a minimal CARTOS system.

---

*UNIX is a trademark of AT&T Bell Laboratories.

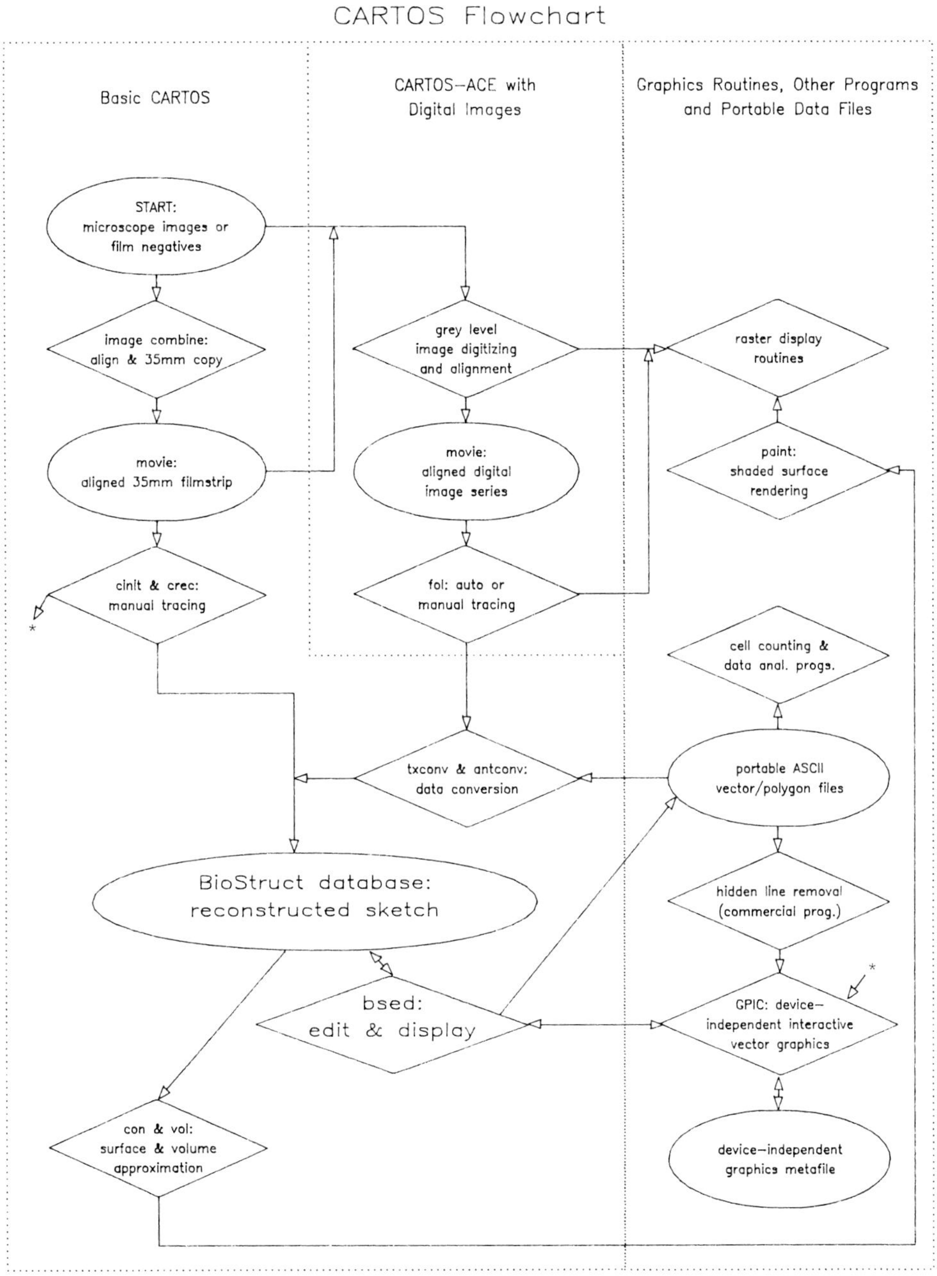

CARTOS Flowchart

Basic CARTOS

CARTOS-ACE with Digital Images

Graphics Routines, Other Programs and Portable Data Files

START: microscope images or film negatives

image combine: align & 35mm copy

movie: aligned 35mm filmstrip

cinit & crec: manual tracing

grey level image digitizing and alignment

movie: aligned digital image series

fol: auto or manual tracing

raster display routines

paint: shaded surface rendering

cell counting & data anal. progs.

txconv & antconv: data conversion

portable ASCII vector/polygon files

BioStruct database: reconstructed sketch

hidden line removal (commercial prog.)

bsed: edit & display

GPIC: device-independent interactive vector graphics

con & vol: surface & volume approximation

device-independent graphics metafile

Figure 4. Flowchart illustrating the overall processing of information in CARTOS. Procedures and programs are shown as diamonds, and stored information inside of ovals. The two left-hand columns represent the essential aspects of the basic manual tracing system, the semiautomatic CARTOS-ACE system, and the BioStruct reconstruction database management system. The right-hand column depicts programs, subroutines, and files that may be used in CARTOS but are also useful in other contexts.

No assembly language routines are built into CARTOS, although some may be desirable in the future to optimize performance of the image display and digitization procedures used in ACE. Various UNIX library routines are used, which are similar though not identical in all UNIX systems.

CARTOS Programs and Data Structures

A Model of the Reconstruction Process

A series of cross-sectional images is the primary source of data. Each image is a two-dimensional distribution of light intensity (or film density) information. In a digital computer each image can be represented as a 2-D array of discrete intensity samples, called pixels.

Each image may represent a thick or a thin slice, and the distance between adjacent slices may vary. In other words the sections (x–y planes) may represent a sparse sampling of the original specimen in the z dimension. For reconstruction, sections must be close enough together to enable a human expert to identify related features from one image to the next.

The main purpose of reconstruction is to reduce the large information content in the input images to a more compact three-dimensional "sketch" embodying the important qualitative and quantitative aspects of structures under study while maintaining their geometrical and topological relationships with precision. Exactly what information is needed in such a sketch depends on the particular field of study. Our laboratory has worked primarily on neuroanatomical problems, and this orientation is reflected in the procedures we use. In all cases irrelevant background clutter is removed and the sketch invariably improves the visibility of structures of interest and permits further analysis to proceed.

CARTOS Methods

Figure 4 illustrates the procedures (diamonds) and forms of information storage (ovals) used in CARTOS. This diagram shows how the tasks of reconstruction and data analysis are divided and provides a starting point for describing the programs and data structures used. The most obvious and the simplest method of obtaining a reconstructed sketch is to have a human operator trace features from the sections and to record the coordinates in the computer memory. This method is used in the manual CARTOS data collection procedure. Alternatively, if digitized images are available, programmed "automatic" techniques (including image thresholding, 2-D boundary following, and heuristics using global 3-D information) can be applied to the images to produce parts of the sketch. In this case the human operator is still needed to guide the program and verify or correct all questionable results. When the "automatic" algorithms fail to produce acceptable results, the operator can trace directly from the digital image display.

In either case we assume that the tracing is done section by section, and is concentrated in the region around the "current structure of interest." The automatic tracing program (FOL) takes advantage of the section-by-section restriction by operating on each digital 2-D image only once for each profile. This is in contrast to the automated 3-D surface-following method of Herman and Liu (1978), which operates on a 3-D array of volume densities.

The recorded features are stacked in the third dimension according to section number and stored in a database called BioStruct on the computer disk. The operator then uses the BioStruct editor program BSED to display one or more of the reconstructed units ("nerves") as a "wire frame" seen from an arbitrary viewpoint and to correct any data entry errors or remaining misalignments in the sketch. BSED can display and manipulate reconstructions from several entities (movie databases) at once, potentially with different magnifications and traced on different data entry systems. In such cases all displays are guaranteed to be to scale. BSED is an interactive command-driven program and it accepts a uniform syntax for specifying which parts of a reconstruction each command applies to (e.g., "all T-cells," "every third contour," "only synapses in sections 2–20").

The graphical output from BSED is routed through GPIC, a set of device-independent subroutines and programs similar to the GSPC Core System (GSPC, 1979). Depending on the capabilities of the selected display device, objects may be differentiated visually by color, line texture (dotted, dashed, bold), or other attributes. When used with a 3-D graphics work station (such as the Evans & Sutherland PS2) the display can be zoomed, rotated, or translated dynamically (in real time) under the control of dials or a joystick. Furthermore, once a satisfactory view is obtained, the corresponding transformation matrix can be read back into BSED and used to plot that view on a hard-copy plotter or to modify the coordinates in the database permanently.

BSED can also output parts of reconstructions as a stream of ASCII characters that can be read by almost any computer program. Such ASCII files contain the coordinates of points and vectors, as well as text labels, but they do not retain much of the associated structural and historical information that is stored in the BioStruct database. Either millimeters or BioStruct "digitizer resolution" coordinate units can be used. Each line in the file defines a point in 3-D space, and consists of four or five items separated by commas. The first item is the digit 1 if the point is an end point of a line segment initiating on the preceding point (DRAW command). Otherwise the first item is the digit 0 (MOVE command). The next three items are the x, y, and z coordinates of the point. The last item, which may not always be present, is a textual label to be associated with the point.

If more sophisticated displays or data analyses are required, several options are available. Assuming that contour outlines have been traced, programs that approximate the volume and surface (VOL and CON) operate directly on the BioStruct contours. Triangulation is used to create a surface model (Fuchs et al., 1977). These programs currently work only on simple (spheroid or tubelike) structures. Hidden line removal (see Figure 2) can be performed on ASCII files containing contours. A program for eliminating duplicate counts of cells and for producing statistics of cell size distribution operates on ASCII files made by CARTOS using center-of-mass plus radius to represent each cell body. Other special-purpose data analysis programs can be written easily to make use of the ASCII representation of the reconstruction.

Basic CARTOS Tracing Programs

Let us look at the actual operation of the programs needed to manually reconstruct objects from aligned images with CARTOS. The CINIT program is run once the first time tracing is made from a movie. In a conversational dialogue,

CINIT requests from the operator and records in the database master file all information necessary to establish a new BioStruct database. First, an *entity name* must be assigned to identify the reconstruction(s) from this filmstrip with some meaningful name. The operator must measure and type in the *scale,* defined as magnification from original tissue to the image projected on the tracing surface. The coordinate units used for reconstruction are the finest resolution units of the tracing device. Once the size of the tracing unit is measured, the scale defines the physical size of the reconstructed objects. The last parameter that must be entered is the distance between sections (not always the same as section thickness). If sectional spacing is nonuniform, some average value must be chosen and corrections to the reconstructed z coordinates made later.

At the next stage the CREC program is used to obtain the vector "sketch" representing the structure(s) of interest. Each structure is called a *unit,* and can represent a single neuron, a cluster of cells (ganglion, glomerulus, or packet), an entire organ, and so on. When CREC is started, the operator selects which entity and then which specific unit is to be traced. He or she then advances the film (or otherwise selects an appropriate image) and begins to enter data points by moving a cursor or stylus on the image. These points are placed and joined together with lines so as to delineate cell profiles, mark synapses or nuclei, or form a "skeleton" or stick-figure representation of a dendritic field. The three types of structural representation that can be traced are made up of the following features:

contour—a set of connected line segments within the section plane;

tree—a set of connected line segments with topological (binary branching) information and an optional "fiber thickness" at each point;

mark—a point in space that may be associated with a textual "label," or a scalar quantity such as the radius of a sphere centered on that point.

When a tree is being traced, the position is recorded at each point that a fiber divides in two. The operator then follows one of the subtrees until it is completed at which time the program automatically returns the film to the previously saved position and prepares to reconstruct the unfinished subtree.

In practice, branched structures are often traced manually as stacks of surface contours that merge together. When this is done, the topological branching structure is visually obvious when the contours are displayed, but is only implicit in the information stored by CARTOS. The alternative, tracing a tree, stores a skeleton, stick-figure sketch. In such a tree, the cell surface may be approximated by a stack of circles with appropriate radii. The operator can trace a tree using previously traced contours as input data. This simplifies recognition of where the center of the fiber is.

CREC can display the reconstruction as it proceeds in vector (wire-frame) form on a graphic screen next to the tracing surface. Exactly what is shown on this display can be controlled by the operator. Normally the structure being worked on is depicted in a projection in the x–y plane (as if viewed looking along the z axis). The operator can specify any angle of view, and any previously traced structure can be called up on the screen as well. A *depth clip window* can be enabled so the display shows only elements within the two or three sections on either side of the current section. These and various other displays help the operator to keep track of the particular structure being traced and to monitor the state of the overall reconstruction. The end result is a completed BioStruct database containing 3-D vectors and points visually delineating or quantitatively defining each of the features and structures of interest.

The BioStruct Sketch Database

Although different parts of CARTOS use different data structures to store different information about the sketch, there is one central representation in which all reconstruction data is stored, which is called *BioStruct*. It serves as a common format for reconstructions entered from different data input *(tracing)* systems, data obtained from outside CARTOS completely, data from different experiments, and for movies with different magnifications. Usually any required editing (e.g., additional alignment) of a completed reconstruction is done using the BioStruct representation. Data analysis and display are performed from this database or representations derived from it.

The data format used in BioStruct databases makes them compact and is sufficiently flexible to permit new types of information to be added. Coordinates are stored as integers, with as much precision as the digitization process provides. Use of floating-point numbers would provide unwarranted precision and would typically increase the memory space used by a factor of 2. Scale information that relates coordinates in the database with real-world distances is recorded separately. One of the design goals was to develop a data structure in which new types of information could be added to a database without interfering with the operation of existing programs. Conversely, programs with new functions should work with old databases.

Since we use the UNIX operating system, CARTOS utilizes the features of the UNIX hierarchical file system. At the top level, users can access both group and private *directories* into which files are placed. One or more BioStruct databases can be located in any one directory. Each database consists of several individual files. Conceptually each database corresponds to a single three-dimensional physical specimen or *entity* (for example, a specific rat brain). Ordinarily there is one movie (series) of images per entity. Each entity consists of a number of units, each of which usually represents a single structure or object such as a nerve or an organ. The unit corresponds to the *nerve file,* which was described earlier and is made up of structural features. The features that CARTOS supports in the sketch database are as follow:

1. *Stack of contours:* Contours are made up of a variable number of connected points in 3-D space. The x and y coordinates at sampled points around the contour are either traced by hand or extracted automatically by the computer. The z coordinates are calculated from the section number and the spacing between sections. The section number is also stored for each contour. Contours are usually placed to delineate the surface boundary of a structure. However, they are not required to be closed polygons and any useful arrangement of line segments can be stored.
2. *Stick-figure tree (skeleton):* A tree consists of an arbitrary number of branches, each of which in turn is made up of some number of connected points. The three space coordinates plus section number are stored for each tree point. In addition, each point may be assigned a "thickness" to indicate the approxi-

mate or average radius of the fiber as it crosses the plane of the section. No more than one tree is permitted in each unit file.

3. *Label points with associated text string:* The coordinates, section number, and a two-character string are stored. Two or more label points can be combined to obtain a text string longer than two characters. Labels are used to mark synapses and other pointlike features, as well as for annotating larger structures.
4. *Spheres:* Each sphere is defined by its center coordinates, section number, and radius. Spheres are useful for indicating the location and approximate size of cells or their nuclei, when the exact shape is unimportant.

BioStruct File Structure and Processing Modules

The layout of information in BioStruct databases is shown in Figure 5. There is a uniform organization (Figure 5a) common to all BioStruct files, within which different types of information are packaged in *blocks.* At the beginning of every file a special code is stored to identify it as a properly constructed BioStruct file. Following the ID code, a series of data blocks is found. Each block begins with a header indicating how long the block is and what type of information it contains (the *type code*). The data contained in the body of the block varies according to the type code.

Figure 5b shows the organization of the BioStruct master file. Each database has only one master file, which contains bookkeeping information necessary to preserve the integrity of the entire database. Currently two separate blocks store this information: a MASTER_HISTORY block and a MASTER_FILES block. In the future, additional blocks can be added with other information without impairing the ability of existing programs to extract the information in these two existing blocks. The MASTER_FILES information functions as a kind of directory of the individual units within the database. It establishes a correspondence between meaningful user-assigned unit names (e.g., OPTIC_NERVE) and file names. More importantly, if a unit file is ever accidentally deleted from the computer disk, this loss will be apparent when the MASTER_FILES block is checked.

Each unit file (Figure 5c) contains a HISTORY block and at least one CONTOUR, TREE, or MARK block. If present, the CONTOUR block simply consists of a series of contour records. The length of the record depends on the number of points in that contour, and the information recorded is just that described previously for contour features: length, section number, and point coordinates. The TREE block contains a series of branch records. Each branch consists of a set of connected points (with optional thickness), plus pointers to left and right "daughter" branches. In the case of a terminal "twig," the daughter pointers are null. The MARK block contains fixed length records, each of which may be either a label or a sphere, as described above.

The CARTOS programs are modular: Each closely related set of functions is performed by a separate program, and the programs communicate by way of files on disk. Also, function routines and subroutines for manipulating BioStruct databases are collected together and then selected as needed from a BioStruct library.

There is a module in this library called the BLOCK I/O module that handles tasks associated with reading from and writing to files at this level of the layout. This set of subroutines is used by all programs that access BioStruct databases,

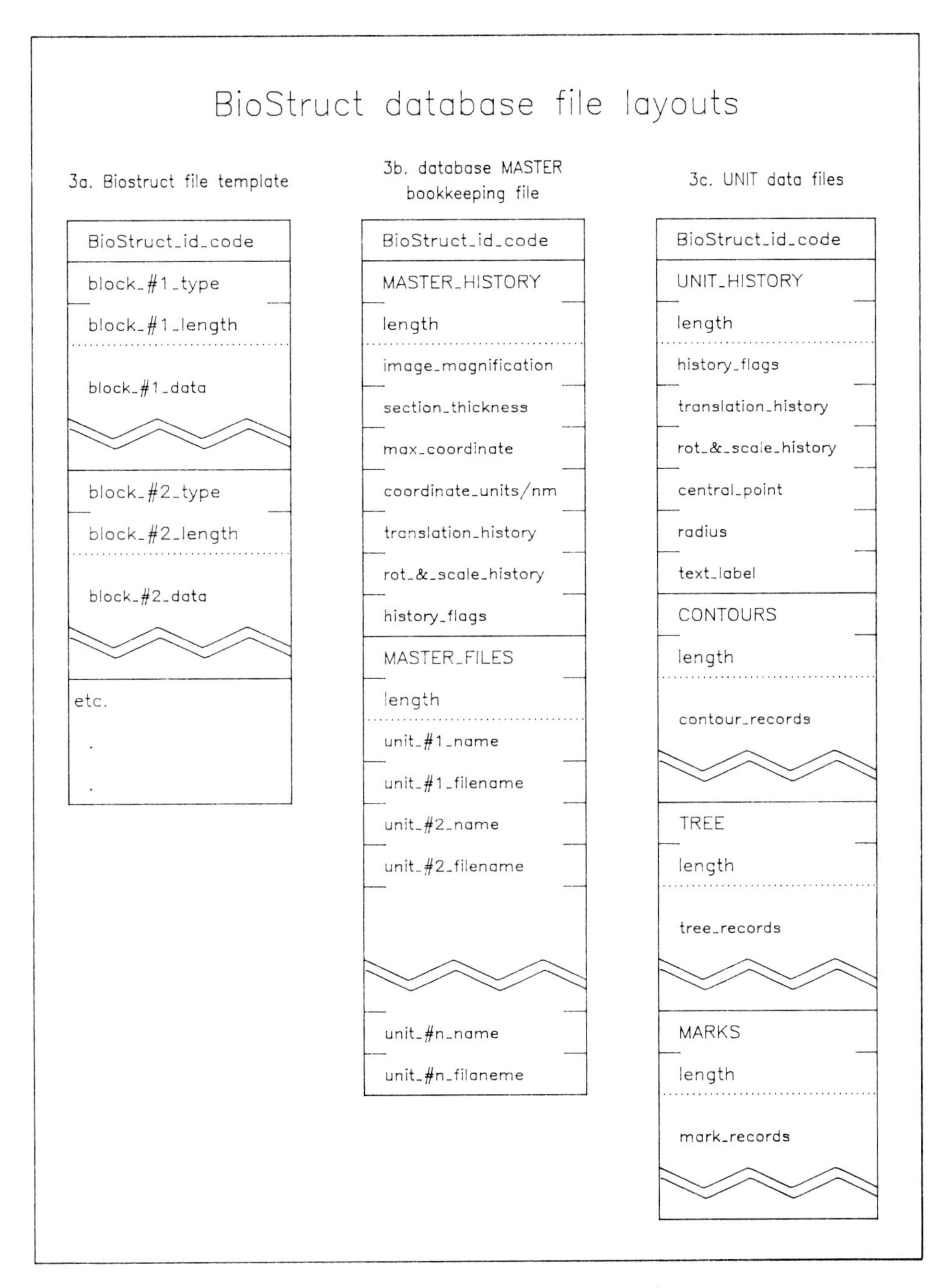

Figure 5. Layout of BioStruct files. The template that defines the structure of all BioStruct files is on the left. Two specific types of file are illustrated: the MASTER file containing bookkeeping information for the database, and the UNIT file, which contains a representation of features in a single nerve or other fundamental structure. Other file types not shown contain alignment and interconnection information.

and it is completely independent of the type and structure of information stored in the blocks. For example, the subroutine BREAD reads an entire block into a buffer supplied by the calling program. BREAD checks to make sure that the block will fit in the buffer space, requests the appropriate amount of data from the file, and verifies that the data was transferred successfully. The calling program can then process the data if it has been programmed to handle that type of data; otherwise it can ignore the entire block.

The UNITSPEC (unit specification) module functions as a layer on top of BLOCK I/O. Given a set of unit names (with "wildcard" characters for pattern matching) it calls on BLOCK I/O to read the MASTER_FILES block into memory, then searches in it for any matching unit names present in the database. For each match, it opens the appropriate unit file and performs some function on the data in that file.

The layer underneath UNITSPEC is called OBJSPEC (object specification). This module operates on, and depends on the layout of, the data records inside CONTOUR, TREE, and MARK blocks. Once one of these blocks has been read into a memory buffer by UNITSPEC, OBJSPEC implements another level of selectiveness. For each data record in the buffer, OBJSPEC routines check whether that feature meets specified criteria relating to section number or sequence number. For example, OBJSPEC can be made to select every third contour and ignore all other data.

The layer that actually does something with objects is supplied by the specific application program. There is no library module for calculating the area enclosed by a contour, for example. Instead, a pointer to the subroutine or function that performs this task is passed to the library routines as a formal parameter. This method of calling a general-purpose data I/O routine and supplying a data processing routine as a parameter is used at several levels, and we have found it to be quite general and powerful.

CARTOS-ACE Overview

The time it takes a human operator to trace a complex nerve cell by hand is long. Using manual CARTOS might require as much as a person-year to reconstruct a large network of extensively branched and interacting nerves. The tracing work is boring, visually strenuous, and tiring. In order to cut down on the amount of human operator time spent on such painstaking tasks, we have developed an extension to CARTOS called ACE (Automatic Contour Extraction). CARTOS-ACE is based on an initial design and implementation by Dr. Irwin Sobel. The main program that implements CARTOS-ACE is called "the boundary follower" or simply FOL. This program permits the computer to "recognize" directly the boundaries of cell profiles and other structures in digitized images of sections, and to exploit the 3-D continuity of the structures in this task. We do not expect this program to be completely automatic. Rather, it is interactive and operator guided. The conceptual model that motivates the design of FOL and the algorithms and data structures used in its implementation are the subject of this section. The way in which we define the problem has had a profound effect on the design of CARTOS-ACE. FOL currently runs on a VAX superminicomputer, but we have the required hardware and are in the process of transporting the program to the Codata microcomputer system.

Our aim was to automate the boundary tracing process without sacrificing confidence in the correctness of the extracted boundary. These are inherently conflicting goals, and we attempt to err on the side of confidence. The important aspects in the design of FOL are as follows: As much boundary recognition as possible is performed by the program but any "difficult decisions" are deferred or presented to the operator for resolution. These interactions are postponed as long as possibile so that the maximum amount of information is available when making decisions, and so that unattended operation can continue for as long as possible. If and when an operator decision is required, the program provides a clear display of available options and useful contextual information. Color graphics are intermixed with display of the input images so as to optimize the operator interaction with the program. We also wish to retain the option of manual tracing for situations where the automatic follower fails.

Image Acquisition Requirements

As with the manual CARTOS system, a series of two-dimensional images of sections provide the input data for CARTOS-ACE, and for the computer to operate directly on an image it must be digitized. In this process the intensity of light (or the density of some medium) is measured at regular intervals across the image. For each sample point (pixel), the light intensity is converted to a number, and this *gray-value* or *gray-level* is stored. The result is a grid or array of values, each of which corresponds to the brightness of the image in a small square. The size of each pixel relative to the total image area defines the spatial resolution of the scanning and digitizing hardware. The signal-to-noise ratio and number of binary digits sampled at each pixel determine the number of distinguishable gray levels. The requirements for these and several other parameters (such as speed) must be determined before the hardware is acquired.

We have found that sufficient resolution of gray levels can be obtained using 256 levels, i.e., one byte of information. It is important that these gray-level numbers be appropriately distributed over the contrast range of the specific images being analyzed. A logarithmic scale may be the best way to achieve this. For the time being, we adjust the range of the digitizer for each movie to obtain acceptable results.

Image Storage and Management

One very challenging problem in neuroanatomy is to reconstruct nerves in detail over substantial distances from serial section electron micrographs. This provides an extreme example of the large amount of storage required for digitized imagery. Assume that a 70-mm film negative can accurately record the electron-generated image. The resolution of such a negative is typically 100 or more line pairs per millimeter. If we wished to retain this resolution, the digitizer would have to sample approximately 1400 × 1400 pixels. Even at one byte per pixel this amounts to 196 million bytes (196 MB) per image! However, after copying to 35-mm film, at most 50 line pairs per millimeter, can be resolved. This corresponds to approximately 3600 × 2400 pixels (rectangular 35-mm format). In our work we have compromised further and accept digitized images of 1800 × 1200 pixels, which requires only 2 MB per image. This permits us to store up to 150 digital section images on a large disk drive. Many applications only require resolution on the order of 512 × 512 pixels. In this case many more sections can be stored in the "image library" on disk.

Once the required image resolution is chosen, one must still devise an efficient method for the computer to access the relevant parts of these images. Memory management as it applies to serial section images is clearly an important aspect of any system such as CARTOS-ACE. Many microcomputers limit program size to far less than 2 MB, so this is not just a question of efficiency, but of feasibility. Moreover, the time it takes to transfer an entire image between peripheral devices and main memory can be prohibitive.

Rather than restrict the maximum image resolution so that a single section could be accessed quickly and retained in main memory in its entirety, we have chosen to break each section image into a mosaic of overlapping tiles. The overlap is two pixels so that a 3 × 3 neighborhood around any pixel is completely contained in some tile. The size and layout of these tiles can be altered by changing small modules within FOL. Optimal tile size for a given type of reconstruction can be ascertained by simulation. Currently we use 32 × 32 pixel tiles, which is the size of the laser scanner's raster, which tesselate (divide up) the image in a brickwork pattern.

The laser scanner can be operated to provide the densities from film directly to FOL. Alternatively, the aligned series can be scanned, microaligned, and stored on disk in bulk. In either case, FOL maintains in memory a cache of the tiles most recently used, while the remainder of tiles within the "current" section can be quickly accessed as needed.

Coherence or locality of references by our algorithm to the image library is exploited to minimize disk accesses. Currently FOL only operates on one section image at a time as it is extracting a boundary for a specific structure in that section. Furthermore, the inner loop only examines a 3 × 3 neighborhood of pixels to calculate a single boundary point.

For efficiency, the image library is organized into three parts: the movie directory, the tile directory, and the tiles themselves. The movie directory contains the names assigned to one or more image series (movies), each of which represents serial sections from a single block of tissue. For each movie, it also indicates which sections are digitized and where each section starts in the tile directory. The tile directory contains one entry for each section, with a 2-D array giving an absolute *tile number* for each tile within that section. This tile number allows single-seek access to the data for that tile in a separate file that contains all of the tiles.

Image Analysis Requirements

Once digitized images are available to the computer, we define the problem in two parts. Boundary following within an image is the low-level part. This requires the computer to identify correctly portions of a closed curve bounding the profile of a cell (or other structure) in the cross-sectional image. The high-level task is to track the structure from section to section, recording where branches occur and guiding and controlling the boundary-following procedure.

This orientation is quite different from most other approaches to automatic analysis of section images. Typically the problem is defined in terms of identifying or measuring features or blobs in a single section without reference to 3-D information from previous to subsequent sections.

The choices of which low-level image features FOL will search for, and what method it uses to find them, are fundamental to the design of both the hardware and software for CARTOS-ACE. The obvious feature that must be recognized is the boundary between a structure and the surrounding tissue. The boundary contour should be a simple (non-self-intersecting) closed curve with maximal resolution. For simplicity we consider a boundary contour to be any ordered, connected set of pixels lying on or adjacent to the visually correct boundary. Normally FOL follows the pixels just inside the cell membrane or extracellular space by tracking the *isodensity contour* for some specific density value called the *threshold.* Isodensity contours in density images are equivalent to height contours in a topographic map. We designed the low-level follow algorithm to be very fast, primarily because it is the "inner loop" of the entire system. This has been achieved in FOL by

1. using isodensity contour following as opposed to more complex image analysis techniques;
2. treating large digitized images as mosaics made up of tiles;
3. requiring that only those tiles needed by the follower be kept in main memory; and
4. using a random-access scanner for on-line digitizing of tiles from film.

The higher-level control is where most of the decisions about the interpretation of images are made in CARTOS-ACE. Two of the main control functions are three-dimensional tracking of the structure (where to expect a profile) and recognition and rejection of incorrect boundary contours. The control level also adjusts the threshold level automatically when this is necessary. This level also "decides" when special heuristics should be invoked to perform tasks such as gap filling and noisy contour filtering, which make use of more complex image analysis techniques.

To illustrate, consider the problem of when to reject as incorrect a contour extracted by the lower level. Assume that at least one boundary contour is known to be correct. Various quantifiable shape features (center of mass, area, perimeter, width, height, minimum and maximum x and y coordinates, and other features derived from these such as circularity and "smoothness") are calculated from the known contour(s). These quantities are extrapolated to obtain the "predicted" features of the contour in the subsequent section. If, in subsequent sections, some such feature of the contour proposed by the extractor differs too much from the expected value, the control level rejects the contour. It may then adjust certain parameters and call the extractor for a reevaluation. Alternatively it might call for human operator intervention, invoke a special heuristic procedure, or simply skip over the section.

Image Processing Functions

We have incorporated into CARTOS-ACE capabilities to process digitized images by filtering and convolution before they are used by the boundary follower. In limited experiments that we have performed to date, we have not found any such analysis procedures that significantly improve the efficacy of the follower. However, we suspect that edge enhancement operations possible with image processing techniques may eventually prove useful in this regard.

The Basic FOL Algorithm

The lowest-level isodensity contour following algorithm is very simple and fast. Some of the subtleties and details of its optimization are described by Sobel (1978). The operator must begin by choosing a threshold gray level that will distinguish the inside of the structure's profiles from its bounding membrane or from the extracellular space. On our system, the operator initially varies that threshold by moving the cursor of the digitizing tablet left and right, causing the program to move the boundary between two pseudocolored regions. In other words, the Grinnell color table is modified in real time so that areas lighter than the trial threshold appear as shades of one color, while areas darker than the trial threshold are displayed in a contrasting color. Conceptually this creates a binary image whose pixels are all either "white" (above threshold) or "black" (below threshold). Assume that the structure of interest is white and its boundary black (the algorithm works equally well if the reverse is true). Given a pixel (the search point) that lies within the structure of interest, the algorithm (illustrated in Figure 6.) performs the following steps:

1. FINDTHRESH (look left for a boundary crossing): Examine successive pixels to the left of the search point until the threshold is crossed. If no such transition is found within a reasonable distance, the procedure fails. Otherwise the white pixel just before the transition is considered to lie on the boundary. Store this pixel's coordinates as the starting point for the contour. This procedure corresponds with pixels labeled 1–11 in Figure 6. The first pixel on the boundary (start point) is number 10.

2. FOLLOW (track the isodensity contour until it closes on itself): Set the environment variable *entrydir* to indicate that a black pixel is located to the left of this point. Then repeatedly perform the ADJ operation (see step 2a), which locates the next neighboring white pixel that lies on the isodensity contour. The *sense* of this search (clockwise or counterclockwise) is set by the higher-level algo-

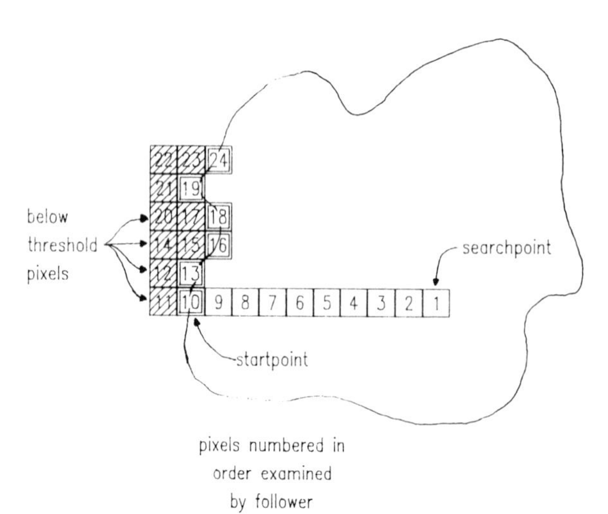

Figure 6. FOL boundary search. The algorithm used for identifying and tracking an isodensity contour boundary is illustrated. The boxes represent pixels (picture elements) that are examined in the first stages of this particular example. See text for explanation.

rithms and communicated via a global program variable (the external environment parameter *sense*). The process terminates when the contour closes on itself, or when a predetermined maximum perimeter length is exceeded (environment parameter *leash length*).

2a. ADJ (find the next neighbor pixel on the isodensity contour): Given a "current" boundary pixel (white), its predecessor boundary pixel (also white), and the rotational sense of spiral searching, examine neighboring pixels in order until a white one is found. This pixel is returned as the next boundary point. To traverse the black border of a white object clockwise as illustrated in Figure 6, neighbor pixels are examined clockwise, starting just beyond a known black pixel. As soon as a white pixel is found, it is taken as the next interior boundary point. Note that following a black object from the "outside" will produce a counterclockwise tracing of its white boundary pixels. The first time ADJ is called, the known black neighbor (number 11) was previously identified by FINDTHRESH (step 1) in order to locate the start point. ADJ examines pixels numbered 12 and 13, and 13 is identified as the next boundary point. Subsequent invocations of ADJ examines pixels 14, 15, and 16; 17 and 18; 19; and 20–24. Each time neighbor pixels to the last known boundary pixel are examined, circling clockwise, and starting just beyond the last known black pixel. Note that the full gray-scale image is available throughout these procedures. The concept of a binary image is used to simplify the description and the implementation. FOL can still calculate quantities such as the local density gradient perpendicular to the boundary, and other image processing can be performed on any part of the image.

The higher-level algorithms that control the isodensity follower are rather complex and will not be discussed in detail. It is worth noting, however, that small black spots in the image caused by dirt specks, "noise," or small intracellular structures will often be encountered by the follower. If the contour returned by the lower level is very short, this is presumed to be the case; the search point is moved just beyond (to the left of) the offending particle, and the procedure is started over from step 1.

A New Control Strategy for CARTOS-ACE

We have developed two new algorithms to handle cases in which no single isodensity contour exists that gives a satisfactory match to the observed or to the previous boundary. These situations arise when there are gaps in the image of a cell membrane, when cell bodies or fibers are in very close contact, or when dirt or cell organelles produce an apparent invagination of the boundary.

In the first method, the boundary contour from a previous section, which is assumed to be correct, is stored in an image array. A *search band* is generated around this contour, indicating the area inside which the current cell boundary is expected to lie. The isodensity follower is then applied to the gray-scale image of the current section, checking whether each point on the isodensity contour is inside or outside of the search band. The points at which the isodensity follower leaves the search band are then investigated to determine whether gaps can be closed by deleting the region outside the search bands.

The second method, which has proven more successful, makes use of a strategy developed by Dr. Peter Selfridge of AT&T Bell Laboratories, who has been helping us with the problem. The original procedure developed by Selfridge operates on images where the interior and exterior of the object under study are uniformly different colors. It requires as input several points that are known to be on the boundary. A description of this original algorithm follows (Selfridge and Prewitt, 1981).

Divide and Conquer Algorithm. Given an initial, incomplete set of boundary points, the boundary is refined by repeatedly applying the following steps to interpose a new point between each pair of adjacent boundary points, until a satisfactory number of points have been obtained:

1. The midpoint of the line segment joining the two adjacent points is taken as the starting point for a boundary search.
2. Choose a direction to search by observing if the image density at this point indicates "inside" or "outside" of the object.
3. Search for a limited distance (determined by the distance between adjacent points) for a boundary in the image.
4. If none is found, use the midpoint from step 1 as a tentative boundary point.

Our modified procedure starts by looking at the image density data moving outward in eight directions from an extrapolated center of mass that is based on previous contours. Eight initial boundary points are selected based on density or gradient calculations. Then the isodensity follower is run from each initial point. If the follower closes to the next point, a correct isodensity boundary is assumed to exist in that segment. If not, a modified version of the Divide and Conquer algorithm is used to refine the boundary *in that segment only.* In step 2, we use the information inherent in the center of mass and eight start points to determine which direction from the midpoint is "in" and which is "out." The search for boundary point begins some distance "in" from the midpoint and continues "outward" somewhat further beyond the midpoint (since the contour is usually convex).

Many special-case tests are used to control the strategy. For many types of light and electron micrograph, the combination of rapid isodensity following supplemented by search bands and modified Divide and Conquer works well. So far it has been applied to electron micrographs of the neuropil of invertebrates and we are exploring its use with other types of neural tissue so that further modification through algorithms can be implemented as needed.

The entire system, including the automatic nerve tracing, is now being implemented on the Codata Motorola 68000-based microcomputer. Since all of the programming to date has been done on a DEC VAX-11/780 running UNIX, it is relatively straightforward to transfer the programs to the microcomputer, which also runs UNIX. A video image digitizing system will replace the laser scanner on the microprocessor. The entire system has been operational since September 1984.

FOL Data Structures

FOL extracts profile boundaries while tracking an object that may be branching in three dimensions, so it stores both contours and a tree for each structure. Within FOL, contours are represented exactly to the resolution of the input images. This is implied by the definition of boundary contours as a connected set of pixels. Thus a single contour could easily contain 400 points, and hundreds of contours contain a very large number. In order to minimize the amount of memory needed to hold this data, contours are represented by a starting point (x, y, and z coordinates) plus a series of *direction codes* or *chain codes* indicating in which of eight possible directions the boundary proceeds. At a later stage contours are recursively smoothed so its direction changes by at most 45° at any point. The result is differentially encoded and stored on disk with only 2 bits required per point. This type of smoothing reduces the effects of spatial quantization noise and removes artifacts such as scratches on the negative. Associated with each contour is a set of measurements used in various ways by FOL. The center of mass is calculated and used to decide where to search for the corresponding contour in the next section. The total contour length (perimeter), area of the enclosed region, area to length ratio, and so on are also calculated and stored as each contour is recorded.

At any given stage of reconstructing with FOL, there may be a stack of "unfinished branches" to which to return. Each such branch point is located in a section in which the structure divided. The algorithm tracks one subtree first and must eventually return to reconstruct the remaining subtree.

Since FOL stores reconstructions in data structures specially tailored for its particular needs, a separate program (ANTCONV) was written whose sole function is to translate the files created by FOL into a BioStruct format database.

Research Applications

A serious question that arises when evaluating systems of the kind described here is whether substantive scientific insights can be obtained with them that could not be obtained without their use. When reconstructions were first produced of the developing optic system of *Daphnia magna,* it became clear that the answer to this question was affirmative. The level at which developmental noise produces differences in the structures of adult nerves in isogenic animals could be determined (Macagno et al., 1973). The observation was made that target neuroblasts wrap around incoming optic axons and that gap junctions are formed in this interaction (LoPresti et al., 1973). This observation would have been extremely difficult to make without the reconstruction since the cell bodies involved are many microns apart when the wrapping takes place. Geometrical and topological information had to be obtained and integrated from many serial electron micrographs before the conclusions could be reached. In addition, results obtained from a reconstruction at one level of detail provide guides that are then used in deciding where in the organism further serial section studies are needed at higher magnification.

The functional implications of these close structural interactions during development have been investigated by E. Macagno and his collaborators in further studies with *Daphnia* (Macagno, 1978). Françoise and Cyrus Levinthal and their collaborators have studied the interactions of neurons in the developing optic system of the small tropical fish *Brachydanio rerio* (the Zebrafish) (Levinthal and Levinthal, 1983). The importance of the interaction of the growth cones on the tips of growing fibers with the fibers of more mature neighboring neurons became clear as the result of these studies (Bodick and Levinthal, 1980).

Reconstructions performed on leeches have facilitated the counting of cells and characterization of cell body distribution in that animal's segmental CNS. These results in turn led to further experiments using microlesion and biochemical techniques with 3-D reconstructions of developing and mature neurons. By perturbing the environment of developing neurons and observing their response over the course of development cellular interactions have been determined and their functional significance evaluated.

Numerous other investigations have been successfully carried out with CARTOS on diverse biological systems including monkey cortex, shrimp muscle fibers, moth antennal sensory apparatus, and bacterial ultrastructure (Nierzwicki-Bauer et al., 1983).

Conclusions

Brief Summary of System

CARTOS has proven its worth over many years of use, for micro- and macroscale neuroanatomical investigations and for reconstruction work in a number of different systems. Noncomputer methods for reconstruction (such as tracing on acetate sheets) become unmanageable when the number of sections grows large, and they do not directly provide numerical data. Stereology has its place as a statistical measurement tool, but fails to provide answers to structural questions. There are a number of computer-based systems that have been implemented for reconstruction, some using methods similar to CARTOS, some designed for problems that CARTOS does not address. For a broad class of reconstruction problems where one starts out with many high-resolution serial sections, CARTOS is a very effective and fairly inexpensive tool.

Availability of Hardware and Software

The basic CARTOS computer system described, based on a Codata microcomputer and complete with a color graphics display terminal and x–y digitizing device (but without the image alignment and 35-mm film transport equipment), can be purchased for about $16,000. This system would permit tracing and reconstruction from aligned micrograph prints.

The cost of upgrading the basic system to handle aligned 35-mm filmstrips is substantial. The Vanguard pin-registered film transport with projection screen is available for about $7,000. The custom interface to the film transport can be built in any good electronics shop from standard parts for about $300. Design drawings are available from us for this interface. The video components of our image combiner cost about $4,500, the Automax camera alone is about $7,000, and a light box, moveable stage, copy lens, and optical bench are also required.

Adding the capability for digitizing video images adds about $13,000 to the cost of the basic system ($2,000 for a good video camera and $11,000 for the Imaging Technologies frame buffer boards). For some applications digitizing hardware may alleviate the need for 35-mm filmstrips. It may, however, require modifying the Codata to strengthen the power supply and to provide more backplane slots. The basic CARTOS programs including GPIC are available for research and educational purposes (as is, and without support) from our laboratory.

Many individuals have played important roles in the development of CARTOS, but we want to acknowledge the special contributions of Eduardo Macagno, Christos Tountas, and Irwin Sobel, all of whom have contributed greatly to the systems described here. C. Tountas wrote the reconstruction programs for the Adage AGT50 and the Tektronix 4051-based systems. R. Bornholdt wrote the GT40 reconstruction programs and much of the GPIC graphics package. N. Kropf designed and programmed BSED, the BioStruct database, reconstruction software for the two loaner systems, and the Evans & Sutherland PS2. I. Sobel designed and programmed much of CARTOS-ACE; subsequently, C. Kim, P. Selfridge, and N. Kropf have made significant contributions to its development. L. Vernooy and I. Sobel wrote CON; J. Jacobs wrote VOL. Robert Schehr and Steven Stern provided helpful editorial assistance with the manuscript. This work has been funded by a grant from the National Institutes of Health (PHS P41RR00442).

References

Artzy, E., Frieder, G., and Herman, G.T. (1983) The theory, design, implementation, and evaluation of a three-dimensional surface detection algorithm. Tech. report MIPG43, Dept. of Radiology, Univ. of PA. Med. Ctr., Phila, PA.

Bloch, P., and J.K. Udupa (1983) Application of computerized tomography to radiation therapy surgical planning. Proc. IEEE 71:351–355.

Bodick, N., and C. Levinthal (1980) Growing optic nerve fibers follow neighbors during embryogenesis. Proc. Natl. Acad. Sci. USA 77:4374–4378.

Engel, L. (1981) CARTOS: Visualizing Nerves in Three Dimensions (16-mm color sound film). NIH Research Resources Information Center, Rockville, MD.

Farrell, E.J. (1983) Color display and interactive interpretation of three-dimensional data. IBM J. Res. Dev. 27:356–366.

Fuchs, H., Z.M. Kedem, and S.P. Uselton (1977) Optimal surface reconstruction from planar contours. Commun. ACM 20:693–702.

Fuchs, H., S.M. Pizer, L.C. Tsai, S.H. Bloomberg, and E.R. Heinz (1982) Adding a true 3-D display to a raster graphics system. IEEE Comp. Graph. Appl. 2:73–78.

Gaunt, P.N., and W.A. Gaunt (1978) Three Dimensional Reconstruction in Biology. Kent, England: Pitman Medical.

GSPC (1979) Status report of the Graphics Standards Planning Committee of ACM/Siggraph. Comput. Graphics 13(3):II-1–II-179.

Herman, G.T. (1979) Detection and display of organ surfaces from computer tomograms. Proc. 6th IEEE Conf. Comp. Applications in Radiology.

Herman, G.T., and H.K. Liu (1978) Dynamic boundary surface reconstruction from planar contours. Comput. Graphics Image Process, 7:130-138.

Levinthal, C., and F. Levinthal (1983) On the Pathways of Neural Development. In S. Kety et al. (eds.): Genetics of Neurological and Psychiatric Disorders. New York: Raven Press.

Levinthal, C., and R. Ware (1972) Three-dimensional reconstructions from serial sections. Nature 236:207–210.

Liu, H.K. (1977) Two- and three-dimensional boundary detection. Comput. Graphics Image Process. 6:123–134.

LoPresti, V., E.R. Macagno, and C. Levinthal (1973) Structure and development of neuronal connections in isogenic organisms: Cellular interactions in the development of the optic lamina of *Daphnia.* Proc. Natl. Acad. Sci. USA 70:433–437.

Macagno, E.R.,V. LoPresti, and C. Levinthal (1973) Structure and development of neuronal connections in isogenic organisms: Variations and similarities in the optic system of *Daphnia magna.* Proc. Natl. Acad. Sci. USA 70:56–61.

Macagno, E.R. (1978) Mechanism for the formation of synaptic projections in the arthropod visual system. Nature 275:318–320.

Macagno, E.R., C. Levinthal, and I. Sobel (1979) Three-dimensional computer reconstruction of neurons and neuronal assemblies. Ann. Rev. Biophys. Bioeng. 8:323–351.

Nierzwicki-Bauer, S.A., D.L. Balkwill, and S.E. Stevens (1983) Use of a computer-aided reconstruction system to examine the three-dimensional architecture of cyanobacteria. J. Ultrastruct. Res. 84:73–82.

Schlusselberg, D.S., W.K. Smith, M.H. Lewis, B.G. Culter, and D.J. Woodward (1982) A general system for computer based acquisition, analysis, and display of medical image data. Proceedings of the 1982 Annual Conference. Assoc. Comput. Mach. New York pp. 18–25.

Selfridge, P.G., and J.M.S. Prewitt (1981) Organ detection in abdominal computerized tomography scans: Application to the kidney. Comput. Graphics Image Process. 15:265–278.

Sobel, I. (1978) Neighborhood Coding of Binary Images for Fast Contour Following and General Binary Array Processing. Comput. Graphics Image Process. 8:127–135.

Sobel, I., Levinthal, C., and Macagno, E. (1980) Special techniques for the automatic computer reconstruction of neuronal structures. Ann. Rev. Biophys. Bioeng. 9:347–362.

Stevens, J.K., T.L. Davis, N. Friedman, and P. Sterling (1980) A systematic approach to rcconstructing microcircuitry by electron microscopy of serial sections. Brain Res. Rev. 2:265–293.

Udupa, J.K., Srihari, S.N., and Herman, G.T. (1979) Boundary detection in multidimensions. Tech. Report MIPG31, Dept. of Radiology, Univ. of PA Med. Ctr., Philadelphia PA.

Whitted, T., and D.M. Weimer (1981) A software test-bed for the development of 3-D raster graphics systems. Comput. Graphics 15:271–277.

13

An Algorithm for Removing Hidden Lines in Serial Section Reconstructions Using MICROS

Cameron H. Street and R. Ranney Mize

Introduction

Serial section reconstruction is an important tool in biology. The technique is useful both for analyzing structures volumetrically and for modeling their three-dimensional molecular configuration. In cell biology, for example, it is possible to view the three-dimensional structure of mitochondria, chromosomes, and other molecular entities by reconstructing them from multiple consecutive thin sections cut through the structure (Moens and Moens, 1981; Perkins et al., 1979; Veen and Peachey, 1977).

In neurobiology, serial section reconstruction is commonly used to reconstruct nerve cells (Glasser et al., 1977; Johnson and Capowski, 1983; Levinthal and Ware, 1972; Llinas and Hillman, 1975; Macagno et al., 1979; Shantz and McCann, 1978; Stevens et al., 1980). We have used the technique to study the neurons of the central visual system (Mize et al., 1982; Street and Mize, 1983). By stacking a series of thin sections in exact register, we can reconstruct cells three-dimensionally and study both their shape, size, and dendritic arborization and their patterns of synaptic input. Slicing the brain into thin sections allows us to examine synaptic morphology at very high resolution with the electron microscope. The reconstruction process allows us to piece together the three-dimensional structure of the cell so that we can examine the spatial distribution and density of the synapses that contact it. The information obtained is of great value both for modeling the output of neurons based upon their synaptic input and for classifying cell types.

In order to appreciate properly both the patterns of synaptic input and the dendritic branching patterns of the cell, it is useful to view the reconstructions from different angles with a three-dimensional perspective. This requires that sections lying behind the frontal plane of the object be hidden from view. This chapter describes computer algorithms for suppressing hidden lines of images that have been drawn from serial sections. The algorithms were developed for use with a microcomputer with only 64 KB of memory. The programs produce high-resolution reconstructions at specified rotations displayed either on a CRT screen or on a digital plotter.

The MICROS Reconstruction System

Our reconstruction system, called MICROS (Microcomputer Reconstruction of Sections), has two programs: one to enter and align the sections (section digitization program), and the other to display the sections at different rotations with hidden lines removed (rotation program). The software for the system is written entirely in an enhanced interpretive BASIC and is run on a Hewlett-Packard 9845B microcomputer [described in Chapter 5]. The system is described in detail in Street and Mize (1983).

The section digitization program is used to enter cell outlines from micrographs placed on a digitizer tablet. The operator first enters data on a keypad to identify the number and thickness of each section. The cell outlines are then traced with a hand-held cursor on the digitizer. Up to 25 separate fragments belonging to a single reconstructed object can be digitized for each section (see Figure 1). The x, y coordinate positions of and a code type for up to 30 separate synaptic inputs to the cells can also be entered from the digitizer (squares in Fig-

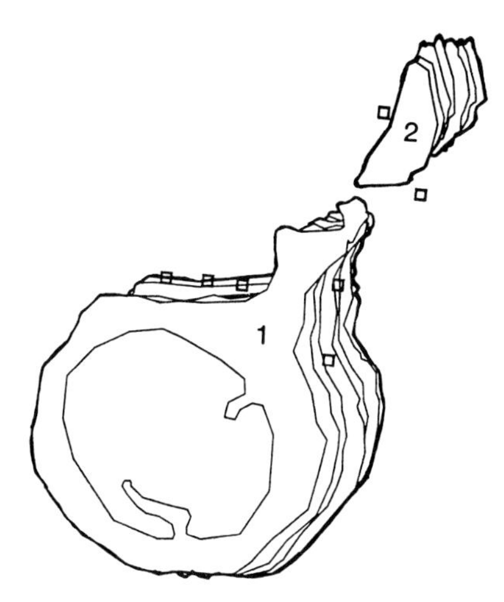

Figure 1. Serial section reconstruction of a neuron in cat superior colliculus. Seven sections have been drawn through the cell. Note the suppression of hidden lines gives the cell a three-dimensional perspective. The hidden areas are defined as all points that lie within the sequence of data points represented by the heavy black line that circumscribes the two-dimensional projection of the cell. In this figure, the cell has two fragments: One fragment is the soma, the other fragment is a dendritic process that will eventually merge with the soma. The open squares represent synaptic inputs to the cell.

ure 1). Sections are aligned by simultaneously displaying two consecutive sections on the graphics CRT screen of the microcomputer. The two sections are roughly superimposed using a centering algorithm that centers both sections around the center of the CRT screen. Precise alignment is achieved by rotating and translating the sections using a reference cursor on the CRT. The reference cursor is manipulated by special arrow keys on the computer keyboard. Sections can be edited using insert and delete subroutines. The aligned sections are currently stored on floppy disks using an HP-9895A dual floppy disk drive. The sections can also be stored on DC-100 magnetic tapes using the tape drives installed on the 9845B computer.

Once all of the sections have been entered and aligned, they can be displayed graphically using the rotation program. The rotation program replots the completed reconstructions on the graphics CRT, a high-resolution four-color digital plotter (HP-9872A), or a thermal line printer (using the dump graphics command to copy the contents of the CRT screen to the printer). The rotation program plots the reconstructions at any specified scale and angle of rotation in the x, y, and z planes of the display. The display of the reconstructions in pseudo-three-dimensional space is achieved by removing hidden lines, which lie behind objects in the line of sight of the viewer. The hidden line suppression algorithms give a three-dimensional perspective to the reconstructions. The algorithms for hidden line removal involve rather complex calculations, which are described in detail below.

The Hidden Line Removal Problem

One of the basic aspects of representation of three-dimensional objects is the removal of lines and surfaces that are hidden from view because they are behind parts of the object nearer the viewer. This task is not so complicated if the entire image can be represented by only a few points. In the case of multiple serial sections cut through an irregular object, however, a tremendous number of points may be required. Information representing a section through a cell is entered as x, y data pairs taken around the periphery of the cell. Using our MICROS system, typically about 100 points are entered into the computer for each section. If 250 sections are digitized, over 25,000 points will be taken, or 50,000 x, y data values. This quantity of data is far beyond the capacity of most small computers. It is simply not possible to load all of this data into computer memory for hidden line analysis.

One advantage of the algorithm described in this chapter is that data for only one section at a time is needed in computer memory. This allows data to be stored on a mass storage device and read into the computer sequentially. The algorithm makes use of the fact that serial sections are ordered in the z direction. This eliminates the need for extensive sorting routines. Sections are drawn successively, from the section nearest the observer to the section farthest away.

A second advantage of the algorithm described here is that hidden line suppression is achieved at the precision of the plotting device. As sections are drawn, the program maintains a data sequence that describes the boundary between regions that are no longer visible because of overlying sections and regions that are still visible. The area described by this data sequence is termed the *hidden area* and is shown by the dark line bordering the reconstructed cell in Figure 1. Since the image is being drawn from front to back, this area represents that part of the image, nearest the observer, that has already been drawn, and behind which the view is obstructed. Any portions of sections that might later fall within this two-dimensional area are actually behind sections already drawn, and so cannot be seen. The plotting resolution afforded by this method allows a high-resolution plotter to produce a drawing in which hidden lines are suppressed at precisely the point of intersection with the region described by the hidden data sequence. The reconstructions are thus produced at the maximum level of resolution. The price paid for this precision, of course, is speed. Where every point is analyzed to determine points of intersection, the algorithms must pass through a very large number of iterations. This slows the program appreciably, particularly when run in interpretive BASIC.

Terminology

The following definitions are used in describing the hidden line algorithms. A *cell* is a structure occupying three-dimensional space and residing within the boundaries of a piece of *tissue.* A cell, of course, could be any object occupying space within the tissue. A cell *process* is a cell specialization, such as a dendrite or an axon of a neuron, that projects from the cell. A *section* is a single slice through the tissue. A section may contain both portions of the cell soma and portions of several of the cell's processes. Each of these is termed a cell *fragment.* All cell fragments are closed in that they have beginning and ending points that are identical. *Serial sections* are a group of sections cut sequentially through a cell. Techniques have been developed for routinely obtaining 400–500 serial sections or more (Macagno et al., 1979; Stevens et al., 1980; Street and Mize, 1983). Each of these sections gives a two-dimensional view of parts of the cell. The third dimension, depth, is obtained by stacking the serial sections on top of one another. The *thickness* of the section is coded so that depth can be realistically represented in the reconstruction.

Reconstruction refers to the procedure of drawing each section sequentially so as to produce a three-dimensional representation of the original cell. Information must be entered into the computer to represent both the location and shape of all fragments in each section. In our system this is done with a digitizer, as described earlier. Essentially, a series of points representing the outer contour of each fragment is entered into the computer. A point is defined by an x and a y coordinate value, describing the location of the point in two-dimensional space. These series of points are stored in data *arrays* (i.e., memory blocks) within computer memory. Each data array is circular in the sense that the last point in the array is followed by the first point in the array, in the same way that all section fragments are circular or closed.

A *plotter* is a device that is able to move a pen or cursor to any specific position on a piece of paper or CRT screen as defined by a pair of numbers representing the x and y positions. The plotter pen can be raised or lowered under program control. The plotter draws only when the pen is lowered. In this way the pen can be moved without drawing. A *three-dimensional image* can be created by drawing each section on the plotter or the CRT screen of the computer, such that the sections are superimposed. Three-dimensional perspective is achieved by not drawing those portions of each section fragment that lie behind fragments closer to the viewer and already drawn. This area is called the *hidden area.* The hidden area is defined by a sequence of points representing the border of the two-dimensional projection of the current three-dimensional image (Figure 2). This sequence of points represents the boundary between the visible area and that region, lying behind sections already drawn, that is not visible from a frontal view.

Display and Rotation Methods

The rotation and display program of MICROS displays the reconstruction on either the CRT screen or a digital plotter. The reconstruction can be displayed at any rotation angle in the x, y, and z planes. The rotation angles are specified by the operator at the beginning of the program, along with the scale and direction of rotation. The program then automatically begins to draw the three-dimensional image one section at a time. The image is formed by drawing the section nearest the observer first, and then sections progressively further from the viewer, with the sections superimposed on one another. When hidden lines are not removed, the dimensions of the cell are rapidly obscured and the three-dimensionality of the cell is lost. When hidden lines *are* removed, a shading effect is

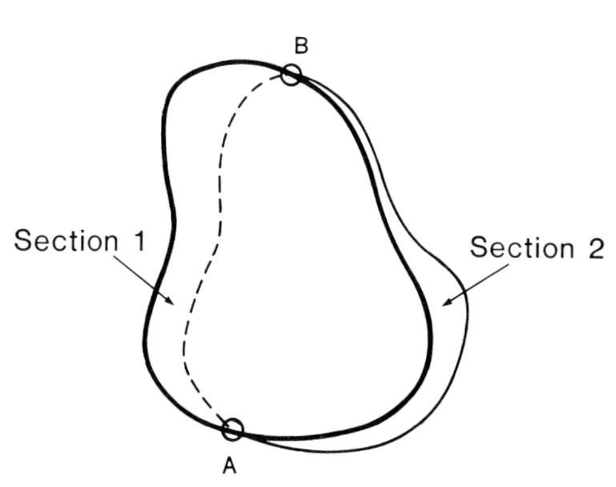

Figure 2. Two sections of a three-dimensional object. The hidden area of the second section has been suppressed. Only those parts of the section that occur outside the hidden area, and that are therefore visible, have been drawn. Portions of the new section that occur inside the hidden area are behind the first section and are represented by dashed lines. The hidden area is updated by inserting the visible portion of the new section, from point A to point B, into the hidden array. The portion of the hidden array that is no longer a border is deleted from the hidden array.

achieved, giving a three-dimensional appearance to the reconstructed cell (Figure 1). In effect, an artist's perspective is added to the reconstruction.

The removal of hidden lines is simply a question of deciding whether a data point of a section lies behind a previously plotted hidden area. If the point lies outside the hidden area, the pen on the plotter is lowered and the point is plotted. If it lies inside the hidden area, the plotter pen is raised and the point is not plotted. The problem of hidden line removal is complicated by the need to merge separate fragments of the cell that are not connected in some planes of sectioning. In practice, many sections will contain several cell fragments, occupying entirely separate locations in space (Figure 1). As additional sections are drawn, the fragments will gradually converge, forming a unified three-dimensional image of the complete process. At some level in the reconstruction, each of the cell processes will also connect with the cell soma. A hidden area must therefore be maintained for each of these section fragments. The hidden area must be updated to include all visible portions of each fragment just drawn (Figure 3).

In addition to removing hidden lines, the rotation and display program also allows the reconstruction to be drawn at any angle in three-dimensional space. Rotation of the cell can occur in any of the three spatial planes. This is easily achieved by setting up a 3×3 matrix for the desired rotations, as described by Hungerford (1978). This matrix is then multiplied by the x, y, z values defining each section point as data is read from mass storage. The z value is obtained by subtracting the section thickness from the current z value each time a new section is read.

The Hidden Line Algorithms

Data Structures

The hidden line algorithm works by calling a single section into memory and comparing it with the currently defined hidden area. In order for the hidden line algorithm to keep track of hidden areas, a number of data values must be stored

Figure 3. Data arrays. Array XY is the section data array. Array PNTR-XY holds values that point to the position of the next section in the array. The hidden area data arrays are HIDE, LINK, and AVAIL. HIDE holds all XY data pairs that describe the outer boundary of the hidden area. LINK holds data values associated with the merging process. AVAIL contains the locations in array HIDE that are available. The pointer arrays for these arrays are described in the text.

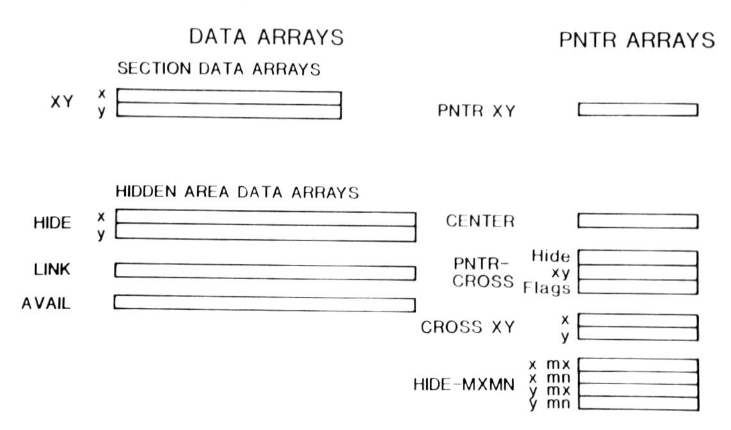

in memory. The two primary data arrays are XY and HIDE. These arrays hold section data and hidden areas, respectively. A variety of smaller arrays are used as pointer arrays. Data arrays are described in Table 1 and diagrammed in Figure 3.

The array HIDE stores data in memory using a linked-list data structure. A *linked-list data structure* is one in which data may be stored in a physical order different from the true logical order of the data. This allows data to be inserted and deleted without rewriting the order of the entire data array.

The *x, y* coordinate data for each section are held in array XY. This array holds data for all fragments of the cell for that section. Data are stored sequentially, and array PNTR-XY keeps track of the data boundaries between each fragment

Table 1. Hidden Line Program Variables

Array Name	Description
XY(DIM+2,2)	Data array for all *x, y* coordinate values of a single section.
OFFSET(3)	Holds the section thickness of each section, which is added to the current *z* value during reconstruction. *x, y* offset values that are set during reconstruction are also stored in this array and are added to each point before plotting.
PNTR-XY(25)	Pointer array that points to each fragment of the new section, found in the array XY.
SYNAPSE(30)	Points to the position in array XY at which coordinates of the first synapse may be found. The remainder of the array contains data values assigned to each synapse digitized, up to a maximum of 29.
DIM	The size (dimension) of array XY. Two additional words of memory must be allocated to the *x* and *y* arrays to hold scaling data.
NO-DATA	Number of XY points in current section.
HID-DIM	The size of array HIDE, for hidden line data.
HIDE (HID-DIM,2)	*x, y* coordinates representing the hidden area. The data are stored in a linked-list structure. All hidden centers are represented in the array.
HIDE-MXMN(25,4)	Maximum and minimum coordinates for each of the possible 25 hidden centers.
PNTR-CROSS(25,3)	Pointers to crossing points in array HIDE (*,1), arrays *x* and *y* (*,2), and a location for flags (*,3).
CROSS-XY(2)	Array for *x, y* intersection points, calculated each time the new section crosses the hidden area.
LINK (HID-DIM)	The linked-list for the array HIDE.
AVAIL (HID-DIM)	Available locations in array HIDE.
AVAIL-PNTR	Pointer to end the AVAIL list.
NO-HID(25)	Number of data values in each hidden area.
CENTER(25)	Pointer to array LINK. If not zero, each element in the array points to an arbitrary position within a hidden center represented in array HIDE and ordered by array LINK.

(Table 1). For example, PNTR-XY(1) points to the end of data for the first fragment of the current section found in array XY. PNTR-XY(2) points to the end of the second fragment of the same section, and so forth.

All hidden areas are represented in the linked-list array called HIDE (Table 1). It is possible to use a single array because each hidden area is circular, with the last point of the hidden area linked to the first point. Thus, different linked-lists within the same array do not interfere with each other. A second array called CENTER contains pointers to each of the hidden areas represented in data array HIDE. In this way, any nonzero element of the array CENTER points to an arbitrary point within a closed data loop in array HIDE. The order to each data loop stored in array HIDE is found in array LINK. Twenty-five hidden areas are allowed, which matches the maximum number of cell fragments allowed in the reconstruction program.

Two arrays are maintained in the process of marking intersections between a new section and a hidden area. These arrays are updated each time an intersection is found. Array PNTR-CROSS is a three-dimensional array that contains the array positions of XY and HIDE at the point of intersection, and the number of the hidden area involved in the intersection. Intersections are always between line segments, and the respective line segments are defined by the array position stored in PNTR-CROSS and the next sequential array element. The precise x, y coordinate of the intersection of the new section and the hidden area is calculated, and this coordinate pair is stored in array CROSS-XY. This is the information that allows hidden line suppression with the precision of the plotting device. At the time array HIDE is updated, the precise point of intersection stored in array CROSS-XY will be inserted into array HIDE, along with visible points from array XY.

Array HIDE-MXMN contains the maximum and minimum x and y positions of each hidden center. This information is used in two places. First, time-consuming calculations used to determine whether a section point lies within a hidden center can be skipped if the section point lies outside the max–min borders of the hidden center. Secondly, the case can be found of a section fragment entirely circumscribing a hidden center. In this case the hidden center is replaced by the new section fragment.

Drawing Serial Sections

The drawing algorithm produces a three-dimensional representation of a cell by drawing the outer border of all serial sections taken through the cell (Figure 1). The sequence of points maintained in array HIDE represents the border between visible space and space hidden from view by fragments already drawn. As sections are drawn, then, the critical question is whether a given line segment of the section lies outside the area circumscribed by the hidden array, and so should be drawn, or inside that area, and so should not be drawn.

A subroutine called INSIDE, described in more detail below, determines whether a given point lies inside or outside the hidden area. As a section is drawn, routine INSIDE is called for each point of the new section. It should be realized that sections are in reality polygons, consisting of many small line segments between adjacent x, y coordinate points. As a section is drawn, the plotter pen moves along the polygon describing the section. Each time the plotter pen is to be moved along a line segment to the next section point, routine INSIDE is called to determine whether that next point occurs inside or outside a hidden area. If inside a hidden area, the pen is in the raised position as it is moved, and the line segment is not drawn. If the next section point is not in a hidden area, the pen is placed in the down position as it is moved, and then the line segment is drawn.

Each time the section being drawn crosses a hidden area border, routine INSIDE calculates a point of intersection. This intersection will be between a line segment defined by two points within the new section array, and a line segment defined by two points in the hidden array. It is important to keep track of precisely where the intersection occurred in both the section array XY and the hidden area array HIDE, for array HIDE will have to be modified later to include those data points of the section that were visible. This is done in array PNTR-CROSS.

Array PNTR-CROSS contains three columns, as shown in Figure 4b. This array is used to keep track of each intersection between a line segment of array XY and a line segment of array HIDE. Each time such an intersection is found, three data values are stored in array PNTR-CROSS. The location of the intersecting line segment in array XY is stored in column 1 of array PNTR-CROSS. The location of the intersecting line segment in array HIDE is stored in column 2. The third item stored is the number of the hidden center in which the intersection occurred. A number of hidden areas may exist, and these are numbered, as defined by array CENTER.

The intersection calculation is the basis on which routine INSIDE decides whether a section fragment has just crossed a hidden border. The intersection point calculated is the precise x, y location at which the fragment intersects with the hidden border. This can be used to advantage in drawing the fragment, producing a high resolution of hidden line suppression. The plotter pen is moved to the point of intersection, its position (up or down) reversed, and then moved to the end of the fragment line segment. In this way, the fragment edge is drawn precisely to the point at which it disappears under overlying fragments.

Updating the Hidden Area

The hidden area array HIDE must be updated after each new section fragment is drawn. The general procedure for this is to replace all segments of array HIDE that lie within the visible regions of the section fragment just drawn with the visible segments. In Figure 4a this would involve deleting segments *AB, CD,* and *EF* (dark line segments) from array HIDE and inserting segments *AB, CD,* and *EF* (light line segments) from array XY into array HIDE. The information needed for this procedure is contained in PNTR-CROSS. Each row of the array PNTR-CROSS contains information about one intersection of the new section with the hidden area (Figure 4b). Columns 1 and 2 are array pointers indicating the array position involved in the intersection. Column 1 contains pointers to array XY, column 2 contains pointers to array HIDE, and column 3 contains the number of the hidden center involved in the intersection. Note that in this case column three contains only 1s, indicating that only hidden center 1 is involved in this updating procedure.

There are six entries in array PNTR-CROSS, pointing to the boundaries of three segments of the hidden area that must be replaced. Replacement of the first segment is carried out as follows. Data points defining the outdated segment of array HIDE (from *A* to *B*) are deleted, and array space is returned to the AVAIL list. The first intersection point is inserted into array HIDE from array CROSS-XY, and then the new data segment (from *A* to *B*) is entered into array HIDE from array XY. Finally, the second intersection point is inserted into array HIDE from array CROSS-XY. Linked-list structure is maintained throughout this updating procedure.

The updating procedure just described can be executed as a loop, repeating until all segments listed in PNTR-CROSS have been replaced. It is crucial, however, that the first pointer used from PNTR-CROSS refer to an intersection in which the new section fragment is leaving the hidden area and entering the visible area. This is not difficult, since the intersections listed in PNTR-CROSS are in circular order (*A* follows *F*). If the first section fragment point appeared inside the hidden area, updating will begin with the first intersection in the PNTR-CROSS list. Otherwise, updating will begin with the second intersection of the PNTR-CROSS list.

It is crucial that all section fragments be digitized in the same direction (counterclockwise in our programs). The reason is that all fragments are circular, and insertion and deletion procedures start from an array position marked in PNTR-CROSS and continue until the next array position listed in PNTR-CROSS is reached. A subroutine exists in our digitizing program to check the direction of digitizing and to reverse the data if necessary.

Merging Hidden Areas

As the image reconstruction process progresses, it is expected that the hidden areas of all fragments will eventually merge. Each hidden area represents a series of fragments that have been taken from a cell process. If such a process never merged with the cell, it would be difficult to claim that the process was indeed part of the cell unless the process was marked with an intracellular tracer such as HRP.

Figure 4. (a) Two hidden centers are shown, one circumscribed by dark, the other by light lines. A new section fragment is superimposed over hidden center 1. Updating the hidden center involves replacing hidden area segments *AB, CD,* and *EF* with the corresponding segments of the new section fragments. (b) The contents of array PNTR-CROSS are illustrated after the new fragment has been drawn. Columns 1 and 2 contain pointers to the array positions of XY and HIDE at each point of intersection. Column 3 contains the number of the hidden center involved in each intersection. (c) Three hidden centers are shown, numbered 1, 2, and 3. A new section fragment is drawn that intersects with all three hidden centers, requiring that the three hidden centers be combined into one hidden center. This is done in three unit procedures as described in the text. One unit operation occurs within the region circumscribed by the dashed line. Here the hidden array segment *CB* will be deleted, and the new section fragment segment *AB* inserted. (d) The contents of array PNTR-CROSS for Figure 4c. Column 3 contains the number of the hidden center involved in each intersection. (e) This figure is slightly more complicated than 4c, in that the new section fragment intersects with hidden center 2 four times, which is reflected in Figure 4f. Column 3 of array PNTR-CROSS shows 4 consecutive intersections in hidden center 2. As described in the text, only the first and last of these points (*B* and *E*) are needed for the merging process. (f) Contents of array PNTR-CROSS for Figure 4e.

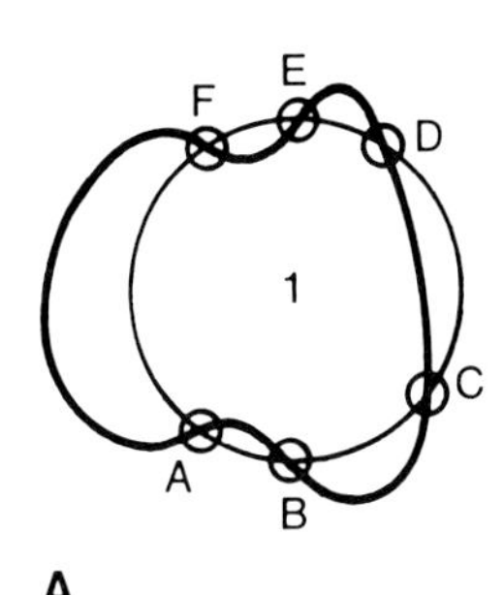

A

ARRAY PNTR-CROSS

1	2	3
A	A	1
B	B	1
C	C	1
D	D	1
E	E	1
F	F	1

B

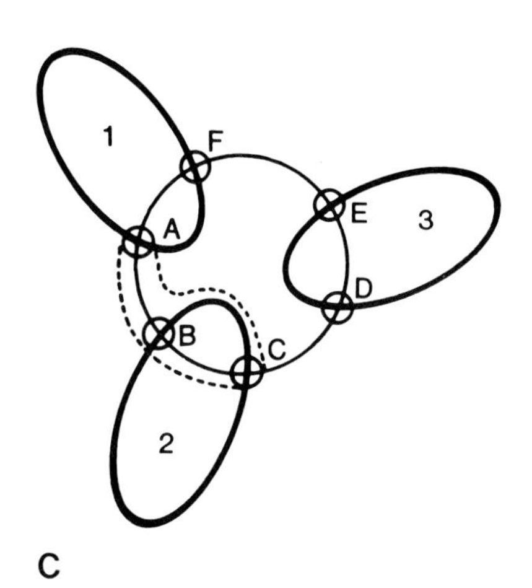

C

ARRAY PNTR-CROSS

1	2	3
A	A	1
B	B	2
C	C	2
D	D	3
E	E	3
F	F	1

(1) Pointers to array XY
(2) Pointers to array hide
(3) Hidden area of intersection

D

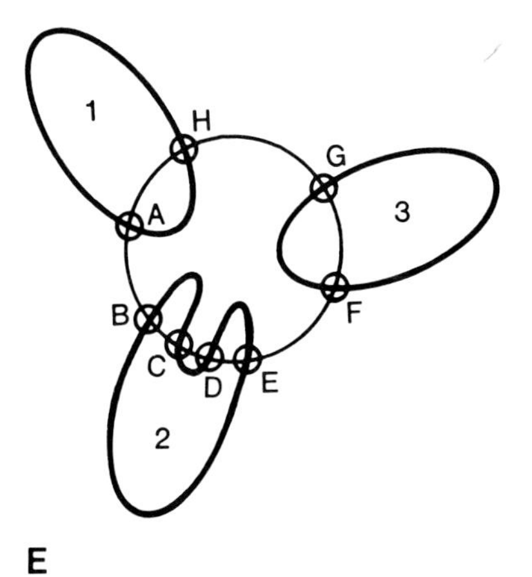

E

ARRAY PNTR-CROSS

1	2	3
A	A	1
B	B	2
C	C	2
D	D	2
E	E	2
F	F	3
G	G	3
H	H	1

F

Any time a single fragment intersects two or more hidden areas, those areas are considered to be merged. At this point, only one list of data points is needed to represent the merged hidden area. Intersection of a fragment with more than one hidden area is determined by looking at the third column of numbers in array PNTR-CROSS. Each time an intersection occurs, the hidden center number in which the intersection occurred is listed in column 3, as can be seen in the diagram of array PNTR-CROSS, Figure 4d. In this case, three different hidden areas are listed, and so three hidden areas must be merged.

The procedure of merging centers needs to be sufficiently flexible to handle any number of centers. This can be done most conveniently if the procedure consists of a basic operation that can be repeated the same number of times as there are centers to merge. Examination of Figure 4c shows that such a basic operation must be performed three times to merge the three centers. One of these has been circled with a dotted line in Figure 4c. The merging routine executes the following procedure. The hidden border from point C to point B is deleted, and visible points of the new fragment are inserted, from point A to point B. The procedure can be repeated for the next unit by deleting segment ED and inserting CD, and so on.

Some care must be taken in setting up this procedure in order to maintain the linked-list structure of array HIDE. Five pointers may be defined for each loop of the merging routine. These will be called HP1, HP2, HP3, SP1, and SP2. For the first merge loop, these pointers will be defined as follows. HP1 is the pointer that points to hidden point A, HP2 to hidden point B, and HP3 to hidden point C. These three values are taken directly from the first three intersections listed in array PNTR-CROSS for array HIDE. Pointer SP1 is the pointer to section point A, and SP2 to section point B. These two pointers are the first two listed in array PNTR-CROSS for array XY.

During each merge loop, one hidden array segment is deleted, and one section segment inserted. Deletion of hidden array points begins with location HP3 and continues until array position HP2 is reached. The reason for deletion in the direction of HP3 to HP2 is that all arrays are linked in counterclockwise direction, as described earlier. Next, section data points are transferred from array XY into array HIDE, starting at SP1 and continuing to SP2. Linked-list structure is maintained in array HIDE for these points. Intersection points from array CROSS-XY may also be inserted, thus maintaining a high degree of precision in the hidden center array. At the completion of the first loop, the linked-list structure is open, for HP3 is not linked to a valid point. This data point will be linked at the end of the next loop, for it will become HP1 and therefore be linked to the newly inserted segment CD.

At the start of the second merge loop, the merge pointers will be defined as follows. HP1 will become the pointer to hidden point C, HP2 to hidden point D, and HP3 to hidden point E. SP1 will become the pointer to section point C, and SP2 to section point D. Execution of the loop proceeds as described in the previous paragraph.

One intersection pointer in array PNTR-CROSS must be modified at the beginning of the merging operation. It will become necessary to delete hidden array points between location A and location F. However, this will not occur until the final loop of the merge routine. By this time, hidden point A has been linked to the section added earlier. Therefore, it is necessary to modify the pointer to hidden point A, located in array PNTR-CROSS, to LINK(A) during execution of the first loop of the merge routine.

A minor complication of the updating procedure is shown in Figure 4e, f. Here the new section has entered and left the hidden area 2 twice. This can be determined from array PNTR-CROSS by looking at column 3. Rows 2–4 show an intersection with hidden area 2. In this case the unit updating operation is simply to delete segment EB from array HIDE and insert segment AB from XY into HIDE. Rows 3 and 4 of PNTR-CROSS are ignored in this operation.

The hidden line algorithm must be able to create new hidden areas as well as to merge hidden areas. Determining whether a new area is present is quite simple. It is indicated by the appearance of a section that is drawn with no intersections. A special case must be checked for, however, in the instance of a section drawn with no intersections. This may be a case in which a hidden area is entirely engulfed by the section. Max–min data stored in array HID-MXMN allow such cases to be found easily.

Routine INSIDE

The problem of determining whether a data point lies inside or outside a hidden area can be tricky and often requires a considerable amount of computer time. One basic approach to the problem follows. First, a section point is found whose position inside or outside a hidden area is known with certainty. For the remaining points in the section, the question then becomes whether or not an intersection has occurred between two line segments. One line segment is formed by the section point and its previous point. The other line segment may be any other consecutive two points that are part of the hidden area.

It is possible to calculate all possible intersections for every line segment of the hidden array with each new section point, but such massive number crunching is neither time efficient nor necessary. A variety of tests can be devised to eliminate the majority of these intersection calculations. The first test to be performed is to determine whether the data point lies within the limits of the hidden area. This can be determined by looking at data values stored in the array HID-MXMN.

Many intersection calculations may be eliminated using the criteria described below. Figure 5 shows a hidden area (dark line) and a new section. Four points, labeled S1–4, are shown on the new section, representing four possible situations. The general idea of this routine is to find all pairs of points in a hidden area lying immediately adjacent to a vertical line through the section point in question. The task is then simply to compare the y value of the section point with the y values of the pairs of hidden array points. Figure 5 shows that with this approach points S1 and S4 are clearly defined as outside the hidden area, and point S2 is clearly inside the hidden area. Intersection calculations will have to be performed for the line segment defined by S3 and the previous section point and the line segment defined by the two hidden array points. If the two line segments intersect, it can be concluded that S3 is outside the hidden area, since the previous section point was inside (Figure 5).

Pairs of hidden area points lying to either side of a vertical line through the section point can be found as follows. The algorithm starts with any point of array HIDE and determines whether its x value is greater or less than the section point. The algorithm then proceeds through the hidden array comparing x values. Each

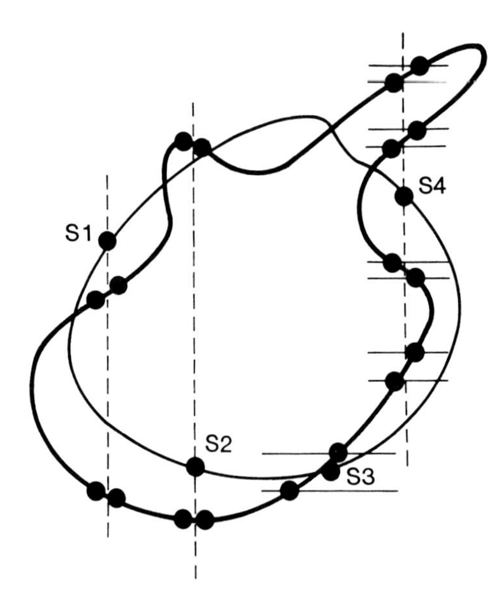

Figure 5. A single hidden area is shown with a single section fragment superimposed on it. Four points along the section fragment have been selected to illustrate conditions checked for by routine INSIDE. These points are labeled *S1–4*. Pairs of points along the hidden area array have been marked, lying on either side of a vertical line drawn through each section point. These points are described in detail in the text. *S1* is outside the hidden area, *S2* is inside, and *S4* is outside. *S3* lies between the *y* value of a pair of points and therefore its location inside or outside the hidden area cannot be determined by routine INSIDE. Intersection calculations are required for point *S3*.

time the comparative values flip, a pair of points has been found, and their location can be stored in a temporary array. After all points of the hidden area have been examined, those points that were selected as pairs are sorted by *y* value.

The procedure now is to loop through the hidden pairs of points, comparing their *y* values with the *y* value of the section point. If the section point lies between a pair of hidden points, it is necessary to perform intersection calculations, as in the case of point S3. If the section point lies below the first hidden point or above the last hidden point, it is outside the hidden area (point S1, Figure 5). Point S2 is above the first pair of hidden points and below the second pair, and is therefore inside. Point S4 is an example of the usefulness of this procedure for a more complicated structure. For this case it is simply necessary to keep track of the number of pairs of hidden points that have been checked. S4 is above the first and second pairs but below the third pair, and is therefore outside the hidden area.

The simple tests described here can require many comparisons, but nevertheless involve far fewer calculations than checking for line segment intersections between every section point and every hidden array point. In fact, generally only a small number of intersection calculations is required. This provides a significant increase in the speed of the hidden line program.

An additional advantage of these comparative criteria is in error detection. The INSIDE routine assumes that as a section is drawn it changes between visible and hidden regions only in cases where an intersection is found. There are situations, however, when limited computational precision results in an erroneous decision, especially in cases where line segments intersect or nearly intersect only at their extreme ends. An incorrect decision can be detected with the next section point because that section point can be absolutely placed inside or outside a hidden area, using the comparative criteria. When such a point is found there will be a conflict between where the point is and where the program thought it was with respect to the hidden areas. In these cases, it is necessary to make corrections by raising or lowering the plotter pen and adjusting array PNTR-CROSS and CROSS-XY.

Conclusions

Reconstruction of cells and other organelles found in biological tissues is an important technique for studying structure–function relationships. The analysis of reconstructions is greatly aided by displaying them three-dimensionally. A pseudo-three-dimensional image can be produced by removing hidden lines that lie behind the viewing angle of the observer. This chapter has described a computer program that removes hidden lines from computer reconstructions. The program is written in enhanced BASIC and can be run on a 64-KB microcomputer.

A number of other hidden line algorithms have been devised for this purpose (Sutherland et al., 1974). One such algorithm utilizes an array of adjacent, parallel line segments of variable length to represent the hidden area (Veen and Peachey, 1977). Another algorithm utilizes the concept of bit-mapped points. This method uses an array that contains bits for each pixel of the plotter device. Bits for each pixel that occur within a hidden area are set to the value 1, and all other bits are set to 0.

The algorithm described in this paper maintains a series of points defining the border between visible space and space hidden behind cell fragments already drawn. This series of points is termed the hidden area and is in linked-list form. The hidden array is modified after each fragment is drawn. This is done by inserting into the hidden array all portions of the fragment that were drawn, for these represent the new boundary between visible space and space occurring behind cell parts previously drawn.

This algorithm offers two advantages over others. First, the hidden array contains the precise boundary between visible and hidden space. This permits extremely accurate calculation of intersection points between a fragment line segment and the hidden border. Visible fragment lines can be drawn to the precise point of intersection with a hidden area and the pen lifted. The resolution of line suppression is therefore very high. A second advantage of the algorithm described here is that the data structure allows very conservative use of memory for data storage. This is essential for operation of the program on a 64-KB microcomputer. The program should operate with modifications on any microcomputer with 64KB of RAM.

The algorithm was designed specifically for use in reconstruction of neurons and synapses and has been used successfully for this purpose for several years. Removal of hidden lines from serial section reconstructions should prove useful in a variety of other areas of biology. In fact, the algorithms we have developed should operate successfully on any reconstruction that utilizes closed sections constructed of Cartesian coordinate data pairs. The methods used in this algorithm are fairly straightforward and can be easily translated to a number of high-level languages. It is hoped that the program logic described here will prove useful to other cell and neurobiologists. A copy of the complete BASIC program for hidden line removal can be obtained for a modest handling charge by writing to Dr. Mize.

We thank Linda Horner and Lee Danley for preparing the illustrations. Discussions with Rob Smith of the University of Pennsylvania provided valuable insights into the problem of hidden line suppression. Funds for this project came from USPHS Grant EY-02973 from the National Eye Institute and a New Faculty Research Grant from the State of Tennessee.

References

Glasser, S., J. Miller, N.G. Xuong, and A. Selverston (1977) Computer reconstruction of invertebrate nerve cells. In R.D. Lindsay (ed.): Computer Analysis of Neuronal Structures. New York: Plenum, pp. 21–58.

Hungerford, J.C. (1978) Graphic manipulations using matrices. BYTE 3:156–165.

Johnson, E.M., and J.J. Capowski (1983) A system for the three-dimensional reconstruction of biological structures. Comput. Biomed. Res. 16:79–87.

Levinthal, C., and R. Ware (1972) Three-dimensional reconstruction from serial sections. Nature (Lond.) 236:207–210.

Llinas, E., and S.W. Hillman (1975) A multipurpose tridimensional reconstruction computer system for neuroanatomy. In M. Santini (ed.): Golgi Centennial Symposium Proceedings. New York: Raven Press, pp. 71–79.

Macagno, E.R., C. Levinthal, and I. Sobel (1979) Three-dimensional computer reconstruction of neurons and neuronal assemblies. Ann. Rev. Biophys. Bioeng. 8:323–351.

Mize, R.R., R.F. Spencer, and P. Sterling (1982) Two types of GABA-accumulating neurons in the superficial gray layer of the cat superior colliculus. J. Comp. Neurol. 206:180–192.

Moens, P.B., and T. Moens (1981) Computer measurements and graphics of three-dimensional cellular ultrastructure. J. Ultrastruct. Res. 75:131–141.

Perkins, W.J., A.N. Barrett, J. Green, and D. Reynolds (1979) A system for the three-dimensional construction, manipulation and display of microbiological models. J. Biomed. Eng. 1:22–32.

Shantz, M.J., and G.D. McCann (1978) Computational morphology: Three-dimensional computer graphics for electron microscopy. IEEE Trans. Biomed. Eng. 25:99–103.

Stevens, J.K., T.L. Davis, N. Friedman, and P. Sterling (1980) A systematic approach to reconstructing microcircuitry by electron microscopy of serial sections. Brain Res. Rev. 2:265–293.

Street, C.H., and R.R. Mize (1983) A simple microcomputer-based three-dimensional serial section reconstruction system (MICROS). J. Neurosci. Methods 7:359–375.

Sutherland, I.E., R.F. Sproull, and R.A. Shumacker (1974) A characterization of ten hidden-surface algorithms. Comput. Surveys 6:1–55.

Veen, A., and L.D. Peachey (1977) TROTS: A computer graphics system for three-dimensional reconstruction from serial sections. Comput. Graphics 2:135–150.

MICROCOMPUTER USES IN IMAGING AND DENSITOMETRY

14

Principles of Computer-Assisted Imaging in Autoradiographic Densitometry

Peter Ramm and Jeffrey H. Kulick

Introduction

Computer-assisted image analyzers are becoming common tools in neuroscience laboratories. Two categories of image analyzer can be distinguished by the types of information made available to the computer. One class of device digitizes positional information while an operator traces features of a microscope image or photomicrograph. The resulting data are used by the computer to provide morphometric descriptions of the features.

A second category of image analyzer, the computer-assisted imaging device (CID), does not require operator interaction during the digitizing process. Rather, a scanning digitizer (Nagy, 1983) transforms the image into numbers containing one intensity (z) and two positional (x, y) image dimensions. These are stored in a dedicated image memory from which the image can be accessed for display or processing. As computing power decreases in cost, CIDs are finding increasing application to cytometry (review in Wied et al., 1982), stereology (review in Weibel, 1979), morphometry (Capowski, 1983; Rigaut et al., 1982, 1983; Wann et al., 1974; Woolsey and Dierker, 1978), and autoradiographic densitometry (Gallistel et al., 1982; Goochee et al., 1980; Yonekura et al., 1983).

This chapter discusses some principles involved in CID densitometry. Categories of host and image processor units are compared as are advantages of various scanner types including scanning microscope photometers, scanning microdensitometers, charge coupled device cameras, and vidicon cameras. Errors associated with the scanners are treated, as are software procedures for error correction.

Densitometry

Densitometry is the measurement of the attenuation of light by a sample (usually film) placed between an illumination source and a sensing element. The densitometric method substitutes an easily obtained optical density (OD) measure for the more tedious point/unit area measures obtained by grain counting. Densitometry is used with the 2-deoxyglucose (2-DG) method for determination of local cerebral glucose utilization (Alexander et al., 1981; Sokoloff et al., 1977), for measurement of local rates of protein synthesis (Kennedy et al., 1981), and for regional localization of other radio-labeled ligands in brain (Herkenham and Pert, 1982; Lewis et al., 1981; Palacios et al., 1981; Penney et al., 1981; Rainbow et al., 1981; Unnerstall et al., 1982; Wooten and Horne, 1982; Zarbin et al., 1983) and whole-body autoradiographs (Som et al., 1983; Yonekura et al., 1983). Densitometry has also been applied to quantitative immunocytochemistry (Benno et al., 1982) and high-resolution cytometry (Wied et al., 1982). Quantitative autoradiographic densitometry is as demanding an application as a CID is likely to encounter. Therefore, a capable densitometric CID will be useful in many applications.

The Density Measure

There are two units in which density is commonly expressed. *Transmittance* is the ratio of the flux passing through a sample to the flux present when there is only air in the sampling aperture. *Optical or transmission density* (OD) is the common logarithm of the reciprocal of transmittance.

Perception of brightness differences is approximately logarithmic. Thus, a region of 2 density units (2 D; 1% transmittance) appears twice as dark as a region of 1 D (10% transmittance). As the scale of density units bears a readily appreciated relation to the perception of brightness, OD is the most common measure of light transmission.

Measuring Density

The OD value obtained varies with the type of density measured. *Specular transmission density* is taken with a small solid angle of collection (the angle over which light from the sample is collected to the sensing element). Not all of the transmitted light is collected. *Diffuse density* is taken with an angle of collection of 180°. All of the transmitted light is collected. Specular and diffuse light collection procedures yield different density values. Most density standards are calibrated to *double diffuse density,* in which both illumination and collection angles are 180°. This rather difficult procedure is not implemented in most densitometers which, therefore, exhibit some degree of specularity. Thus, the "semispecular" density value obtained will vary between devices reading a sample with different collection angles.

A correction for specularity involves calibrating the densitometer to a diffuse density reference. As diffuse density references are not available for many films, densitometers are often calibrated with carbon filters. This calibration will not yield accurate diffuse density measurement from a photographic film, because carbon films and various photographic films scatter incident light to various degrees. We have found that calibration to a photographic wedge yields a system response that is linear to within 1% in reading of that wedge. If the same wedge is read after calibration to a range of Wratten #96 (carbon) neutral density filters, system response is linear to within only 8%.

A measure of scatter is the Callier *Q* factor, defined as the ratio between specular and diffuse density. Photographic films exhibit various *Q*s, typically about 1.4. A dispersion of fine carbon particles does not appreciably scatter light and exhibits near unity *Q*. If *Q* were constant for a given film, a single calibration back to diffuse density in that film would be adequate. However, in most of the density range found in autoradiographic material, the *Q* factor is density-dependent [see Altman (1977) for a discussion of this]. For accurate diffuse densitometry, the densitometer should be calibrated back to a full range of diffuse density readings made on the particular film in use.

Fortunately, autoradiographic OD is expressed with reference to the effects of given radioactivity exposures upon film. Diffuse or double diffuse density values have little intrinsic meaning and it is sufficient that the CID measure density consistently. The usual practice is to normalize density values relative to some reference density (e.g., brain mean OD), or to calibrate the densitometer with reference to a known set of activity standards. In quantification of deoxyglucose autoradiographs, for example, ^{14}C standards calibrated in $\mu Ci/g$ of tissue are exposed with each film. Density readings are transformed to ^{14}C values by reference to the film densities produced by the standards.

The Densitometric CID

General Principles

A common method of densitometry is to move an autoradiograph under the aperture of a photometer mounted on a microscope and to record a spot reading. The autoradiograph is then moved until the next area of interest lies under the aperture, and another reading is then taken. Although this technique has been implemented at relatively low cost (Bryant and Kutyna, 1983; Dauth et al., 1983; Haas et al., 1975), it has two major disadvantages.

1. Sampling speed is very slow, limiting the number of samples that can be taken and the adequacy with which sampled regions are defined.
2. Visualization of the autoradiograph and placement of the sample aperture is difficult.

Because of these disadvantages, manual densitometry is particularly cumbersome in comprehensive mapping studies, for example, the labeling of glucose utilization throughout the nervous system.

A CID is well suited to large-scale densitometric mapping because it permits easy visualization of regions in the autoradiographs and rapid sampling from those regions. The entire autoradiograph is generally held in the CID's image memory as an array of picture elements (pixels). The image formed by the array is viewed on a monitor. Any portion of the image can be sampled from image memory as if directly from the autoradiograph. Density sampling can be performed with computer-generated apertures (windows) of various sizes and shapes. The stored image can be contrast-enhanced and color-coded without actually changing the data values. Thus, the visibility of subtly delineated regions can be improved for display. Finally, because the entire image is directly accessible to the computer or to a specialized image processor, the CID can subject large image databases to image processing algorithms.

A CID usually consists of four major components:

1. an image acquisition device, such as a vidicon camera;
2. a general-purpose processor (GPP);
3. an image processor (essentially a peripheral computer dedicated to imaging tasks); and
4. image display monitors.

As shown in Figure 1, not all CIDs include all of these devices. The simplest and least expensive form of CID can be built without the image processor or display monitors (Figure 1a). In this case, positional drive signals generated by the GPP direct the stepping stage of a scanning microphotometer. Analog density data are read following each step of the stage, converted to digital format by the digitizer and stored, with x, y coordinates, in the GPP data memory. Because there is no video generator or image memory, the image is not displayed.

More flexible CIDs (Figure 1b,c) include an image refresh memory (abbreviated to image memory). *Image memory* is a memory bank in the image processor dedicated to containing an image. To fill image memory, CIDs may use either

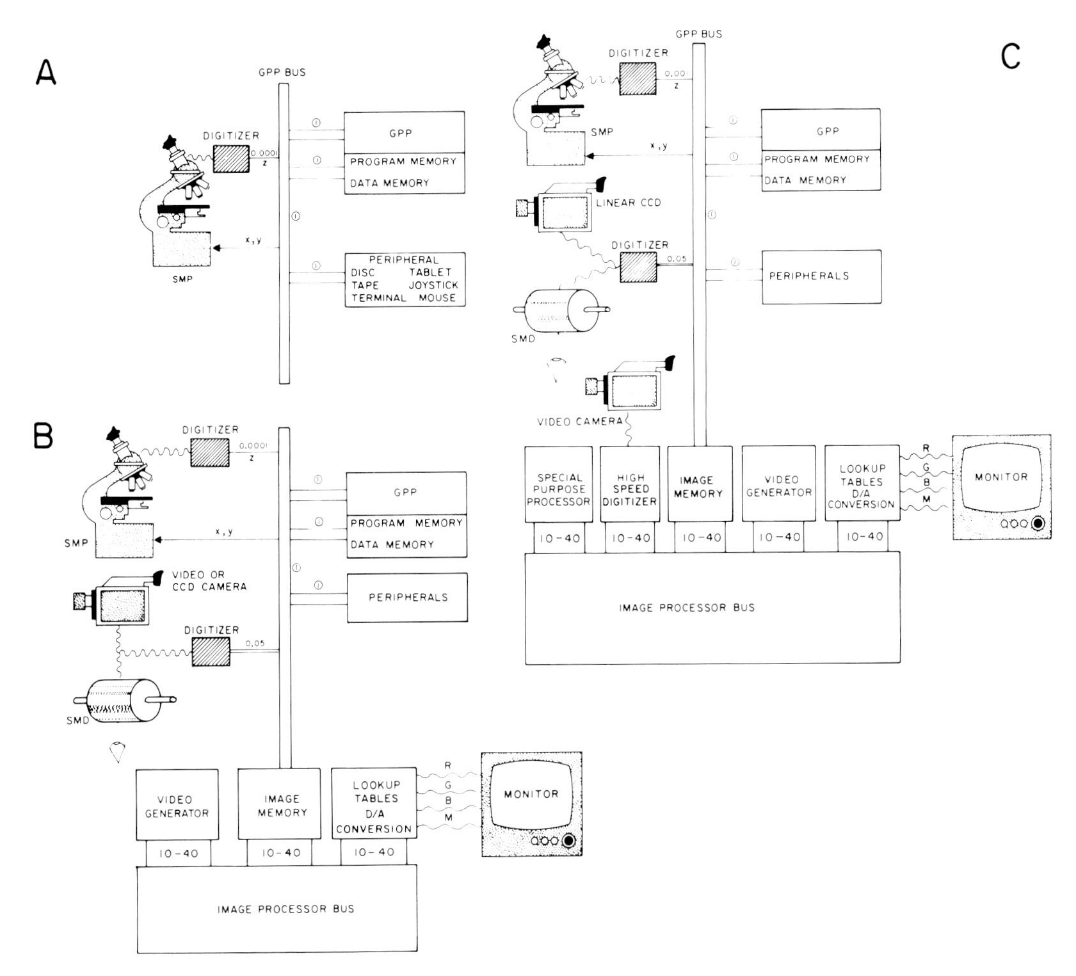

Figure 1. Configurations of CIDs. (A) The computer-assisted densitometer/photometer. Positional drive signals (x, y), generated by the general-purpose processor (GPP) direct the stepping stage of the scanning microscope photometer (SMP). Density data (z) are read following each step of stage movement and are stored, with x, y coordinates, in the GPP data memory. As there is no image refresh memory, the image is accessible as numerical data only. Numbers associated with the bus indicate approximate bandwidth (million operations/sec). Analog signals are represented with wavy lines, digital signals with straight lines. (B) The basic CID, including an image refresh memory. Data signals from any scanner can be placed on the GPP data bus. Image data are stored in image memory, where they are accessible to the data memory of the GPP at a rate limited by the GPP bus bandwidth. (C) A CID capable of real-time digitization and rapid image processing. A video camera (vidicon or array CCD) and high-speed digitizer provide data at a rate that permits real-time digitization and storage of the video signal. Similarly, the special-purpose processor (SPP) is connected to image memory by the image processor bus, permitting rapid manipulation of the contents of image memory.

slow scanners (scanning microscope photometers, scanning microdensitometers, line-array charge coupled device cameras) or fast scanners (area-array CCDs or vidicon cameras). Data signals from a slow scanner are placed on the GPP data bus and then transferred to image memory. In this case, the rate-limiting step of data transfer to image memory is either:

1. the read rate of mechanical/photometric components of the scanner;
2. the digitizing rate;
3. the rate at which the scanner interface can transfer data; or
4. the bandwidth of the GPP bus.

Typically, a 512 × 512 × 8-bit image is acquired in 1–10 sec. The contents of image memory are regularly overwritten with the image (refreshed), usually at the video frame rate (30/sec). The rate at which data in image memory are refreshed and displayed is dictated by the video generator, in which synchronizing and drive signals for all imaging components originate. Lookup tables map density data into color (R,G,B) or monochrome (M) codes, and permit rapid manipulation of color and contrast. Digital-to-analog converters (DACs) provide an analog signal to modulate the guns in the display monitor. All these functions of the image processor are carried by an independent bus of wide bandwidth. However, image data are accessible to the data memory of the GPP at a rate limited by the GPP bus bandwidth.

A CID incorporating a video digitizer and sophisticated image processor (Figure 1c) is capable of both real-time digitization and rapid image processing. The vidicon camera (or array CCD) and high-speed digitizer can provide data at rates much faster than the GPP bus can accept them. Therefore, these devices are interfaced directly with image memory via the image processor bus. This feature permits real-time digitization and storage of the video signal. Similarly, the special-purpose processor is connected to image memory by the image processor bus, permitting rapid manipulation of the contents of image memory. With this device, such image processing procedures as frame averaging can be performed very rapidly. Each of the basic components of a CID are described in detail in the sections that follow.

Image Acquisition

Image acquisition devices (scanners) include the vidicon camera and its relatives (Gallistel et al., 1982; Wied et al., 1982; Yonekura et al., 1983) and mechanically driven scanners such as the scanning microscope photometer, scanning microdensitometer (Goochee et al., 1980), and line-array CCD camera (e.g., Eikonix 78/79). CCD cameras employing area arrays are becoming available, but have yet to be implemented in a CID. The scanner is coupled to a digitizer, which converts the analog image signal to a digital one. The mechanical scanners do not require high rates of digitization. Typically, their scanning rate is limited both by the rate at which a sensing element can be positioned and the rate at which it can perform a stabilized measurement. In contrast, the video camera (vidicon or area-array CCD) can take advantage of a high-speed ADC (typically 30 Msamples/sec) that digitizes pixels at rates commensurate with the rapid video scan.

It is critically important that an appropriate scanner be selected for the task at hand. Factors dictating scanner performance include the following:

(a) *Spatial resolution:* Spatial resolution of the CID is dependent upon the resolution of all components in the system. The image scan is represented as a gridwork of digitized pixels in image memory. There must be sufficient pixels in the memory grid so that, at a convenient magnification, the smallest area of interest in the image can be uniquely represented. Scanner resolution should match the storage capabilities of image memory. A scanner exhibiting resolution of 1,000 lines/image is of dubious value if image memory contains only a 256 × 256 pixel array. Similarly, a scanner capable of resolving only 400 lines/image (a typical $\frac{3}{4}$-inch vidicon) will not efficiently fill a 1,024 × 1,024 pixel image memory. Most 1-inch vidicon scanners exhibit resolution of 700–900 lines and are appropriate to 512 or 1024 pixel image memories.

(b) *Dynamic range:* Scanners have a dark threshold below which no output is generated and a saturation threshold corresponding to maximum output. The difference in OD between the two thresholds is the dynamic range of the scanner. The dynamic range should be as wide as possible to contain the range of densities found in material to be analyzed. Autoradiographic material typically lies in the range of 0.05 to 1.5 D, which is reproduced fairly well by most vidicons.

(c) *Sensitivity:* The ability of the scanner to discriminate gray level is dependent upon its noise characteristics (see the subsection, Noise and Sensitivity, below). For quantitative autoradiographic densitometry, a scanner should be capable of providing at least 6-bit (64 gray levels) densitometric resolution (using an integrative noise-reduction procedure).

(d) *Shading:* Shading is a measure of photometric variability over the image, expressed as a proportion of the dynamic range of the scanning device. The success of shading correction procedures ultimately dictates the densitometric accuracy of vidicon and CCD scanners.

CIDs in common use incorporate optical microdensitometers or camera scanners (Table 1). Optical microdensitometers such as the scanning microscope photometer (SMP) and scanning microdensitometer scan an object with a photometer aperture. In the case of the SMP, the object lies on a stepping microscope stage. The position of the stage is usually controlled by and registered within a microprocessor. At each step of the scan a reading of aperture position and of the integrated OD under the aperture is taken and stored. SMP systems (e.g., Boyle and Whitlock, 1977; Wann et al., 1974; Zeiss ZONAX; Leitz DADS-560) have wide dynamic range, high sensitivity, and high precision. They are, however, very slow (typically about 20 min. to scan a 512 × 512 pixel image). Furthermore, because the mechanical and optical components of an SMP are expensive, the image processing and display capabilities of a moderately priced (<$50,000) unit are rather limited. The DADS, for example, can only provide an eight-color display of a maximum of 16,384 data points (128 × 128 pixels). Sophisticated scanning microscope systems (e.g., Cambridge Quantimet 720; Leitz TAS; Zeiss IBAS) incorporate vidicon scanners rather than photometer heads. This arrangement provides the densitometric characteristics and flexibility of a vidicon-based CID.

A scanning microdensitometer (SMD) contains a rotating drum (e.g., Optronics P1700) or a flat plate (e.g., Joyce-Loebl Microdensitometer 6) upon which the film sheet is mounted. In the drum scanners, a light source and detector mounted on a lead screw move past the rotating film clamped to the drum. On the flatbed scanners, the film is moved between the fixed illumination source below and the collector above. Various sizes of aperture can be selected to scan the film at a fixed resolution. Scanning microdensitometers are accurate, have wide dynamic range, high sensitivity, and excellent spatial linearity. Their major disadvantages are high cost (>$40,000), a limited set of available resolutions (aperture settings), a

Table 1: Scanner Manufacturers

Data rate:	high (<5 sec) medium (6 sec–3 min) low (>3 min)
Spatial resolution:	high (1,000 × 1,000 pixels or better) medium (about 500 × 500 pixels) low (256 × 256 pixels or less)
Density resolution:	high (7 bit or better with software correction) medium (6–7 bit) low (<6 bit)
Cost:	high (>$15,000) medium ($5,000–$15,000) low (<$5,000)

Manufacturer	Scanner Type	Data Rate	Spatial Resolution	Density Resolution	Cost
Cohu Inc. San Diego, CA	vidicon	high	medium–high	medium	medium
Dage-MTI Inc. Michigan City, IN	vidicon	high	medium–high	medium	low
Datacopy Corp. Palo Alto, CA	CCD	medium	medium–high	?	low–medium
Eikonix Corp. Bedford, MA	CCD	medium	high	high	high
Fairchild Palo Alto, CA	CCD	medium	low–high	?	low–medium
Hamamatsu Systems Inc. Waltham, MA	vidicon	high	medium	medium	medium
Joyce-Loebl Malden, MA	SMD SMP	medium low	high high	high high	high high
E. Leitz Inc. Rockleigh, NJ	SMP	low	high	high	high
Optronics International Chelmsford, MA	SMD	medium	high	high	high
Sierra Scientific Corp. Mountain View, CA	vidicon	high	medium–high	medium	medium
Carl Zeiss Inc. Thornwood, NY	SMP	low	high	high	high

Note: This partial listing includes only companies manufacturing their own scanners, and whose equipment appears to be suitable for incorporation into a CID. Most scanner manufacturers offer a range of resolutions and prices within their price category. This is particularly true of vidicons. Scanning microscope photometers include the microscope and stepping stage as essential elements. The data rate is the time required to acquire a 500 × 500 pixel scan of an image approximately the size of a rat brain.

relatively slow scan rate, and the fixed ultimate resolution dictated by the scanning step accuracy of the device.

The line-scan charge coupled device (CCD) camera uses a specialized integrated circuit consisting of a linear array of individual photodiodes. A lens forms a focal plane image across which the photodiode array is carried by a lead screw. The line-scan CCD camera exhibits wide dynamic range, good response to low light levels, and a near-linear response to incident illumination. Typically, it scans at about the same speed as the scanning microdensitometer, requiring up to several minutes for a high-resolution scan of a section of rat brain. CCD area arrays (e.g., Fairchild CCD3000, Table 1), which scan the image electronically and can provide data at video frame rates, are a more recent development. The resolution of such cameras is at present limited because of the difficulty and expense involved in producing large, error-free photodiode arrays. In the near future, however, area CCD cameras may replace vidicons in many applications.

Although CCD cameras can be very cheap (e.g., Fairchild CCD 1200C, Table 1), those that have been packaged as precision scanners incorporate manufacturer-supplied diode-response compensation and software, which increase their cost. Such devices (e.g., Eikonix EC 78/79, Table 1) are relatively expensive (>$20,000). The line-scan Eikonix EC 78/79 may perform almost as well as a scanning microdensitometer (Ramm et al., 1984), provides some of the convenience of a video system (e.g., continuous magnification selection), and does not require user software correction for scanner nonlinearities. Its price lies midway between that of the scanning microdensitometer and of the vidicon camera.

Because scanning microscope photometers are slow, scanning microdensitometers expensive, and most CCD devices unproven, vidicon camera scanners are used in many CIDs. Their characteristics are discussed in a later section.

The General Purpose Processor

The most common form of image display is the cathode ray tube (CRT). Most CRTs are designed to display a new image (refresh) 30 times a second (the standard video frame rate). This refresh rate allows the perception of a continuous display rather than a succession of discrete images. If we consider a typical image to consist of 500 × 500 pixels, then a monochromatic, 30 frame/sec display system must be supplied with data at a rate of 7.5 million bytes/sec. A real-color image with independent red, green, and blue data valves can require a byte transfer rate of over 20 million/sec. These data transfer rates far exceed the capabilities of almost all computers. Therefore, imaging systems have two parts: a computation part and an imaging part. The imaging hardware is constructed to transfer massive amounts of data at high speed. The GPP has responsibility for running the imaging programs and system utilities, and sometimes for processing the stored image.

Ideally, a computer used for imaging should have the following features:

1. There should be a large, directly addressable memory store, preferably 1 MB or more.
2. There should be a complete and fast instruction set for integer arithmetic (arithmetic execution time on the order of 1 μsec). There should be a compatible floating-point arithmetic processor, preferably supported by the processor itself rather than requiring use of an add-on array processor that may not be easily integrated into existing software.
3. The processor should have virtual memory capability. Thus, memory expansions are handled automatically by the operating system and are transparent to the user. With nonvirtual systems, each expansion in system capabilities requires major adjustments in the use of image storage devices (such as disks) and other system facilities.

Although the above criteria define a powerful imaging computer, practical requirements for a GPP vary with the funds available and with the type of imaging to be performed. General-purpose processors can be large computers (e.g., IBM 370, VAX 11/780), capable of very fast program execution. The advantage of this arrangement is that, because such computers are already in place at many institutions, it may be possible to avoid purchase of a dedicated imaging computer. The disadvantage of mainframe computers is that they are costly and, therefore, must function as time-sharing machines. Thus, their several million bytes of memory are distributed among many users and calculations requiring large databases or extensive interactive procedures (such as autoradiographic imaging) are performed very poorly.

If it is only planned to image and sample density data from autoradiographs, particularly at low resolution, a GPP can be selected from the large family of single-user microprocessors such as the DEC PDP 11/23 or IBM PC. The *Q* bus on the 11/23, in particular, can be interfaced to a very wide range of peripherals including video digitizers and frame buffers. Both the PDP 11 and IBM PC have adequate memory and moderately fast internal buses (*Q* bus and IBM PC, approximately 1 MB/sec), but limited directly addressable memory. This limitation makes access to the maximum memory of the computer difficult. These machines also have limited facilities for memory management, so that expansion beyond 1 MB (IBM PC) is not possible. It is, however, possible to perform competent imaging with machines such as the PDP 11 or IBM PC. In contrast, lower-cost home computers are not suitable for imaging tasks. They use a relatively slow bus (e.g., S100), which degrades performance for even the least demanding autoradiographic imaging. Further, there are very few high-quality peripherals available for these computers.

If complex imaging tasks, such as image overlay by intercorrelation, cell reconstruction, or feature detection and processing, are likely to be required, it is best to select a highly capable processor. These processors speed imaging calculations because they incorporate buses of wide bandwidth (e.g., Multibus, Unibus, and Versabus). Some very fast computers incorporate multiple bus structures on a single backplane (e.g., Multibus II), which permit multiple simultaneous data transfers concurrent with processor operations. A further advantage of these high-performance computers is that they can be interfaced with a wide range of quality peripherals.

Fortunately, a new generation of 32-bit processors is becoming available. These processors (e.g., Motorola 68020, National Semiconductor 16032/32032, DEC Microvax) have three attributes that make them particularly attractive for imaging applications.

1. They have large address space and can directly access several million bytes of real or virtual memory.
2. They are fast. Like the larger mainframes, these devices (with the exception of the Microvax) have wide bandwidth bus structures (e.g., Multibus; 10 MB/sec) and can execute integer computations at the rate of about 1 million/sec or floating-point computations (16,032/32,032) at about 250,000/sec.
3. They have virtual memory capability.
4. Low-cost computer systems using these processors are available in the $10,000–$20,000 range, placing them in price competition with personal computers.

The latest generation of image processing systems is distinguished by the inclusion of a high-speed GPP based on one of these microprocessor units. Thus, it is becoming more common to purchase an image processor as a stand-alone unit containing a general-purpose computer.

Two other developments in the microcomputer area hold particular promise for imaging. First, inexpensive disk subsystems suitable for single-user computers and capable of storing large numbers of high-resolution images are becoming available. A 50-MB disk subsystem for one of the new generation processors will cost about $4,000. Thus, a very respectable computer with image storage capability can be secured for less than $20,000. Secondly, manufacturers are developing array processors designed for use in single-user computer systems. These array processors share resources with the local processor and, although most lack the software support available for the more expensive derivatives of mainframe array coprocessors, they are much less expensive (<$5,000). Performance improvements with these array processors should be up to an order of magnitude. Therefore, a reasonably priced single-user system can be configured to meet any computational load now serviced by a mainframe computer like the DEC VAX-780.

The Image Processor

The imaging portion of the system generally contains a memory array for image storage and a high-speed translation table placed in the data path between the image memory and the display screen (Figure 1b,c). By changing values in the translation table, the visible image may be rapidly changed. Such operations as pseudocolor display and contrast manipulation operate via the translation table. A device containing only these capabilities is usually called a *display controller.* Once the separate image data path has been established, however, it is possible to place more computational facilities (special-purpose processors) in the data path. The limiting factor for real-time operation is the requirement that the hardware process image data at the bandwidth of the display process. The placement of special-purpose processors in the data path typically allows real-time operations including image averaging (a noise reduction procedure).

The simplest form of CID (Figure 1a) includes a scanner, digitizer, and GPP with peripherals. Because a display controller is not included, image data are held in the data memory space of the general purpose processor. There are serious limitations to this approach:

1. Since data memory is limited, the number of pixels that can be stored in most GPPs is also limited. Manipulation of pixels in larger arrays requires time-consuming disk read-write operations, which degrade system performance.
2. A typical microcomputer has a modest bus bandwidth, and cannot acquire a high-resolution image in any reasonable time.
3. Processing of the image pixels on an individual basis (such as is required for gray-level histograms and contrast manipulation) is very slow.
4. Since an image memory and associated hardware (lookup tables, DACs) are not present, the image cannot be displayed.

A more powerful configuration (Figure 1c) includes the image memory and a special purpose processor tied together on the wide-bandwidth image processor bus, and operating independently of the GPP. The performance benefit of this approach is to shift the load of data acquisition and display to a specialized subsystem, allowing the CID to achieve performance levels higher than those of stand-alone GPPs. A configuration that includes a special-purpose processor is almost mandatory if operations requiring massive arithmetic throughput (e.g., image averaging) are required.

Selection of the image processor component requires careful consideration of the end use. Whether selection is made from commercially packaged processor systems or from individual components, the following device capabilities should be considered when configuring a system:

(a) *Size of image memory:* Usually, imaging systems are designed to work with images of about 256 × 256 to 2,048 × 2,048 pixels. A 256 × 256 array is adequate for most autoradiographic densitometry.

(b) *Number of memory planes:* Although an image can be stored in and analyzed from a single memory plane, multiple image memory planes (or a single large plane that can be segmented to hold a number of images) are useful for overlay of images or simultaneous multiple displays. In double-label autoradiography, for example, one memory plane could contain a ^{131}I-labeled autoradiograph, a second plane a ^{14}C-labeled image, and a third plane an image of the stained tissue. If the three images are made to lie in the same memory locations of each memory plane (by manual overlay or intercorrelation), it is possible to sample the same regions from all three planes at once. Thus, the operator can adjust the placement of the sample window on the stained section while obtaining data from the corresponding portions of the ^{131}I and ^{14}C autoradiographs.

(c) *Bit density:* Most vidicon cameras provide about 4-bit precision on a single scan, and scanning microdensitometers 7–8. To attain 7–8-bit densitometric precision, vidicon CIDs require image storage capabilities of 12–16 bits. This permits noise reduction by image averaging.

(d) *Arithmetic capabilities:* Arithmetic operations are performed by the special-purpose processor at a rate much faster than is possible for the GPP. The arithmetic capability of most importance for video densitometry is that of image averaging. Other data transformations may be necessary for tasks such as automated feature extraction or surface modeling. For these purposes, card-mounted special-purpose processors or extremely capable external units can be interfaced with most image processors.

CID Costs

A number of expensive ($90,000 and up) commercial systems offer software support for such applications as cell typing, stereology, particle analysis, grain counting, karyotyping, and x-ray digitization. Typical devices are marketed by Bausch and Lomb, Eikonix, Leitz, Joyce-Loebl, Zeiss, and others [see Chapter 9]. The high price of such CIDS results partly from a lag between the development of new hardware and its implementation in a marketable system with specialized software support. Although imaging hardware is rapidly decreasing in price, it will take time for this decrease to filter through to commercial CIDs. Therefore, one can pay a premium price for a CID with a packaged capability or one can select a very reasonably priced (and often more capable) system from the newer generation of imaging devices. The disadvantages of this "ground-up" approach to

CID construction are that the user must have an excellent conception of the end use, must carefully select hardware appropriate to that use, and must have access to programming expertise for algorithm construction.

CIDs are most cost effective in a multiuser environment. With relatively minor modifications, a device capable of imaging autoradiographs can also be used for video-enhanced contrast or video intensification microscopy (Reynolds, 1980), for automated cytometry (Wied et al., 1982), for fluorescence histochemistry (Cowen and Burnstock, 1982), and for many other tasks in which morphometry or processing of the stored image are required (Dytch et al., 1982; Gardette et al., 1982; 1977; Kressner et al., 1980; Lindsay, 1977; Miller et al., 1979; Nyssen et al., 1982; Piper, 1982; Reddy et al., 1973; Rigant et al., 1983; Schlusselberg et al., 1982; Wree et al., 1982). Potential flexibility should, therefore, be a prime requisite of any new CID.

Major costs of the CID lie in the host computer, scanning device, the image processor, and display monitor. A computer, image processor, and display monitor system can cost $10,000–$80,000. Therefore, the input scanner is a very significant portion of the total CID cost. A high-quality vidicon camera costs $2,000–$15,000, a CCD camera about $1,000–$30,000 (with necessary accessories), and a scanning microdensitometer $40,000–$150,000.

The Vidicon CID

Advantages of vidicon scanners include:

1. their low cost;
2. real-time image viewing so that selection, illumination, contrast, and position of images can be conveniently optimized;
3. continuously variable image resolution selected by optical magnification; and
4. availability of a wide variety of vidicons for various applications.

The continuously variable resolution available with vidicon (and CCD) cameras is a major advantage. The desired resolution of any scan invariably seems to fall between settings of a scanning microdensitometer (SMD) aperture. Furthermore, the spatial resolution of an SMD is ultimately limited by its mechanical precision. This limitation precludes SMD applications requiring resolution better than the minimum aperture size or step interval of the lead screw. In contrast, vidicon cameras do not have fixed resolution apertures or lead screws and it is possible to vary the effective spatial resolution continuously by using lenses or microscopes.

Of the wide choice of vidicon imaging elements, three are in most common use. These are the low-light vidicons, and vidicons with linear or curvilinear response to incident illumination. A silicon-target vidicon is particularly useful for low light-level work such as video intensification or fluorescence microscopy. The characteristics of a linear response vidicon (Chalnicon, Newvicon) are suited to densitometry. Lower-priced antimony trisulphide vidicons are perfectly adequate in applications in which precise densitometry is unnecessary.

There are also disadvantages of vidicon scanners. These include densitometric and spatial nonlinearities and lack of dynamic range. Because the practical effects of these errors upon autoradiographic densitometry have not been described, vidicon densitometry has often been limited to procedures in which within-subject normalization is used (Gallistel et al., 1982). Before evaluating the feasibility of quantitative vidicon densitometry, we discuss some sources of vidicon error and some error-correction procedures.

Vidicon Camera Basics

The sensing face of the vidicon (called the *raster*) forms a Cartesian coordinate system with each discrete x, y coordinate point or pixel representing a specific position in the image memory array. The array size (number of pixels), characteristics of the vidicon and lens, and bandwidth of the video amplifier all limit the spatial resolution of the system.

In addition to positional assignments, pixels have a density value. The amplitude of the video signal generated at any pixel is related to the intensity of illumination falling upon that pixel. The analog voltage value associated with each pixel is digitized into one of a set number of levels (usually 256) and the resulting digital value is stored in the pixel position within the memory array. A fast digitizer makes an entire raster scan available to the memory array at the video frame rate.

Reasonably priced 8-bit imaging systems in the range of 256–512 pixels/picture height have a resolution commensurate with that of a 1-inch vidicon. Larger memory arrays (e.g., 1,024 $\times$ 1,024 pixels) are relatively expensive, as are vidicon cameras with the necessary bandwidth and photoconductor elements for 1,000-line resolution. Fortunately, perfectly adequate imaging can be performed with moderate sized memory arrays (e.g., 512 $\times$ 512) and a high-quality 1-inch vidicon.

The justification for large memory arrays is to provide high-resolution representation of the image. For example, if 1 cm^2 of film occupies 256 $\times$ 256 pixels, each pixel represents a film area of about 40 $\times$ 40 μm. With a 512 $\times$ 512 array each pixel represents 20 $\times$ 20 μm while the entire 1 cm^2 of film is present in memory and is displayed on the monitor. By doubling optical magnification of the image in the 256 $\times$ 256 pixel array, each pixel is made to represent an area of 20 $\times$ 20 μm. Thus, resolution is made equivalent to that of the 512 $\times$ 512 pixel array operating at lower magnification, but now only a portion (0.25 cm^2) of film is present in memory (and visible) at any one time (Figure 2).

Shading

Equal levels of incident illumination do not produce equal signal values from all points of the raster. The deviations from uniform output are typically less than 10% of the dynamic range of the vidicon. Sources of this shading error include the illumination source, camera lens, and camera electronics.

To what degree must shading be controlled? The typical vidicon camera can provide 256 gray levels (with image averaging). To sample 256 levels without quantitizing errors would require that shading be controlled to within 0.4%. Such fine correction is difficult to implement and may be unnecessary for autoradiography. Density resolution is practically limited by technical errors and film characteristics. Therefore, we have defined an adequate shading correction as one which results in the presence of less than $\pm 1.5\%$ error, through the density range in which the data lie.

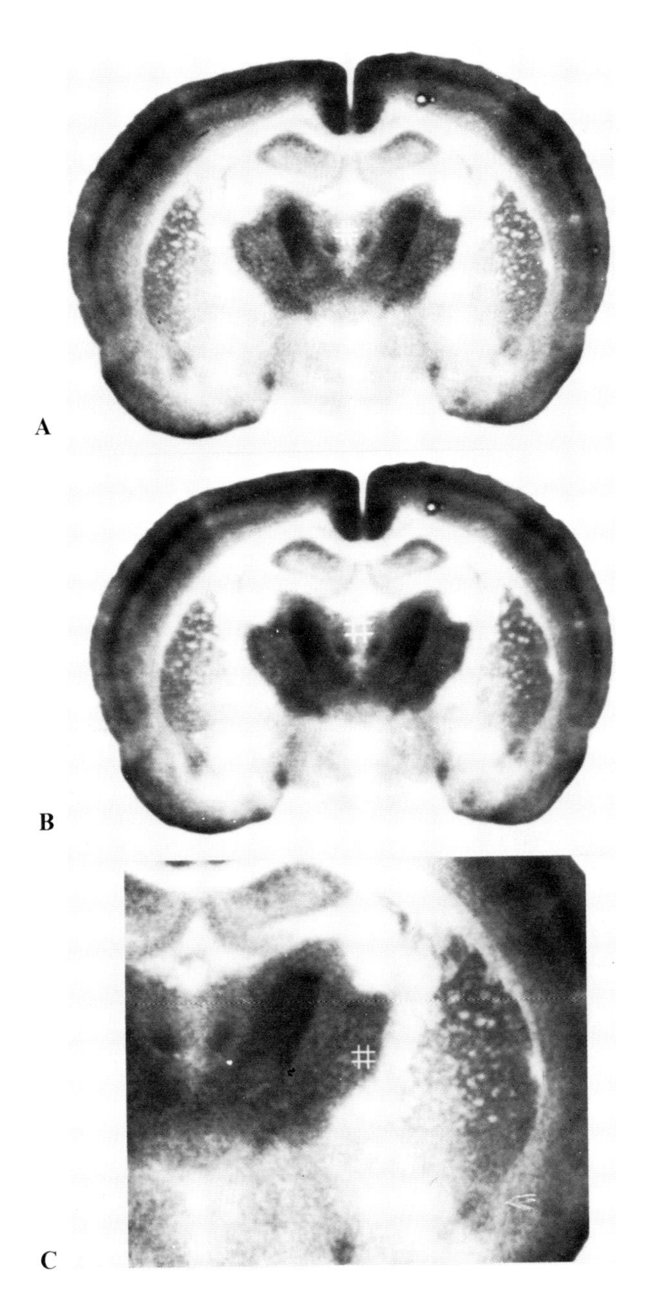

Figure 2. Digitized autoradiographs of rat brain at resolutions of (A) 30 μm/pixel (512 × 512 pixel array) and (B) 60 μm/pixel (256 × 256 pixel array). The 256 × 256 pixel array can contain image data at equivalent resolution to the larger array [a high-resolution image in the smaller array is shown in (C)], but less of the image is stored. The window visible in each image is 10 × 10 pixels in size.

The Shading Correction Algorithm

A simple shading correction is applied by first digitizing and storing a defocused blank field of medium luminance (the error field). Each area of the error field is then compared to the mean density value of the whole field. Any deviation from reference value is expressed as a proportion of full scale, and this proportion is stored as the shading correction factor for that portion of the field. On every subsequent sampling of a given region, the actual value read is multiplied by the shading correction factor for that region. If time is available, the entire image can be stored, corrected, and redisplayed in corrected form. The simple shading correction works moderately well in that worst case shading errors can be brought to under ±1% with densities close to that at which the error field was established. If the edges of the field are avoided, shading error is typically less than ±1.5% across a range spanning most autoradiographic data. This level of correction permits (with frame summing) practical sampling at 6–7-bit densitometric resolution.

There are nonlinearities associated with the simple shading correction. An error field established at a given luminance is not appropriate for deriving correction factors to be applied at other luminances. In practical terms, if shading error is less than ±1% at 50% transmission, values in the 10% and 90% range will retain up to ±3% shading error. There are two ways to decrease this error:

1. A better correction can be obtained by using two error fields, established at low and high transmission values. Each sampled area in the autoradiograph is then corrected by interpolation between the error fields. This procedure may bring shading across the entire image, and across the density range to about ±1%.
2. An area of low shading can be determined. As shading error is minimal in this region (usually midraster), the error correction and its associated nonlinearity are also minimal. Extreme data values can now be sampled in this area of the raster.

Nonlinear Response to Illumination

Two classes of vidicon are defined by the dependency between signal output and intensity of incident illumination. The relation is expressed as $I = KL^{\gamma}$, where I is signal current, K is a constant, and L is the level of incident illumination. The power γ varies from about 0.65 for the antimony trisulphide vidicon to unity for vidicons using most other photoconductors (Figure 3). The 0.65-γ vidicon has wide dynamic range (approximately 350:1) but requires a relatively high level of incident illumination and exhibits compressed response to high luminance levels. Densitometry with this vidicon requires a hardware or software correction for nonlinearity. For these reasons, it is best to select a photosensor whose signal output is more directly proportional to intensity input. Although cadmium selen-

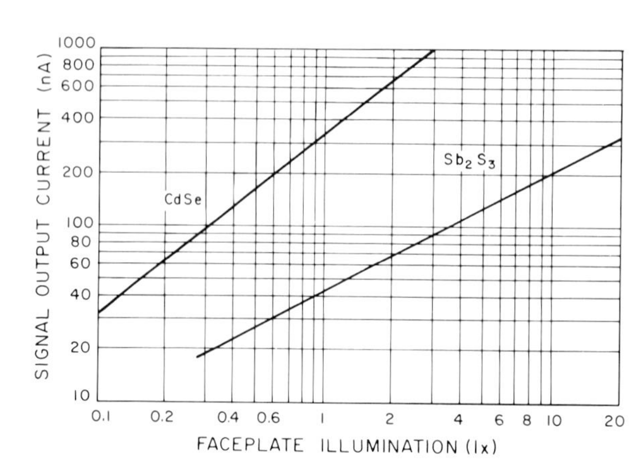

Figure 3. Light transfer characteristics of antimony trisulphide and cadmium selenide (Chalnicon) vidicons. The trisulphide vidicon (γ = 0.65) exhibits progressively compressed response at high levels of incident illumination. The unity-γ Chalnicon exhibits superior sensitivity and linearity at all levels of faceplate illumination. [Adapted from data provided by Hamamatsu and Hitachi.]

ide and cadmium–zinc telluride vidicons such as the Chalnicon and Newvicon have a dynamic range of only 100:1, they are characterized by unity γ, low dark current, high sensitivity, low lag, and negligible burn-in. The combined advantages, we feel, outweight the dynamic range limitations of these vidicons.

Noise and Sensitivity

The analog video output contains both the signal and a white noise component contributed by the camera. As the number of levels comprising the digitized signal (bit density) increases, the size of the interlevel increment decreases relative to the noise component. At some point, the interlevel increment becomes so small that there is error in the discrimination between adjacent shades of gray. For accurate quantification, signal amplitude must be large enough to negate the effects of noise.

A typical vidicon camera operating at a bandwidth of 8 MHz (signal-to-noise ratio decreases with increasing bandwidth) can reliably discriminate 16 levels of gray (4 bits). To increase density resolution, the camera's effective signal-to-noise ratio must be increased by *averaging frames* or by spatial averaging across pixels. As frames or pixels are averaged N times, the root-mean-square amplitude of noise is decreased by a factor of $1/\sqrt{N}$. Theoretically, averaging of 100 frames (or pixels) acquired by a vidicon with a 40 db S/N ratio allows a 60 db S/N ratio to be achieved. This S/N ratio is appropriate to 8 bits (256 levels) of gray level resolution. Frame averaging requires a special-purpose processor to add new image data rapidly to the memory bank contents. Therefore, image memory must have sufficient depth to hold multiple frames of image data. Fifteen bits are required for 128 frames and 16 bits for 256 frames.

An alternative to frame averaging is *spatial averaging*. A lower limit is placed upon the size of the sample window used. Spatial averaging of 256 pixels (a 16×16 pixel window) yields a density resolution (8 bit) equivalent to that obtainable in a 1×1 pixel window with averaging of 256 frames. Thus, accurate densitometry is possible with a cost-efficient 8-bit memory plane.

Modulation Transfer Function

The response of any densitometric system (including the SMP and SMD; see Gallistel and Nichols, 1983) is a function of the spatial frequency of the input. A vidicon presented with successively finer sine wave frequencies on a luminant field would ideally produce a sine wave voltage that varied in frequency but not in amplitude. In fact, however, as spatial frequency increases, modulation of the video output decreases in a nonlinear fashion (Figure 4). The practical consequence of this effect is a decrease in the visible contrast and measured density of very small regions in the autoradiograph.

To avoid nonlinearities due to the modulation transfer function (MTF), higher optical magnifications or smaller microdensitometer apertures may be necessary in densitometry of small regions. In vidicon densitometry, the measured area should be sufficiently wide to fall in the spatial frequency range where close to 100% contrast is obtained. The Hamamatsu N1453 vidicon (Chalnicon), for example, exhibits in response to a line with a thickness equivalent to $\frac{1}{100}$ the height of the photoconductor surface (100 to line resolution). The densities of narrower

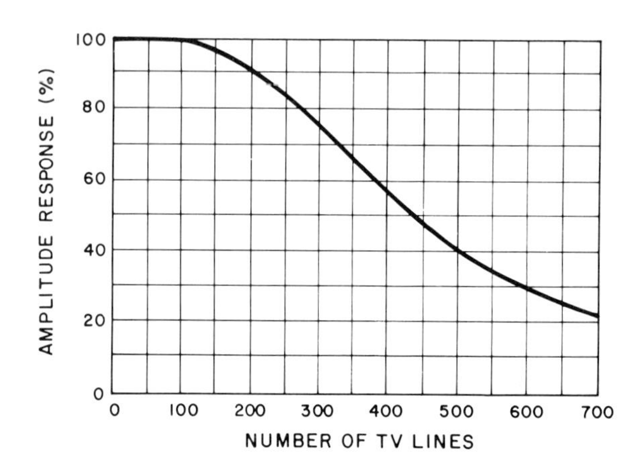

Figure 4. Contrast transfer of the Hamamatsu N1453 Chalnicon. As image spatial frequency increases, there is decreased output from the camera. However, there is a range in which image contrast is faithfully rendered. Most autoradiographic features lie in this range. [Adapted from Hamamatsu bulletin SC-5-3.]

lines are underestimated. As our imaging system represents the raster with a 512 × 512 pixel array, 100-line resolution is equivalent to features approximately 5 pixels in width. Therefore, we select optical magnification so that the smallest features occupy at least five pixels in the image. At the magnification of 30 μm/pixel (Figure 2), convenient for displaying an entire section of rat brain, very few features are less than 5 pixels wide.

Nonlinearities Resulting from the Autoradiographs

Although corrected video densitometry can yield data unaffected by idiosyncratic characteristics of the imaging system (Figure 5), there are also autoradiographic sources of nonlinearity. First, the relation between film OD and radioactivity in the underlying tissue is nonlinear (Goochee et al., 1980; Sharp et al., 1983). Therefore, the relation between the radioactivity and density scales must be flexible to permit use of various films and exposure ranges.

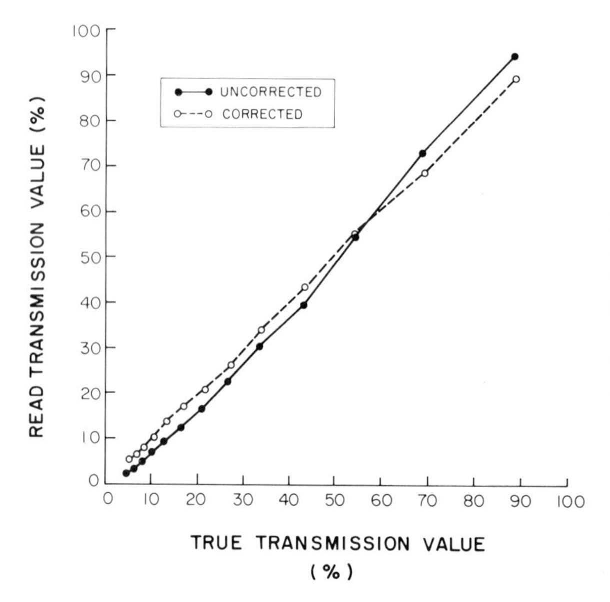

Figure 5. Linearity of an uncorrected (●———●) and corrected (○– –○) vidicon-based CID scanning a photographic step tablet spanning the density range contained in most autoradiographic data.

A simple correction for film nonlinearity is applied by setting the system to the exact illumination and magnification levels to be employed in the analysis session. A set of film density standards spanning the data range is then read. These standards are areas of film exposed to radioactive materials calibrated in units of radioactivity per equivalent weight of tissue, or steps on a calibrated photographic wedge. Any change in illumination or other conditions requires entering the entire standard set again. In our hands, this procedure results in a system response that is linear to within 1% across the dynamic range of our camera.

An alternative procedure allows minor changes in illumination and magnification, and allows periodic corrections for any drift in system response. First, linearity is established on a photographic wedge spanning the range of autoradiographic data in 0.05-D steps. Data are subsequently read in units of percent transmission, which can be converted to μCi/g or to levels of cerebral metabolic rate for glucose (μmol/100g/min). The system is then given a densitometric reference point using a SET TO STANDARD command. To set to standard, the CID is again presented with the calibration standard. It is instructed to sample a mid-density step and to derive a factor by which the sampled value must be multiplied to equal the calibration value. Every subsequent image data value is multiplied by the adjustment factor. If the data continue to lie in the range for which the densitometric scale has been made linear, the multiplicative correction is appropriate. For example, a 10% increase in illumination level followed by set to standard results in a system response linear to within 1.5%. A change in magnification from 60 μm/pixel to 30 μm/pixel leaves system response linear to within 2%.

Film Noise

A uniformly exposed area of film contains randomly dispersed, opaque silver halide crystals ranging from about 0.2 to 2 μm in size. These cannot be individually resolved at low magnifications. The granularity evident at low magnifications results from random groupings of grains, which introduce noise into the densitometric procedure. Lower magnifications and larger apertures integrate spatially and result in lower granularity and noise.

Our scans of the film used for precision step wedges suggest a 6–7-bit precision over an area of 100 × 100 μm. The x-ray film used for most autoradiography does not approach even this accuracy. We have found that on SB5 film, radioactivity standards differing by 1% or 2% require averaging of film areas larger than many brain regions for reliable discrimination (Ramm et al., 1984). Scans (25 μm/pixel) of film exposed to methacrylate ^{14}C standards showed pixel-to-pixel variability of about ±2.5% in XT-L and of about ±6% in SB-5. These scans were performed in regions of moderate film density. Film noise becomes an even more serious problem in regions in which exposure is nonoptimal. Areas of low contrast (resulting from underexposure), in particular, exhibit increased noise, whereas areas of film saturation show poor densitometric resolution due to a decrease in film sensitivity at high exposure levels. Thus, film noise imposes a practical limit upon the ability of the CID to quantify density in high resolution images. The effects of film noise must be minimized by spatial averaging with larger apertures (SMD), lower magnifications (camera), or windows containing more pixels (CID). Large film areas rarely contain homogeneous density information and are difficult to assign to individual brain regions. Therefore, it is rare that high-precision density data are available from film autoradiographs. Enthusiasm for any high-precision analog scanner must be tempered by this fact.

A Typical CID Configuration

We now describe hardware comprising a CID presently in routine use, and summarize features of the software used for densitometry. We then briefly evaluate performance of this CID and make specific recommendations regarding construction of a low-cost densitometric system.

Scanning Devices

The Hamamatsu N1453-01 video camera (Table 1) is fitted with external drive, an N1453 1-inch Chalnicon, and a 55-mm micro-Nikkor lens set at *f*8. Horizontal resolution of the camera is better than 700 lines. Amplitude response (modulation transfer) is 60% at 400 lines. Shading (before software correction) is 9.7%, and spatial distortion is $\pm 1\%$ using EIAJ test chart C2. Corrected shading is less than 1.5% in the central 200 $\times$ 200 pixel region of the raster across the range of 10%–90% transmission. Shading at 50% transmission is less than 0.5% in the central portions of the raster and less than 1.5% over the entire raster.

Illumination is provided by an Olympus PMT macroilluminator fitted with an opal glass diffuser mounted 2 cm below the stage surface, or by a light table of our own design. The illuminators are powered by a variable voltage, highly stable DC power supply (our construction) fitted with digital readout of lamp voltage.

The SMD, an Optronics P1700, scans with square apertures of 12.5, 25, 50, 100, 200, or 400 μm. Its integral digitizer provides 8 bits (256 levels) of gray-scale discrimination on a 0–2-D logarithmic scale. The SMD is interfaced to an independent imaging system consisting of a VAX 11/780 and NORPAK graphics system running under the UNIX operating system. Scans performed on this system are stored on tape for transfer to the PANDA system.

The Image Processor and Host Computer

The analog video output, or the digitized SMD output, is fed to a Grinnell Systems (San Jose, CA) GMR 270 Display System (512 $\times$ 512 visible resolution, 30-Hz refresh format). The display system contains an 8-bit digitizer (bypassed with SMD input), which can fill one of the two independent memory planes in $\frac{1}{30}$ sec. A processor card allows for true pipelined operations on one or more display channels in $\frac{1}{30}$ sec. Addition, subtraction, and various logical functions can be directly executed on 8-bit quantities. There are facilities for zooming images by factors of 1, 2, 4, or 8. Also present (selectable) is one 12-bit monochrome lookup table or three 8-bit DACS with color lookup tables. Image data is routed through the lookup tables and combined with video synchronization to provide three 8-bit video signals to drive the RGB color monitor, or one 12-bit signal to drive a B/W monitor.

A sample window is generated and moved by a quad independent cursor card and joystick control, unit or Summagraphics (Fairfield, CT) mouse. Grayscale or pseudocolor images are displayed on an Hitachi HM-2719 19-inch color monitor with resolution of 720 pixels horizontal and 540 pixels vertical.

The host computer, a PDP 11/23 running under RT11, is fitted with dual floppy and dual RL02 hard disk drives, 224 KB of memory, and a direct memory access interface to the GMR 270 unit. The control terminal is a DEC VT100.

Data are analyzed using the PANDA (Program for Analysis of Digitized Autoradiographs) routines developed in our laboratory. PANDA is written in Whitesmith's C to run under RT11, but could easily be transferred to a UNIX system and standard C compiler. PANDA's functional modalities include the following:

1. select input mode (video or SMD);
2. manipulate image, including digitization, summing, storage, and retrieval from disk;
3. apply shading and linearity corrections to the video input;
4. manipulate display (gray scale, pseudocolor, thresholding, contrast);
5. manipulate a variable sample window under joystick or mouse control;
6. convert digitized levels to units of percent transmission, optical density, μCi/g or glucose utilization;
7. express data in normalized or ratio form;
8. create data files and perform statistical summary;

A detailed evaluation of the PANDA system is presented elsewhere (Ramm et al., 1984). Briefly, we obtain a comparable level of densitometric accuracy with the SMD or vidicon (Table 2). We suggest that this level of accuracy is appropriate to autoradiographic densitometry. However, the SMD can be made to yield better ultimate accuracy and repeatability than the vidicon scanner. For example, SMD accuracy could be improved by use of the calibration procedures, which work well with the vidicon camera. Although we do not have a CCD camera in the system, our limited experience with the Eikonix unit (Ramm et al., 1984) leads us to believe that the CCD camera is a high-precision alternative to the SMD at a more reasonable cost.

Conclusions

An autoradiograph is used as an analog record of radioactivity distribution in underlying tissue. The most important characteristic of an autoradiographic densitometer is its ability to perform consistent measurements at a level of precision

Table 2: Response of a Vidicon Camera and SMD to Wratten No. 96 Neutral Density Filters

True Density		Camera Density (±S.D.)	SMD Density (±S.D.)
D	%	(% transmission)	(% transmission)
0.10	80.0	80.1 (1.06)	78.8 (0.16)
0.20	63.0	63.8 (0.67)	63.2 (0.10)
0.30	50.2	50.4 (0.56)	49.0 (0.08)
0.50	31.7	31.5 (0.49)	32.2 (0.13)
0.70	20.0	20.2 (0.55)	19.8 (0.09)
1.00	10.0	10.9 (1.20)	10.5 (0.04)
1.50	3.2	3.9 (0.52)	3.2 (0.03)
2.00	1.0	0.8 (0.31)	0.8 (0.00)

Note: Both scanners read the filter series five times and were turned off between each scan of the series. In addition, the camera error corrections were reestablished between each scan. Thus, the standard deviations provide an estimate of session-to-session repeatability attainable with a corrected vidicon camera and uncorrected SMD. Note that accuracy of the scanners is comparable. Repeatability of camera is deemed acceptable, that of the SMD is excellent.

at least equal to that of the analog record. The individual investigator must decide whether his or her autoradiographs and preparation techniques provide an analog record that can tax the 6–7-bit precision of vidicon densitometry. We suggest that in a well-configured CID, vidicon performance is superior to the sensitivity of the autoradiographic method. Methodological sensitivity is limited by aspects of the autoradiographic procedure (e.g., variation in section thickness, calibration of radioactivity standards) and properties of autoradiographic materials (e.g., film noise and sensitivity).

In sum, a low-cost imaging system can be constructed using a vidicon scanner, a display controller (or more complex imaging system), and any of a variety of host processors. Such a system is capable of accurate densitometry and is applicable to many other imaging tasks. We estimate the total cost of basic systems to be in the $20,000 to $35,000 range, which is less than the cost of many technologically inferior commercial systems and includes a powerful general-purpose computer. For example, our uses of the PANDA hardware range from text editing to the presentation of electronically generated stimuli during experiments in visual psychophysics or single-cell physiology.

The major difficulty of the user-implemented approach lies in construction of appropriate algorithms. As most laboratories write their own programs, a great deal of time can be wasted in implementing new capabilities. We suggest that the portability of algorithms will be maximized if neuroscientists using imaging equipment standardize on an operating system (e.g., UNIX) and programming language (e.g., C). If this were done, a library of routines for such purposes as densitometry, cell reconstruction, and analysis of electron microscope images would rapidly accumulate. Minor program changes could then allow transfer between CIDs. These changes are a trivial task in comparison to the construction of an entirely new program. Further, the ability to easily implement multiple tasks would encourage cooperative purchasing and would bring the benefits of imaging to many more investigators.

References

Alexander, G.M., R.J. Schwartzman, R.D. Bell, J. Yu, and A. Renthal (1981) Quantitative measurement of local cerebral metabolic rate for glucose utilizing tritiated 2-deoxyglucose. Brain Res. 223: 59–67.

Altman, J.H. (1977) Sensitometry of black-and-white materials. In T.H. James (ed.): The Theory of the Photographic Process. New York: Macmillan, pp. 481–516.

Benno, R.H., L.W. Tucker, T.H. Joh, and D.J. Reis (1982) Quantitative immunocytochemistry of tyrosine hydroxylase in rat brain. I. Development of a computer-assisted method using the peroxidase–antiperoxidase technique. Brain Res. 246: 225–236.

Boyle, P.J.R., and D.G. Whitlock (1977) A computer-controlled microscope as a device for evaluating autoradiographs. In R.D. Lindsay (ed.): Computer Analysis of Neuronal Structures. New York: Plenum Press, pp. 1–19.

Bryant, H.J., and F.A. Kutyna (1983) The development and evaluation of a low-cost microdensitometer for use with the 2-deoxy-D-glucose method of functional brain mapping. J. Neurosci. Methods. 8: 61–72.

Capowski, J.J. (1983) An automatic neuron reconstruction system. J. Neurosci. Methods. 8: 353–364.

Cowen, T., and G. Burnstock (1982) Image analysis of catecholamine flourescence. Brain Res. Bull. 9: 81–86.

Dauth, G. W., K.A. Frey, and S. Gilman (1983) A densitometer for quantitative autoradiography. J. Neurosci. Methods 9: 243–251.

Dytch, H.E., P.H. Bartels, M. Bibbo, F.T. Pishotta, and G.L. Wied (1982) Computergraphics in cytodiagnosis. Anal. Quant. Cytol. 4: 263–268.

Gallistel, C.R., and S. Nichols (1983) Resolution-limiting factors in 2-deoxyglucose autoradiography. I. Factors other than diffusion. Brain Res. 267: 323–333.

Gallistel, C.R., C.T. Piner, T.O. Allen, N.T. Adler, E. Yadin, and M. Negin (1982) Computer assisted analysis of 2-DG autoradiographs. Neurosci. Biobehav. Rev. 6: 409–420.

Gardette, R., R. Joubert, and J.C. Bisconte (1982) A standardized automatic procedure to evaluate cell numbers in low cell density tissues by image analysis. Microsc. Acta 86: 105–116.

Goochee, C., W.R. Rasband, and L. Sokoloff (1980) Computerized densitometry and color coding of [^{14}C] deoxyglucose autoradiographs. Ann. Neurol. 7: 359–370.

Haas, R.A., D.M. Robertson, and N. Meyers (1975) Microscope densitometer system for point measurement of autoradiograms. Stain Technol. 50: 137–141.

Herkenham, M., and C.B. Pert (1982) Light microscopic localization of brain opiate receptors: A general autoradiographic method which preserves tissue quality. J. Neurosci. 2: 1129–1149.

Kennedy, C., S. Suda, C.B. Smith, M. Miyaoka, M. Ito, and L. Sokoloff (1981) Changes in protein synthesis underlying functional plasticity in immature monkey visual system. Proc. Natl. Acad. Sci. USA 87: 3950–3953.

Kressner, B.E., R.R.A. Morton, A.E. Martens, S.E. Salmon, D.D. Van Hoff, and B. Soehnlein (1980) Use of an image analysis system to count colonies in stem cell assays of human tumours. In S.E. Salmon (ed.): Cloning of Human Tumour Stem Cells. New York: Alan R. Liss, pp. 179–193.

Kriss, M.A. (1977) Image Structure. In T.H. James (ed.): The Theory of the Photographic Process. New York: Macmillan, pp. 592–635.

Lewis, M.E., M. Mishkin, E. Bragin, R.M. Brown, C.B. Pert, and A. Pert (1981) Opiate receptor gradients in monkey cerebral cortex: Correspondence with sensory processing hierarchies. Science 211: 1166–1169.

Lindsay, R.D. (1977) The video computer microscope and A.R.G.O.S. In R.D. Lindsay (ed.): Computer Analysis of Neuronal Structures. New York: Plenum Press, pp. 1–19.

Macagno, E.R., C. Levinthal, and I. Sobel (1979) Three-dimensional computer reconstruction of neurons and neuron assemblies. Ann. Rev. Biophys. Bioeng. 8: 323–351.

Miller, A.K.H., R.L. Alston, and J.A.N. Corsellis (1979) The practical application of image analyzing systems to neuropathology: Principles and possible scope. In W.T. Smith and J.B. Cavanaugh (eds.): Recent Advances in Neuropathology, vol. 1. Edinburgh: Churchill-Livingstone, pp. 113–128.

Nagy, G. (1983) Optical scanning digitizers. Computer 16: 13–24.

Nyssen, M., A. Cornelis, L. Maes, and R. DeZanger (1982) A system to acquire and process scanning transmission electron microscope images. Ultramicrosc. 8: 429–436.

Palacios, J.M., D.L. Niehoff, and M.J. Kuhar (1981) Receptor autoradiography with tritium-sensitive film: Potential for computerized densitometry. Neurosci. Lett. 25: 101–105.

Paldino, A., and E. Harth (1977) A measuring system for analyzing neuronal fiber structure. In R.D. Lindsay (ed.): Computer Analysis of Neuronal Structures. New York: Plenum Press, pp. 59–71.

Penny, J.B., K.A. Frey, and A.B. Young (1981) Quantitative autoradiography of neurotransmitter receptors. Eur. J. Pharmacol. 72: 421–422.

Piper, J. (1982) Interactive image enhancement and analysis of prometaphase chromosomes and their band patterns. Quant. Anal. Cytol. 4: 233–240.

Rainbow, T.C., W.V. Bleisch, A. Biegon, and B.S. McEwen (1981) Quantitative densitometry of neurotransmitter receptors. J. Neurosci. Methods 5: 127–138.

Ramm, P., J.H. Kulick, M.P. Stryker, and B.J. Frost (1984) Video and scanning microdensitometer-based imaging systems in autoradiographic densitometry. J. Neurosci. Methods 11: 89–100.

Reddy, D.R., W.J. Davis, R.B. Ohlander, and D.J. Bihary (1973) Computer analysis of neuronal structure. In S.B. Kater and C. Nicholson (eds.): Intracellular Staining in Neurobiology. New York: Springer-Verlag, pp. 227–253.

Reynolds, G.T. (1980) Applications of image intensification to low-level fluorescence studies of living cells. Microsc. Acta 83: 55–62.

Rigaut, J.P., P. Berggren, and B. Robertson (1983) Automated techniques for the study of lung alveolar stereological parameters with the IBAS image analyzer on optical microscopy sections. J. Microscopy 130: 53–61.

Schlusselberg, D.S., W.K. Smith, M.H. Lewis, B.G. Culter, and D.J. Woodward (1982). A general system for computer-based acquisition, analysis and display of medical image data. Proc. ACM: 18–25.

Sharp, F.S., T.S. Killduff, S. Bzorgchami, H.C. Heller, and F. Allen (1983) The relationship of local cerebral glucose utilization to optical density ratios. Brain Res. 263: 97–103.

Sokoloff, L., M. Reivich, C. Kennedy, M.H. Des Rosiers, C.S. Patlak, K.D. Pettigrew, O. Sakurada, and M. Shinohara (1977) The [^{14}C] deoxyglucose method for the measurement of local cerebral glucose utilization: Theory, procedure and normal values in the conscious and anesthetized albino rat. J. Neurochem. 28: 897–916.

Som, P., Y. Yonekura, Z.H. Oster, M.A. Meyer, M.L. Pelletteri, J.S. Fowler, R.R. MacGregor, J.A.G. Russell, A.P. Wolf, I. Fand, W.P. McNally, and A.B. Brill (1983) Quantitative autoradiography with radiopharmaceuticals, part 2: Applications in radiopharmaceutical research: Concise communication. J. Nucl. Med. 24: 238–244.

Unnerstall, J.R., D.L. Niehoff, M.J. Kuhar, and J.M. Palacios (1982) Quantitative receptor autoradiography using [^{3}H] ultrafilm: Application to multiple benzodiazepine receptors. J. Neurosci. Methods 6: 59–73.

Wann, D.F., J.L. Price, W.M. Cowan, and M.A. Agulnek (1974) An automated system for counting silver grains in autoradiographs. Brain Res. 81: 31–58.

Wann, D.F., T.A. Woolsey, M.L. Dierker, and W.M. Cowan (1973) An on-line digital computer system for the semi-automatic analysis of Golgi-impregnated neurons. IEEE Trans. Biomed. Eng. BME-20: 233–247.

Weibel, E.R. (1979) Stereological Methods, vol. 1. Practical Methods for Biological Morphometry. London: Academic Press.

Wied, G.L., P.H. Bartels, H.E. Dytch, F.T. Pishotta, and M. Bibbo (1982) Rapid high-resolution cytometry. Anal. Quant. Cytol. 4: 257–262.

Woolsey, T.A., and M.L. Dierker (1978) Computer assisted recording of neuroanatomical data. In R.T. Robertson (ed.): Neuroanatomical Research Techniques. New York: Academic Press, pp. 47–85.

Wooten, G.F., and M.K. Horne (1982) A new autoradiographic approach for imaging forebrain dopamine distribution. Ann. Neurol. 12: 163–168.

Wree, A., A. Schleicher, and K. Zilles (1982) Estimation of volume fractions in nervous tissue with an image analyzer. J. Neurosci. Methods 6: 29–43.

Yonekura, Y., A.B. Brill, P. Som, G.W. Bennet, and I. Fand. (1983) Quantitative autoradiography with radiopharmaceuticals, part 1: Digital film-analysis system by videodensitometry: Concise communication. J. Nucl. Med. 24: 231–237.

Zarbin, M.A., R.B. Innis, J.K. Wamsley, S.H. Snyder, and M.J. Kuhar (1983) Autoradiographic localization of cholecystokinin receptors in rodent brain. J. Neurosci. 3: 877–906.

15

A Simple Microcomputer System For Microscope Fluorometry

Ingemar Rundquist and Lennart Enerbäck

Introduction

Quantitative Cytochemistry

The use of biochemical and biophysical techniques in the analysis of single, intact cells has become increasingly important in cell biology and cell pathology. Information obtained by biochemical methods on the concentration of cellular substances is strictly limited to the mean value for a population of cells. The amount of information can be greatly increased by presenting the results from individual cell analyses as a distribution function for the population. The basic principles of quantitative cytochemistry were first described by Caspersson, who developed instruments and methods for microspectrophotometric quantitation of nucleic acids and protein in single cells (Caspersson, 1936). His methods were based on the native absorption of ultraviolet light by nucleic acids and protein. Quantitative cytochemistry has since advanced by the development of a great number of specific cytochemical methods based on selective binding of dyes or chemical reactions that result in selectively colored cellular material.

Cytofluorometry

The introduction of quantitative fluorescence techniques applied to microscopic objects (Mellors and Silver, 1951) was another important step forward. The use of microfluorometry for measurement of single, living cells was introduced by Chance and Thorell (1959). Cytofluorometry has proved especially advantageous owing to its high sensitivity and freedom from distributional errors. Instruments have also improved concurrently with the technical evolution, and many instruments for microscope fluorometry, often designed for specific applications, have been described (e.g. Caspersson et al., 1965; Chance and Legallais, 1959; Cova et al., 1974; Jotz et al., 1976; Kohen et al., 1975; Olson, 1960; Ploem et al., 1974; Rost and Pearse, 1971; Ruch and Leeman, 1973). Considerable effort has also been devoted to the construction of systems for flow cytometry (see Horan and Wheeless, 1977). Flow cytometric methods have the obvious advantage of speed and permit analysis of large cell populations within a short time. Microscope fluorometry has some unique advantages, however:

1. The cells can be identified visually before measurement.
2. Repeated measurements can be made on the same object.
3. Dynamic fluorescence variables can be analyzed in individual cells.

The limited speed of microscope fluorometry results in a statistical precision lower than that of flow cytometry because fewer cells are measured.

Application of Computers

Recent developments in electronics and computer technology have opened up new approaches for automation in many different kinds of measurement systems. Such automation can as a rule be performed by hardware electronics only, but the integration of a computer system offers great advantages in making a measurement system more flexible to use. Laboratory automation has thus been improved by the use of computers (Shapiro et al., 1976), and the introduction of microcomputers has initiated a new era of computer technology characterized by computer systems that are cheap, compact, and easy to handle (Doerr, 1978; Shipton, 1979).

Cytofluorometry can be greatly improved by using the computer system in following main tasks:

1. to control the measuring procedure;
2. to perform data acquisition and processing; and
3. to present the results in a well-arranged manner.

This chapter describes the design and use of such instruments.

Hardware

Microscope Fluorometer System

Over a period of more than a decade we constructed three systems for microscope fluorometry, based on commercially available microscope photometers (Leitz MPV I, MPV II, and MPV III) equipped for fluorescence microscopy with incident excitation illumination. The fluorometer described in detail in this chapter consists of a Leitz Orthoplan microscope stand combined with the MPV II photometer (Figure 1), but the other two microscope photometers either work in the

Figure 1. The principle of the microscope part of the cytofluorometric system. [Reprinted from I. Rundquist, Histochemistry 70:151–159, 1981, with permission of the publisher, Springer-Verlag, Heidelberg.]

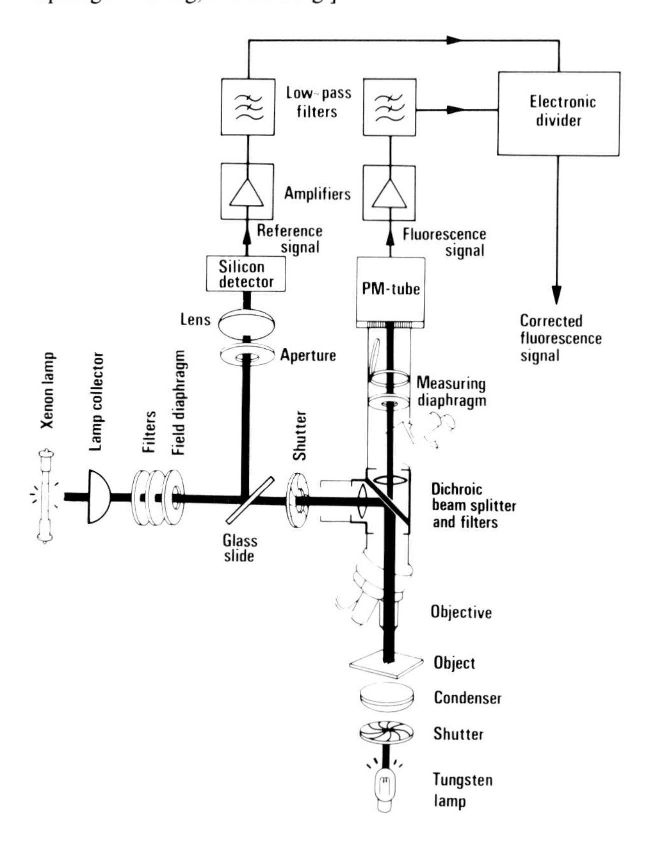

same way (MPV III) or have been modified to do so (MPV I) (Rundquist, 1981). The same principle for computer control is applied in all three instruments. There are other instruments on the market that can easily be adapted to the same type of computer control, notably the Zeiss UMSP 100 universal microscope photometer. The instruments are equipped with two illumination sources, a tungsten bulb for bright-field or phase-contrast illumination for selection and focusing of the objects to be measured, and a stabilized arc (Xe or Hg) lamp for fluorescence activation. Both illumination light paths are controlled by electromagnetic shutters, and the emission light path contains a prism, also controlled by an electromagnet that is used to make the measuring diaphragm visible in the binoculars. The entire measuring procedure can thus be controlled by electric signals. Spectral conditions can be selected by means of interchangeable excitation filters, dichroic beam splitters, and barrier filters in the emission light path. A motor-driven monochromator, containing a continuous interference line filter (Veril S-60, Schott), can be placed in the emission light path for recording of emission spectra. The principle for the optical components in microscope fluorometers has been described in detail by Ploem (1977). Instability of activating gas-discharge lamps can sometimes be a serious source of error in quantitative microfluorometry. A reference channel was therefore designed using the principle described by Ploem et al. (1974) for the automatic correction of such variations. An ordinary glass slide was used as a beam splitter, and an image of the field diaphragm was formed on the surface of a silicon detector (EG & G, type PV-215). The signal from this detector was found to be well correlated to the signal from a fluorescent standard. A permanent standard that could easily be swung into the light path of the microscope was prepared as follows. A fluorescent uranyl glass (Schott GG21) was mounted at the bottom of a hollow metal cylinder which was screwed onto a standard objective (Leitz Apo 25×, N.A. = 0.65), replacing its outer covering (Figure 2). In our experience a reference channel is not often needed today, possibly owing to improvement in lamp quality during recent years.

Fluorescence light is converted to current by a photomultiplier (PM)-tube with S-20 cathode (EMI 9558B). The current is further converted to voltage by a fol-

Figure 2. Uranyl glass mounted in the bottom of a metal cylinder (right), which can be screwed on an objective (Leitz Apo 25×, N.A. = 0.65) (center), replacing its original covering (left). In the foreground, two neutral density filters that may be inserted in the upper part of the objective to reduce the intensity of fluorescent light from the standard.

lower connected to the PM-tube within the same metal housing. This fluorescence signal is then amplified and low-pass filtered. The reference signal is treated in a similar way, and both signals are then connected to an analog divider (Burr-Brown 4291). The resulting corrected signal is amplified by an amplifier providing calibrated gain steps of 1, 2, 5, 10, 20, 50, or 100 before being connected to the analog-to-digital converter (ADC) in the computer. It is also sensed by a sample/hold amplifier, and displayed on a digital voltmeter to allow off-line measurements. The calibrated amplifier is also provided with a coded digital output of its gain, which can be read by a computer. An electronic shutter control unit is included in the MPV II and MPV III systems and has been added to the MPV I system. This unit controls the measuring sequence when the instrument is used off-line. The shutter in the fluorescence activation light path can be controlled to provide four different illumination times between 20 msec and 1 sec. Both shutters can also be controlled manually.

Computer Hardware

Microcomputer technology has developed extensively during recent years, and many small computer systems of high capacity combined with low price are now available. For quantitative cell analysis a computer system should meet four general requirements:

1. on-line data acquisition in real time;
2. high-level computer language (BASIC, PASCAL) for programming;
3. immediate presentation of results; and
4. mass-storage device for programs and recorded data.

Many such computer systems are now on the market. We have used the Swedish computers ABC 80 and ABC 800 for integration with our microscope fluorometers.

The ABC 800 Microcomputer System. The MPV II microfluorometer was connected to an ABC 800 microcomputer system as shown in Figure 3. The ABC 800 basic unit (Luxor AB, Motala, Sweden) consists of a Z80A microprocessor, 32 KB ROM, 32 KB RAM, a keyboard, and a monochrome CRT display. The keyboard includes eight function-keys and a separate numeric key set. A graphic memory of 16 KB, which shares a part of the ROM-area, permits use of medium resolution graphics of 240 × 240 pixels. The ABC 800 also has two RS 232 communication ports and a connector for the ABC bus.

The DataBoard 4680 Bus. The ABC bus is largely compatible with the DataBoard 4680 bus (DataSweden, Täby, Sweden). This general microcomputer bus is divided into an I/O bus for connection to I/O peripherals, and a memory bus for addition of extra memory. A large number of interfaces are commercially available for connection to the bus.

Peripherals. A dual floppy disk (2 × 160 KB, DataDisc 82, DataSweden) was used for storage of programs and recorded data. The cabinet to the floppy disk acts also as a DataBoard 4680 bus backplane for connection of interface cards, which are built up on standardized Europe cards (size 100 × 160 mm). The inter-

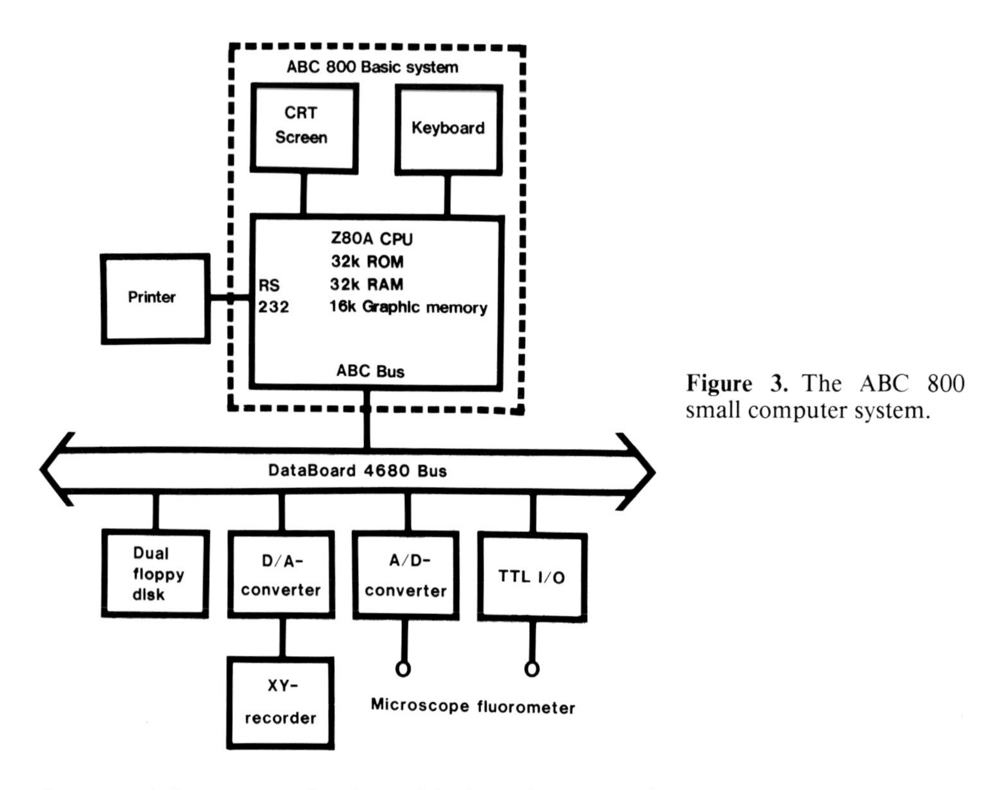

Figure 3. The ABC 800 small computer system.

faces used for communication with the microscope fluorometer consist of a 16-channel, 12-bit ADC (MTD ADC 12 × 16, Multitech Data, Göteborg, Sweden), and a TTL-interface card (4006, DataBoard 4680, DataSweden) providing 4 × 8 digital TTL outputs with external tristate control and 2 × 8 digital TTL inputs. A 4 × 8-bit DAC (4084, DataBoard 4680, DataSweden) is used for output of graphs on an XY-recorder (7015A, Hewlett-Packard). A matrix printer (Epson MX80) is connected to one of the RS 232 ports in the ABC 800 basic unit. This printer also has a graphics capability.

Program Development

Programming Languages

The ABC 800 computer contains a BASIC interpreter located in a 24-KB ROM. The system is also provided with floppy-disk control firmware (4-KB ROM) and a combined printer and terminal handler (4-KB ROM). The BASIC interpreter includes such features as simple statements for I/O-communication (INP, OUT) and communication with machine-coded subroutines (POKE, PEEK, CALL). Our assembler routines were kept very short, so it was quite easy to make direct translations from Z80 mnemonics to machine code. If more extensive machine-coded sequences are needed, several assemblers are available for this computer.

Software Design

Program development was divided into three levels:

1. machine-coded special functions for control of the instrument and primary data acquisition;

2. subroutines written in BASIC linking low-level functions to complete measuring sequences including primary data processing; and
3. main programs written in BASIC for specific biological applications including lower-level subroutines, statistical data processing, and presentation of results.

Main Programs. Main programs were developed for three categories of biological application: (1) cell population analysis; (2) recording of corrected emission spectra; and (3) recording of time-dependent fluorescence variables. Different versions are available within each group to fulfill various biological prerequisites and/or different requirements for the presentation of the results.

When the computer is started, an automatic start routine (MPVINIT) is initiated. This loads the machine-coded subroutines in the RAM area, where they later can be reached by the different programs. The user can also set the real-time clock and select a main program from the program library.

All main programs include a command menu, which is displayed on the CRT when the program is started. The different commands, which are written as subprogram modules, are executed by a single character command (Table 1). The character "#" on the screen shows that the computer is in command mode and is ready to accept a command. When the computer is in operating mode, (i.e., when it is busy performing a subprogram), it cannot be reached via the standard characters on the keyboard. Instead, the set of function keys is used. A special-function key handler was written for communication with these keys. The main programs contain various routines for the immediate presentation of the results during the measurements.

Subprogram Modules. An example of a second-level subroutine is the complete sequence for measurement of fluorescence during brief illumination, which is used in the main program for cell population analysis. The aim of this routine is to minimize the manual steps for such measurements, and to permit easy changes of the different variables. A measurement starts with manual focusing of a cell within the limits of the measuring diaphragm. The measuring process controlled by the subprogram is then initiated by pressing a button or pedal. The shutter in the transmitted light path is closed, and the light path to the PM-tube is opened. Next, the dark current from the PM-tube is measured by a machine-

Table 1: Command Menu in Main Program for Cell Population Analysis

Group	Command	Implication
Measurements	C	Cells
	U	Uranyl standard
	B	Background
Display	L	List of measured values
	Z	Statistics
	H	Histogram
Documentation	S	Save on floppy disk file
	P	Printout of results
Miscellaneous	R	Reset
	M	Show menu
	T	Set the real-time clock
	E	Exit

coded ADC routine and stored. The shutter in the activating light path is then opened for the illumination period, which is usually 40 msec for the lower gain ranges. This period, which is determined in the program, was selected to minimize the influence from rapid, initial, fading phenomena (Rundquist and Enerbäck, 1976) and from conventional fading (Enerbäck and Johansson, 1973). For higher gain ranges the fluorescence signal is further low-pass filtered to reduce noise, and consequently the illumination time is increased to 200 msec. Before shutter closure, the fluorescence signal is measured by the ADC routine (Figure 4). The amplifier gain is sensed by the computer, and the measurement value is divided by the gain. This routine permits changing of gain from one measurement to another in order to increase the resolution for weakly fluorescent cells. Primary data processing includes subtraction of dark current, subtraction of a previously measured background, and normalization to the previously measured standard. When the measurement is completed, the transmitted light is switched on again, and the instrument is ready for the next cell measurement.

A variant of this subroutine is used for the automatic measurements of the uranyl standard. When the standard routine is activated by the selection of a "U" in the command menu, a sequence of 10 measurements of the standard is performed. The mean standard value and the coefficient of variation are displayed on the screen. Measurement of the variation of the standard is a convenient means of checking the stability of the illumination system. The standard subroutine can be used in all main programs, and it can also be used as often as necessary during the study of a cell population.

Special Functions. A machine-coded subroutine for rapid communication with the ADC is common to all programs. This subroutine performs 16 analog-to-digital (A/D) conversions with a sampling frequency of about 4 KHz. The mean of these A/D conversions is returned as the measurement value.

Program Operation

The programs for cell population analysis are based on repeated use of the subroutine for recording of fluorescence from individual cells. These programs are designed to permit a rapid stream of such measurements. A typical cell population analysis starts with measurement of the standard. A routine for measurement of backgrounds is also included. The following individual cell measurement values are corrected for background fluorescence, and are expressed in "uranyl units." In programs of this kind, where it is important to be able to get a quick

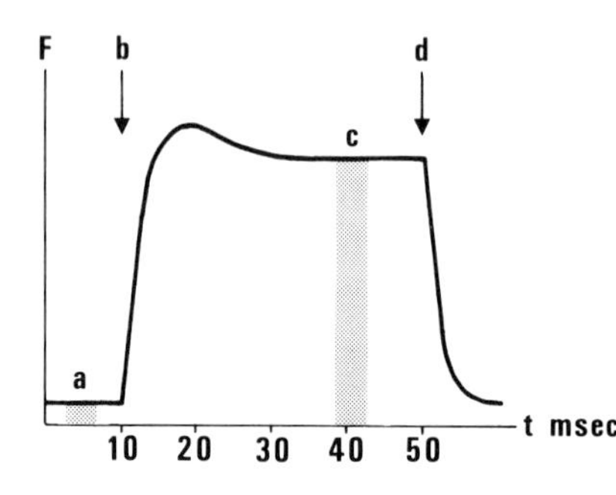

Figure 4. The principle for rapid measurements of fluorescence signals: (A) recording of dark current; (B) opening of shutter; (C) recording of fluorescence intensity; (D) closure of shutter. [Reprinted from I. Rundquist, Histochemistry 70:151–159, 1981, with permission of the publisher Springer-Verlag, Heidelberg.]

glance at the individual values, each cell value and cell number is displayed by a graphic routine that uses the medium-resolution graphics to draw clearly legible 4-cm-tall digits on the screen. Another routine handles error messages. For example, when the signal amplitude is too high, the message "overflow, reduce gain" will appear on the screen. Error messages are also combined with an audible signal to catch the attention of the operator.

Emission spectra from single cells are recorded with the aid of the motor-driven interference wedge whose position is sensed by the computer, but the same principle can be used for other devices, such as the grating monochromator now available in the Leitz MPV III system or in other comparable systems. Such uncorrected spectra can be corrected for the spectral characteristics of the instrument, which were determined by recording a spectrum with known distribution, in this case from a tungsten lamp (Ritzén, 1967; Ploem et al., 1974). Spectral correction functions for different optical configurations were stored in the computer, which automatically corrected the unknown emission spectra. The recording time for one spectrum is about 30 sec, and consequently most spectra are distorted by fading. To obviate this, such spectra can be sampled at intervals of 5 or 10 nm using the routine for brief illumination (20 or 40 msec), thus reducing the effective illumination time of each cell to about 1 sec.

Programs for recording time-dependent fluorescence variables resemble in many respects those for recording emission spectra, except that time is used as the independent variable instead of wavelength. Two modes are also available here, continuous illumination or interrupted illumination. The illumination period and the dark intervals are determined in the program, and they can readily be changed within wide margins.

Data Analysis

After termination of the measurements in the different main programs, several routines for analysis of the collected data can be selected from the command menu. The first group of commands concerns documentation of the results on the screen and/or printer. The individual values can be listed in a table and/or presented as a histogram. The histogram has a resolution of 205 channels, which is one-tenth of the resolution of the ADC (11 bits for positive signals). The gain of the amplifier in the system is again read by the computer and used to determine the scale of the x axis. The scale on the y axis (number of cells per channel) is automatically calculated by the computer. After presentation of the histogram on the screen, the operator can easily change variables such as class width and scale factors. When the operator is satisfied, the histogram can be printed using the graphics mode of the printer.

A small package of statistics is available for the main programs. It contains routines for calculation of mean, S.D., and S.E.M. Some versions include the estimation of the median, which can be very useful because the quantities of many cellular components are log-normally distributed. Furthermore, the median can be employed for simple estimations of the fraction of cells in S+G2 phase of the cell cycle.

All data can be stored as files on floppy disks. The routines for recording emission spectra and time-dependent fluorescent variables also include functions for storing the curves from individual cells as separate files. Such files are automatically given updated file names when new cells are measured. Stored data can readily be recalled for further analysis by other programs in the library.

Performance of the Measurement System

Sensitivity and Accuracy

The unsystematic errors of the system were checked by repeated measurements of the uranyl standard. The relative gain was varied by varying the supply to the PM-tube and by using the available ranges of the amplifier. The fluorescence intensity was also adjusted with the aid of different sized measuring diaphragms and filters to achieve the same final signal amplitude. The results are shown in Table 2. For comparison, the DNA in a rat lymphocyte is measured after staining with Hoechst 33258 (Enerbäck and Rundquist, 1981) with a relative gain of ~5, and the endogenous content of serotonin in a rat peritoneal mast cell (~0.3 pg) is measured after formaldehyde condensation (Enerbäck et al., 1977) with a relative gain of ~50. Most fluorescence measurements on whole cells are performed with a relative gain not exceeding 70; the total unsystematic error therefore seldom (<1%) exceeds 2.3% for a measurement with an illumination period of 40 msec. An additional low-pass filter for noise reduction can be used for gain ranges exceeding 10, and the illumination time is then prolonged to 200 msec. The weakest fluorescence intensity used in this experiment was barely perceptible by a thoroughly dark-adapted eye. Such signals could be measured with a coefficient of variation (CV) below 4%.

It is probably less important to determine the total accuracy of a cytofluorometric method than its overall precision, owing to the relative nature of the method. Cytofluorometric measurements of the DNA content in cells from a nonproliferating population are useful for studying the overall precision because these cells are assumed to have a constant DNA content. Rat lymphocytes stained with Hoechst 33258 were used for this purpose. The results, obtained over a period of 5 months, showed that the total imprecision was about 7% (CV) when a standardized staining procedure and a standardized measuring routine (uranyl standard) were used (Rundquist, 1981). The variations within the populations ranged

Table 2. Precision of the Instrument after Different Settings of Relative Gain and Illumination Time[a]

Instrument settings			Measurement variables	
Supply to PM-tube (V)	Range	Illumination time (msec)	Relative gain[b]	CV[c]
600	1	40	1	0.38
600	10	40	10	0.60
1,035	1	40	100	1.2
600	10	200	10	0.12
600	100	200	100	0.30
1,035	10	200	1,000	0.98
1,035	100	200	10,000	3.8

[a]Based on 100 measurements of the uranyl standard. Mean fluorescence values were about 1,000 units.
[b]Derived from V and range.
[c]Coefficient of variation.

from 2.3 to 5.3%, reducing the total long-term variations to about 6%. However, when such lymphocyte populations were used as biological standards the total variation in the measurements of mast cell DNA was reduced to 2.4% (Enerbäck and Rundquist, 1981). The short-term variations due to noise, about 0.4%, are in this case small enough to be neglected. These results suggest that variations in the staining method together with variations in the optical components of the instrument represent the greater part of the overall imprecision of this cytofluorometric method.

Standardization

Careful standardization of the cytochemical method is most important in cytofluorometry. Many cytofluorometric methods are based on a staining procedure by which a fluorochrome is selectively bound to the substance of interest. Other methods involve chemical reactions resulting in fluorescent reaction products. All methods should be standardized by detailed examination of the various steps, such as cell preparation, fixation, and staining. The final step should include such variables as dye or reagent concentration, staining or reaction time, pH and ionic strength during the staining or reaction, and temperature. In the case of staining methods it is imperative that equilibrium conditions be obtained and maintained throughout the period of measurement. The simplest way to obtain a staining equilibrium is to use a low dye concentration and a staining time long enough to ensure a constant fluorescence intensity, and to examine the cells mounted in the dye solution (Liedeman and Bolund, 1976) under a cover glass sealed with varnish (Enerbäck, 1974; Enerbäck and Rundquist, 1981). Suitable staining conditions are then characterized by stable measurement values for small changes in the staining variable.

After establishing the cytochemical conditions, the method usually has to be checked to confirm the absence or presence of quenching phenomena. This can be done by model experiments or by referring to a biochemical method. In model experiments the relevant substance is mixed with a carrier such as protein (Deitch, 1964; Enerbäck, 1974; Kelly and Carlson, 1963; Ritzén, 1966) or cellulose or polyacrylamide (van Duijn and van der Ploeg, 1970) and positioned on a slide in the form of a smear, a film, or microdroplets. Such models can be used for checking the stoichiometry of a histochemical method, but they are less suitable for calibration purposes. Microdroplets can be used for calibration, but the preparation and measurement of dry weight and fluorescence of such droplets may introduce variations that are difficult to control, and the simple cytofluorometric technique is made more complicated. Furthermore, many fluorophores may be structurally oriented within cells, leading to concentration conditions that cannot easily be reproduced in simple model systems.

Calibration is preferably performed by comparison with a biochemical method. In many cases this is a difficult procedure owing to the lack of suitable reference methods sensitive enough to permit analysis of small samples. Methods for the preparation of pure cell suspensions must also be available. An example of a highly sensitive biochemical method is high-performance liquid chromatography, (HPLC) followed by electrochemical or fluorescence detection. We used such a calibration to establish the relationship between formaldehyde-induced fluorescence and dopamine content in rat mast cells (Rundquist et al., 1982).

When cytofluorometric measurement results are compared, two measurement situations should be considered: (1) results obtained from the same instrument but at different times; and (2) results obtained from different instruments. In both cases, the use of a fluorescent standard as reference object is the simplest method of standardization. The second situation is especially troublesome as it requires special qualities of the standard; the spectral characteristics of the standard must, for instance, be almost identical to those of the object (Ploem, 1970).

The uranyl standard can only be employed if the long-term variations affect the standard and the object in a corresponding way. If the standard and the object have different spectral characteristics (which is usually the case), all variations that affect the spectral characteristics of the instrument itself will reduce the capacity of the standard. The two diaphragms of the instrument, a field diaphragm that limits the illuminated area and a measuring diaphragm that limits the measured area, will also affect the standard and the object differently, because the standard has macroscopic and the object microscopic dimensions. Fixed diaphragms should therefore be used instead of continuously variable iris diaphragms. Interchangeable pinhole diaphragms are preferred for this purpose.

The uranyl standard compensates for long-term changes in the excitation intensity and for variations in the amplification level. Acceptably small total variations, adequate for most experimental purposes, have been obtained over periods up to 5 months by careful control of the measuring situation. However, it is difficult to keep the optical conditions of the instrument absolutely constant for more than a few days. If higher precision is required, it will therefore be necessary to include a chemical standard with the same size and spectral characteristics as the objects. The use of a microvolume of a fluorochrome solution has been proposed (Jongsma et al., 1971; Sernetz and Thaer, 1970), but this has limited value in cytofluorometry. A biologic standard such as the DNA content of a well specified and appropriately stained cell population (Enerbäck and Rundquist, 1981; Ruch, 1973) is probably the best alternative.

Research Applications

Cytofluorometric methods are available for many applications in cell biology. For example, the fluorescent dye acridine orange (AO), can be used not only for the quantitation of nucleic acids in intact cells but also for the quantitative study of the conformation of nucleic acids in situ. The latter type of study exploits the metachromatic properties of the AO-nucleic acid complexes (Darzynkiewicz, 1979; Liedeman and Bolund, 1976; Rigler, 1966). Other cytofluorometric methods can be used for the determination of monoamines such as serotonin and the catecholamines dopamine and noradrenaline (Enerbäck et al., 1977; Jonsson, 1971; Ritzén, 1967; Rundquist et al., 1982).

The weakly fluorescent dye berberine forms fluorescent complexes with certain glucoseaminoglycans (GAG). Of the naturally occurring GAGs, heparin forms a complex with by far the highest quantum yield of fluorescence. This property can be used for the detection and quantitation of heparin in mast cells and individual mast cell granules (Enerbäck, 1974; Enerbäck et al., 1976; Gustafsson and Enerbäck, 1978). Recently, fluorescent berberine binding has been used to monitor secretory activity in living mast cells (Berlin and Enerbäck, 1983).

An interesting application of cytofluorometry concerns the study of reaction kinetics in living cells. Intracellular and exogenous fluorochromes can be used to monitor both energy metabolism and biosynthetic or metabolic activity by spectral or multisite topographic analysis (see Kohen et al., 1978).

Measurement of DNA in Mast Cell Populations

Study of the DNA distribution of cell populations is probably the most common application of cytofluorometry, owing to current interest in cell proliferation and tumor pathology. The DNA can be quantitated with the aid of fluorescent Schiff reagents (Kasten et al., 1959), but this reaction is less well understood than are conventional Feulgen reactions, and tends to give unspecific dye binding. However, this problem can be circumvented by extraction procedures, and a useful technique with acriflavine as fluorescent Schiff reagent has recently been described (Tanke and van Ingen, 1980).

DNA can also be measured by DNA specific dye binding, using for example AO, ethidium bromide, propidium iodide, mithramycin, or the bibenzimidazole dye Hoechst 33258 (Hilwig and Gropp, 1972; Johannisson and Thorell, 1977; Krishan, 1975; Le Pecq and Paoletti, 1967; Rigler, 1966). Ethidium bromide and propidium iodide in particular have been widely used in flow cytofluorometry, but unspecific dye binding requires extensive enzymatic treatment before measurements can be made. Consequently, the measurements will in principle be performed on nuclear ghosts rather than intact cells. It follows that these methods are not appropriate for applications in microscope fluorometry where structural information is required.

The dyes mithramycin and Hoechst 33258 are eminently suitable for microscope fluorometry. They show highly selective base-specific DNA binding and negligible unspecific fluorescence. The staining of mast cell DNA is complicated because the strongly charged dye binding sites in the cytoplasmic granules may interact with the dye. However, we found that Hoechst 33258 was also suitable for measurement of mast cell DNA. After establishing suitable staining conditions, diploid cells were measured with a CV or 3%–4%. The method has been described in detail (Enerbäck and Rundquist, 1981).

The technique has been used to study the degree of proliferation of peritoneal mast cells in growing rats (Figure 5). Diploid nonmast cells from each animal served as a biological standard, which resulted in total long-term variations of 2%–3%. The proportion of mast cells in the hyperdiploid region of the DNA distribution, estimated by a simple statistical method, was found to decrease in relation to body weight. The concomitant increase in the total number of mast cells in the peritoneal cavity suggests that the total number of proliferating cells remains constant during the growth period. We concluded that peritoneal mast cells of adult rats increase in number by mitotic proliferation of differentiated cells.

Quantitation of Cell Protein

Measurement of protein in intact cells or cell organelles is of considerable value in many fields of cell biology. Available methods are often based on dye binding and cytophotometry of the resulting absorbance or fluorescence. Fluorometric methods for measurement of arginine (Rosselet, 1967) and lysine (Rosselet and

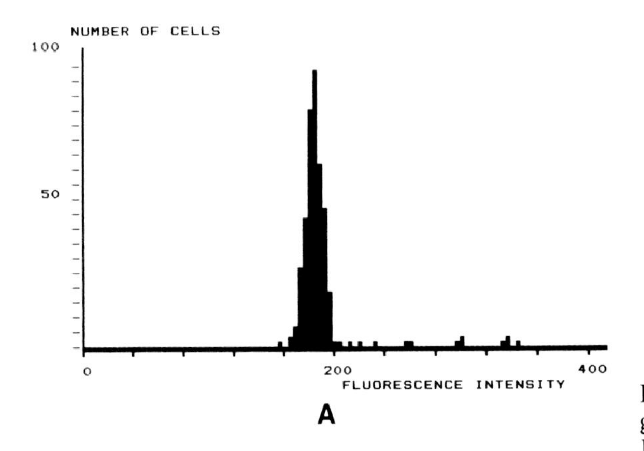

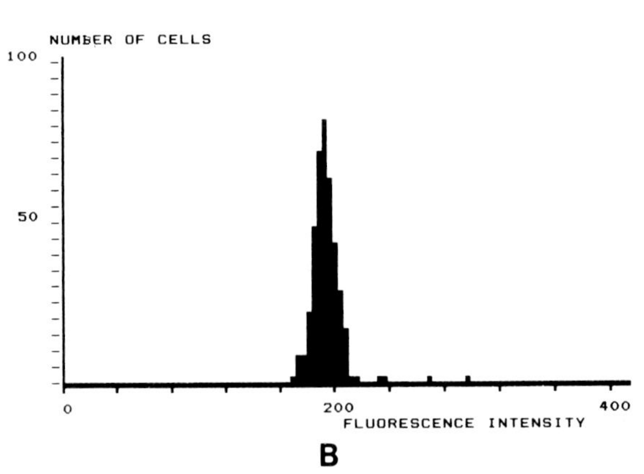

Figure 5. Characteristic histograms showing distributions of DNA, measured after staining with Hoechst 33258, for (A) a young rat, and (B) an old rat.

Ruch, 1968) and for the determination of basic and total proteins with sulfaflavine (Leemann and Ruch, 1972) have been published. Six fluorescent protein stains were evaluated for use in flow microfluorometry (Freeman and Crissman, 1975). Dye binding methods have some important limitations because charged groups are often not available for dye binding in the intact biological structure. For example, nuclear basic proteins cannot be measured by dye binding methods unless DNA is first removed.

We have been engaged for some time in the search for new cytofluorometric protein methods, and have successfully employed a new protein reagent containing o-phthalaldehyde (OPT), which has a number of interesting properties. It measures the total protein, and the results are strongly correlated to cellular and nuclear dry mass. In addition to OPT, the reagent contains mercaptoethanol, and can also be used in the determination of protein by gel electrophoresis and in solution (Weidekamm et al., 1973) and for the detection of amino acids in HPLC systems (Roth, 1971). The mechanism of the reaction is not fully understood, but it is clear that the fluorescent complex is formed by a covalent binding to OPT. For technical details see Mellblom and Enerbäck (1979). The method has been evaluated by combined cytofluorometry and microinterferometric determination of dry mass. Excellent proportionality was demonstrated between fluorescence intensity and dry mass on dried protein microdroplets of cell size. In experimental studies close correlation was also shown between fluorescence intensity and

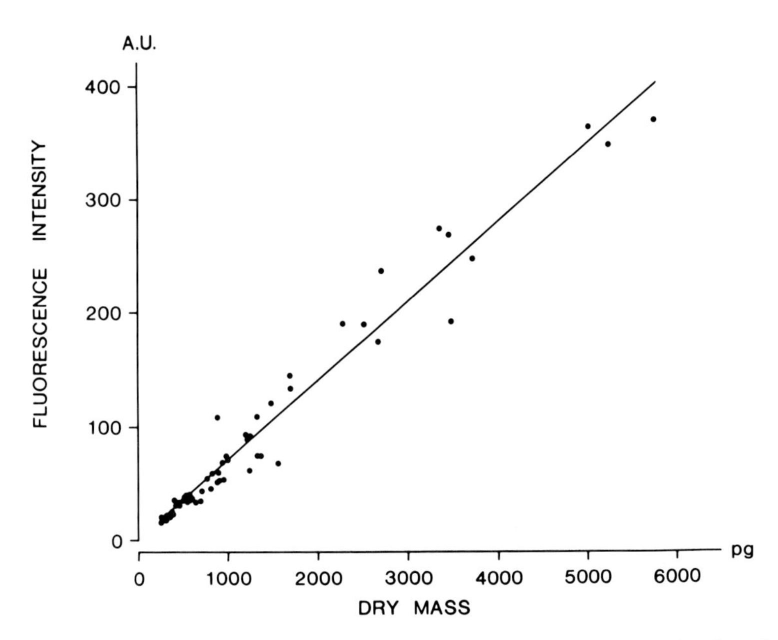

Figure 6. Fluorescence intensity measured after the OPT reaction in relation to dry weight in mouse neurons.

dry mass of liver cell nuclei, mast cells, and neurons. The fluorescent OPT method is thus a reliable means of weighing cells or nuclei. Figure 6 shows the correlation between OPT fluorescence and dry weight in neurons.

Conclusions

Determination of the sensitivity and total accuracy of cytofluorometric measurements is fraught with uncertainty. In principle, the relationship between the measured value and the quantity of fluorescent substance in the cell is exponential, but for low (<0.05) absorbance values it is roughly linear. Inevitably, the sensitivity and accuracy will be to a great extent influenced by the histochemical method. The errors can be characterized as short-term variations caused by different kinds of noise in the system, or as long-term variations caused by variations in the staining procedure or by optical variations in the instrument. Errors caused by rapid changes in the illumination can be corrected for by using the reference channel. Other errors can be compensated for by using the uranyl standard. However, it is difficult to keep the optical conditions of the instrument absolutely constant for more than a few days. According to our experience, long-term variations contribute far more to the total error than do noise or other uncorrected short-term variations. Long-term variations can be kept reasonably small, however, if a standardized measuring routine is used. The S.D. of 7% for measurements over a period of 5 months (Enerbäck and Rundquist, 1981) would be acceptable for many long-term studies, but if greater precision is needed a chemical or biological standard of the same size and spectral characteristics as the objects to be measured must be included (Jongsma et al., 1971; Ruch, 1973; Sernetz and Thaer, 1970).

To obtain satisfactory statistical precision of microfluorometric data, a rapid measuring procedure is essential. This can be achieved by automation. Automation involving hardware electronics only, however, would be rather rigid and difficult to handle. Computer technology offers unique possibilities of rendering a measuring system flexible by placing determinations and decisions in a readily modifiable program. The systems described here can be used for different applications with a minimum of hardware changes. Further, about 500 cells can be measured in 1 hour on well-prepared specimens, so that most cell populations can be analyzed within reasonable time and with sufficient precision. On-line statistical data processing and presentation of results also save time.

Small computers started to appear on the market in 1976, and were originally mainly used as toys for entertainment and education. In 1978, Doerr defined personal computers as "desk-top or table-top user programmable systems with CPU, RAM, alphanumeric keyboard, display, and high-level language." At about the same time, single-board computers were introduced for industrial applications, but these required programming on an additional microcomputer development system. Today, the gap between the two categories has been spanned by more sophisticated small computer systems, of which a number are now commercially available and can be used in cytofluorometry. The value of such systems is determined more by the availability of peripheral devices than by the actual computers. A system suitable for professional use must include a random-access mass-storage device, such as a floppy disk, and a printer. These components are often more expensive than the computer itself, but cheap peripheral devices designed for the small computer market are becoming increasingly available.

Many computer technologists look upon BASIC as a second-class computer language owing to its lack of structure. It has two great advantages, however: It is easy to learn and it is interactive. These features are often important in biomedical research applications, where the user is usually not a technologist and where programs have to be modified at frequent intervals. Such versatility can be difficult to achieve with a compiling language.

References

Berlin, G., and L. Enerbäck (1983) Fluorescent berberine binding as a marker of secretory activity in mast cells. Int. Arch. Allergy Appl. Immunol. 71:332–339.

Caspersson, T. (1936) Über die chemische Aufbau der Strukturen des Zellkernes. Scand. Arch. Physiol. 73(Suppl. 8).

Caspersson, T., G. Lomakka, and R. Rigler (1965) Registrierender Fluoreszenzmikrospektrograph zur Bestimmung der Primär und Sekundärfluoreszenz verschiedener Zellsubstanzen. Acta Histochem. (Jena) Suppl. 6:123–126.

Chance, B., and V. Legallais (1959) Differential microfluorimeter for the localization of reduced pyridine nucleotide in living cells. Rev. Sci. Instrum. 30:732–735.

Chance, B., and B. Thorell (1959) Localization and kinetics of reduced pyridine nucleotides in living cells by microfluorometry. J. Biol. Chem. 234:3044–3050.

Cova, S., G. Prenna, and G. Mazzini (1974) Digital microspectrofluorometry by multichannel scaling and single photon detection. Histochem. J. 6:279–299.

Darzynkiewicz, Z. (1979) Acridine orange as a molecular probe in studies of nucleic acids in situ. In M.R. Melamed, P.F. Mullaney, and M.L. Mendelsohn (eds.): Flow Cytometry and Sorting. New York: Wiley, pp. 285–316.

Deitch, A.D. (1964) A method for the cytophotometric estimation of nucleic acids using methylene blue. J. Histochem. Cytochem. 12:451–461.

Doerr, J. (1978) Low-cost microcomputing: The personal computer and single-board computer revolutions. Proc. IEEE 66:117–130.

Enerbäck, L. (1974) Berberine sulphate binding to mast cell polyanions: A cytofluorometric method for the quantitation of heparin. Histochemistry 42:310–313.

Enerbäck, L., and K.-A. Johansson (1973) Fluorescence fading in quantitative fluorescence microscopy: A cytofluorometer for the automatic recording of fluorescence peaks of very short duration. Histochem. J. 5:351–362.

Enerbäck, L., and I. Rundquist (1981) DNA distributions of mast cell populations in growing rats. Histochemistry 71:521–531.

Enerbäck, L., G. Berlin, I. Svensson, and I. Rundquist (1976) Quantitation of mast cell heparin by flow cytofluorometry. J. Histochem. Cytochem. 24:1231–1238.

Enerbäck, L., B. Gustafsson, and L. Mellblom (1977) Cytofluorometric quantitation of 5-hydroxytryptamine in mast cells: An improved technique for the formaldehyde condensation reaction. J. Histochem. Cytochem. 25:32–41.

Freeman, D.A., and H.A. Crissman (1975) Evaluation of six fluorescent protein stains for use in flow microfluorometry. Stain Technol. 50:279–284.

Gustafsson, B., and L. Enerbäck (1978) Cytofluorometric quantitation of 5-hydroxytryptamine and heparin in individual mast cell granules. J. Histochem. Cytochem. 26:47–54.

Hilwig, I., and A. Gropp (1972) Staining of constitutive heterochromatin in mammalian chromosomes with a new fluorochrome. Exp. Cell Res. 75:122–126.

Horan, P.K., and L.L. Wheeless (1977) Quantitative single cell analysis and sorting. Science 198:149–157.

Johannisson, E., and B. Thorell (1977) Mithramycin fluorescence for quantitative determination of deoxyribonucleic acid in single cells. J. Histochem. Cytochem. 25:122–128.

Jongsma, A.P.M., W. Hijmans, and J.S. Ploem (1971) Quantitative immunofluorescence, standardization and calibration in microfluorometry. Histochemie 25:329–343.

Jonsson, G. (1971) Quantitation of fluorescence of biogenic monoamines. Progr. Histochem. Cytochem. 2:299–344.

Jotz, M.M., J.E. Gill, and D.T. Davis (1976) A new optical multichannel microspectrofluorometer. J. Histochem. Cytochem. 24:91–99.

Kasten, F.H., V. Burton, and P. Glover (1959) Fluorescent Schiff-type reagents for cytochemical detection of polyaldehyde moieties in sections and smears. Nature 184:1797–1798.

Kelly, J.W., and L. Carlson (1963) Protein droplets, especially gelatin, hemoglobin and histone, as microscopic standards for quantitation of cytochemical reactions. Exp. Cell Res. 30:106–124.

Kohen, E., J.G. Hirschberg, C. Kohen, A. Wouters, A. Pearson, J.M. Salmon, and B. Thorell (1975) Multichannel microspectrofluorometry for topographic and spectral analysis of NAD(P)H fluorescence in single living cells. Biochim. Biophys. Acta 396:149–154.

Kohen, E., C. Kohen, J.G. Hirschberg, A. Wouters, and B. Thorell (1978) Multisite topographic microfluorometry of intracellular and exogenous fluorochromes. Photochem. Photobiol. 27:259–268.

Krishan, A. (1975) Rapid flow cytofluorometric analysis of mammalian cell cycle by propidium iodide staining. J. Cell Biol. 66:188–193.

Le Pecq, J-B., and C. Paoletti (1967) A fluorescent complex between ethidium bromide and nucleic acids. Physical–chemical characterization. J. Mol. Biol. 27:87–106.

Leemann, U., and F. Ruch (1972) Cytofluorometric determination of basic and total proteins with sulfaflavine. J. Histochem. Cytochem. 20:659–671.

Liedeman, R., and L. Bolund (1976) Acridine orange binding to chromatin of individual cells and nuclei under different staining conditions. I. Binding capacity of chromatin. Exp. Cell Res. 101:164–174.

Mellblom, L., and L. Enerbäck (1979) Protein content, dry mass and chemical composition of individual mast cells related to body growth. Histochemistry 63:129–143.

Mellors, R.C., and R. Silver (1951) A microfluorometric scanner for the differential detection of cells: Application to exfoliative cytology. Science 114:356–360.

Olson, R. (1960) Rapid scanning microspectrofluorimeter. Rev. Sci. Instrum. 31:844–849.

Ploem, J.S. (1970) Standards for fluorescence microscopy. In E.J. Holborow (ed.): Standardization in Immunofluorescence. Oxford: Blackwell, pp. 63–73.

Ploem, J.S. (1977) Quantitative fluorescence microscopy. In G.A. Meek and H.Y. Elder (eds.): Analytical and Quantitative Methods in Microscopy. Cambridge: Cambridge University Press, pp. 55–89.

Ploem, J.S., J.A. de Sterke, J. Bonnet, and H. Wasmund (1974) A microspectrofluorometer with epi-illumination operated under computer control. J. Histochem. Cytochem. 22:668–677.

Rigler, R. (1966) Microfluorometric characterization of intracellular nucleic acids and nucleoproteins by acridine orange. Acta Physiol. Scand. Suppl. 267.

Ritzén, M. (1966) Quantitative fluorescence microspectrophotometry of catecholamine–formaldehyde products. Model experiments. Exp. Cell. Res. 44:505–520.

Ritzén, M. (1967) Cytochemical identification and quantitation of biogenic monoamines. A microspectrofluorometric and autoradiographic study. MD Thesis, Stockholm.

Rosselet, A. (1967) Mikrofluorometrische Argininbestimmung. Z. Wiss. Mikrosk. Tech. 68:22–41.

Rosselet, A., and F. Ruch (1968) Cytofluorometric determination of lysine with dansylchloride. J. Histochem. Cytochem. 16:459–466.

Rost, F.W.D., and A.G.E. Pearse (1971) An improved microspectrofluorimeter with automatic digital data logging: Construction and operation. J. Microsc. 94:93–105.

Roth, M. (1971) Fluorescence reaction for amino acids. Anal. Chem. 43:880–882.

Ruch, F. (1973) The use of human leucocytes as a standard for the cytofluorometric determination of protein and DNA. In A.A. Thaer and M. Sernetz (eds.): Fluorescence Techniques in Cell Biology. Berlin: Springer, pp. 51–55.

Ruch, F., and U. Leeman (1973) Cytofluorometry. In V. Neuhoff (ed.): Micromethods in Molecular Biology. Berlin: Springer, pp. 329–346.

Rundquist, I. (1981) A flexible system for microscope fluorometry served by a personal computer. Histochemistry 70:151–159.

Rundquist, I., and L. Enerbäck (1976) Millisecond fading and recovery phenomena in fluorescent biological objects. Histochemistry 47:79–87.

Rundquist, I., S. Allenmark, and L. Enerbäck (1982) Uptake and turnover of dopamine in rat mast cells studied by cytofluorometry and high performance liquid chromatography. Histochem. J. 14:429–443.

Sernetz, M., and A. Thaer (1970) A capillary fluorescence standard for microfluorometry. J. Microsc. 91:43–52.

Shapiro, M.B., A.R. Schultz, and W.H. Jennings (1976) Computers in the research laboratory. Ann. Rev. Biophys. Bioeng. 5:177–204.

Shipton, H.W. (1979) The microprocessor, a new tool for the biosciences. Ann. Rev. Biophys. Bioeng. 8:269–286.

Tanke, H.J., and E.M. van Ingen (1980) A reliable Fuelgen-acriflavine-SO_2 staining procedure for quantitative DNA measurements. J. Histochem. Cytochem. 28:1007–1013.

van Duijn, P., and M. van der Ploeg (1970) Potentialities of cellulose and polyacrylamide films as vehicles in quantitative cytochemical investigations on model substances. In G.L. Wied and G.F. Bahr (eds.): Introduction to quantitative cytochemistry—II. New York: Academic Press, pp. 223–262.

Weidekamm, E., D. Wallach, and R. Fluckiger (1973) A new sensitive, rapid fluorescence technique for the determination of proteins in gel electrophoresis and in solution. Anal. Biochem. 54:102–114.

16

Computerized Radioautographic Grain Counting

James A. McKanna and Vivien A. Casagrande

Introduction

In recent years, radiolabeling techniques have become fundamental assays in physiology and biochemistry experiments. They also have assumed increasingly important roles in morphologic studies. Characteristically, radioautographic analysis of structure has been qualitative rather than quantitative; however, microcomputers have opened the door to several methods for quantifying grain counts and density. The overall goal of this chapter is to describe grain counting using the Bioquant, an image analysis package based originally on the Apple II+, and now available for several popular microcomputers. We discuss our image analysis procedures by applying them to a study of development in the central nervous system (CNS).

The Biological Problem

Radioautography has been especially valuable in qualitative studies of CNS development and organization. For example, monocular injection of tracer in primates reveals a laminar distribution of the ganglion cell terminals in the dorsal lateral geniculate nucleus (LGN), their primary target in the CNS. In the mature brain, the neurons of the LGN also are organized in layers.

We have studied the development of the LGN laminae in the tree shrew, *Tupaia belengeri,* a highly visual mammal that is quite immature at birth (Brunso-Bechtold and Casagrande, 1982, 1984). The LGN in newborn tree shrews displays a homogeneous population of neurons, allowing experimental examination of the factors involved in cell layer formation.

The retinal afferents to the tree shrew LGN, however, are segregated in layers at birth, apparently predicting the adult topography. In immature animals, the label seems qualitatively more dense in layers 2 and 4, and less dense in layer 3. All three of these layers (2, 3, and 4) are innervated by the contralateral eye. These differences have been observed using tritiated proline and tritiated wheat germ agglutinin (WGA) radioautography, and using WGA coupled to horseradish peroxidase (HRP) demonstrated histochemically (Casagrande and Brunso-Bechtold, 1983). Qualitatively, the label density seems slightly different for the three tracers. We sought to determine quantitatively whether the labeling differences were consistent from one slide to the next and whether they persisted throughout maturation. We have used computer-assisted methods to quantify the label density in identified regions of radioautographs from animals at several developmental stages. These studies provide an experimental framework around which we describe the advantages and drawbacks of computerized morphometry.

Quantitative Radioautography

As discussed with considerable insight elsewhere (Salpeter, 1978), quantification of radioautographs is a risky pursuit on both theoretical and practical grounds. The potential for sampling errors argued for large numbers of experiments with full complements of controls and radiolabeled standards, while the tedium and difficulty of manual counting made large samples unworkable.

The advent of microcomputer-assisted morphometry has diminished the technical drawbacks of radioautography grain counting, allowing scientists to overcome many original limitations on data interpretation. The technical assistance from computers is available at several levels of complexity and cost.

Manual Methods (Touch Counting)

In some experimental situations, the label is associated with distinct structures of comparable size, justifying a strategy of simply counting each labeled object. For instance, proliferating cells labeled by tritiated thymidine incorporation into DNA could be counted at a low magnification. At higher power, the individual grains could be counted, providing "grains per nucleus" data.

In practice, several options are apparent for gathering the counts into the computer. Some investigators prefer to mentally tally the labeled cells at the microscope and then type the value at the computer keyboard. This format may be approximate, but it provides rapid estimates of large samples.

Other investigators prefer to count images on a digitizing tablet (described in the section, Computer Hardware and Interfaces). Counts are recorded by moving the cross hairs or stylus to each cell and pressing the cursor button or stylus (Fig-

Figure 1. Combined area measurements and "touch counting" using the IBM PC version of the Bioquant. An image projected from above is outlined with the ball-pen stylus to provide a permanent record and measurement simultaneously. Labeled cells are counted by touching the tablet.

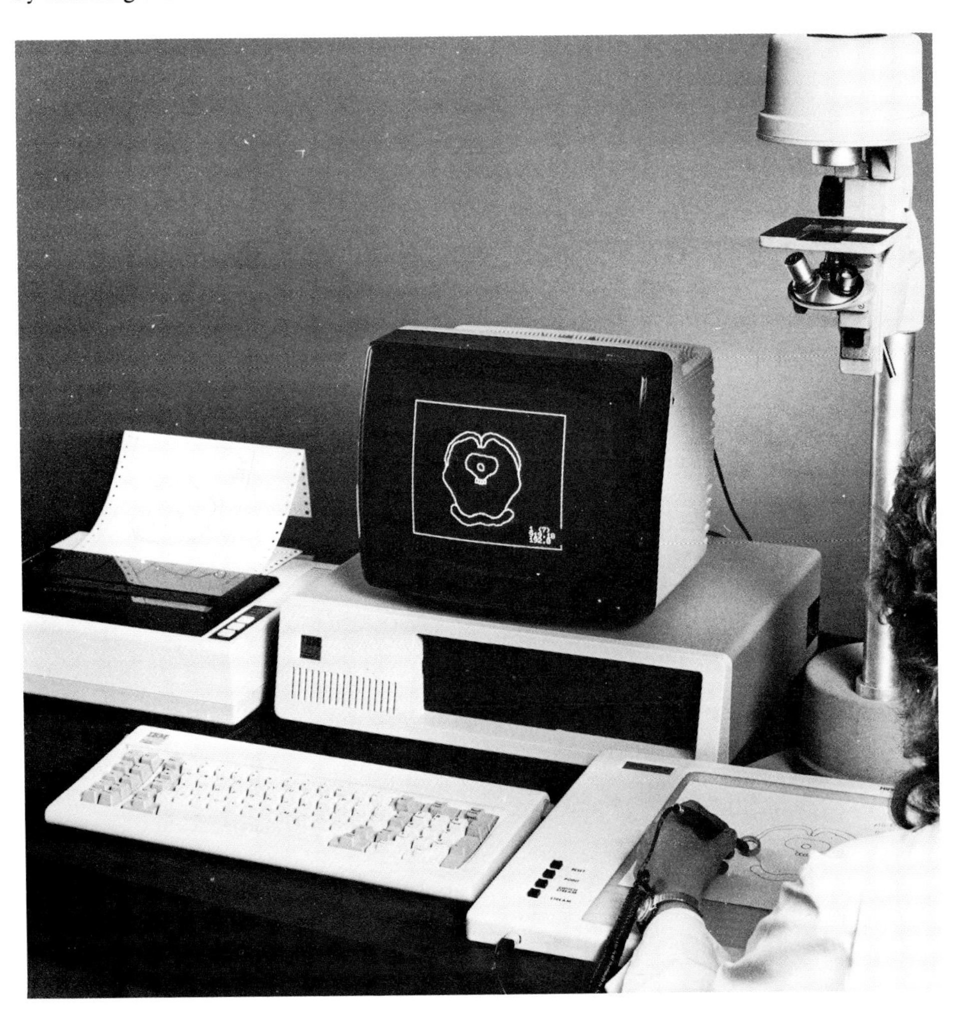

ure 1). This format, which provides a dot on the video image to identify positively each object that has been counted, frequently is most accurate but also time consuming.

Automated Methods (Grain Counting and Microdensitometry)

In order for the computer actively to analyze an image, it must have access to a numeric representation of the image. Each point in the image, referred to as a pixel (picture element), is represented by a value in an array corresponding to its brightness. The array may be generated either from consecutive values recorded by a photodiode as the specimen is moved or from digitization of the signal from a video camera. The diode method is very linear because the sensor and the illumination are constant, but it is slow (several minutes per scan) and quite expensive to fit the microscope stage with stepper motors and controllers.

The video method is fast but cannot provide the absolute accuracy of the diode. The readings are usually compared in a relative rather than absolute sense. Video digitization involves interfacing the microcomputer with appropriate accessories such as an analog-to-digital converter (ADC) and video camera (Figure 2). With most microcomputers, the digitizing screen is approximately 200–400 pixels square; and the numeric values for brightness (gray scale) span 64 or 256 gray levels.

As the computer digitizes the image, it can utilize the pixel values to generate several complementary sets of data:

The values may be summed to indicate total brightness of the image.

The average pixel brightness (sum/number) may be correlated with the area digitized as described below.

Each pixel may be compared to a threshold and scored as brighter or darker than the threshold.

Groups of contiguous pixels may be tested to see if they meet the computer's criteria for identifying a grain.

The task of automated counting is made easier by attention to illumination. Radioautography grains or other labels can be made to appear quite different from the specimens they identify. The specimen contrast can be enhanced by either dark- or bright-field illumination. Especially in dark field, the grains appear bright while the tissue is invisible (Figure 3), and there is little question that the brightness of a field is directly proportional to the density of label.

Magnification is another important variable. Depending on the magnification of the image, one pixel may correspond to a single grain or to many grains. At low powers, the brightness of a region in the specimen may be correlated with readings from radiolabel standards. This approach is often used in protocols such as 2-deoxyglucose uptake, where the tissue has been used to expose x-ray film. Either photodiode or video microdensitometry would be valid in such cases because the readings are taken relative to standards embedded and sectioned with the tissue.

In other cases, the specimens are examined at higher magnifications to allow closer correspondence between the number of pixels above a certain brightness and the number of grains. The computer could count the number of these pixels

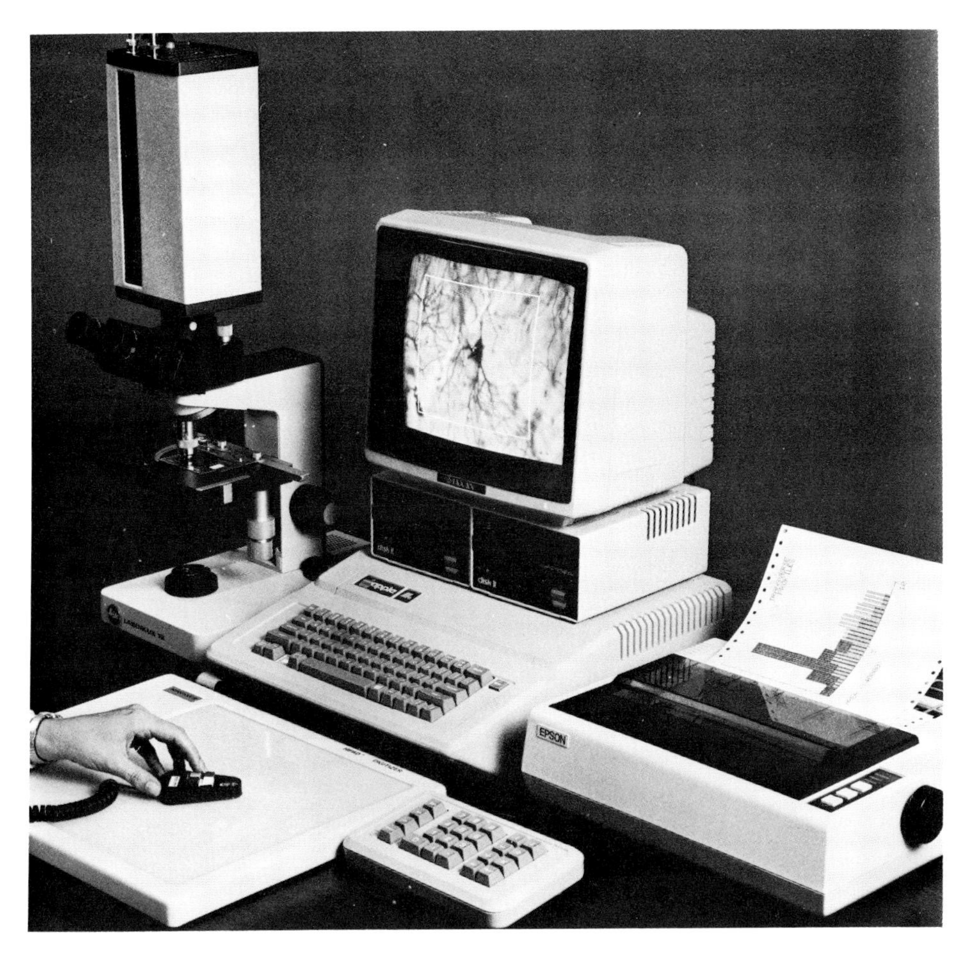

Figure 2. Apple IIe version of the Bioquant fitted with the video camera overlay option. High-resolution image of the microscope specimen is displayed on the screen along with a cross hairs generated by the computer (positioned right on the soma in the photo), corresponding to the cursor position on the tablet.

meeting the criterion and thereby directly estimate the number of grains present. The efficiency and power of this approach are augmented considerably by the associative methods discussed below.

At even higher magnifications, several pixels would be lit for each grain, and the computer could examine their relationships to determine whether one or more grains were represented by each cluster of pixels. This type of analysis, most familiar to us in photos from satellites, requires a great deal of computing power and time. It is generally not available in microcomputer systems.

Associative Counting

Frequently, experiments require quantification of label in irregular structures, and/or need to determine the density of label (counts per unit area). Some specimens might also require associated counts of adjacent regions for background label corrections.

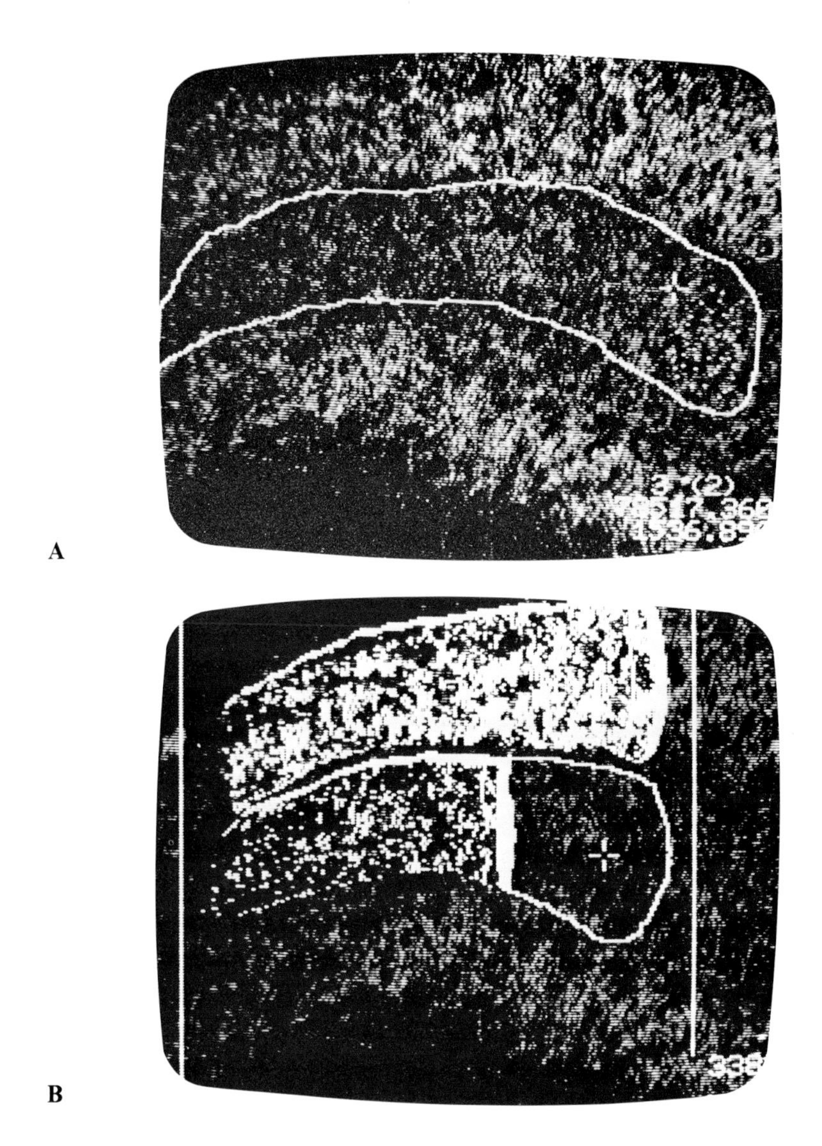

Figure 3. Photographs of the video monitor during various grain counting steps. **(A)** Associative touch counting: the area and perimeter of a labeled region in layer three have been measured; values are displayed in lower right corner; manual counting procedure is underway as indicated by dots in right quadrant of the layer and cross hairs positioned there. **(B,**

In manual mode, the object of interest would be measured as a standard area, and either keyboard entry or touch counting is performed as described above. The computer may be programmed to accept the area and count values as corresponding elements in adjacent arrays (e.g., the fifth area measurement would be correlated with the fifth count value). Such correlation would facilitate later calculation of the count density because the values in one array could be divided by the other.

In the automated mode, the operator would measure the area in standard fash-

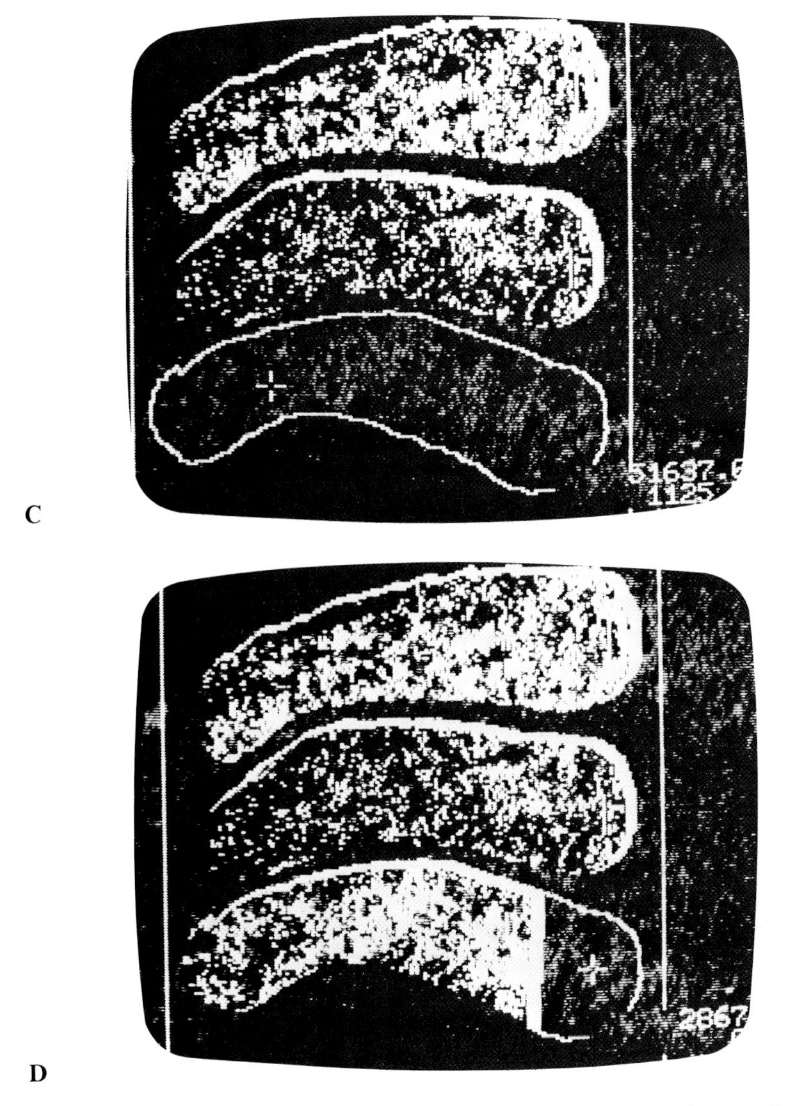

C, D) Automated counting: layer 4 (top) has been measured and counted; layer 3 has been measured and is in the midst of counting **(B)**; layer 3 counting finished and layer 2 measured **(C)**; counting underway in layer 2 **(D)**.

---

ion, and then designate the area to be counted by placing the video cursor inside the outlines of the area. Only the regions within the outlines, regardless of how convoluted, would be counted. This point is especially important for biomedical scientists, since most specimens do not conform to the circular or rectangular masks usually offered by automated digitizers. As demonstrated in Figure 3, our system easily restricts counting to curved lamellae and other obtuse shapes common to specimens in the life sciences.

Computer Hardware and Interfaces

The Bioquant is a comprehensive image analysis system developed for our research using off-the-shelf microcomputers interfaced to data entry and display devices. Currently, programs are available for the Apple II+ and IIe and the IBM PC and PC XT. The Apple LISA, Macintosh, and DEC Rainbow programs will be available soon. Standard systems include dual floppy disk drives, at least 64 KB RAM, a Houston Instrument Hi-Pad digitizing tablet and an Epson dot-matrix printer. The software, developed by R&M Biometrics, Inc. of Nashville and available from E. Leitz microscope dealer representatives, is written in BASIC, C, and ASSEMBLER. Source code is provided to the end user, facilitating modification and customization.

In the simplest formats, the system functions as an electronic planimeter with advanced data storage and analysis capabilities. For instance, a micrograph may be placed on the digitizing tablet and structural dimensions recorded by tracing with a cross-hairs cursor or a ball-pen stylus. Areas, curved-line lengths, and straight-line distances may be determined. The measured values automatically are entered into the computer's memory in separate arrays, allowing concurrent differentiation of more than 100 types or classes.

Although it might be possible to perform our grain-counting analysis from micrographs, the methodology described concentrates on a video camera accessory that obviates the intermediate photographic steps and provides a direct image for digital discrimination as described below. A high-resolution camera (Dage-MTI model 65) with vidicon tube is standard. More sensitive models including newvicon and silicon-intensified tubes can be used also. By using the timing from the computer to synchronize the camera, the video image of the microscope specimen may be displayed on the same monitor (mixed) with the computer's graphics.

Since the computer can graphically generate a cross hairs corresponding to the position of the cursor cross hairs on the digitizing tablet, it is possible to utilize the resolution of the tablet (± 0.1 mm) to measure the video image. As the cursor on the tablet is moved to circumscribe the video objects, it provides measurement coordinates to the computer.

Program Operation

In the experimental examples below, we have quantified the label in the tree shrew lateral geniculate nucleus following eye injection. Six specimens included ages from newborn to adult, and tracers included tritiated proline and tritiated WGA visualized using standard radioautography methods and WGA–HRP visualized with tetramethyl benzidine (TMB) histochemical localization of HRP. All three labels may be visualized easily with dark-field light microscopy and quantified by several methods. Using the Bioquant, we have asked these questions:

1. Are the different tracers transported at the same relative rates in newborn animals?
2. Does the relative density of label in identified LGN laminae change throughout maturation?

Below is a step-by-step description of the data-gathering and analysis procedures we followed, commencing immediately after turning on the computer.

Morphometry Protocol

Upon receiving instructions that we intend to measure up to five different classes of structures using micrometers (μm; UM on screen) as units of measurement at a magnification of 290×, the computer offers us the following menu:

```
HIPAD MEASUREMENT

CHOICES:
   1 — GRAIN COUNTING (BRIGHT)
   2 — LENGTHS
   3 — CONSECUTIVE DISTANCES
   4 — INDIVIDUAL DISTANCES
   5 — AREAS & PERIMETERS
   N — NAME & COMMENT
   M — MAG CHANGE ×290
   X — THRESHOLD CHANGE >30
```

Although measurement could commence at this point, experience has taught us that entering appropriate identifiers facilitates orderly access to the various data types. For instance, we intend to study the density of label in layers 2, 3, and 4 of the LGN contralateral to the eye injected with tracer. In order to simplify association, we decide to assign AREA array A2 to hold area measurements from layer 2. Such assignment is accomplished by pressing the "N" key, which directs the program to the NAME & COMMENT routine as indicated on the menu above. We enter the name LAYER2 for array A2. Similarly, arrays A3 and A4 will accommodate the areas from layers 3 and 4, and are named accordingly.

At this time it is not necessary to enter comments for these arrays. However, up to 255 characters may be entered at any time, providing a permanent notepad to serve as a reminder of any peculiarities relevant to the specimen or data. The names and comments are printed on hard copies for reference, and are stored along with the data on disks for later retrieval.

We next locate the region of the specimen to be measured, center it in the microscope field, and indicate our intention to measure AREAS by selecting "5" from the menu. The screen responds by listing the names we have entered for the area arrays (A2 = "LAYER2," etc.).

```
DIGITIZATION

GRAIN COUNTING  (UM)

0  D0
1  D1
2  LAYER2
3  LAYER3
4  LAYER4

PLEASE BEGIN MEASUREMENT
```

During the measurement routine, we need access to two types of information that, especially when using the video camera system, do not mix very well. For this reason, the computer has been programmed to display two "pages." The screen shown above is the text page display. The computer uses this page to list words and numbers (text) related to the measurements as they are being accomplished. The other screen display used for measurement is the graphics screen (Figures 1, 2, 3). It provides a combination (overlay) of the video image from the camera on the microscope and the computer's graphics. The graphics screen is used to perform the actual measurements. By remembering the codes "T" for text and "G" for graphics, the operator can switch between these screens at will.

Upon examination of the graphics image and centering the specimen in the field, we proceed with the following steps.

Area Measurement

Press "4" to indicate that we will begin with layer 4.

Move the cursor on the tablet, thereby moving the cross hairs on the screen to the edge of layer 4.

Press the cursor button and circumscribe the region of interest.

Upon reaching the vicinity of our starting point, the computer beeps, indicating that we should terminate the measurement by letting up on the button.

Release the button and the computer closes the perimeter line.

The measurement values for area and perimeter appear on the screen.

The values also are entered in the computer's memory as the first elements in A4 and P4 (the fourth AREA and PERIMETER arrays, respectively).

Manual Counting (Figure 3a)

Press the "P" key to indicate POINT entry.

Move the cross hairs to the first grain and press the cursor button.

A dot appears on the screen at the first point entered.

Move to other grains and press the button once for each grain.

Upon counting the last grain, press the "/" key to terminate the counting.

The total number of counts is displayed on the screen and is entered into the associative "TC-LAYER4" array.

Next Areas

Press the "3" key to indicate that layer 3 is next.

The screen displays "A3(0) LAYER3," confirming that we are ready to measure layer 3, but currently have no measurements of this type.

We proceed as above, and repeat the procedure also for layer 2.

At this point the values for the labeled area and number of grains counted in each of the three layers of the first section are stored in separate arrays. We are ready to proceed to another section. After centering the LGN in the field, we press "4" to indicate that we intend to measure layer 4, and proceed exactly in the sequence starting under Area Measurement above. This time, however, the values measured and counted are entered into the arrays as the second value, corresponding to the fact that we are measuring the second section.

After measuring and counting all three labeled layers in the second section, we proceed to another then to another until we have obtained a representative sample. (The Bioquant Statistics options can help decide whether the sample is adequate.) The values continue to be stored in orderly arrays and are available for further manipulation according to routines selectable from the master menu (Procedure Options) as described below.

Automated Counting

In the automated mode, the Bioquant displays a vertical line on the video image and the line moves from left to right as the pixels adjacent to it are digitized. The entire field is covered in approximately 4 sec, and the computer graphically places a dot on the screen adjacent to each pixel that meets the criterion for counting (either bright or dark pixels may be counted).

In order to perform accurate and repeatable assays, the automated grain-counting apparatus must be aligned. Sometimes it also is desirable to calibrate the instrument using standards. For instance, experimental designs using 2-deoxyglucose as a metabolic tracer require that a series of known concentrations of isotopes be embedded and sectioned with the labeled specimens. In other cases such as our developmental studies, internal comparisons and standards make absolute counts unnecessary for determinations of relative activities. In all situations, however, it is necessary to set the threshold for counting. For automated counting, the following steps are performed.

Alignment. The image from the video camera must be aligned with the computer graphics to ensure appropriate placement of the graphics dots that identify counted pixels. For this purpose, the Bioquant video control box has circuitry for generating points on either side of the field. The computer generates two similarly placed points. The operator adjusts a potentiometer to align the points on the right side of the screen so that they are coincident, and then adjusts the left-hand points using arrow keys from the keyboard. The steps are repeated until satisfactory alignment is achieved.

Calibration. This procedure provides steps for standardizing the illumination, brightness threshold, and contrast response of the digitizer. The procedure provides real-time microdensitometry (brightness) readings from a small region in the center of the field and follows these steps:

Set the brightness and contrast scales to a reproducible midpoint.

With the microscope focused but moved to an empty field, set the illuminator to the desired level (the level determined from previous standardization).

Move the specimen to the darkest level anticipated (or to the most dense standard specimen) and set the brightness control so that the digitizer reads your standard low value (usually 2–5, never 0).

Move the specimen to the brightest region anticipated (or to your least dense standard) and adjust the contrast control to achieve a standard reading near but less than the maximum (63 or 255).

In the above alignment and standardization, the BRIGHT control is used to set the *y*-intercept of the digitizer sensitivity curve. The CONTRAST control is used to adjust the slope of the sensitivity. In protocols with intermediate standards such as 2-deoxyglucose studies, the density of the intermediates may be determined after setting the limits of the curve as described above. The intermediate densities may be plotted against concentration of tracer in the standards, and the resulting curve used to normalize the experimental data.

Threshold. Following alignment, a representative region of the specimen is centered in the field and the operator enters a midrange value such as 30. The system scans the field, digitizing the pixels and displaying dots for each grain that meets the criterion. The operator inspects the scan and judges whether grains have been distinguished from background. Usually, one tries several other thresholds (e.g., 20, 40) and chooses the value most suited to the present specimen or field. Although, ideally, one could find a single threshold that would suffice for an entire day of grain counting, in practical use it may be necessary to adjust the threshold when moving from one field to another. In our experience, adjusting the threshold does not invalidate the standardization; measurements are reproducible so long as the illumination, brightness, and contrast have been set appropriately and are not changed during the assay.

Counting the Specimen (Figure 3B–D). Following alignment and calibration, it is easy to quantify grain density in many fields of different types. From the menu above we select "1—GRAIN COUNTING," and the names of the arrays are again displayed on the text page. We press "G" for graphics, and center the first specimen field. The sequence below is followed for each field:

Press the numeric key to indicate the array for the first measurement ("2" for layer 2).

Move the video cross hairs to the edge of the region we wish to measure.

Press the cursor button and encircle the region of interest, thereby measuring its area and perimeter.

Following release of the cursor button, the area and perimeter values are displayed in the lower right corner of the graphics screen, as well as on the text page.

Place the cross hairs in the center of the encircled area and press the "F" key, indicating that we wish to perform a "fill" count in which only the area inside the line will be assayed.

The computer fills the area counted with solid white, and erodes holes in the solid figure to indicate pixels exceeding the threshold.

The count is displayed on the graphics screen and its value is automatically stored in a data array corresponding to the area measurement.

We are now ready for another measurement (e.g., press "3" to indicate that layer 3 is next).

A permanent copy of the graphics image is available at any juncture by pressing "H." Similarly, the screen may be saved to disk and subsequently reloaded to the video for future reference.

Although the current protocol requires grain counts, it should be noted that microdensitometry is also available in the "fill" mode. The total brightness value for all the pixels inside the line and their average brightness may be determined and entered into arrays exactly like the grain counts.

In a fashion similar to that described for manual methods above, the measurement continues until the investigator feels that the sample is adequate. At this point it is simple to request a statistical analysis from the Procedure Options. If the data do not yet meet your criteria for significance, it is equally simple to return to the measurement routines and gather more data. The Apple-based systems hold 3,800 values in RAM; however, their analytical routines may be configured to handle virtual data from the disk, providing limitless data arrays. The 16-bit micros such as the IBM PC may be configured for limitless virtual arrays in digitizing measurement as well as analysis.

Data Storage and Analysis

Experience has shown that one of the best ways to make the power of a microcomputer available to novice users is by providing lists (menus) of the selections available at each step of the program. The master menu of the Bioquant morphometry programs may be reached at any point in any program simply by pressing the RETURN key. By pressing a number from 1 to 8, any of the procedures may be initiated.

BIOQUANT II

PROCEDURE OPTIONS

1 DATA FROM DISK
2 DIGITIZING MEASUREMENT
3 DATA LISTING
4 STORE DATA ON DISK
5 STATISTICS
6 DISTRIBUTION & PERCENTAGES
7 CALCULATIONS
8 CORRELATION

Although the eight Procedure Options are available at any point in the program, a normal sequence of data gathering and analysis follows in relative order. Thus, at the conclusion of measuring (option 2), the data would be listed (3) and then stored on disk (4) for safekeeping. Data from disk (1) may be retrieved at any time for coanalysis with current data.

Data Listing

These routines allow the investigator to format video and hard-copy displays for maximum clarity (Figure 4). For instance, we would wish to list the areas for layer 2 adjacent to the grain counts for layer 2. Furthermore, if we had measured areas in square millimeters, it would be desirable to display several decimal places since most values would be fractional; whereas if the units were square micrometers,

Figure 4. Hard copy of data listing may be formatted according to the requirements of the data (e.g., names and comments, number of columns, number of decimals).

```
D1     NAME - LAYER2%
       COMMENT -

D2     NAME - LAYER3%
       COMMENT -

D3     NAME - LAYER4%
       COMMENT -


UNITS OF MEASUREMENT = UM
```

	D1	D2	D3
Ø	Ø	Ø	Ø
1	46.354	11.456	42.188
2	51.215	3.483	45.3Ø1
3	47.Ø45	4.488	48.466
4	47.Ø78	3.7Ø9	49.211
5	29.361	23.235	47.4Ø3
6	27.4Ø5	23.486	49.1Ø8
7	36.894	19.75Ø	43.354
8	38.37Ø	21.554	4Ø.Ø74
9	29.273	9.624	61.1Ø2
1Ø	26.491	12.971	6Ø.536
11	39.511	27.923	32.564
12	48.223	32.698	19.Ø77
13	57.478	14.176	28.344
14	47.372	11.7Ø1	4Ø.925
15	43.744	24.779	31.476
16	29.562	22.418	48.Ø19
17	46.282	3.5Ø8	5Ø.2Ø8
18	45.177	3.4Ø6	51.416
19	61.93Ø	5.891	32.178
2Ø	62.335	4.235	33.428
21	46.5ØØ	14.466	39.Ø33
22	49.365	11.916	38.718
23	52.38Ø	14.711	32.9Ø8
24	61.Ø4Ø	3.1Ø7	35.851
25	7Ø.288	5.626	24.Ø84
26	7Ø.643	6.741	22.614
27	53.826	12.283	33.89Ø
28	56.Ø56	12.7Ø2	31.24Ø
29	57.178	11.892	3Ø.928
3Ø	57.115	8.651	34.232
31	42.575	22.4Ø4	35.Ø2Ø
32	59.494	12.4Ø3	28.1Ø2
33	41.365	24.543	34.Ø91
34	5Ø.61Ø	17.3Ø8	32.Ø8Ø

integers would suffice. Subsequent to the video listing, simply pressing "H" activates the printer to produce a hard copy.

Store Data on Disk

The data values, along with name and comment, may be written to the disk in drive 2. The name entered for the array becomes the file name; multiple arrays may be stored in a single file at the discretion of the operator. The disk files are standard DOS text files, allowing the data to be accessed by other programs.

Statistics

Statistical parameters, including number, total, mean, sum of squares, variance, standard deviation, and standard error of the mean, may be determined for up to 10 arrays simultaneously. Up to 100 arrays may be analyzed simultaneously using the separate STATISTICS RETRIEVAL program. These arrays may be further characterized using the Student t test and one-tailed ANOVA (F-value). Median and Mann–Whitney U-tests are also available. Following video display, the data may be printed (Figure 5).

Distribution and Percentages

A frequency distribution is available in both tabular and graphics form for raw data or percentages determined by normalizing each array to 100%. The operator may specify the following:

- linear or logarithmic scales;
- up to 100 bins;
- up to four arrays simultaneously;
- bin size (linear) or exponentiation (log);
- lower limit and y scale.

Both line and bar graphs are available, and the graph labels may be specified to facilitate preparation of illustrations for posters and publication (Figure 6).

Calculations

A valuable option allows manipulation of data arrays by algebraic and trigonometric operators. Thus, we are able to determine the number of grains per unit area for each of the measurements. A new array is created in which each element is the quotient of the corresponding elements in the grain count array divided by the area array. In order to normalize the data with regard to varying levels of background, we have set the total count density for all three layers from each section equal to 100% and then calculated the percentage contributed by each layer. These calculated values have been used for the statistics, distribution, and correlation analyses.

```
1/D1     NAME  - LAYER2%
     COMMENT  -

2/D2     NAME  - LAYER3%
     COMMENT  -

3/D3     NAME  - LAYER4%
     COMMENT  -
```

FILE	NAME	COMMENT	NUMBER	SUM	MEAN	SQ SUM	VARIANCE	STD.DEV.	S.E.M
1/D1	LAYER2%		34	1629.552	47.928	82529.051	134.175	11.583	1.986
2/D2	LAYER3%		34	463.260	13.625	8505.088	66.455	8.152	1.398
3/D3	LAYER4%		34	1307.186	38.446	53605.242	101.462	10.072	1.727

```
T-VALUE (1<>2) = 14.121125   (66 D.F.)

T-VALUE (1<>3) = 3.60153268   (66 D.F.)

T-VALUE (2<>3) = 11.1690578   (66 D.F.)



F-VALUE = 105.94553

     DEGREES OF FREEDOM = 99, 2
```

Figure 5. Standard hard copy of statistical values including t test and ANOVA.

Correlation

As illustrated in Figure 7, the values from two arrays may be displayed in a scattergram overlayed with the linear regression line. The coefficient of correlation, slope, and y intercept are listed also.

Research Applications

Although the figures in this chapter have been presented as examples of computer technique, they represent actual experimental data that may deserve comment. For these specimens, we have concentrated on automated counting because the large areas displaying label contained large numbers of grains, too large to count manually except for illustration. Also, because the labeled laminae are contiguous, we were confident in comparing consecutive samples.

Comparison of Label Density in LGN Layers 2–4

Development. Six tree shrews, spanning the ages of newborn to adult, received injections of tracer into the vitreous humor of one or both eyes and survived for 6–24 hr. In all cases the label density in layer 3 was significantly lower than in layers 2 and 4, indicating that whatever factors are responsible for decreased

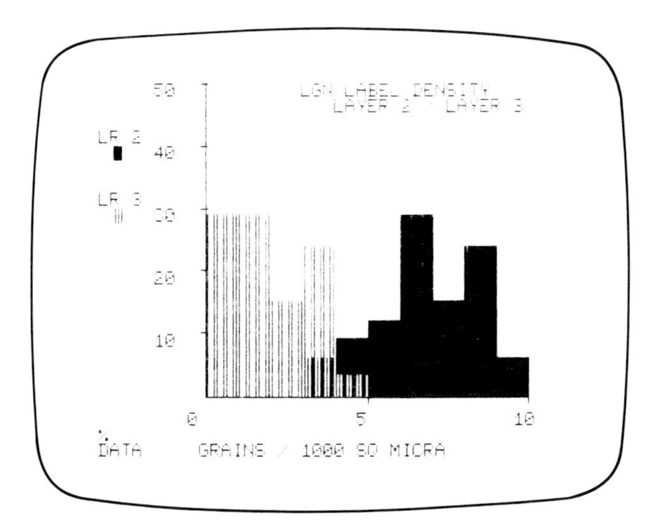

Figure 6. Bar graph of frequency distribution provides ability to add labels. Data, which have been normalized to percentages to compensate for variation in the background, demonstrate magnitude of differences in label density for layers 2 and 3.

transport to layer 3 in the adult are already present at birth. These factors could include a lower level of uptake in the retina by the ganglion cells projecting to layer 3, a slower rate of transport by these cells, a smaller caliber of fiber or less axonal arborization resulting in less retinal afferent axoplasm in layer 3, or fewer fibers projecting to that layer. Correlation with ultrastructure studies may indicate which factors are responsible for the reduced label density.

Figure 7. Correlation analysis provides graphic scattergram with linear regression line and other numeric parameters. Data demonstrate that layers 2 and 4, although they seem more similar to each other than to layer 3 (which always has less label), are not especially well correlated either (coefficient of correlation = .583).

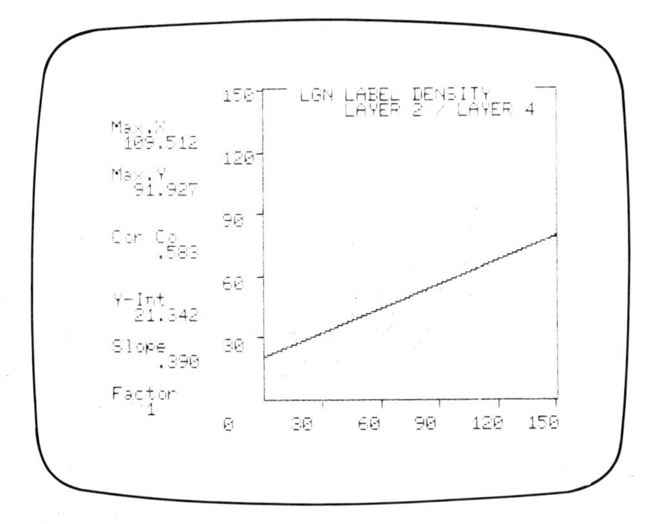

Different Tracers (Proline, WGA, WGA-HRP). At least in the newborn, our data show the same pattern of label density in layers 2, 3, and 4 for each of the three tracers studied. The label in layer 3 is significantly less than layers 2 and 4 in all cases. These data indicate that the retinal ganglion cells exhibit similar transport characteristics for dissimilar tracers, one of which is incorporated into proteins synthesized de novo and the other an exogenous lectin.

Conclusions

The Bioquant programs provide powerful tools for image analysis that may be added to popular microcomputers already situated in many laboratories. The modular design permits starting at any level, from manual measuring to automated video digitizing, with capabilities added as required. Other special accessory programs include those for point-count stereology, 3-D serial reconstruction and rotation, and bone morphometry.

Thorough documentation is provided, including both step-by-step "tutorial" and reference sections for each program. A "Help Disk," which may be summoned by entry of "?" at any point, provides a description of the current procedures and suggestions for their use.

The Bioquant uses the operator-interactive approach, depending on the superior discriminative powers of human vision to decide where and what to measure. The program design fine-tunes the computer's capacity for accurate measurement, data storage, calculation, and graphics display, allowing investigators at any level of computer expertise to use it with ease.

To those not familiar with biomedical image analysis, the fully automated video systems may seem like more sophisticated tools to attack the same problems. Frequently, however, such is not the case. Regardless of how expensive the system, it cannot match the human eye for detecting the objects of interest.

The fully automated systems excel in two areas. Very expensive and powerful ones are great for enhancing images, subtracting noise or augmenting contrast. Less expensive ones are fast and accurate for measuring high-contrast images where the objects do not overlap, such as the masks used in manufacture of integrated circuit chips. However, rarely are our specimens so distinct.

Good advice for anyone intending to purchase an image analysis system is, "Use it in your own lab with your own specimens!" Make sure that it does what you need before committing the funds. Only by trial, will you be sure that the system meets your needs.

Costs of the Bioquant depend on the capabilities required and hardware available already. The basic measuring program is approximately $1,000, with additional data storage and analysis modules $895 each. The complete morphometry and stereology programs, ideal for a multiuser facility where various projects are underway, is packaged at $6,995. The special accessories, including video, grain counting, 3-D reconstruction and rotation, and bone morphometry, are priced at $2,000–$4,000. In the United States, Canada, and many other regions of the world, the Bioquant system is available through E. Leitz microscope dealer representatives, who gladly provide demonstrations, trials, and advice.

Research supported by grants EY-05038 and EY-02221 from the National Eye Institute.

The authors are grateful to Tim Lempka, Cliff Batson, and Linda Brown for excellent assistance in preparation of the manuscript, and to James Ratzlaff for outstanding program support.

References

Brunso-Bechtold, J.K., and V.A. Casagrande (1982) Early postnatal development of laminar characteristics in dorsal laternal geniculate nucleus of the tree shrew. J. Neuroscience 2:589–597.

Brunso-Bechtold, J.K., and V.A. Casagrande (1984) Development of layers in the dorsal lateral geniculate nucleus in tree shrew. In W. D. Neff (ed.): Contributions to Sensory Physiology, vol. 8. New York: Academic Press.

Casagrande, V.A., and J.K. Brunso-Bechtold (1983) The relationship between afferent laminar development and cell layer formation in the lateral geniculate nucleus (LGN) Soc. Neurosci. Symp. 9:25 (abst.).

Salpeter, M.M., F.A. McHenry, and E.E. Salpeter (1978) Resolution in electron microscope radioautography. IV. Application to analysis of autoradiographs. J. Cell Biol. 76:127.

17

Quantitation of Two-Dimensional Gel Electrophoretograms by Computerized Analysis

Cary N. Mariash and Steven Seelig

Introduction

Developments in electrophoretic technology over the past several years (O'Farrell, 1975) have permitted the examination of multiple proteins from a given sample simultaneously. The ability to perform two-dimensional (2-D) gel electrophoresis of protein mixtures enables clinicians and basic scientists to uncover changes in specific proteins from various biologic fluids (Young and Anderson, 1982). Other developments in molecular biology, such as in vitro mRNA translational assays (Pelham and Jackson, 1976), when combined with two-dimensional gel electrophoresis, offer the same advantage in examining the expression of multiple genes. Thus, one can examine the pretranslational effect of hormonal or other physiologic manipulations on any tissue of interest by isolating the mRNA from that tissue, translating the mRNA in an in vitro translational assay, and separating the translated products by two-dimensional gel electrophoresis. By adding radioactive amino acids to the translational assay, one can visualize the separated proteins by exposing the two-dimensional gel to x-ray film to create an autoradiogram.

We initially applied these techniques to study the influence of thyroid hormone on gene expression in rat liver (Seelig et al., 1981). There were well over 200 individual translated products that were visualized on the 2-D gel autoradiograms. Many of these products changed with alterations of the hormonal status. Some of these changes involved apparent alterations in the concentration of the products rather than complete appearance or disappearance. In order to study the kinetics of response, and to classify the various responses, it therefore became necessary to quantitate the precise amount of radioactivity in each of the visualized spots.

The traditional method for quantitation of radioactivity in electrophoretograms is to cut the appropriate area on the electrophoretogram, solubilize the gel, and measure the radioactivity by scintillation counting. For two-dimensional gel electrophoretograms, one must first localize the spot of interest on the dried gel, precisely cut out the area corresponding to that spot, cut out another area of equal size in which there are no visualized spots to determine an appropriate gel background, and finally solubilize and count each cutout. Because of the inherent difficulties in this technique, it became obvious that alternative methods for analysis of the 2-D gel autoradiograms were necessary. Stimulated by several recent articles describing the utilization of large computers for the analysis and quantitation of two-dimensional gels (Bossinger et al., 1979; Garrels, 1979), we felt that such techniques could be applied to microcomputers. Although it had been amply demonstrated that computer technology could be adapted for the rapid quantitation of two-dimensional gel electrophoretograms, our goal was to utilize microcomputer technology to decrease the cost required for computerized analysis of 2-D gels by several orders of magnitude (Mariash et al., 1982).

Materials and Methods

Gel Preparation

The two-dimensional gel autoradiograms were prepared by standard techniques (O'Farrell, 1975; Seelig et al., 1981). In brief, isolated mRNA is translated in an in vitro reticulocyte lysate translational assay system with ^{35}S-methionine as the radioactive amino acid. The radioactive proteins whose synthesis was directed by the mRNA added to the lysate are separated by isoelectric focusing in the first dimension, and sodium dodecyl sulfate (SDS) electrophoresis on a slab gel in the second dimension. Proteins are therefore separated by their unique charge in the first dimension and their molecular size in the second dimension. The gel is placed in fixative, and if necessary impregnated with a fluor, dried, and exposed to standard x-ray film for an appropriate length of time at $-80°$ C. The developed x-ray film is used for quantitative analysis of the two-dimensional gel electrophoretogram. The basic assumptions in quantitation are that the density of the individual spot is proportional to the amount of radioactivity in the gel, and the radioactivity is proportional to the amount of the mRNA species coding for that protein.

Hardware

The overall hardware design is depicted in Figure 1. The computer used is an Apple II+ (Apple Computer, Inc., Cupertino, CA). Current configuration of the Apple II+ consists of 64*K* RAM (Apple Language Card) and two $5\frac{1}{4}$-inch floppy disk drives. The computer is also configured with a parallel interface for printer output. The upper 16*K* of RAM memory space is shared with the 12*K* of ROM by software controlled bank switching. The only program occupied by this 16*K* RAM is the disk operating system (Diversi-DOS, DSR, Inc.). The autoradiogram is placed on a standard x-ray viewbox, or a photographer's light box, for backlight illumination. A black and white video camera is used to obtain the data. The video image is digitized by a video digitizing circuit installed within the microcomputer and the digitized array is stored and analyzed within the microcomputer. To complete the hardware interfacing, two video monitors are required. One video monitor is used for display of the video image, the other for display of the computer output. Last, we utilize a dot-matrix printer to obtain a hard copy of the results.

Figure 1. Schematic drawing of gel analysis hardware. The autoradiogram to be analyzed is placed on the x-ray viewbox for backlighting. (CCTV = closed-circuit television). [Reproduced with permission from Mariash et al. (1982).]

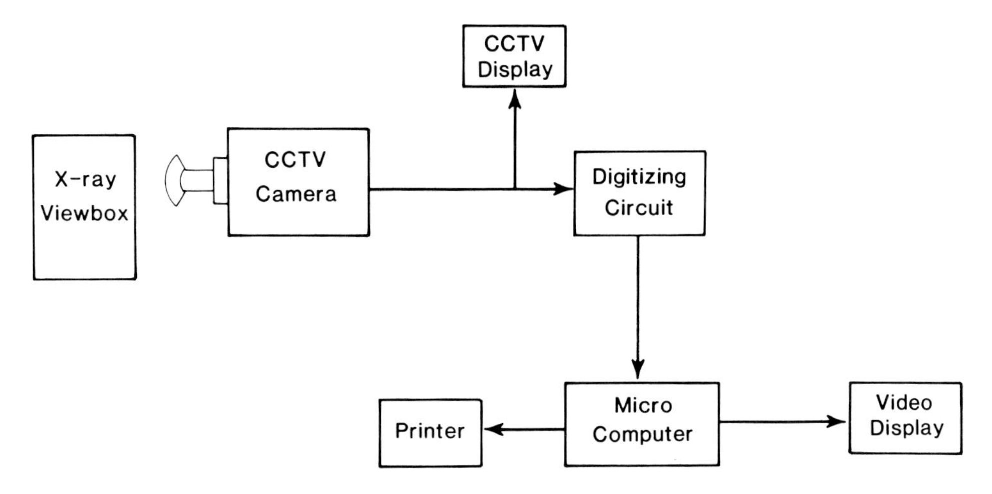

A major requirement of the x-ray view box is uniform light emission. It is important that shading across the image be minimized for accurate quantitation. Nevertheless, a correction for a small amount of shading by the digitizing circuitry is described later.

Although there are a variety of techniques for obtaining densitometric data from an autoradiogram, we elected to utilize videodensitometry because of the ease of interfacing and the rapidity of scanning a 2-D gel. Any video camera can be utilized in this system. However, we suggest using a high-quality video camera in order to maximize the signal-to-noise ratio. As described below, the greater the signal-to-noise ratio, the fewer times an area must be scanned in order to ensure accurate digitized data.

For greatest flexibility, we also recommend obtaining a zoom lens with either macro capabilities or an additional close-up attachment. This zoom lens allows one to alter the resolution to the magnification required for optimum separation of individual spots. The equipment we currently use is an RCA TC1001/05 black and white video camera, which provides 800 lines of horizontal resolution. It utilizes a C mount lens to which we have attached a Cannon V/6 zoom lens with a Cannon 450 close-up lens.

The most important piece of equipment in the 2-D gel analysis system is the video digitizing circuit. Although there are many video digitizers available, our choice was dictated both by the Apple computer and by the price/performance ratio. Of the few digitizers available for the Apple computer, only one had the capabilities required—the Digisector-65 (DS-65; The Microworks, P.O. Box 1110, Del Mar, CA). The Digisector is supplied as a single peripheral board that fits into any available slot in the Apple computer. The input to the board is the composite video signal originating from the video camera. The board converts the video signal to a 6-bit digital value. It divides the video signal into 256 horizontal by 256 vertical directly addressable pixels. Thus, one can obtain a gray scale of 0–63 at any of 63,536 pixels in a given video image.

The DS-65 uses both the horizontal and vertical sync pulses for timing. To digitize a pixel one sends the x and y coordinate of that pixel to the DS-65. The DS-65 determines the correct time to digitize the voltage of the video signal. This gray-scale value can be read by the computer from the DS-65. The timing circuits, the analog-to-digital (A/D) circuits, and the peripheral interface adapter are all integrated on the Digisector. Timing of the A/D conversion is such that only one pixel can be digitized per horizontal scan line, and of the 256 scan lines which can be sampled, only every other scan line can be sampled sequentially. It takes approximately 8 sec to digitize the entire $65K$ pixels available.

Calibration

One characteristic of a video signal is that the peak-to-peak voltage across the entire video image is relatively constant over a wide range of total light presented to the video pick-up tube (videcon tube). This permits generation of pleasing video images while changing the lighting from low to high intensity. Unfortunately, this characteristic prevents a correlation of a digitized gray-scale value with the actual light input to the video camera. However, this problem can be overcome by appropriate adjustment of the sensitivity and threshold levels of the A/D circuitry. There are two potentiometers on the DS-65 for this purpose. Because these potentiometers must be adjusted for each video image, we have found it most convenient to remove the potentiometers of the DS-65 and replace them with equivalent multiturn potentiometers on long leads that lie outside the Apple computer.

The threshold voltage is adjusted by placing a black border around the video image. A pixel on this border is repetitively scanned and the threshold potentiometer adjusted to yield a gray-scale value just under the maximum of 63. Next, the sensitivity of the A/D circuitry is adjusted by repetitively scanning a blank area of the gel for background. The potentiometer is adjusted so that this gray-scale value is just above 0. The process of electronically adjusting the A/D circuitry for each video image is analogous to establishing a dark-field voltage and blank optical density in a standard spectrophotometer. The gray-scale value of any digitized pixel subsequently obtained is directly related to the light intensity of the pixel. By applying this electronic correction procedure, one can scan gels of varying overall densities and yet maintain a constant relationship between gray-scale value and optical density of any given pixel.

In order to provide the Digisector with the correct x and y coordinates, we have overlaid the video monitor with a transparent film ruled off in 25-pixel increments. Because the DS-65 also outputs an intensified cursor to the video monitor at the pixel location digitized, one is ensured that the area requested for analysis is included in the area actually digitized. Moreover, the combination of the intensified cursor and the overlay eliminates much guesswork in defining the appropriate x and y coordinates for digitization.

Standard Preparation

One must be able to relate gray-scale values to actual counts per minute per square millimeter (cpm/mm^2). Since the relationship between gray-scale value and cpm/area is not linear, a standard curve must be established for each experiment. Thus, the standard curve allows comparison between experiments and corrects for changes in exposure time, film composition, and film development. A standard curve is prepared by uniformly dissolving radioactive proteins in 12% acrylamide. The acrylamide solution is serially diluted in nonradioactive 12% acrylamide. Each serial dilution is allowed to polymerize at the same thickness as the slab gel utilized for the second dimension. The polymerized standards are then cut into 0.5×0.5 cm squares and processed simultaneously with the two-dimensional gels. Following development of the x-ray film, these standard squares are solubilized and counted by liquid scintillation techniques to determine the cpm/area in each standard chip. One can then directly relate the gray-scale value of that chip to a known cpm/mm^2. A standard curve can be established using the equation proposed by Rodbard (1974). Since the gray-scale value/pixel is independent of the magnification used, the cpm per pixel can be obtained by multiplying the cpm/mm^2 obtained from the standard curve by the magnification of the image expressed as mm^2/pixel. The magnification of the image is easily obtained by determining the number of pixels contained in an image of known size.

Program Operation

Setup

Several compromises were necessary to implement the image analysis package in APPLESOFT (BASIC). Although the program development was written in BASIC, most of the program as finally implemented is written in 6502 ASSEMBLER. This was necessary to optimize memory utilization and minimize execution time.

The first step in the image analysis program is to obtain the x and y coordinates for setting up the threshold and sensitivity of the ADC. The coordinates are sent to the DS-65 and the results displayed on the terminal. This process is continuously repeated until the electronics are appropriately set. Next, the area of the gel to be scanned is obtained by requesting the starting and ending x and y coordinates. In addition, the program requests the number of times this area is to be scanned. As previously pointed out, repetitive scans are required to maximize the signal-to-noise ratio. It can be shown that the reduction in noise is proportional to the natural log of the number of scans (Williams, 1980). We have found that approximately 32 scans of each pixel are required to reduce the 95% confidence interval of the gray-scale value to less than one unit out of the 63 gray-scale values available.

Scan

Program control is then passed to the subroutine SCAN. In order to minimize the time required for scanning the area of interest, the entire subroutine SCAN was implemented in assembly language. Because each pixel in the video image appears only once each 16.66 msec, it is important to optimize the order in which the pixels are accessed, and minimize the instructions between each access. This is best accomplished if columns are accessed sequentially, and every other row is accessed in each column. The only code between pixel digitizations is that required for summing the digitized data at the appropriate place in memory where the scan data is stored. Thus, to obtain all the pixels in a given column, each column must be accessed twice, odd rows followed by even rows. It takes approximately 1 min to complete the data acquistion of a 16,000 pixel area; 2 sec for each of the 32 scans. Since a video image contains 65,536 pixels, approximately 10 min are required to scan an entire video image fully.

Because the 6-bit gray-scale values are summed 32 times, each pixel requires two 8-bit bytes for data storage. This substantially reduces the memory available in the Apple for further analysis. Therefore, the data is divided by 4 to produce a 1-byte gray-scale value (0–252) for each pixel. Following completion of scanning, the data can be represented in memory as a two-dimensional matrix (Figure 2A).

Control is then passed back to the BASIC program where the operator has the option of storing the primary data on disk for later analysis, or immediately completing the analysis of the area scanned. If analysis is to be completed, control is passed to the subroutine MINIMUM. This subroutine is also implemented in assembly language to decrease execution time.

Minimum Map

MINIMUM examines the data matrix to produce a second two-dimensional matrix containing the identification of the individual spots. The minimum

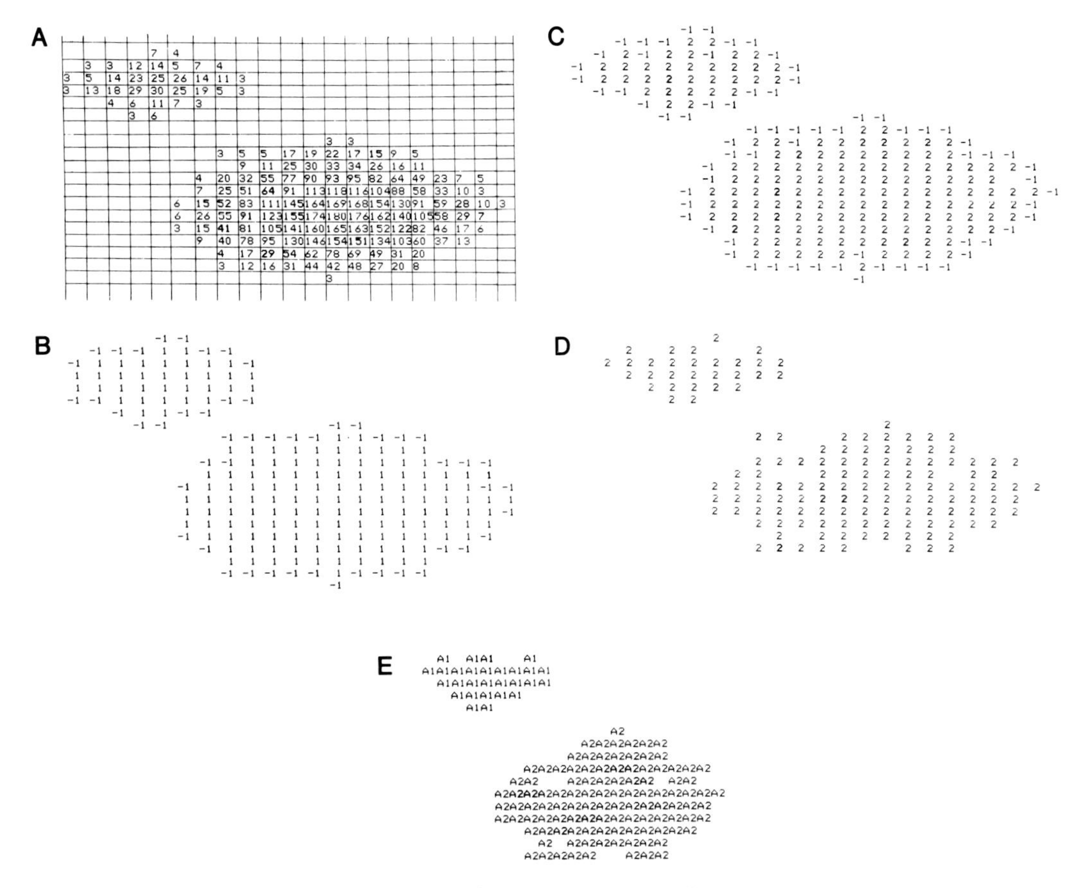

Figure 2. Step-by-step analysis of an area on a two-dimensional gel autoradiogram. The area scanned contains two spots. Panel A represents the original data matrix. Each value is the gray-scale value at that pixel. Panel B is the minimum matrix after all columns are searched for minima. Panel C is the minimum matrix after rows are searched for minima. In panels B and C a −1 represents a minimum. Panel D is the minimum matrix just prior to assignment of individual spot numbers. In panels A–D all empty pixels contain a value of 0. Panel E represents the final printout of the minimum matrix. Al is spot 1 and A2 is spot 2.

matrix is mapped one-to-one with the data matrix. Because of memory constraints, each pixel in the minimum matrix is also represented by a single byte. Therefore, the maximum number of spots that can be individually numbered is 255.

A spot is defined by MINIMUM as a set of contiguous pixels that are bounded in at least two orthogonal axes by minima. This definition makes no assumptions regarding the shape of a spot and only implies that the edges of a spot have minimum intensities. In brief, MINIMUM operates by definition of absolute and relative minima. An absolute minimum occurs when a negative or zero slope is followed by a positive slope; a relative minimum is located when a zero slope occurs between slopes of identical signs. As described in detail below, the location of both absolute and relative minima are assigned a value of −1 in the minimum matrix. The locations between the minima are incremented by 1. We utilize the x and y axes to search for minima.

There are five different subroutines within MINIMUM. After the minimum matrix is set to 0, the first subroutine searches the data matrix by columns for minimum intensities. An initial nonzero slope is obtained, its location marked in the minimum matrix by a -1, and its sign saved at variable $S1$. If $S1$ is negative (decreasing intensities) the search is continued until the slope changes sign (either 0 or positive). This location is also marked in the minimum matrix by a -1 and all values between these two minima in the minimum matrix are incremented.

If the initial slope was positive, then two additional slopes, $S2$ and $S3$, are located. $S2$ is the next changing slope, and $S3$ is obtained according to the following table:

		$S2$	
		0	−
	−	continue	continue
subsequent	0	continue	$S3$
slope	+	$S3$	$S3$

$S3$ is marked by a -1 in the minimum matrix, and all pixels between $S1$ and $S3$ are incremented by 1. All columns in the data matrix are searched sequentially by this algorithm. Upon completion, the minimum matrix contains values of either -1 or $+1$ (Figure 2B).

The second subroutine utilizes the same minimum algorithm to search the data matrix by rows (x direction). After completion of this subroutine, the minimum matrix contains values from -1 to $+2$ (Figure 2C). However, only those pixels between minima in both the x and y axes will contain the value $+2$.

Since a spot must have minima in all axes examined, spot separation can be improved further by searching in more than two axes. However, we have found adequate resolution by continuing the search for minima in the axis at 45° from the x axis. These minima are marked in the minimum matrix with a -1, but the pixels between minima are not incremented. In order to improve the resolution further, a pixel immediately between two minima on the diagonal axis is also marked with a -1. All values that are not $+2$ are then removed from the minimum matrix (Figure 2D). The process of making diagonal minima improves spot resolution but, as demonstrated in Figure 2D, may lead to minor deterioration of contour definition.

By definition, an individual contour contains all pixels whose values are $+2$ and are adjacent to other values of $+2$ in either the x or y axis. Since there is a limit of 255 to the number of uniquely identified spots, it is necessary to remove contours that contain only one or two pixels before assigning a unique number to each contour. Therefore, all $+2$ pixels in the minimum matrix that have at most one other $+2$ pixel in either the adjacent x or y axis are assigned the value 0.

The final subroutine in MINIMUM replaces all the $+2$'s within a contour by a unique number from 1 to 255. Thus, all values of 1 in the minimum matrix represent contour 1, all values of 2 represent contour 2, and so on. Upon completion of the MINIMUM routine, two two-dimensional matrices exist in memory:

1. the data matrix containing all pixel intensities, and
2. the minimum matrix, a one-to-one map of the data matrix, containing the spot identification numbers.

Analysis

Control is passed back to the BASIC program for completion of the image analysis. At this point the minimum matrix is sent to the printer. The contour number is converted to a unique two-digit alphanumeric value, and each pixel is represented by two spaces on the printout (Figure 2E). The operator can examine the printout to ensure that all spots were separated correctly. An example of a two-dimensional gel and a corresponding analysis is presented in Figure 3.

Both the intensity of each spot and, based on the standard curve, the cpm/mm^2 for each spot, is summed pixel by pixel. The program then multiplies the summed cpm/mm^2 for each spot by the magnification factor (mm^2/pixel) to give the total cpm in each spot. Finally, the program prints out the alphanumeric spot identification number with the total intensity and the calculated cpm for that spot. The entire process of scanning, analyzing, and printing a 16,000 pixel area with this system takes about 10 min. We use a magnification of about 0.175 mm^2/pixel to obtain adequate resolution of nearly all spots on a 180 $\times$ 100 mm gel. However, at this resolution, 4–6 scans are required per gel. Therefore, it takes approximately 1 hr to perform a complete analysis on a single gel.

Limitations

As presently designed, the program utilizes all available memory in the Apple computer. Therefore, an area comprising more than 16,000 pixels cannot be scanned at once. Another limitation to the maximum area that can be scanned at once is the number of spots in that area. In order to assign more than 255 spots, each pixel in the minimum matrix would have to be represented by 2 bytes. Since the current program already uses all available memory in the Apple, it would be impossible to expand the minimum matrix to 2 bytes per pixel. Therefore, one must be sure that the area scanned contains fewer than the maximum 255 spots allowed.

Although the system described in this chapter has limitations both in its hardware and software implementation, we have found the system to be highly accurate and reproducible. As shown in Figure 4, comparison of traditional methods for spot analysis with our system yields a straight-line function with a correlation coefficient of 0.98. Another test of quality control is to compare the results obtained on the same gel by two different people at two separate times. Such data are reproduced in Figure 5 and again demonstrate a high degree of reproducibility with a correlation coefficient of 0.99.

Research Applications

As stated in the Introduction, the ability to quantitate individual spot densities permits detailed kinetic analysis of the responses of individual gene products to hormonal and other physiologic manipulations. After development of this system, we were able to show that thyroid hormone administration led to an

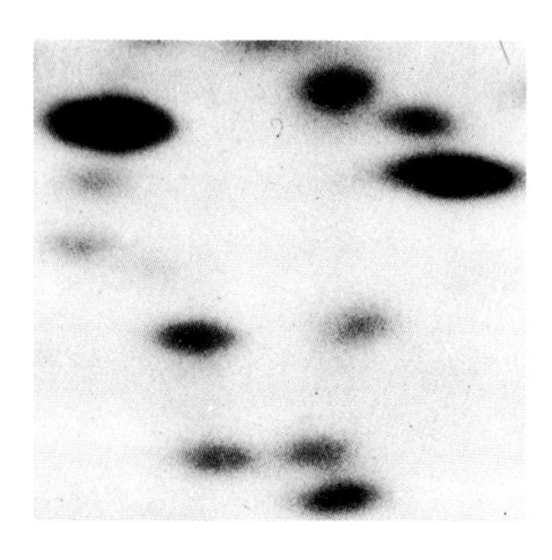

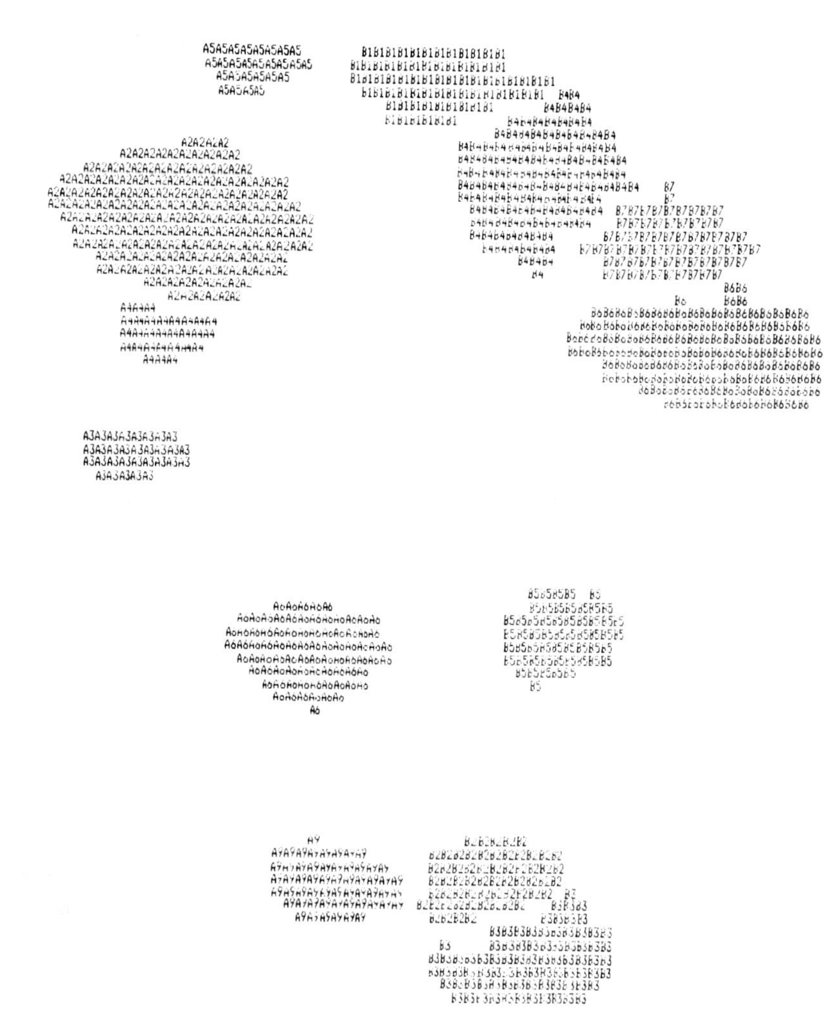

Figure 3. A representative two-dimensional gel electrophoretogram and associated analysis. The upper panel is the middle of a two-dimensional gel. The radioactive proteins were separated by pH from left (basic) to right (acidic) and by molecular size from top (M_r approximately 35,000) to bottom (M_r approximately 20,000). The bottom panel is the printout of all the spots identified by the analysis program. The total intensity ranged from 18 (spot A5) to 987 (spot A2). [Reprinted with permission from Mariash et al. (1982).]

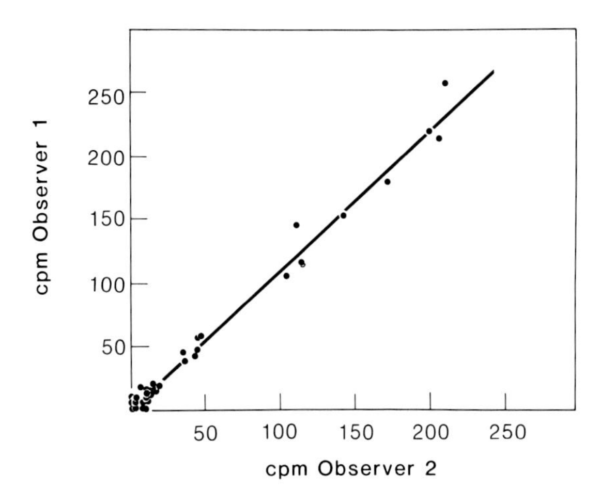

Figure 4. Correlation between the cpm calculated by videodensitometry and cpm measured by cutting each spot from the gel as described in the text. Sixty-seven spots were measured with cpm ranging from 4 to 955. The correlation coefficient is r = .981. [Reproduced with permission from Mariash et al. (1982).]

extremely rapid induction of one specific mRNA ($mRNA_{S14}$) with virtually no time lag following the administration of thyroid hormone (Seelig et al., 1982). Applying similar techniques, we subsequently demonstrated that many of the mRNAs altered by thyroid hormone were mediated by a thyroid hormone induced increase in pituitary growth hormone with growth hormone the proximate stimulus for several specific mRNAs (Liaw et al., 1983). Lastly, we have taken advantage of the ability to quantitate the entire 2-D gel pattern to classify the responses to multiple hormonal stimuli and physiological states. In these studies, the level of mRNA for each changing spot under each condition studied was analyzed by multivariate analysis techniques. We were able to show that the hepatic mRNA profile was unique for each of the 10 states studied (Carr et al., 1984). In addition, we were able to determine the degree of similarity between the different states. Such analysis would not have been possible without the quantitative determination of spot intensities of an entire two-dimensional gel electrophoretogram.

We have utilized the system to quantify 2-D gels stained with the silver technique, single-dimensional gels stained with Coomassie brilliant blue R250, and

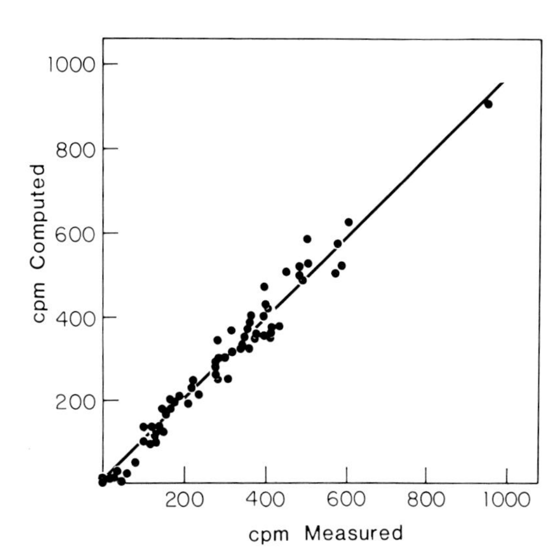

Figure 5. Correlation of computed cpm on one gel between two individuals. On separate occasions 44 spots were quantitated as described. The correlation coefficient was r = .99 and the slope 1.0.

autoradiograms produced by RNA–cDNA dot hybridizations with a ^{32}P-cDNA probe. The latter has allowed us to quantitate the actual mass of specific mRNAs present under varying physiologic conditions. Because of the flexibility built into the system, it should be possible to quantitate nearly anything which can be visualized.

Conclusions

This chapter has described an Apple-based densitometer system for analyzing two-dimensional gels. Many other investigators have used either videodensitometry or other densitometric methods for analysis of two-dimensional gel electrophoretograms (Bossinger et al., 1979; Garrison and Johnson, 1982; Kronberg et al., 1980; Lemkin et al., 1979). Moreover, many different algorithms for image analysis have been developed. The ability to analyze much larger arrays with much more effective algorithms is available by attaching a data acquisition system to a mini or mainframe computer. Additionally, other techniques of data acquisition such as drum scanning (Bossinger et al., 1979) or microscopic photometry (Garrels, 1979) can yield true optical densities as opposed to arbitrary gray-scale values obtained with videodensitometry.

With the rapid advancement in microcomputer technology, as well as the advancement in the electronics industry, it should be possible to build a low-cost system for the analysis of two-dimensional gels with much greater capabilities than the system described in this chapter. For example, there are several microcomputer systems available now that can directly access megabytes of memory and operate at 10–20 times the speed of an Apple computer. Additionally, there are newer video digitizers on the market that have 4–16 times the resolution of the Digisector with the ability to digitize the entire video image in 30 msec. Very recent video technology has led to the development of digital video cameras that consist of large arrays of photosensitive detectors. These cameras produce a digital output signal rather than an analog output signal and therefore do not require any special A/D conversion hardware in the computer system.

With the information provided in this chapter, it should be possible for any researcher to develop a video image analysis system for the quantitation of two-dimensional gel electrophoretograms for well under $4,000. All necessary equipment described in this chapter is commercially available. Moreover, the software we have developed for the Apple is available from The Microworks. If one wishes to develop a more sophisticated system, software image analysis techniques are available in the text by Castleman (1979).

We thank Dr. Jack H. Oppenheimer for his encouragement, support, and helpful discussions during the development of this project. The secretarial assistance of Hobie Pharis and Lori Spangler is also greatly appreciated. This work was supported in part by NIH Grants 1-RO1-AM19812, 1-K08-AM00800, and National Research Service Award AM06478.

References

Bossinger, J., M.J. Miller, K.-P. Vo, E.P. Gerduschek, and N.-H. Xuong (1979) Quantitative analysis of two-dimensional electrophoretograms. J. Biol. Chem. 254:7986–7998.

Carr, F.E., C. Bingham, J.H. Oppenheimer, C. Kistner, and C.N. Mariash (1984) Quantitative investigation of hepatic genomic response to hormonal and pathophysiologic stimuli by multivariate analysis of two-dimensional mRNA activity profiles. Proc. Natl. Acad. Sci. USA 81:974–978.

Castleman, K.R. (1979) Digital Image Processing. New Jersey: Prentice-Hall.

Garrels, J.I. (1979) Two-dimensional gel electrophoresis and computer analysis of proteins synthesized by culture lines. J. Biol. Chem. 254:7961–7977.

Garrison, J.C., and M.L. Johnson (1982) A simplified method for computer analysis of autoradiograms from two-dimensional gels. J. Biol. Chem. 257:13144–13149.

Kronberg, H., H.G. Zimmer, and V. Neuhoff (1980) Photometric evaluation of slab gels: I. Data acquisition and image analysis. Electrophoresis 1:27–32.

Lemkin, P., C. Merril, L. Lipkin, M. Van Keuren, W. Oertel, B. Shapiro, M. Wade, M. Schultz, and E. Smith (1979) Software aids for the analysis of 2D gel electrophoresis images. Comput. Biomed. Res. 12:517–544.

Liaw, C., S. Seelig, C.N. Mariash, J.H. Oppenheimer, and H.C. Towle (1983) Interactions of thyroid hormone, growth hormone, and high carbohydrate, fat-free diet in regulating several rat liver messenger ribonucleic acid species. Biochemistry 22:213–221.

Mariash, C.N., S. Seelig, and J.H. Oppenheimer (1982) A rapid, inexpensive quantitative technique for the analysis of two-dimensional electrophoretograms. Anal. Biochem. 121:388–394.

O'Farrell, P.H. (1975) High-resolution electrophoresis of proteins. J. Biol. Chem. 250:4007–4021.

Pelham, H.R.B., and R.J. Jackson (1976) An efficient mRNA-dependent translation system for reticulocyte lysates. Eur. J. Biochem. 67:247–256.

Rodbard, D. (1974) Apparent positive cooperative effects in cyclic AMP and corticosterone production by isolated adrenal cells in response to ACTH analogues. Endocrinology 94:1427–1437.

Seelig, S., D.B. Jump, H.C. Towle, C. Liaw, C.N. Mariash, H.L. Schwartz, and J.H. Oppenheimer (1982) Paradoxical effects of cycloheximide on the ultra-rapid induction of two hepatic mRNA sequences by triiodothyronine (T_3). Endocrinology 110:671–673.

Seelig, S., C. Liaw, H.C. Towle, and J.H. Oppenheimer (1981) Thyroid hormone attenuates and augments hepatic gene expression at a pretranslational level. Proc. Natl. Acad. Sci. USA 78:4733–4737.

Williams, T. (1980) Digital storage of images. Byte 5:220–238.

Young, D.S., and N.G. Anderson (eds.) (1982) Special issue: Two-dimensional gel electrophoresis. Clin. Chem. 28:737–1092.

18

Microcomputer Systems for Analyzing 2-Deoxyglucose Autoradiographs

C. R. Gallistel and Oleh Tretiak

Introduction

The 2-deoxyglucose technique detects alterations in the metabolically coupled functional activity of neural tissue. Nerve cells preferentially use glucose for their energy requirements. Changes in the functional activity of a neuron (for example, its impulse firing rate) often change its rate of energy utilization and hence its rate of glucose uptake. The 2-deoxyglucose (2DG) molecule is a mutilated form of the glucose molecule. It is made radioactive and injected into the circulatory system (or into the peritoneum, from which it enters the circulatory system) in tracer amounts. It is taken up by neurons (and other cells) in proportion to the cell's rate of glucose utilization. Like glucose, it is phosphorylated to a 6-phosphate form, but it cannot then be further metabolized. Hence, the amount of it that has accumulated in a neuron during the period when the tracer injection has cleared from the circulatory system is an index of the neuron's rate of glucose utilization. By sacrificing the animal, slicing the neural tissue, and exposing the slices to x-ray film, one obtains an autoradiographic image. The darkness of the film at a given point in the image reflects the accumulation of 2DG in the underlying tissue. Localized changes in the functional activity of neural structures change the relative rate of glucose utilization, hence the relative gray level of the corresponding region of the autoradiograph. The technique is superb for revealing localized alterations in tissue activity anywhere in the brain of an animal. The animal can be freely behaving during the period of isotope incorporation. The technique reveals the spatial pattern of activity generated by an experimental stimulus or experimental situation. It serves primarily to highlight areas whose activity may be of special relevance to a behavioral phenomenon and that merit further investigation by other techniques.

The 2-deoxyglucose technique cannot achieve its full potential in the absence of computer-assisted analysis of the autoradiographs. Systems suitable for analyzing these images have recently been described (Goochee et al., 1980; Gallistel et al., 1982). The systems so far described were implemented on minicomputers. The implementation involved a cost in both hardware acquisition and software development that placed them beyond the means of most individual laboratories. To meet the needs of a large community of users, a Biotechnology Resource Center devoted to the computerized analysis of 2-deoxyglucose autoradiographs (and other neurobiological images presenting similar analytic problems) has been established at Drexel University. The Center has as part of its mission the development of a low-cost microcomputer-based system that will perform the critical analytic functions now performed by the minicomputer systems. In this chapter, we describe those analytic functions. We also survey the rapidly emerging microcomputers and the specialized image-processing hardware to which they must be mated, calling attention to the aspects of their architecture and their operating systems that render them more or less suited to the implementation of these functions. The Center will develop and distribute software for one of the configurations to be described.

Analytic Functions

The analytic functions that are essential to the analysis of 2-deoxyglucose autoradiographs for localized changes in metabolically coupled functional activity are the following:

1. the rapid fine-scale digitization of the autoradiographic image;
2. enhancement of the operator's ability to detect small localized changes in the darkness of the image;
3. overlaying the histological image on the autoradiographic image;
4. the outlining of the features to be quantified;
5. the conversion of gray-scale values (the digital representations of local image darkness) either to rates of glucose utilization or to normalized indices of relative activation;
6. the computation of descriptive statistics;
7. the creation of data files suitable for computerized statistical analysis of the descriptive statistics;
8. the creation of analytic displays suitable for black and white (B/W) publication.

Digitization

The first step in the computer-assisted analysis of an image is digitization of the gray levels at each point in the autoradiographic image. The individual points in an image are called picture elements or *pixels.* The digitization of the pixel gray level allows storage of the image as a two-dimensional array in a digital memory *(picture memory).* The numerical value of each pixel in the array represents the gray level of the corresponding point on the image. This numerical value is called the pixel's *gray-scale value.* It is usually a 6-bit or 8-bit number. The analytic functions are performed using these gray-scale values. The image stored in picture memory is displayed on B/W and color monitors, so that the operator can direct the analysis.

Enhanced Detection of Changes in Relative Gray Level

The detection of regions in the autoradiographic image that are darker or lighter than would be expected from control readings is facilitated by the use of color windows to make false-color images. A color window is created when the user selects a color to represent a range of gray-scale values. All those pixels whose gray-scale values fall within the specified range (the window) are given the selected color in the display shown on the color monitor. The resulting image is called a *false-color image.* The operator may select several different colors, specifying a different range of gray-scale values for each color, to create a false-color image in which the approximate gray-scale value of any region can be read from its color.

The unaided eye cannot make accurate comparisons of the gray level of nonadjacent regions of an image or of the gray level of the same region in two different images. Color windows make this judgment easier. If a two-dimensional graytone image is likened to a mountain range, with variation in elevation corresponding to variation in gray level, then the investigator's task is analogous to the task of judging whether two peaks or two valleys have the same elevation. Figure 1 shows how color windows facilitate such judgments. We have found that the window is most useful for detecting asymmetries in the activation of homologous contralateral structures if its width and elevation can be rapidly and independently varied. One way to achieve this is to have a joystick (or other graphics input device) control the upper limit of the window while the width of the win-

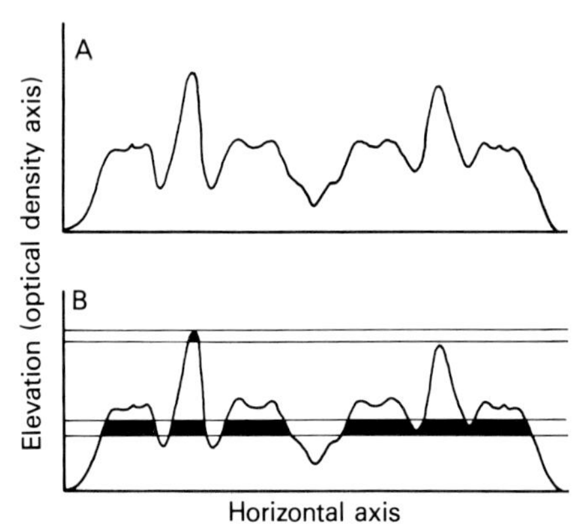

Figure 1. Schematic illustration of the principle behind the use of a color window to detect subtle differences in optical density. (**A**) It appears that the peak on the left may be higher than the peak on the right and the valleys on the left lower, but it is hard to be sure. (**B**) Lighting up (coloring in) narrow bands at user-choosable gray levels (optical densities) makes the judgment easy. The peak at left is illumined by the upper window whereas the peak at right is not; the valleys at right are illumined by the lower window, whereas the corresponding valleys at left are not. [Reproduced from Gallistel et al. (1982) by permission of ANKHO International.]

dow (the interval between the upper and lower limit) is increased or decreased by keystrokes.

Overlaying the Histological Image

A suitable system must permit the operator to alternate between the display of the autoradiographic image and the display of the histological image of the section from which the autoradiograph was made, with provision for bringing the two images into precise registration. The overlaying and registration of the histological image permits the operator to outline features (hot or cold spots) on the autoradiographic image and transfer the outline to the histological image to determine what structures are showing the effect. It also permits the outlining of structures on the histological image and the transfer of these outlines to the autoradiographic image for the purpose of quantifying the activation of histologically specified structures.

The registration of the histological and autoradiographic images is accomplished in one of several ways. In all of them, the autoradiographic image is kept in picture memory while the stained section is moved beneath the video camera. In one method, the source for the image displayed on the monitor is flickered back and forth between the autoradiograph in memory and the stained section beneath the video camera. When the images from these two sources are out of alignment on the monitor, there is an apparent motion. The operator adjusts the position of the stained section until there is no detectable apparent motion. In another method, the computer generates an outline of the autoradiographic image (see Figure 2A). The outline is generated by setting a threshold gray level darker than the background but lighter than the image. At the edge of the autoradiographic image of a brain section, there is an abrupt decrease (darkening) of gray values. Thus, the edge of the image can be detected by this simple threshold procedure. A contour following algorithm follows the threshold transition from one

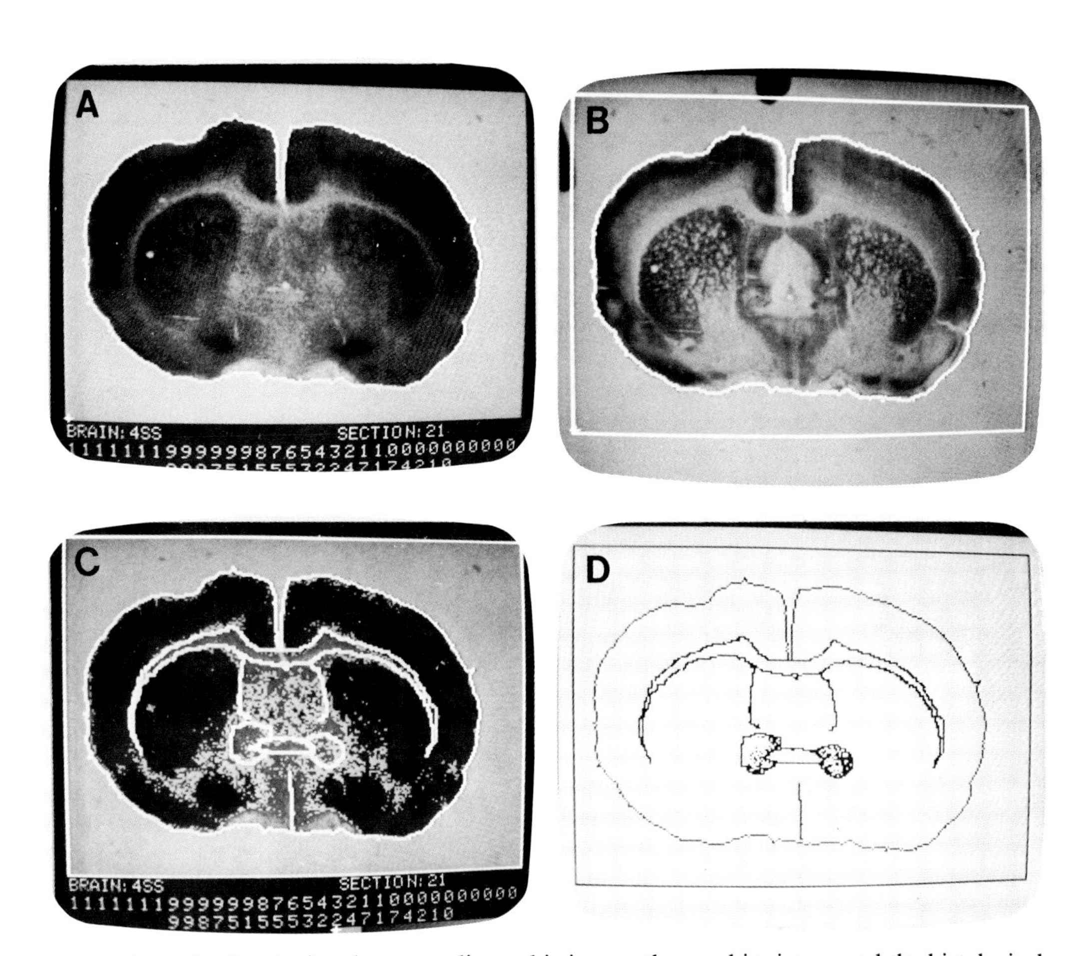

Figure 2. Overlaying the autoradiographic image, the graphics image and the histological image. (**A**) An automatic outlining routine makes an outline of the autoradiographic image. This display overlays the graphics image and the autoradiographic image, both of which are stored in picture memory. (**B**) The autoradiographic image is suppressed and the stained section is placed beneath the video camera and positioned so that the histological image falls within the outline. This display overlays the graphics image (from graphics memory) and the video image (from the camera). (**C**) The operator uses the cursor to outline various structures seen in the histological image, then overlays the outlines on the autoradiographic image and establishes a yellow color window, which colors all pixels whose RODs fall between 37 and 24. The colored pixels appear light gray on this photograph. The approximate position (ROD range) of the color window is indicated by the light gray patch on the gray-scale bar at the very bottom of the picture (with the cursor on the left of the patch). (**D**) The operator has the system map all the colored pixels inside the outlines of the bed nuclei of the stria terminalis into the graphics image, so that when the graphics image alone is viewed (after reversing black and white) one sees the differential stippling of these two homologous structures.

row of pixels to the next, generating the outline. The outline, which is stored in a subcomponent of picture memory called *graphics memory,* remains on the monitor when the operator switches the source of the display to the video camera. The operator moves the stained section under the camera until its image fits within the outline (Figure 2B). We have found that precise alignment is achieved most quickly by using the image-outline method first, then fine-tuning the alignment with the flicker method.

The overlaying of the histological image requires that the system be capable of mixing the sources for the image displayed on the monitor. It requires that some of the pixels in the image on the monitor come live from the video camera while adjacent pixels come from the autoradiographic and graphics images in picture memory. Some image-processor components for low-cost image processing systems do not have this crucial capability. Those that do have it may not be compatible with the bus on the microcomputer one intends to use. Attention to this issue is important when assembling a system.

Identifying Features

Features are regions of the autoradiographic image for which the system is to compute a descriptive statistic of some kind (e.g., the mean rate of local glucose utilization or the mean relative activation). Alternatively, features may be hot spots or cold spots on the autoradiograph, regions whose borders are to be mapped onto the histological image. Features are identified by means of outlines. The operator draws the outlines on the image displayed on the monitor, using a graphics input device, such as a joystick, trackball, mouse, bit-pad, or light pen. (These are more or less interchangeable devices; hereafter, "joystick" will be used to mean "graphics input device.") The joystick controls the position of a cursor as it is moved around the border of the feature the operator wishes to delineate. The path of the cursor is displayed on the monitor as a white line overlaid on the image. This white overlay is part of the so-called graphics image, which is stored in the graphics subcomponent of picture memory. (The outline of an image, generated by the automatic outlining routine, is another constituent of the graphics image.)

An automatic fill-in routine may be called to whiten all the pixels within the outline, creating a graphics image of the entire feature. If a statistic is to be computed for the feature, then the gray-scale values of the pixels in the graphics image of the feature are read into a register for use in subsequent computations. The system prompts the operator for a label, so that the statistics may be linked to a structure name in the data file.

When the features are to be defined by reference to the histology, the operator displays the histological image (having already aligned it with the autoradiographic image), locates the landmarks that are to be used to establish the boundaries of the structures, and draws in all the structures for which statistics are desired. When the structures are drawn, the system prompts for labels. The feature corresponding to a given label is specified by placing the cursor inside its outline. When the label is supplied, the computer fills in the outline and reads the gray-scale values of the corresponding autoradiographic pixels into the statistics register. The outlines of the bilateral bed nuclei of the stria terminalis shown in Figure 2C were made while the histological image in Figure 2B was displayed, then the histological image was replaced by the autoradiographic image. To obtain a descriptive statistic (e.g., a mean isotope concentration) for each bed nucleus, the operator places the cursor inside the outlines and calls the statistics routine.

Converting Gray-Scale Values

Before the descriptive statistics for a feature are computed, the operator generally wants to convert the gray-scale values of the pixels to values that represent either an isotope concentration, an optical density, a rate of glucose utilization, or a normalized index of relative activation. This conversion is generally not a linear one; therefore the conversion must be done pixel by pixel before the means are computed.

In order to convert gray-scale values to rates of glucose utilization, the operator must show the system the images of a series of standards of known ^{14}C concentration, which are exposed to the film along with the tissue sections. The system fits a polynomial equation to the gray scale and concentration values to obtain the function relating gray scale to concentration for that sheet of film. In order to convert concentration values to rates of glucose utilization, the operator must supply the system with data on arterial ^{14}C concentrations during uptake of the isotope. The data are derived from samples of arterial blood taken at frequent intervals following the injection of the isotope. The system fits a concentration-versus-time curve to the data points supplied, computes the integral of this curve, and uses this integral and several constants taken from the literature to compute a scalar factor that converts concentration to rate of glucose utilization [for details, see Sokoloff et al. (1977)].

The conversion to rate of glucose utilization is only desirable when the investigator is interested in metabolism per se, or in sleep or drug studies, where global changes in brain activation are both anticipated and interpretable. The majority of investigators now use the technique as a marker of localized changes in the functional activity of neural tissue. They are not interested in metabolism per se and would not know what to make of global changes in brain metabolism if they were to encounter them. When the purpose is to detect localized (rather than global) changes in metabolically coupled functional activity, then a normalized index of relative activation is and ought to be preferred [for an extensive discussion of why, see Gallistel et al. (1982)].

The most commonly used normalized index of activation is the ratio between a structure's mean gray-scale value (or derivative thereof, such as optical density or mean concentration) and the mean value of the white matter in that section (usually the corpus callosum). This is not the best normalized index. Its value depends somewhat on the overall darkness of the image (Kelly and McCulloch, 1981). Unpublished measurements of our own have consistently shown this statistic to be less "robust" (insensitive to global factors, factors that affect the darkness of the image as a whole) and less sensitive than another normalized statistic, the relative optical density (ROD).

The ROD of a pixel is the rank order of its gray-scale value in the distribution of the gray-scale values for all the pixels in the image. The automatic outlining technique, which enables the computer to outline the autoradiographic image, enables it to identify the pixels that belong to the autoradiograph rather than the background. The entire autoradiographic image is treated as a feature, a feature distinguished by the fact that its pixels all have gray levels darker than the threshold value used to separate image from background. Using this threshold procedure to isolate the image pixels from the background pixels, the system constructs the gray-scale histogram of the image. The gray-scale histogram gives for each gray-scale value the number of pixels in the image that have that value. The cumulative gray-scale histogram is the integral of this histogram; it gives the number of pixels having a given gray-scale value or lighter. The ROD of a pixel derives directly from the cumulative gray-scale histogram of the image pixels; it is the percentage of the pixels in the image having a gray-scale value as light or lighter than the gray-scale value of the pixel in question.

The robustness of various normalized statistics was assessed by exposing the same sections for three different durations (5, 10, and 20 days). A statistic was robust if its mean value for a variety of structures, some light, some dark, was not significantly affected by exposure duration (overall image darkness). Of several statistics tried, including the gray/white (G/W) concentration ratio, only the ROD was unaffected by overall image darkness. Sensitivity was assessed by repeatedly determining the same descriptive statistic for the same structures in the same images. From these repeated determinations, we calculated a pooled standard error of measurement. We divided this standard error of measurement by the difference between the darkest and lightest structures measured. For example, the mean RODs for the medial geniculate (a dark structure) and for the optic tract (a light structure) were determined twice in each of the three images of the same section. The difference between the mean RODs for these two structures was divided into the pooled standard error of measurement to get the index of sensitivity. The ROD was the most sensitive of all the statistics evaluated, both on this test and on a similar test involving images from comparable sections through different brains. For the ROD statistic, the average error of measurement in the first test was 5.6% of the difference between the mean for the medial geniculate and the mean for the optic tract; for the G/W ratio, 5.9%. In the second test of sensitivity, the measurement error in the numerator reflected the between section and between brain variance. For ROD, this measurement error was 16% of the range; for G/W ratio, 26%.

If one wants a statistic that captures *localized* changes in the functionally coupled metabolic activity of neural tissue, then the relative optical density statistic makes sense. Any localized change in metabolism will translate into a change in the rank order of the optical density in that locale relative to the optical density in all the other locales. It is possible to contrive simultaneous changes in another locale that prevent the change in rank order that would ordinarily occur, but these changes are indeed contrived—unlikely to occur in fact and unlikely to pass unnoticed if they do occur. If a great many structures in the same section show a change from control values of about the same magnitude, then one needs to check whether this might not be an artifact of a change in some single large structure elsewhere in the image. We have so far not encountered such a case. Very few structures occupy a large enough percentage of the total pixels in an image to cause such artifactual changes.

Computing Descriptive Statistics

When the gray-scale values have been suitably converted, the system must compute the mean and standard deviation of the resulting numbers. It should also record the number of pixels in each feature, since areas can be derived directly from this statistic. (This is particularly useful in determining the extent of lesion effects.)

Creating Data Files

The efficiency of computer-assisted image analysis permits the investigator to obtain data on 10–20 features (neural structures) in 50–500 sections in the course of the analysis for one experiment. If these statistics are recorded only on printouts, they will have to be reentered by hand into a computer when it comes time to do the overall data analysis. A well thought out system should obviate this by creating data files that permit the data to be handled by the many statistical programs now available for microcomputers, or by statistical packages that run on mainframes to which the microcomputer can talk. The best way to do this is to have the system write the labels and associated descriptive statistics to files set up in the Data Interchange Format. The Data Interchange Format is an emerging industry-wide standard format for data that are to be moved from one software package to another (for example, from a database package to a spreadsheet package). The advantage of storing the data in this general format is that the investigator does not have to commit herself in advance to a particular form of data arrangement or a particular kind of data treatment. If the data files are in Data Interchange Format, they can be called into a database management program or spreadsheet program and restructured at will.

Creating Analytic Displays

The system should enable the investigator to create images that give quantitative information about the autoradiographs in a form suitable for publication in archival journals, which will not print color plates or will charge high fees to do so. The ability to overlay a graphics image on the autoradiographic and histological images is central to the creation of these displays. Figure 2D is an example of a black on white analytic display. Various stages in the creation of that display have already been described: The autoradiographic image was outlined (Figure 2A); the histological image was positioned within this outline (Figure 2B); and the bed nuclei of the stria terminalis were outlined. As an aid to the recognition of the plane of section, the ventricles, the corpus callosum, and the anterior commissure were also outlined. These outlines were superimposed on the autoradiographic image (Figure 2C). Computation of the descriptive statistics for the bed nuclei showed that the nucleus on the right was slightly darker than the nucleus on the left. (This reflects a slight but highly reliable activation of this nucleus by the rewarding brain stimulation delivered to the right posterior medial forebrain bundle, which contains fibers going to and from the bed nucleus.) The operator, wishing to portray where in the bed nuclei this activation occurred, set the upper (darker) limit of a yellow color window at the level that first colored in a pixel within the bed nuclei (an ROD of 37), and extended the lower limit down far enough to color in about half the pixels on the more activated side (down to an ROD of 24). The pixels colored yellow by this window appear light gray in Figure 2C. The density of yellow pixels is equal along the medial boundary of the two bed nuclei; the heightened density on the stimulated side (the right) is only seen in the lateral half of the nucleus. To make an image that communicates this clearly, the operator placed the cursor first inside the outline for the nucleus on the right, then inside the outline for the nucleus on the left, and, in each case, called a *stipple routine*. The stipple routine scans within the indicated outline and reads colored pixels (pixels whose gray-scale value falls within the limits of the window) into the graphics image. After stippling the graphics image of the bed nuclei in this way, the operator asked the system to display only the graphics image, in black on white. The photograph of the resulting display (Figure 2D) shows the location of the activation in the bed nuclei in an objectively arrived at image suitable for B/W publication: One can readily see the increased stippling in the more lateral portions of the bed nucleus on the stimulated side. This subtle activation would not be apparent in any printed reproduction of the original autoradiograph.

This concludes the description of the essential functions in a system designed to assist the analyses of 2DG autoradiographs. We turn now to a description of the hardware components needed to create a system and to the specification of those combinations of currently available components that would realize such a system at a cost of less than $20,000.

The Components of a System

The hardware required for an image-analysis system is quite similar to that in a video game. Sad to say, a system for the acquisition and analysis of biological images currently costs about $20,000, rather than $200. The difference between these prices is due, primarily, to economics of mass production. In the future, there is every reason to suppose that the prices of image-processing equipment will decrease. In this section, we describe the components of an image-processing system for the analysis of autoradiographs. We then concentrate on the various hardware components and investigate some of the currently available alternatives. Finally we describe some configurations that, we feel, are good combinations of currently available equipment. The discussion that follows attempts to provide an overview of the equipment and techniques for image analysis. There are many different ways of assembling such a system. While attempting to describe all of the currently viable alternatives, we concentrate on the most practical system implementations. The less attractive options are covered briefly.

The components of a system for the analysis of autoradiographs are shown in Figure 3. Images are converted to electrical signals with a *scanner,* and a digital image is stored in an *image processor.* The information in the image processor is displayed on the *monitor.* The picture is controlled by the *microcomputer,* which can produce false color information and graphics. The operator indicates the parts of the image to be analyzed with a graphics input device *(joystick).* The operation of the system is controlled from a *terminal,* and the results of the analysis are reported on a printer (not shown). A disk is used for loading programs, and to store results for further analysis.

Figure 3. The components of an image-processing system.

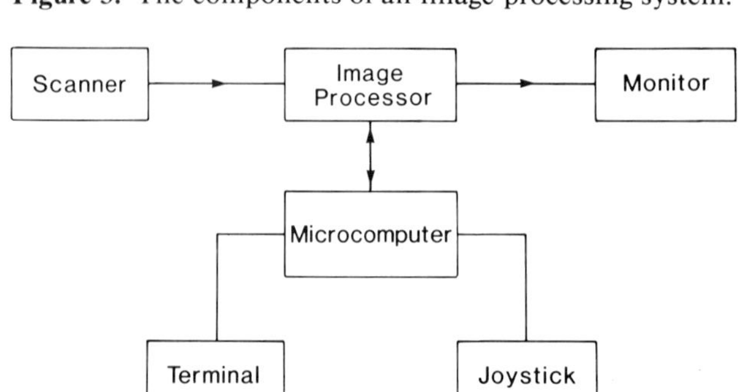

The Choice of a Scanner

This choice is somewhat independent of the choice of other components in the system. There are three options: a television camera, a line-scan camera, and a drum scanner. The drum scanner, when carefully maintained and used, will produce quite accurate digitization of autoradiograms, but the price (about $50,000) prohibits their use in a system whose total price is intended to be about $20,000. Also, these devices are quite slow and are not suited to the histological overlaying function. Line-scan cameras can provide quite high spatial resolution and good photometric accuracy, but these units are also slow and quite expensive ($25,000). A scanner built with a television camera is currently the most attractive. A television scanner consists of an illuminator, optics, and the camera. We have used a fluorescent light box as an illuminator, a conventional *F*/1.4 25-mm lens, and a vidicon. The optics are suitable for images larger than about 10 mm (50 μm per pixel). Improved resolution can be obtained with a macro lens, though manual image positioning is awkward for small images, and at higher magnification (2-mm images) light box brightness is inadequate. Under these conditions low-power microscope optics and condenser illumination should be used. We are currently using a vidicon televison camera (Spatial Data Systems) designed specifically for computer input. Though vidicons are notoriously inaccurate, we have found that this camera has adequate stability and photometric uniformity. It is our opinion that new systems should be constructed with solid-state cameras, which promise greater accuracy, stability, and repeatability than vidicons. General Electric manufactures a CID (charge injection diode) camera for about $2,100. CCD (charge coupled device) cameras are available from Fairchild Semiconductor ($3,500), NEC ($1,500), and have been announced by RCA ($1,300) and Sony ($800). Hitachi manufactures an MOS camera ($985). According to current information, the General Electric CID camera is the best choice: It is relatively free of defects, and does not suffer from the *blooming* (overloading in the presence of high light levels) that occurs with CCD cameras. If the promise of solid-state cameras is realized, they should provide a very attractive replacement to vidicons.

The Choice of a Monitor

Inexpensive computers use television receivers for image and alphanumeric display, but the process of converting an image to broadcast frequency introduces serious distortions. A television *monitor*—that is, a unit driven by unmodulated video signals—is the appropriate device for viewing images. There is a choice between monitors driven by composite video (one signal containing color and brightness information) and monitors that accept three signals (RGB signals for the red, green, and blue information). The latter produces better image quality. In selecting a monitor one should avoid the units designed primarily for color graphics since they do not provide adequate gray-scale displays. [See Powell (1984) for a review and price/manufacturer information.]

The Image Processor

The image processor is the component of an image-processing system that digitizes and stores the image and provides the display signal to the monitor. Figure 4 is a block diagram showing the typical subcomponents of the image processor.

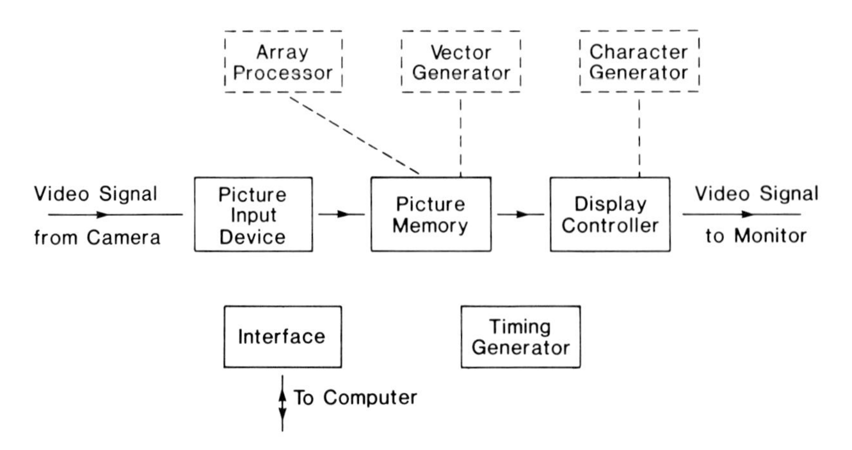

Figure 4. The subcomponents of the image processor in an image-processing system.

Many firms manufacture such equipment. This industry is fairly young, the terminology is not settled, and there is very little product standardization. In what follows we describe the function of the various subcomponents and also define the terms to be used in further descriptions. It is important to distinguish between *image processing* and *graphics* products. Though there is some overlap, we refer to systems for working with continuous gray-scale images as *image processors,* whereas systems that are intended for drawings (black and white or color) will be called *graphics* devices. This distinction is important because some of the simpler graphics systems cost much less than the kind of equipment we are considering, but the applications we have in mind require fairly high gray-scale resolution. In digital image processing, image brightness (density, luminance, transmittance) is represented by digital numbers. In autoradiographic applications each pixel must be represented by six or more bits (64 or more levels). Graphics applications use only one to three bits per pixel. Also, graphics systems often have no provision for the direct input of video pictures (see subsection System Specifications).

The central subcomponent of the image processor is the *picture memory.* This solid state read-write memory contains the digital image to be viewed on the monitor. Nominally, the memory is organized in pictures containing R rows by C columns by B bits. It may contain more than one such picture. Some systems have *configurable* memories, allowing different values of R, G, and B to be obtained under program control.

One group of R by C memory elements is called a *bit plane.* A *graphics memory* is a bit plane used to store binary information such as drawings or alphanumerics (the graphics image). The graphics memory may be separate from the picture memory, or may be included among the picture memory planes, but its function is to superimpose or *overlay* drawings on the continuous tone image.

The *picture input* device takes the signal from the scanner (such as a television camera), digitizes it, and stores it in picture memory (or in the computer). It is an analog-to-digital converter (ADC) with suitable gain control and appropriate timing circuits. A picture input device that can digitize a television image in one-frame time ($\frac{1}{30}$ or $\frac{1}{60}$ sec) is called a *frame grabber,* and it stores images in picture memory.

The image on the monitor is displayed 30 to 60 times per second, and this cyclic process is called *image refreshment* so that the picture memory is sometimes called an *image refresh memory.* The *display controller* converts the digital data in the picture memory and the graphics plane into analog signals that produce an image on the monitor. When the display controller is programmable, the image on the monitor can be changed rapidly by transferring appropriate data from the controlling computer. The display controller implements the false color function. It may also allow an image from an analog source (television camera) to be combined with the image from picture memory. This capability, called *video overlay,* is used to overlay histological, autoradiograhic, and graphics images.

A gray-scale/color map, which is a subcomponent of the display controller, establishes a correspondence between the digital codes stored in picture memory and the brightness/color values shown on the monitor. It contains a *lookup table;* a small digital memory whose addresses are generated from the digital numbers (gray-scale values) stored in picture memory. The contents of the lookup table (the binary strings stored at the addresses) are applied to digital-to-analog converters (DACs), which produce video signals driving the monitor. If only one signal is generated, the image is monochromatic. If three different signals are produced, they are used to specify three color components. For example, if the picture memory stores 6-bit gray-scale values for each pixel (so that there are 64 gray-scale values), one might use a lookup table with 64 addresses, with each address containing a string (word) 24 bits long. The 24-bit word is split into three bytes (8-bit miniwords), which are applied to the three 8-bit DACs that supply the R, G, and B signals to a color monitor. This allows simultaneous display of 64 colors from a "palette" of 2^{24} or about 16 million possible hue–brightness combinations. Some systems implement the color map in a read-only memory, but to be effective the lookup table should be under control of the computer.

The *timing generator* produces clock and synchronizing signals that control the refresh functions. It should be synchronizable with an external video source. The *interface* allows the computer read/write access to picture memory, and permits the computer to control the other image processor operations.

One might ask, why not store the image in the computer memory? Why use a separate picture memory? The display of standard television signals requires that pixels be refreshed at about 5–10 MHz (0.1–0.2 μsec per pixel). Common computer memories do not operate at this speed, and typical *very simple* image processing operations, such as reading pixel values from memory to the central processing unit, require several microseconds. Thus frame refreshing and direct image digitization require specialized memories and digital circuits whose operation must be somewhat independent of any associated computer. The picture memory, timing, interface, and a minimal display generator are the irreducible subcomponents of a true image processor. In contrast to this, for bit-plane graphics applications, eight pixels are stored in each memory byte, so that it is possible for the computer and the image processor to use the same memory. With some elaborate memory architectures (multiport, multimodule) it is possible for the computer and the image processor to share memories, but these memory configuration techniques are not (currently) employed in microcomputers. Currently, simple image processors operate with autonomous memories and the computer accesses this memory through a special port.

The units described above are the minimal subcomponents of the image processor component. The functions described below are available in some image processors, and are useful additions. The *cursor generator* produces a cursor, a visible mark on the display whose location can be controlled either by the computer or by the joystick (directly, without computer intervention). It permits the operator to "point" to things in the image. It may also be used to control the computer by choosing among alternatives displayed on a screen.

Vector and *character generators* are special-purpose computers that allow lines or symbols to be shown on the monitor. Typically, vectors are written in the graphics plane, whereas standard alphanumeric characters are generated dynamically.

Zoom is a feature that allows the picture on the monitor to be expanded, showing only an enlarged portion of the image stored in picture memory. Typically, the system implements only *integer zoom,* where pixels are displayed repeatedly. *Roam* is a feature that operates when the image on the display depicts only a portion of the picture memory contents, such as with zoom. The displayed image frame may move or roam across picture memory. Since this is done with the display controller, roaming occurs instantaneously, under control of the joystick.

An *array processor* is a special-purpose programmable digital computer that operates on the data in picture memory or from a frame grabber, and stores results in picture memory or in an auxillary memory. Array processors perform certain standard image processing functions, such as:

1. picture averaging (adding a number of input pictures point-by-point to improve signal-to-noise ratio);
2. picture subtraction (to detect changes between pairs of pictures, as, for example, in situ binding studies where one subtracts the "nonspecific image" from the "specific image");
3. picture filtering (image enhancement, edge detection);
4. extraction of stereological features;
5. distance transforms, skeletonization; and
6. histogram calculation.

Assembling a System

Image-processing systems are available in roughly three kinds of package: (1) integrated systems, (2) backplane systems, and (3) board products. This diversity allows users to purchase various combinations of hardware and software, appropriate to their fundamental needs and budgets. An integrated system consists of a computer, image processor, peripheral equipment, and (one hopes) applicable software. There are two integrated image-processing systems for biomedical applications: the IBAS, marketed by Carl Zeiss, Inc., and the Magiscan 2, made by Joyce-Loebl and marketed by Nikon. Both units cost about $80,000, so they fall outside the scope of our discussion. An integrated system is available for about $20,000 from Digital Graphics, though this system is not well suited for the applications we are considering.

Backplane systems consist of a special chassis, bus, and modular components. Only those components required for a given application need to be purchased. The prices for these systems start at about $20,000 (for the image processor component alone), so that they too are outside our scope.

In a board system, all or most components of the image processor are placed on one circuit card designed to operate on standard computer buses (Q-bus, Multibus, Versabus, S100, and others). Board systems are available from Datacube, Digital Graphics Systems, Imaging Technology, and Matrox. They provide very impressive price–performance combinations. However, the architecture of these devices limits the performance that can be achieved: The size of picture memory is restricted, array processors are not available, and the degree of reconfigurability is limited. On the other hand, the units have a very low overhead, since they can be placed in the computer chassis. The performance currently available is equal to that of backplane systems of a few years ago, at one-fifth the price. The board systems are the systems we will consider in our design.

System Specifications

A low-cost system for autoradiographic image analysis should meet the following specifications:

1. The image should contain at least 200 × 200 independent pixels. The video data should be digitized to at least 6 bits per pixel.
2. The system should have a color display. It should supply at least 64 simultaneous colors through some sort of a color map.
3. The system should have a graphics input device for interaction with video data. The position of the device (hence, the location of the cursor on the monitor) should be determined by a program in less than 16 msec.
4. The system should be capable of combining for display on the monitor the video image and the images in picture and/or graphics memory.

The specification of resolution is based on experience with our present equipment. A substantially higher resolution (400 × 400 pixels) entails higher equipment cost for the picture memory and for a monitor that will display this image. It will also slow down the system, since the computer must process more data. Also, scanners that realize the higher resolution are more expensive and more difficult to operate. A substantially lower resolution (100 × 100 pixels) produces a relatively coarse image, which makes it more difficult to delineate image features when one wishes to measure activities over arbitrary regions.

The color map is used to create color windows for enhanced detection of intensity differences in the autoradiograph. There is no great need for many distinct colors, but the color maps must be able to assign colors to arbitrary intensity levels (arbitrary gray-scale values). The capacity of an appropriate image processor to combine the video image and the image in picture memory is used to overlay the histological image, graphics image, and autoradiographic image. The graphics input device is used to identify features in the image and to generate graphics (drawings). It is also useful in other system operations, such as controlling the color map.

Although the above characteristics are necessary, there are other features that can be useful in system operation. One of these is a graphics plane. If the picture memory stores more than 6 bits per pixel, one can simulate the graphics plane with the color/brightness map. Zoom and roam are useful when one is identifying small features. It is also desirable to have an image processor that can compute histograms of the pixel brightness values.

The image processors made by Digital Graphics and by Imaging Technology can produce overlays of the stored image and video. It is also possible to perform overlays with software, by storing the images in computer memory and producing a combined image with a program. The most difficult operation is bringing the histological image into registration with the autoradiographic image. In our current system, one of the ways of doing this is by superimposing a graphics outline of the autoradiograph and the image of the histology. One can keep the outline in computer memory and "move" it with the joystick by continually modifying the display of the graphics outline in response to the joystick coordinates. We estimate that the outline can be updated five to ten times a second, which is an acceptable speed. To "move" a full gray-scale image with the same technique would require more than half a second, which is too slow. Clearly, it is better to perform overlay registration with hardware; that is, it is better to choose an image processor component that has the capacity to combine signals from picture memory and from the video camera.

Choice of a Microcomputer

So far, we have not discussed the choice of the computer. The computational requirements for the system are modest. Our current program, which runs on a Digital Equipment Corporation PDP11/34, occupies about 80 KB of memory. No storage is allocated for pictures: all images are kept in picture memory. Current generation microprocessors will take about the same amount of code. The best choice is a computer based on a Motorola MC68000 because it is the most popular, most readily available, and best proven of the new generation of CPU chips. One very valuable characteristic of this microprocessor is the large memory page. This allows one to write long programs and process large arrays of data without resorting to onerous and laborious paging and/or overlaying. Other processors such as the Intel 8088 (IBM PC) are not out of the question, but they are slower and their small memory page makes them more difficult to program for these applications. However, Sritek Inc. manufactures a co-processor using the MC68000, which operates in the IBM PC. Another possibility is the LSI/11 processor in the Micro-11 package, a version of the PDP/11 computer, which has been a laboratory workhorse for more than a decade. This computer can run our current software with minimal modification, but it is slower than the MC68000 and also has a limited memory page.

The most important factors in the choice of computer are compatability with the image processor, reliability and maintainability, and the availability of a good software development environment. Image processors are available for the Q-Bus (Digital Equipment Corporation), and the Multibus and S-100 buses, which are industry standards, and for the IBM PC bus. (See the Appendix for a list of computer manufacturers and bus compatibility.)

Since the video overlay requirement is an essential feature of the system we are concerned with, and since image processors with this capability are available for the Multibus and the S-100 bus, the Q-Bus and the IBM PC bus (announced by Datacube for August, 1984, delivery), we are concentrating on computers that can operate with these buses. ERG and Cromemco make S-100 bus MC68000-based computers, and an inexpensive Multibus-based MC68000 is manufactured by Callan Data Systems. It is also possible to use the Multibus image processors with a Versabus computer (available from IBM Instruments division and Charles River Systems), because a Multibus-to-Versabus adaptor is available from Hal-Versa Engineering. BIT-3 Computer Corporation manufactures an adaptor between the IBM PC bus and the Multibus, so the IBM PC may be mated to a Multibus image processor.

The software for autoradiographic image processing can be written in any good high-level language, such as FORTRAN, PASCAL, or C. We feel that the UNIX operating system with the associated C language is the best choice for software development. UNIX is becoming a de facto standard for sophisticated microcomputers, and C has a rich range of data types and operations, so that many of the bit manipulations required for this software can be performed without recourse to assembly language. A system written in C under UNIX will be relatively portable to other microcomputers, and we feel that this is a valuable characteristic.

We can currently propose the following hardware configuration.

1. Computer: IBM PC/XT with a black and white terminal ($5,675) together with a MC68000 coprocessor made by Sritek and with Xenix operating system ($2,995).
2. Image Processor: Datacube IVG-128 ($2,995).
3. Camera: GE TN2505 CID solid state camera ($2,100).
4. Graphics Input Device: Hitachi HDG-1111 graphics tablet ($728).
5. Monitor: Matrox Electronic Systems MCM-14L RGB monitor ($1,395).
6. Printer: Epson FX-80 ($539).

Total system cost is $16,427. Those users who would apply this system only to image analysis and do not plan to do extensive programming can choose an IBM PC (without a hard disk) and further reduce the price by about $3,000. The IBM PC is a very popular computer used for home, business, and small-scale scientific calculations. The PC has a 64 KB memory and a dual-sided floppy disk. It can be expanded to PC/XT by adding a 10 MB hard disk and another 64 KB memory. There are numerous software packages available in various applications, for example, spreadsheets, word processing, accounting, graphics, and communications. With the 68000 coprocessor made by Sritek Inc., this system can be upgraded to a 16/32 bit computer with additional 512 KB memory. Thus the full power of the 68000 can be utilized. The operating system, Xenix, developed by Microsoft, is a version of UNIX, which supports many software packages.

The specified image processor is currently being developed by Datacube. Its features include pan, scroll, two-fold zoom, and bit plane write protection. The memory size is 512 lines and 384 pixels per line. Each pixel has 8 bits so that a 256 gray level image can be displayed. There is an input video source switch for two camera inputs and four lookup tables (LUT), one for input and three for output. The input LUT may be used for gain and offset controls on the input signal before the digitized data are stored in picture memory. The output LUTs implement the color map function. One or more of the bits can be used to generate graphics overlays. Picture memory can be subdivided into two parts to store two lower resolution image frames, each having 256-line resolution vertically.

It is interesting to explore the possibility of an even less expensive image-processing system. Tecmar manufactures an image digitizer and a picture memory with a color map for the IBM PC. The digitizer operates at about 15 kHz, so the image would be scanned in about 6 sec. The picture memory stores only four bits. To use this system the gray-scale value (six or more bits per pixel) image must be stored in computer memory (rather than in the picture memory of the image processor component). Reduced gray-scale images can be produced by software. This system would have a slow response, and it would be difficult to develop an effective software system for this configuration. Interactive Video Systems produces a picture memory with a 6-bit frame grabber for the IBM PC, but this unit has no false color capability. However, new products are continually being announced. It is safe to predict that within a year or two it will be possible to configure a system for about $10,000.

Software

The Autoradiographic Image Processing Resource software (described in the first part of this chapter) is a mature system, which has been used by a large number of investigators. We plan to move this package to the above proposed microcomputer system by the end of 1984. This software will be available free of charge and the Resource will provide training and manuals to interested individuals.

Conclusions

We have described a system for the analysis of autoradiographs, and have discussed the implementation of this system with a microcomputer. The contemplated microcomputer system will cost $13,500–$16,500. It is important to consider the future of this methodology. We can anticipate that the evolution of software and data analysis techniques will produce systems that allow the investigator to extract more data from a given experimental design. In part, this will be accomplished by developing methods that produce more repeatable data. Experience with the results obtained from image analysis, combined with appropriate statistical data reduction techniques, will allow more subtle effects to be observed and will permit more sophisticated experiments to be designed. For this it is necessary to couple the image-analysis system with suitable database management and data reduction software. Further development in instrumentation will permit the analysis of higher-resolution images, such as tritium autoradiograms and HRP-labeled preparations.

Computer technology promises to provide more processing power at a cost affordable by most investigators. To exploit this evolving technology it is necessary to plan for software portability. The UNIX operating system promises to provide this. The development of new techniques and systems will require close collaboration by investigators in the biological sciences and in the computer and information disciplines.

The preparation of this chapter was supported by a Biotechnology Resource Center Grant from NIH (Grant No. 3P4 IRR O1638 Oleh Tretiak, C. R. Gallistel, & Norman Adler, CoPIs).

Appendix

Manufacturers of Integrated Image-Processing Systems

Magiscan 2 Image Analysis System, Joyce-Loebl, Marquisway, Team Valley, Gateshead, NE11 OQW, England. In the USA: Vickers Instruments Inc., Riverview Business Park No. 27, 300 Commercial Street, Malden, MA 02148.

IBAS Image Analysis System, Carl Zeiss, Inc., One Zeiss Drive, Thornwood, NY 10594.

Manufacturers of Image Processor Components for Microcomputer-Based Systems

Datacube, Inc., 4 Dearborn Road, Peabody, MA 01960. Image processor and graphics products, Multibus and Qbus.

Digital Graphic Systems, Inc., 935 Industrial Avenue, Palo Alto, CA. Image processing and computer graphics products, S100 bus.

Interactive Video Systems, 358 Baker Avenue, Concord, MA 01742. Image processors for Multibus and Qbus.

Matrox Electronic Systems Ltd., 5800 Andover Avenue, T.M.R. Quebec, Canada H4T 1H4. Board level products for image processing, Multibus, Qbus, Unibus, S100 Bus, STD-BUS (Mostek), Excorciser bus (Motorola).

Microcomputer Manufacturers

These firms make MC68000-based microcomputers that are compatible with image processors, support UNIX, and are relatively inexpensive.

Callan Data Systems, 2645 Townsgate Road, Westlake Village, CA 91361. Unistar 100: 68000, 512-KB main memory, 616-KB floppy, 21-MB Winchester, Multibus and UNIX system 5.

Charles River Data Systems, 983 Concord Street., Framingham, MA 01701. UV68/05-A: 12.5-mHz 68000, has 256-KB RAM, 1.25-MB floppy, 10-MB Winchester, self diagnostics, Versabus and UNOS (real-time, ASSEMBLER).

Cromemco, 280 Bernardo Ave., P.O. Box 7400, Mountain View, CA 94039. 8052. CS-1HD2: 8 mHz 68000 and 4-mHz Z-80A, 256-KB RAM memory, 390-KB floppy, 21-MB hard disk, with S-100 bus and Cromix (UNIX), 8-slot card cage.

Empirical Research Group, Inc., P. O. Box 1176, Milton, Washington 98354 ERG-11: 12-mHz 68000, 384-KB RAM, 1-MB floppy, 21-MB Winchester, S-100 bus and Idris, C, 10-slot backplane (5 free) with 20-A power.

IBM Corporation, P.O. Box 152560, Irving, TX 75015. PC/XT: 8088, 128 KB RAM, 10MB fixed disk, 1 dual-side floppy disk.

IBM Instruments Inc., P.O. Box 332, Danbury, CT 06810. CS-9000: 68000, 128-KB RAM, 256-KB expansion, floppy, 10-MB hard disk, real-time, multitasking Operating System Extensions, and Xenix, ASSEMBLER and Versabus, IEEE-4888 and CRT, keyboard, and printer.

Sritek Inc. 6615 Snowville Road, Cleveland, OH 44141. 68000 coprocessor: 8MHz 68000, 512 KB RAM, Xenix multiuser operating system, IBM PC bus.

Versabus–Multibus Interface

Hal-Versa Engineering, Inc., 322 Saratoga Avenue, Los Gatos, CA 95030.

IBM-PC to Multibus Interface

BIT-3 Computer Corporation, 8120 Penn Avenue South, Minneapolis, MN 55431.

References

Gallistel, C.R., C.T. Piner, T.O. Allen, N.T. Adler, E. Yadin, and M. Negin (1982) Computer assisted analysis of 2-DG autoradiographs. Neurosci. Biobehav. Rev. 6: 409–420.

Goochee, C., W. Rasband and L. Sokoloff (1980) Computerized densitometry and color coding of (^{14}C)-deoxyglucose autoradiographs. Ann. Neurol. 7: 359–370.

Kelly, P.A.T., and T. McCulloch (1981) Errors associated with modifications of the quantitative 2-deoxyglucose technique. J Cereb. Blood Flow Metab. 1:60–61.

Powell, David (1984) Monitors—Buyer's Guide. Popular Computing (February): 122–135.

Sokoloff, L., M. Reivich, C. Kennedy, M. DesRosiers, C. Patlak, K. Pettigrew, O. Sakurada, and M. Shinohara (1977) The (^{14}C)-deoxyglucose method for the measurement of local cerebral glucose utilization: Theory, procedure and normal values in the conscious and anesthetized albino rat. J. Neurochem. 28: 897–916.

MICROCOMPUTER USES IN ELECTROPHYSIOLOGY

19

A Completely Digital Neurophysiological Recording Laboratory

Melburn R. Park

Introduction

This chapter describes the design and construction of a neurophysiological laboratory that uses digital rather than analog recording technology. Most of the separate and special-purpose instruments used for data gathering, recording, and stimulation in a conventional laboratory are replaced by a few digital components, including a computer, a digital oscilloscope, and a few peripheral devices. The computer is used not so much as a computational machine analyzing and synthesizing data but as a device for the handling and routing of data and processes. This use is common outside the biological laboratory where microprocessors have been used for some time as process controllers in the physical sciences, in automated industrial machinery, and even as controllers in consumer appliances.

Even though neurophysiology as a discipline has been responsible for some of the pioneering efforts in the development of small computers and their operating systems (Clark and Molnar, 1964; Schoenfeld, 1983), their use is still not common in the recording laboratory. Investigators have been discouraged by the high cost of laboratory computers and the specialized knowledge required to use them. Most neurophysiologists are not aware that the performance of a properly implemented digital system is superior to that of analog recording techniques using a conventional oscilloscope and camera. This alone would argue for the computer-based system even if costs had not dropped so much as to make this the most economical way to build a laboratory. Specialized knowledge is still required to set up a digital laboratory. One purpose of this chapter is to impart some of that information.

The chapter describes the author's laboratory, which is being rebuilt at the time of this writing (1984), so that the information and thinking is as current as it can be. Most of the hardware and software techniques, however, have been developed and used in two previous recording laboratories and thus are well tested. One goal of this chapter is to provide a sort of tutorial for the scientist perhaps not wanting to duplicate the system described precisely but who does need to be instructed in the use of laboratory computers in general. Thus, although specifics are thoroughly discussed, some general principles important to the author are also presented as a guide through the many pitfalls encountered in adopting digital computer technology. Considerable space is given to a discussion of the command language devised by the author for laboratory applications and the core set of FORTRAN routines that implement it. This form of interactive control has proven so versatile and produces such orderly and structured program code that it has been used in all the data processing programs used in my laboratory. These include the aquisition and data analysis programs for physiological data (called LAB and ANLSIS) described in this chapter, the iterative voltage clamp previously described (Park et al., 1981), and a set of programs for morphometry and serial reconstruction (see Park et al., 1982). Having explained the overall function of the digital laboratory and familiarized the reader with the command language, the step-by-step design of one part (the stimulator) of the system is presented as an example. In particular, the decisions made in the interest of efficiency and rapid development time are explained.

Advantages of Digital Recording

The initial impetus for adopting digital data handling in my recording laboratory came from the desire to eliminate photographic film as the means of enregistering physiological data. Film is expensive, both in cost and in the time required to process the data contained on 35-mm film, through several photographic steps, into a finished publication figure. Data loss is unavoidable using film, for example, in the faint, underexposed depolarizing limbs of action potentials. Also, a single event must generally be photographed a number of times at different sweep speeds as rapid events cannot be analyzed from photographs of slow oscilloscope sweeps and vice versa. To be effectively visualized, low amplitude events, such as postsynaptic potentials, require high-gain AC coupling resulting in the loss of DC information. By contrast, the digitized records produced by the system I will describe preserve DC information and are of sufficient resolution that they can be expanded in horizontal and vertical axes to provide excellent records of both small and fast events (Figure 1). The data can be stored in a compact form on magnetic media and, with a plotter interfaced to the computer, transferred directly to paper in a form suitable for the final layout of a figure. This eliminates two steps in the preparation of plates for publication.

The complete system is not expensive (see Appendix 1), being less than the total cost of the conventional recording and stimulating devices it replaces (analog oscilloscope, oscilloscope camera, camera mounting hardware, and digital stimulus timer). Once implemented as a data gatherer, the digital recording system can easily be made to substitute for two other expensive laboratory devices,

Figure 1. Digitally recorded intracellular traces from neurons of the rat dorsal raphe nucleus. Numerical differentiation of a spontaneously occurring action potential (**A**) produces the trace in **B**. Spontaneous activity originally recorded at a slow sweep speed (**C**) can be expanded in the time axis to show a single action potential and the hyperpolarization that follows it (**D**). Also in **D**, the extracellular control trace has been plotted as a superimposed series of dots to provide a measure of membrane potential and overshoot amplitude. Synaptic events recorded at a gain appropriate for action potentials (**E**) can be magnified in both time and voltage with no defect in resolution (**F**). The depolarizing postsynaptic potential is an evoked IPSP reversed by the injection of hyperpolarizing current. Traces **A–D** are from the same neuron; the voltage calibration of **A**, **C**, and **D** are identical.

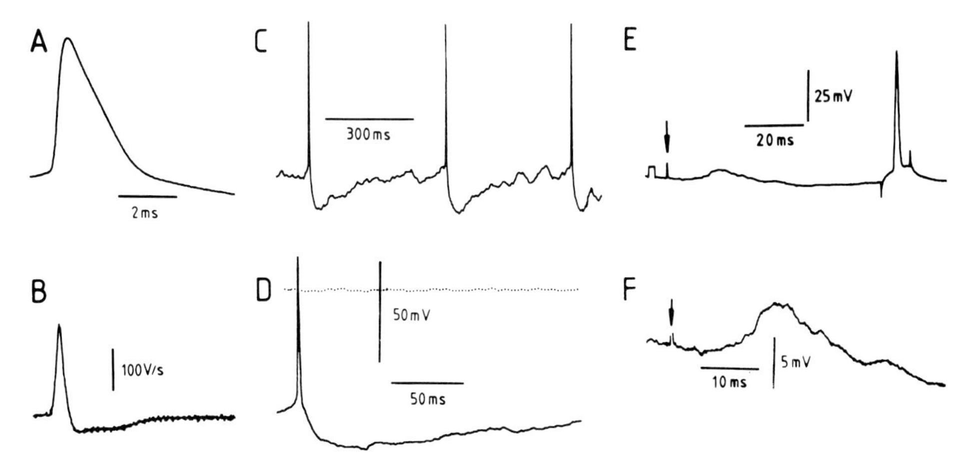

the signal averager and the stimulus timer. The end of this process results in a highly sophisticated, productive, and maintainable laboratory.

Design

In the interest of efficiency, a laboratory is best implemented in a stepwise fashion, whereby a basic, functioning system is first built and then expanded and improved as time and funds are available.

The first step in designing the laboratory is to designate the tasks to be performed by the completed system. In the most general sense, a neurophysiologist is interested in observing the electrical activity of a single neuron or population of neurons. In my experiments, the recordings are intracellular from single neurons and the responses of interest are evoked by electrical stimulation of afferent or efferent pathways. For the time that a neuron is impaled by the recording electrode, which can vary from a few minutes to several hours, the experimenter is continuously changing stimulus parameters, such as stimulus intensity and polarity, giving multiple stimuli combined in various ways, or passing controlled pulses of current through the electrode and into the neuron. The data is collected as discrete records (also called *sweeps* or *traces*) of electrical activity beginning just prior to an electrical stimulation and lasting long enough to capture the relevent evoked activity (usually 5–200 msec). Sweeps are taken at intervals of 1–2 sec so that 10 min of recording could produce a maximum of 600 traces. In practice, only a portion of the total is permanently stored, perhaps 200 for a single day's experiment.

From this brief description of the procedures of a recording experiment, it should be possible to list the tasks that the digital laboratory system will have to perform. These tasks appear below, together with notes [in brackets] as to their priority and whether the task is to be implemented in software, hardware, or a mixture of the two.

During a recording experiment:

1. The system must provide stimulus timing and trigger pulses to the oscilloscope, calibrator, stimulus isolation units (SIUs), and current injection devices. [This is done with some custom hardware; the stimulator software can be developed in stages, as is described below.]
2. Traces are acquired on a digital oscilloscope. [A commercial hardware item that is essential for the trace quality desired.]
3. The traces are moved from the oscilloscope, passing through computer memory, to permanent storage on a floppy disk. [Custom hardware and software must be made for the interface between the oscilloscope and computer; custom software handles the remaining data shuffling.]
4. Signal averaging or other numerical processing of the traces may be performed. [This involves only software and may be added later.]
5. A record of stimulus and recording parameters is kept, substituting for a handwritten protocol sheet [easy software implementation].

All of the above must be under the constant control of the experimenter, through commands given from a keyboard or other controls (implemented by means of a command language, to be described in detail, below).

After an experiment:

1. The traces must be retrieved from permanent storage and displayed [combined hardware and software].
2. Software routines must be available for manipulating the traces. These include routines to:
   Magnify in the x or y dimensions
   Differentiate or integrate
   Add or subtract traces
   Shift traces horizontally or vertically
   Smooth or otherwise filter traces
   [These routines can be added in stages, according to need.]
3. A hard copy of the traces must be provided. [This requires a digital plotter and custom software to drive it.]

These tasks are all done with the equipment schematically represented in Figure 2. The components include a high-quality digital oscilloscope (Nicolet 2090), a general-purpose digital computer (Digital Equipment Corp. PDP-11), a small amount of custom circuitry to interface the oscilloscope to the computer, a console terminal, a printer, and a digital x–y plotter. The software consists of two programs, LAB and ANLSIS, running in conjunction with a commercially available operating system, RT-11. These two are, respectively, for running the labo-

Figure 2. Block diagram of the digital recording laboratory. The biological amplifier and stimulus isolation units (SIUs) are conventional and commercially available. The current command unit is a source of DC voltage gatable by external signals. An SIU can be used. The computer is equipped with serial line interfaces to connect the console terminal, printer, and plotter. A printer equipped with keyboard could substitute for the first two devices. Two DRV11 parallel interfaces connect the digital oscilloscope and stimulator. A programmable clock (KWV11-C) is needed for timing the stimulator pulses. An inexpensive x–y monitor oscilloscope driven by a digital-to-analog converter (DAC) is a great convenience.

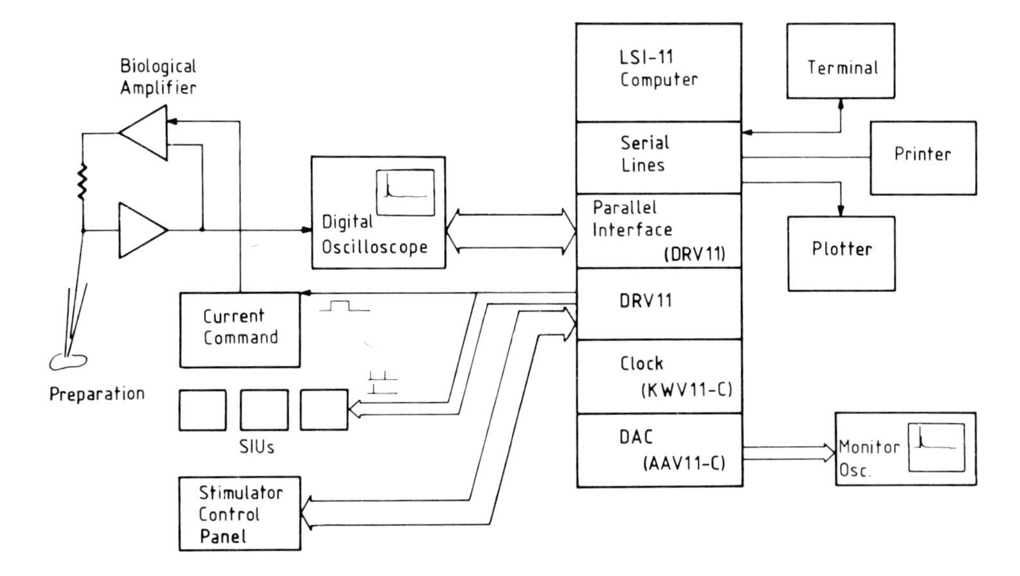

ratory during an experiment and for the subsequent analysis of the data. The operating system, RT-11, has been developed by Digital Equipment Corporation (DEC) for PDP-11 computers operating in just the kind of real-time application that this system is used for. RT-11 is sophisticated enough to allow two programs to run at the same time so that one person can be analyzing and plotting data, or performing some other task with the computer, while another is running an experiment.

Choice of Hardware

Data is acquired with a high-quality digital oscilloscope (Nicolet 2090). This digitizing instrument is built as an oscilloscope and has conventional controls, such as sweep speed and vertical gain. Its bandpass is quite appropriate for biological signals. Thus, the one in my laboratory can digitize a point as fast as every 500 nsec, corresponding to a 0.5-msec sweep. Most digitizing systems advertised for the neurophysiological laboratory have maximum acquisition rates (called *dwell time*) of approximately one point every 40 μsec. Although this is fine for recording postsynaptic potentials at sweep durations of 50 or 100 msec, action potentials require faster sweep speeds, say of 2–5 msec total duration. Also, the Nicolet 2090 has 12-bit resolution, i.e., one part in 4,096, which is uncommonly high for digitizers with such rapid acquisition times. This means that traces can be vertically expanded many times and still be quite presentable. In my intracellular recording experiments, all records are taken at a single gain, with the full screen being equivalent to 200 mV. Records of low amplitude events are made by vertical expansion (see Figure 1).

The Computer

PDP-11 refers to a family of computers produced by Digital Equipment Corporation (DEC). There are two subdivisions within the family, UNIBUS and Q-Bus computers. The former are descendent from the first PDP-11s and include the largest, most powerful, and most expensive members of the family. The second group all utilize Large-Scale Integration (LSI) technology and are in fact often referred to as LSI-11 computers. They are in a sense microcomputers with a minicomputer architecture. They are uniformly less expensive than UNIBUS PDP-11s and yet are software compatible with the larger machines.

The decision to use a Q-Bus PDP-11 (hereafter referred to as an LSI-11) was made because of its computing power, maturity as a product, and manufacturer support and commitment. LSI-11s will not disappear from the market in the foreseeable future nor will the product be changed so drastically that existing units will no longer be compatible with the new. This consideration is important for a laboratory that will be expanding and developing over a number of years. Moreover, LSI-11 products are fully responsive to market pressure so that the purchase price for components and complete systems have been steadily decreasing. LSI-11s are in themselves a family of products, with several variously priced CPUs, memories, and mass storage devices to choose from.The components making up two possible configurations are listed in Appendix 1 together with some suggestions for further economies.

Computer–Oscilloscope Interface

Nicolet offers three interface options with its 2090 oscilloscope. Two conform to international standards, an RS-232C serial interface and an IEEE-488 instrument bus interface. For neither is the data transfer rate sufficiently fast for on-line recording work, where the requirement is to make a round-trip transfer of a 1,024-word trace to the computer and back to the oscilloscope in the time between successive traces (1–2 sec). Therefore, I have always purchased the third interface option, which is a special parallel interface peculiar to this and some other Nicolet instruments (see the Operation Manual for the 2090 digital oscilloscope). Very high data transfer rates can be attained with this interface, at the cost of having to design the circuitry and software to drive a nonstandard interface.

In two previously built installations, the computer was always some distance from the oscilloscope so that the transfer had to be serial. Quite complicated digital circuitry had to be designed, built, and debugged. The results were, however, excellent, with two separate designs (to a PDP-12 and an LSI-11) being able to transfer 1,024-word traces within 30 msec. In the present application, a much simpler solution is available because the Nicolet oscilloscope and computer are close to each other and the approximately 40 lines constituting the parallel interface can simply be strung between them.

Two interface circuits are shown in Figure 3. Both utilize an inexpensive DEC parallel interface card (DRV11; for details see the PDP-11 Microcomputer Interfaces Handbook) and both use a minimum of other hardware. Circuit 1 requires just two integrated circuits, half that of circuit 2. Functions performed by hardware in circuit 2, in particular the generation of synchronizing pulses for the Nicolet, must be done by software for circuit 1. The assembly language routines appropriate for both circuits are to be found in Appendix 3. It is largely a matter of taste and familiarity with a given technology that determines whether a designer leans toward hardware or software solutions of particular problems. Both circuits can be improved by adding inverting buffers (not shown) for all data lines and changing the software so that the inverting commands [i.e., COM (R1)+] can be omitted. In addition, because of the limited size of the output buffer (NICOUT) neither interface can send normalization data (bit N) back to the oscilloscope. One of the free bits from the other DRV11, used for the stimulator (see below), could be used for this purpose. It would then be sensible to reassign the DRV11 output bits by moving WRITE to the stimulator output buffer and putting AC1 into bit 14 and N into bit 15 of NICOUT.

Software

During an experiment, the system is run by the user-written applications program LAB. This program continuously monitors the console keyboard for commands from the experimenter, controls the programmable clock of the LSI-11 in order to give the stimulus timing pulses, and upon command, moves data from the digital oscilloscope to computer memory and from there to permanent storage on disk. This program will be described in some detail. Data analysis is performed with another applications program, ANLSIS, which can retrieve data from disk, display it on a monitor oscilloscope (or the Nicolet 2090) for inspection, and plot it on an x–y plotter for a hard copy.

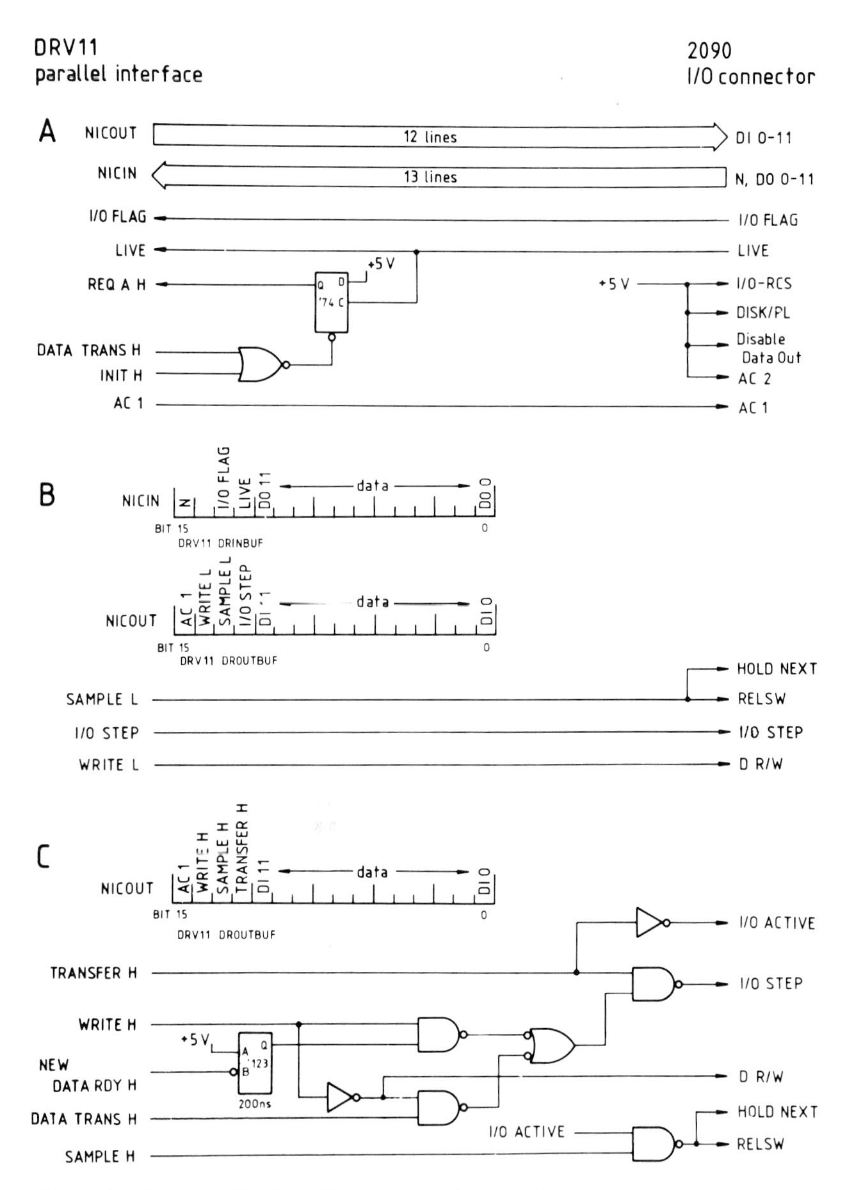

Figure 3. Two possible interface circuits for the Nicolet 2090 digital oscilloscope. An interface using minimum hardware would consist of the elements shown in **A** and **B**. **A**, which is common to both interface circuits, contains the data lines running unbuffered between the digital I/O connector of the 2090 oscilloscope and the connector of the DRV11 card. A single D-type flip-flop ($\frac{1}{2}$ of a 7474) and a NOR gate ($\frac{1}{4}$ of a 7402) constitute the interrupt request circuitry suggested by DEC (PDP-11 Microcomputer Interfaces Handbook). The remaining interconnections and diagrams of the use of the input (NICIN) and output (NICOUT) buffers of the DRV11 are shown in **B**. The suffix H indicates that that bit or line is asserted when it is high (+5 V) or on. The L suffix is for a bit or line that is asserted when it is low (0 V) or off. A more sophisticated interface would consist of the elements of **A** and **C**. The functions of some of the DRV11 output buffer bits (NICOUT) have been changed. The input buffer (NICIN) bit assignments are unchanged. A few additional integrated circuits in **C** provide the pulses required by the Nicolet interface, removing the responsibility for generating them from software.

The bulk of the software is written in FORTRAN, a high-level language of ancient lineage. The reasons for selecting FORTRAN are given under the next subheading. Time-critical input–output (I/O) routines are written in assembly language and are callable from the FORTRAN main program. The program runs under RT-11, a single-user operating system obtainable, as is the FORTRAN compiler and all pertinent documentation, from DEC.

Choice of Language

The choice of FORTRAN as the programming language may appear strange to many readers. The language is old and in many ways outdated. It was developed long before the current software principles of structured programming were devised, principles that, if adhered to, considerably shorten the time required to develop a program and make the finished program more comprehensible to other programmers. Despite these drawbacks, for now and some time to come, FORTRAN has the advantage that it is the one language that every computer science student has studied. The investigator who is not willing to do his or her own programming (and in most cases this would be an unwise use of his time) probably can also not justify hiring a permanent, full-time computer programmer. The programming needs of most laboratories, then, will be met by a succession of part-time people, often students, who must be able to understand the computer language used and work with the programs their predecessors have produced. FORTRAN lends itself well to these circumstances because it is so universally understood. FORTRAN is also well developed in the RT-11 environment. DEC has over the years produced an extensive library of FORTRAN-callable routines that provide the user full access to system resources, such as in the handling of input and output (RT-11 Programmer's Reference Manual 1983). It performs well in the foreground/background environment (a form of time-sharing where two or more programs can run simultaneously), which is required in most laboratory installations. The most time-critical routines (such as controlling the programmable clock and exchanging data with the Nicolet 2090) have to be written in assembly language, but the interaction of the few assembly language routines and the FORTRAN code has been trouble free. FORTRAN is not inherently structured, particularly the version supplied with RT-11, but with care quite legible and maintainable code can be written in it. This is done by enforcing strict programming guidelines, including a modular organization, standard format for subroutines (see Appendix 2), and coding that mimics more structured languages.

An applications program consists of a set of relatively short subroutines, each performing a particular task. These modules are called by another set of routines, which I shall refer to as the *core,* and which is nearly identical in all applications programs. The core provides the interface between user and the remainder of the program. It receives commands from the console keyboard and acts to call the appropriate subroutine to perform the task requested. As the system grows, additional subroutines can be added or existing ones improved. This modular design and the core set of routines, which should serve as a model for interactive programs, allow a functioning program to be written and be running in a very short period of time [for a different approach, see Schoenfeld (1983)].

Command Language

The user controls the operation of the program by means of a command language, and not the inconvenient control by interactive dialog that evolves in most laboratories. In the command language system, all the subroutines constituting an applications program are at the immediate disposal of the user and may be called up instantly and directly by a single command through the keyboard, much as the individual notes of a piano can be played by a single keystroke. An all too common pitfall for programmers and software designers developing an interactive system is to construct their program as an interactive dialog, where the computer leads the user through a matrix of choices by posing a series of questions to which the user responds. In producing this kind of structure, the programmer has done little more than transpose the design discussion with the software designer into code, without any deep inquiry into the eventual facility of the program's use. The result is a program that is easy to learn but unhandy, slow, and burdensome to use. As the interactive dialog is obligatory, the change of even one parameter generally requires that the entire dialog be repeated. Interactive dialogs also come under the guise of menus, a form that is highly touted as being user friendly but just as slow and restrictive to use as is any interactive dialog.

In the laboratory command language I have designed, commands are made up of strings of alphanumeric characters. The first character of a command string must be a letter and it specifies the function to be performed (see Appendix 2 for the list of commands and functions for the program LAB). If any characters (numbers or letters) follow the initial letter, then they are termed modifiers and serve to modify the action specified by the first letter. A command string made up of the command E followed by two modifiers might be EAN. A command string is ended by a separating character, such as a space, comma, or period. Spaces embedded within a command string are not permitted. Several command strings, divided by separators and terminated by a carriage return, may be typed on a single line, making up a command line, such as

GN,A,EAN.

This command line is interpreted by ANLSIS as follows: Get the Next trace, plot it (A is not mnemonic here, but P was already used for another function), and then Erase it All without double-checking with the operator (Noquery).

The core set of routines evaluates the command and calls the subroutine specified by the first letter of the command string. Modifiers function by the simple mechanism of passing the entire command string to the called subroutine, which then has the responsibility of interpreting them. Subroutines can call other subroutines directly or they can replace the command string with another, before returning to the core, as an alternative means of calling other subroutines. Because it is unstructured, the author has avoided the use of this latter technique, except in implementing the HELP feature (see below).

In its most basic form, the core does not respond to a command until a carriage-return character has been typed, terminating the line. Because of this, a single line can contain several commands. They will be properly acted upon, sequentially from left to right, provided that they are separated from each other by one or more nonalphanumeric separators. The command line A,EAN would call first the routine specified by A, then that specified by E, with the characters AN sent to the subroutine as modifiers. The string AEAN, however, would call just routine A with EAN sent along as modifiers.

The heart of the core is the following loop, located in the main program segment.

```
      BYTE CHAR(80)
      .
      .
10    TYPE 200
200   FORMAT ('$Command: ')
      ACCEPT 300, (CHAR(I),I=1,78)
300   FORMAT (78A1)
25    IF (CHAR(1).EQ.'A') CALL AVRAGE(CHAR)
      IF (CHAR(1).EG.'E') CALL ERASE(CHAR)
      .
      .
      IF (CHAR(1).EG.'S') CALL STORE(CHAR)
115   IF (SHIFT(CHAR)) GOTO 25
      GOTO 10
      .
      .
      END
```

Line 10 is the top of the loop. Execution comes to it any time that the program is ready to respond to a new keyboard command from the user. The string "Command:" is typed on the console device to prompt the user to enter a command. Execution proceeds to the ACCEPT statement. If no characters have been typed, the program waits here in a suspended state. When one program is suspended, RT-11 allows another program to run, so that the computer may be performing two entirely different tasks at once. One program is said to be in foreground and the other in background. Real-time data acquisition programs, such as LAB, spend most of their time in suspended state and are usually run in foreground. Once a command is given, the program immediately leaves its suspended state and processes the command. As written, a command line can contain up to 78 characters. Only the first one, contained in CHAR(1), however, is looked at by the succession of IF statements beginning on line 25. These check for a match to CHAR(1) and call the appropriate subroutine when one is found. By their nature, subroutines, upon completion, return execution to the line following this call. Presuming that no subroutine changes CHAR(1), then at most only one subroutine can be called on each pass through the IF statements. Thus, execution always reaches the bottom of the loop, line 115, where the subroutine SHIFT is called.

SHIFT is a LOGICAL PROCEDURE, a form of subroutine, which is also part of the core. Much of the power and sophistication of the command language structure stems from the action of SHIFT on the command string CHAR. Its principal action is to remove the current command from CHAR and check whether another command follows, within CHAR. If one does, then this second-in-line command is placed at the head of the array, that is, beginning at CHAR(1), and SHIFT returns with the Boolean value of TRUE. With SHIFT being TRUE, program execution proceeds to line 25 for another pass through the IF statements, bypassing the ACCEPT statement at the top of the loop. If there is no further command, SHIFT returns FALSE and control passes to line 10 for the acceptance of a new command line.

Besides allowing the interpretation of multiple commands on a single line, SHIFT performs several other useful parsing functions. If one or several commands on a line are enclosed in angle brackets (⟨ and ⟩), SHIFT recognizes that fact and treats the enclosed command string as a sequence to be repeated. The number of repetitions can be specified by following the closing angle bracket (⟩) with a repetition counter. The syntax for the counter is an asterisk (*) followed by a two-digit numeral (e.g., ⟨B,EAN⟩*05). If there is no counter, then the enclosed sequence will be repeated endlessly, until stopped by some other means (see next paragraph). Repeating strings can be nested to any level and in any way, with parsing being from left to right. Thus, the sequence

K,⟨EAN,GN,A⟩*04,K,⟨⟨EAN,GN⟩*15,A⟩

would execute K once, the sequence EAN,GN,A four times, another single K, then the sequence of EAN,GN repeated 15 times followed by a single A, which would be repeated indefinitely.

Indefinite loops must be ended by an external event, such as stopping program execution entirely (this can be done from the console keyboard as a feature of the RT-11 operating system) or by having one of the called subroutines actually destroy the command string contained in CHAR. This is most economically done by having the routine insert a numerical zero into the command string. The value of zero (the ASCII null character and not the ASCII character standing for 0) is commonly used to mark the end of a character string and SHIFT adheres to this convention. Thus, for example, if CHAR(2) is assigned the value zero (i.e., CHAR(2) = 0) then SHIFT recognizes this as the end of the command line and on its next call returns with a value of FALSE. Execution returns to line 10, ready for another user command. As is commonly said, control returns to the user.

This exit feature is used by the author in subroutines that read and write data onto disk. An example that is useful to examine is contained in the subroutine GET, a routine that reads one trace (1,024 data points) from disk to memory. GET is called by the command G. Without modifiers, GET first queries the user as to which trace is wanted (they are numbered sequentially) and expects that an integer will be typed in response:

G

Which trace? 109

(Underlined characters are typed by the user.) Now trace 109 would be retrieved. GN instructs GET to omit the query and to get the next sequential trace. That would now be 110 in the present example. If an attempt is made to get a nonexistent trace, that is, too high a trace number is given, then an error message is typed, stating that the end of the data file has been reached. During this error condition, if in addition the original command had contained the modifier X (as GX, GNX, or GXN, the last two being equivalent), then GET will also break the command line by assigning CHAR(2) = 0. If there is no X, then no further action is taken and the command string is not broken. Repeat sequences, with or without count, and the exit (X) feature have proven very useful in plotting data. Here the task is repetitive and would be time consuming for the user if a command sequence could not be set up to perform the task without user attendance. An indefinitely repeating sequence to get the next data trace, plot it, then erase it suffices, with the X feature being included in the call to GET. The command line for this would be ⟨EAN,GNX,A⟩.

The command language concept has proven to be very powerful. It is easy to use and fosters very rapid program development. It enormously reduces the task of the programmer. Subroutines are kept simple since complex tasks can be performed by stringing these primitive functions together at run time. In this way, the programmer is freed of the major task of having to predict and provide for every possible choice that the user might make. This approach also retains as much versatility in the program as is possible. In the ANLSIS program, for example, the same program can plot traces in just about any form imaginable: for publication, for analysis, as points or continuous lines, with or without axes, superimposed or single.

Further functions have been added to the core, each implemented by calls to subroutines. A HELP function reminds the user which commands perform which functions, serving as a sort of table of contents for the commands of the particular applications program. It works by filling the entire CHAR array with the command string AH,BH,CH, . . . ,ZH so that successive passes through the core's main loop will call each possible subroutine, and in alphabetical order according to their calling command. By convention, all subroutines check for the modifier H in CHAR(2) and, if present, do not execute normally but instead type a short informatory message and then return to the main program (see Appendix 2). Two features have been added to the core to reduce the number of keystrokes necessary to execute commands or command lines in special situations, as during the breakneck activity of an intracellular recording. These can be optionally enabled or disabled. A special command mode can be entered that switches the organization of command console input from line-oriented to character-oriented. In special mode, the program responds immediately to a single letter command instead of only after a carriage return has been typed. The second of these features, a subroutine called ORDERS, allows the user to enter and save in memory an entire command line and have it executed upon the typing of a single special character. This is arbitrarily the ASCII character "|", whose code is 174 octal, and which was chosen because of the convenient location of that character's key on the keyboard used in the author's recording laboratory.

Data Storage

The data is stored on floppy diskette as 1,024(1*K*)-word records in files named for the experiment. FORTRAN is quite flexible in allowing different data storage formats, but, for physiological traces, only one, direct access and unformatted, is appropriate. Data are stored in their most compact (unformatted) form and in a way that any trace can be directly accessed without having to read through the ones preceding it in a file. Each 16-bit word contains up to 15 bits of voltage information. This is three more than the 12 bits produced by the digital oscilloscope, but allows for the summing of at least eight traces without the risk of arithmetic overflow. Overflow, which produces wraparound of the trace, is tolerated in many signal-averaging systems, but is clearly a nuisance of which it is well to be rid. The least significant bit of each data word encodes the trace parameters (i.e., sweep speed, voltage setting, number of active channels) that have been received from the oscilloscope, using the same encoding conventions adhered to by Nicolet.

Stimulation

Timed pulses for oscilloscope triggering, current injection, nerve stimulation, and so on can be delivered entirely by the PDP-11. The only additional hardware required is a KWV11-C, or equivalent, programmable clock and a DRV11 parallel interface card. The KWV11-C is a standard DEC device, inexpensive, and easily installed by the user. There are many ways to program such a device and to manage the interaction between it and the user. The simplest, in terms of programming, is to have all control of the programmable clock be through the console keyboard. A more elaborate solution is to build a dedicated control panel for timing functions and have the PDP-11 scan that panel, interpret its switch settings, and set the programmable clock to deliver the appropriate sequence of pulses. The former purely software system requires no new equipment or nonstandard hardware and so can be implemented quickly. It will, predictably, not be easy to use during an experiment, requiring multiple keystrokes to execute simple changes in stimulus parameters as well as asking that the user memorize an even more extensive set of command language commands. In contrast, a dedicated control panel together with its controlling software can be designed to mimic a digital stimulator, a device with which most users are familiar and skilled. As subroutines used in the pure software system are also used in the dedicated panel system, a wise course would be an orderly progression from the simple, quickly implemented one to the final more elaborate system. The first experiments might be a headache to run, but at least they will be running. It is also possible that some aspects of pure software control will prove so efficient or desirable that they will be retained in the final system.

Development of the stimulus package, then, occurs in three stages. In the first, a simple routine for loading the stimulator parameters and performing other basic functions, such as turning the stimulus pulses on and off, is written. These routines can still figure in the succeeding configurations, such as the loading of stimulus parameters at the beginning of an experiment. In the second stage of development, routines are developed that take full advantage of the power of the command language to alter individual stimulus pulse delays and durations quickly and with as few keystrokes as possible. Finally, a hard-wired control panel can be added, offering the most convenience to the operator. The computer is made to scan the switch settings on that panel and update its internal values for delay and duration accordingly.

In the author's system, eight lines from the DRV11 parallel port are apportioned for stimulation. Each channel should be able to deliver a variable number of pulses of differing duration and delays. As a physical control panel is eventually to be built, it would seem efficient to store the values for delays and pulse durations in arrays defined to give a one-to-one correspondence between the panel switches and array elements (Figure 4). Thus, we use an array DURAT(CHAN,N), which contains the duration in milliseconds of the *N*th pulse on channel CHAN ($1 < \text{CHAN} < 8$) and DELAY(CHAN,N) the time, also in milliseconds, after the start of each stimulus cycle at which that pulse is to begin. Additionally, PULSE(CHAN) contains the number of pulses (six or less) to be delivered for each channel. PERIOD is the length, in seconds, of each stimulus cycle. Limiting values in PULSE to six or less keeps the size of the DELAY and DURAT arrays reasonable. With up to six individually controllable pulses in any of eight channels, 48 array elements in each is sufficient. Long series of pulses are better implemented by programming for an alternative interpretation of the DELAY, DURAT, and PULSE settings, like the pulse-train mode of conventional stimulators. In TRAIN mode, then, DELAY(CHAN,1) and DURAT(CHAN,1) specify the starting point and duration of the first pulses, as normal, but DELAY(CHAN,2) specifies the total duration, in milliseconds, of the pulse train and DURAT(CHAN,2) the interval between pulses in the train. All

Figure 4. Data flow within the software routines for generating stimulus pulses. The raw parameters specifying each pulse delay and duration are first entered into two two-dimensional arrays, DELAY and DURAT. As shown, each of eight stimulus channels can have as many as six individually programmed pulses. Provision for pulse trains is not diagrammed. This data must be converted into a form suitable for direct loading into the buffers of the programmable clock (KWV11-C) and DRV11. This final numerical form resides in the one-dimensional arrays ITIME and ICHAN. The process of this conversion is explained in the text.

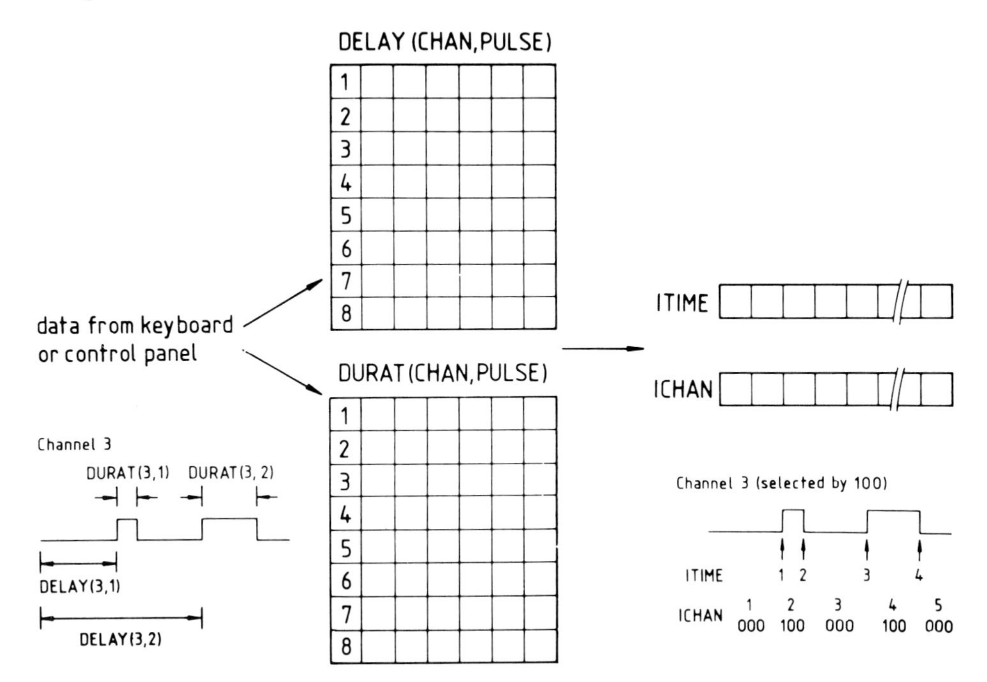

pulses in the train have duration DURAT(CHAN,1). TRAIN mode is specified for a channel by arbitrarily letting PULSE(CHAN) = −100.

Programming the Clock and Parallel Interface

It is not possible to use these parameters to directly control the KWV11-C programmable clock or drive the DRV11 parallel interface. The clock requires a series of negative integer values representing the number of clock ticks in the intervals between discrete events (PDP-11 Microcomputer Interfaces Handbook, 1983). Discrete events would be the beginning and end of a pulse. If the KWV11 has been programmed with a 100-kHz clock rate, then each tick is 10 μsec. At each discrete event, the DRV11 output register is loaded with a byte (8 bits) whose pattern of bits equals the pattern of on and off channels desired. The program code necessary to convert the DELAY/DURAT data into these forms represents an elementary programming exercise that proceeds as follows.

Let ITIME be an integer array, to be filled with the interval values for the KWV11, and ICHAN is an array to contain the bit patterns for the DRV11. Each of the following steps is executed as a loop in which all the values of DELAY and DURAT are evaluated once.

1. For each DELAY/DURAT pair, two values each are placed into ITIME and ICHAN. Into the pair of ICHAN array elements go the times, in clock ticks from the start of the stimulus cycle, at which the pulse is to begin (ITIME(N)) and end (ITIME(N+1)). Into ICHAN(N) goes a single bit corresponding to the channel (1 of 8) to be turned on. This is determined numerically, ICHAN(N) = 2**(CHAN−1). ICHAN(N+1) receives its logical complement, signifying for now that at ITIME(N+1) that channel is to be turned off.
2. ITIME and ICHAN are sorted according to the values in ITIME, the earliest events being ranked first.
3. The values in ICHAN are converted to their final form, in which a set bit is propagated through all subsequent array elements until the complement of that set bit is encountered, indicating that it is to be turned off. The procedure for this propagation is simple. For any pair of adjacent elements in ITIME, if both are positive, indicating only bits to be turned on, then the second array element is replaced with the disjunction (logical OR) of the two. If one is negative, indicating a bit to be turned off (and in this algorithm, only the second of the pair can be negative), then it is replaced with the conjunction (logical AND) of the two, which will always be positive.
4. Next, identical values of ITIME are found and merged by ORing the corresponding ICHAN values.
5. Finally, the absolute time values in ITIME are converted to intervals and negated. For each element in ITIME, its value is replaced by that of the opposite (two's complement) of the difference between it and its predecessor.

Taking full advantage of the command language structure, a set of single-letter commands for rapidly changing timing parameters can be produced so that even the pure software implementation can be made easy to use. Each of the last eight letters of the alphabet can be made a command designating one of six channels and numerical and letter modifiers used to signify parameter values and the identity of a parameter. Thus UW10 would set the pulse width of channel U to 10 clock ticks. The parameter most often changed during an experiment, which in my experience is pulse delay, could be accepted without a letter modifier, as U1000. From commands entered in this way on the keyboard, the software would fill the appropriate array elements, DURAT(CHAN,N), DELAY(CHAN,N), and so on, with the correct values. Additional modifiers could control the switch to train mode, or turn a channel entirely off.

The software implementation described so far provides six controllable delays and pulse durations for each of eight channels. Substituting a digital control (digiswitch) for each of these would be excessive, so a mixed system is envisaged where some lesser but adequate number of digiswitches is used with the remaining positions continuing to be controlled by software only. The control panel is read by a multiplexing circuit controlled by the DRV11. Figure 5 shows the outlines of this circuitry. Six of the unused output lines address the multiplexing circuitry, which in turn selects one set of tristate drivers to place the switch settings on the 16 input lines of the DRV11, to be read. Before each stimulus cycle begins, a software routine can step through the digiswitch addresses, read each switch setting, and update the appropriate DURAT or DELAY array elements, if necessary. Reading the control panel takes less than 5 msec if all the 64 four-digit parameter switches that this circuit can service are used.

Figure 5. Schematic diagram of the multiplexing scheme for sampling the stimulator control panel. The 16 input lines of the stimulator DRV11 are used as a control panel bus and are sufficient to read four digits of binary-coded-decimal (BCD) information at a time. The multiplexing circuit must decode an address in the range of 0–63 present on six DRV11 output lines into one of 64 enable lines. Only one of these enable lines is asserted at a time, in order to turn on just one set of 16 tristate drivers, allowing the signals from just one set of four digiswitches to appear on the control panel bus.

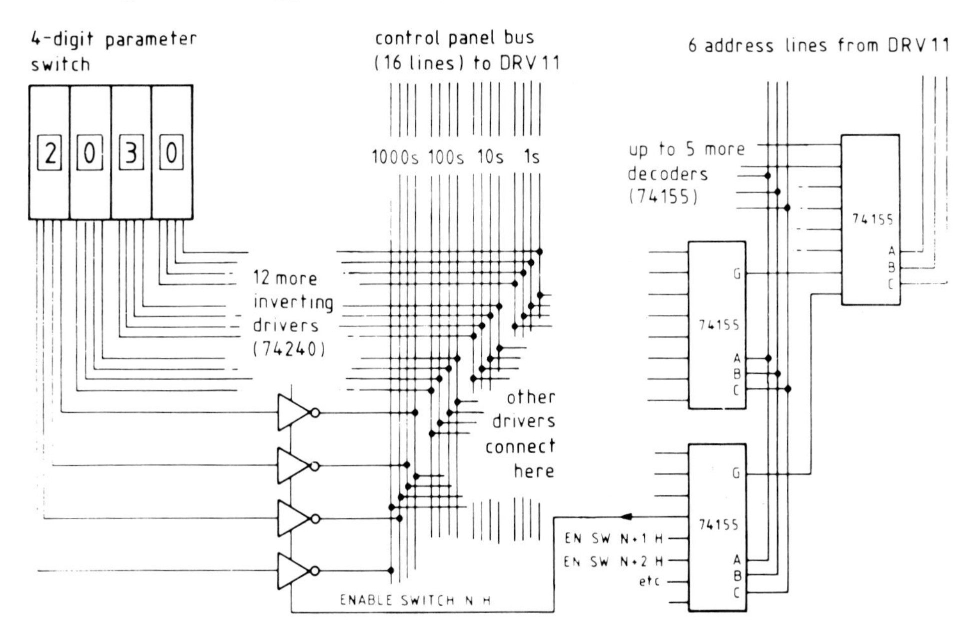

Conclusions

The advantages of the present design for a completely digital recording laboratory are numerous. A well-known and established general purpose computer has been used. The LSI-11 has an advanced architecture, a sufficiently wide data path (16 bits), and adequate speed for this application. Peripherals, operating system, and language processor are all available and appropriate. The implementation is stepwise so that experiments do not have to wait upon the ultimate system being completed. A key element of this efficiency in implementation as well as efficiency in final use is the command language structure. The command language allows modular and logical software development. The resulting programs do not become encumbered with the elaborate dialogs and complex mazes of program logic that I have seen develop all too quickly without it. Quality recording devices have been chosen, particularly with regard to the digital oscilloscope. The data are digitized and stored with very high resolution and fidelity in both the vertical (voltage) and horizontal (time) axes. Data transfer between oscilloscope and computer must be fast and this requires a custom interface. This is far from being a foreboding task, however, and can be realized in any laboratory.

Future developments will most certainly alter the choice of hardware. Viewed another way, however, the addition of new and faster processors to the LSI-11 line will probably reduce the cost of the existing ones, which are entirely adequate for this application. New microprocessor families are also being introduced and becoming competitive with the LSI-11s. They have no significant differences in architecture that would recommend them, and they do not as yet have the software base of the LSI-11s, but this latter point will certainly change. High-speed analog-to-digital boards with their own data buffering are beginning to appear for the Q-Bus (e.g., from Data Translation) and may provide an alternative to the digital oscilloscope, which is a major expense.

Although the largest part of the development work has been done, constructing the laboratory described in this chapter is still a major and highly technical undertaking. No one has ever developed a turnkey recording laboratory so that technical competence and a knowledge of instruments have always been required of neurophysiologists. It appears unlikely that this will change.

This work was supported by USPHS Grant NS 20841.

Appendix 1. System Cost

Two possible combinations of components for a full-featured digital laboratory, as described in this chapter, are given below:

Author's system
- Micro-PDP-11 computer, which includes:
  - LSI-11/23+ CPU
  - 256-KB RAM
  - 900-KB floppy disk storage
  - 10-MB hard disk
  - 4 serial line ports
- KWV11-C programmable clock
- AXV11-C analog I/O board

DRV11-J parallel interface
VT-102 terminal
LA-100 printer
Nicolet 2090 oscilloscope
Hewlett-Packard 7475A plotter
Monitor oscilloscope

Minimal system
LSI-11 CPU
32-KB RAM
2 serial line ports
KWV11-C programmable clock
DRV11 parallel interface (two required)
RX02 floppy disk
Housing and power supply
Decwriter (LA-34, LA-36, or LA-50) printing terminal
Nicolet 2090 oscilloscope
Hewlett-Packard 7470A plotter

The equipment listed under the author's system can be purchased for about $23,000, taking full advantage of substantial university discounts. The smaller system can be obtained for about $14,000, of which about $9,000 is for the oscilloscope and plotter. Great economies can obviously be made if a substitute for the digital oscilloscope (about $8,000) can be found. The reasons for choosing a quality digitizing device have been given in the text. Nonetheless, the author's AXV11-C analog I/O board and KWV11-C could perform the functions of the digital oscilloscope at some loss of time resolution. Minimum interval between samples would be 40 μsec, which is adequate for recording postsynaptic potentials but not fast sweeps of action potentials. The vertical resolution is 12 bits. The sample rate is as good as the best digitizing boards available for the popular 8-bit microcomputers (e.g., Apple) but the vertical resolution is clearly better. The author has in the past written programs that allow a PDP-11 (or a PDP-12) to simulate a digital oscilloscope. CPU overhead is very high so that considerable thought and skill would be required to implement simultaneous data sampling and stimulation. The cost of the monitor oscilloscope is only a few hundred dollars so that this is not a fruitful source for reducing cost. Nor can savings be realized by taking advantage of the ability of the Nicolet 2090 to drive an analog x–y plotter as digital plotters of excellent resolution and quality are now available for less than $2,000.

The cost of software licenses can vary from about $500 to $2,000 depending upon the degree of support required or it may be part of a complete package if the computer is purchased through an original equipment manufacturer (OEM). Most universities and research institutes will already have several PDP-11 installations running RT-11 and FORTRAN. In this case, only a license-to-copy or a general software license may be required. The vendor will be well informed in these matters.

Appendix 2. Outline of a Standard Subroutine

The FORTRAN subroutines that make up an applications program are coded according to a standard form. The form is efficient in memory use and speed of execution and includes a device for providing HELP messages. The design also permits easy implementation of overlays should this be necessary.

```
      SUBROUTINE ERASE(CHAR)
      LOGICAL*1 CHAR(8)                         (Note 1)
      INTEGER XBUF(1024),YBUF(1024)
```

```
        COMMON/PLTDAT/NPTS,IPOINT,XBUF,YBUF              (Note 2)
        IF(CIIAR(2).NE.'H') GOTO 5
        TYPE 900,CHAR(2)                                 (Note 3)
    900 FORMAT(1X,A1,T13,'ERASE',|,
        +' %A',T13,'Erase All',|,
        +' %N',T13,'Noquery')
        RETURN

    5   .
        .
        Body of subroutine                               (Note 4)
        .
        .
        RETURN
        END
```

Note 1: The array CHAR is the only parameter passed to a subroutine and only a portion of it is used in order to save space and operating time.

Note 2: All subroutines operate on data contained in COMMON blocks according to the command contained in CHAR. This is because of the efficiency in both time and space with which FORTRAN handles COMMON blocks. To encourage clear program code, however, the COMMON blocks are treated as a subroutine parameter list. They are named mnemonically and organized in a logical fashion according to function. Their declaration is at the head of a subroutine, in a position analogous to the parameter declarations of ALGOL or PASCAL. COMMON blocks are also accessible to assembly language routines (see Appendix 3).

Note 3: This code implements the HELP dialog. For clarity, FORMAT statements appear where used and not grouped together at the end of a subroutine. The HELP printout can contain any text that the programmer feels would be helpful to the user. The percent sign (%) is meant as a substitute for the calling command letter, which should be E for ERASE, and is part of the shorthand notation used to indicate modifiers that are allowed for a subroutine. For example, the HELP output for LAB is as follows:

```
A     AVERAGE initialize
B     SPECIAL (toggle)
%S    Set special mode
E     ERASE
H     HELP
O     ORDERS
P     PUT on disk
Q     QUIT
R     READ from Nicolet
S     SEND to Nicolet
```

Note 4: The body of the subroutine begins with line 5. The return to the calling routine is standard.

Appendix 3. Assembly Language Routines

The assembly language routines for driving the Nicolet–computer interface (Figure 3) are given below. The full code accompanying routine 1 also illustrates some of the conventions for joining assembly language routines to FORTRAN code, including access to COMMON blocks (RT-11 Programmer's Reference Manual, 1983). This code may be of some informative value in itself but the significance of the control signals (e.g., AC1, D W/R, I/O Active) and interface conventions of the Nicolet oscilloscope should be studied in the 2090 Operation Manual (1981) by the reader wishing to duplicate the author's laboratory.

```
;       Macro program to drive Interface 1 between
;       a DRV11 parallel port and a Nicolet 2090
;       oscilloscope.
        .TITLE NICUTL STARTED 14-JAN-81                     M.R. Park
        .PSECT USER$I,RW,I,LCL,REL,CON                      ;This creates a program
                                                            ;according to FORTRAN
                                                            ;conventions.
        NICOUT=171772                                       ;Output IO register, to Nicolet
        NICIN=171774                                        ;Input IO register
        NICCSR=171770                                       ;CSR for DRV11
        NICVA=330                                           ;Vector addresses
        NICVB=334
        IOSTEP=10000                                        ;Define values to be used as
        NSAMPLE=20000                                       ;as commands
        READ=40000
        AC1=100000
;       Read routine (from Nicolet) follows
RENIC:: MOV     #<READ!NSAMPL!IOSTEP>,@#NICOUT              ;NSAMPLE & IOSTEP must remain
                                                            ;high. Set to read.
        MOV     @#NICIN,RO                                  ;Check state of Nicolet
        BIC     #147777,RO
        BEQ     ERROR                                       ;Not ready for transfer.
        MOV     #YBUF,R1
        MOV     #<READ!NSAMPL!IOSTEP!AC1>,@#NICOUT
        MOV     #<READ!NSAMPL!IOSTEP>,@#NICOUT              ;AC1 line pulsed.
        MOV     #1024.,RO                                   ;This is the counter. Now
                                                            ;start the loop.
```

```
1$:     MOV     #<READ!NSAMPL>,@#NICOUT              ;Pulse IOSTEP to begin a
        MOV     #<READ!NSAMPL!IOSTEP>,@#NICOUT       ;transfer.
        MOV     @#NICIN,(R1)                         ;Get and store the data point.
        COM     (R1)+                                ;All words inverted. Bump
                                                     ;pointer.
        SOB     RO,1$                                ;If not done, loop again.
                                                     ;1024 data points are now in
                                                     ;YBUF. to finish, NPTS is
        MOV     #1024.,NPTS                          ;loaded with the number of
        RETURN                                       ;and the subroutine exited.
;       Send routine (to Nicolet) follows
SENNIC:: MOV    #<NSAMPL!IOSTEP>,@#NICOUT            ;The coding of this routine is
        MOV     @#NICIN,RO                           ;almost identical to RENIC.
        BIC     #147777,RO
        BEQ     ERROR
        MOV     #YBUF,R1
        MOV     #<NSAMPL!IOSTEP!AC1>,@#NICOUT        ;READ is not asserted, of
        MOV     #<NSAMPL!IOSTEP>,@#NICOUT            ;course.
        MOV     #1024.,RO                            ;Counter
1$:     MOV     (R1)+,R2                             ;Data from YBUF
        COM     R2                                   ;is still upside-down!
        BIC     #<AC1!READ>,R2                       ;NSAMPL & IOSTEP remain high
        MOV     R2,@#NICOUT
        BIC     #IOSTEP,@#NICOUT
        BIS     #IOSTEP,@#NICOUT                     ;Pulse IOSTEP.
        SOB     RO,1$                                ;If not done, loop again.
                                                     ;1024 data points have now
                                                     ;been sent.
        RETURN
ERROR:  CLR     NPTS                                 ;Come here if an error is
        RETURN                                       ;detected, return immediately.
```

```
;
;       The following provides access to the FORTRAN COMMON block PLTDAT.
;
        .PSECT  PLTDAT,RW,D,GBL,REL,OVR
NPTS::  .WORD   0
IPOINT:: .WORD  0                                   ;SENNIC RENIC only use NPTS
XBUF::  .BLKW   1024.                               ;and YBUF, but the other
YBUF::  .BLKW   1024.                               ;variable in PLTDAT must also
        .END                                        ;be declared.
```

SENNIC and RENIC appear as follows for Interface 2.

```
;
;       Read routine (from Nicolet) follows
;
        TRANSF=10000
        SAMPLE=20000
        WRITE=40000
RENIC:: CLR     @#NICOUT                            ;WRITE & TRANSF false (L) and
                                                    ;AC1 is true (L).
        MOV     @#NICIN,RO                          ;Check state of NICOLET
        BIC     #147777,RO
        BEQ     ERROR                               ;Not ready for transfer
        MOV     #YBUF,R1                            ;Pointer.
        MOV     #TRANSF,@#NICOUT                    ;I/O Active true and will
        MOV     #<TRANSF!AC1>,@#NICOUT              ;remain so. AC1 line pulsed.
        MOV     #TRANSF,@#NICOUT
        MOV     @#NICIN,RO                          ;Dummy transfer to send first
                                                    ;I/O pulse.
        MOV     #1024.,RO                           ;This is the counter. Now
                                                    ;start the loop.
1$:     MOV     @#NICIN,(R1)                        ;Get and store the data point.
        COM     (R1)+                               ;All words inverted. Bump
                                                    ;pointer.
        SOB     RO,1$                               ;If not done, loop again.
                                                    ;1024 data points are now
```

```
        MOV     #1024.,NPTS             ;YBUF. To finish, NPTS is
                                        ;loaded with the number of
        RETURN                          ;and the sub-routine exited.
;
;       Send routine (to Nicolet) follows
;
SENNIC:: CLR    @#NICOUT                ;See above.
        MOV     @#NICIN,RO
        BIC     #147777,RO
        BEQ     ERROR
        MOV     #YBUF,R1
        MOV     #AC1,@#NICOUT           ;Pulse AC1.
        MOV     #WRITE,@#NICOUT         ;Clear AC1; set WRITE.
        MOV     #1024.,RO               ;Counter
1$:     MOV     (R1)+,R2
        COM     R2
        BIC     #<AC1!SAMPLE>,R2        ;I/O Active and D W/R (Nicolet
                                        ;signals) must be set to pulse
        MOV     R2,@#NICOUT
        SOB     RO,1$                   ;If not done, loop again.
        CLR     @#NICOUT                ;1024 data points have now
                                        ;been sent.
RETURN
```

References

Clark, W.A. and C.E. Molnar (1964) The Linc: A description of the laboratory instrument computer. Ann. N.Y. Acad. Sci. 115: 653–658.

Operation Manual for Series 2090 Digital Oscilloscopes (1981) Nicolet Instrument Corp., Madison, WI, pp. VII-1 to VII-28.

PDP-11 Microcomputer Interfaces Handbook (1983) No. EB-23144-18, Digital Equipment Corp., Maynard, MA.

Park, M.R., H. Imai, and S.T. Kitai (1982) Morphology and intracellular responses of an identified dorsal raphe projection neuron, Brain. Res. 240: 321–326.

Park, M.R., W. Leber, and M.R. Klee (1981) Single electrode clamp by iteration. J. Neurosci. Meth. 3: 271–283.

RT-11 Programmer's Reference Manual (1983) No. AA-5378A-TC, Digital Equipment Corp., Maynard, MA.

Schoenfeld, R.L. (1983) A programming discipline for laboratory computing. IEEE Trans. Biomed. Eng. BME 30: 257–270.

20

Real-Time Analysis of Visual Receptive Fields Using an IBM XT Personal Computer

Walter H. Mullikin and Thomas L. Davis

Introduction

Much of what we know about the operation of the visual brain has grown from our ability to characterize the responses of visual neurons to the presentation of different patterns of light on the retina. Stephen Kuffler was the first to describe the discharge patterns of cat retinal neurons to small circular spots of light (Kuffler, 1953). From analysis of the ganglion cell responses to different patterns of light, Kuffler defined two cell types that differed in their response to the onset and offset of light. These different responses were found to depend upon the spatial organization of the excitatory and inhibitory regions of the two cell types. This spatial organization is called the *receptive field* of the cell. An innovation that made Kuffler's study possible was the ability to project one or two spots of light on the retina, varying their size and location independently, while at the same time capturing the responses of neurons. So powerful was this approach that Kuffler was able to reveal, for the first time, the spatial organization of excitation and inhibition in the receptive fields of retinal ganglion cells. Subsequently, Hubel and Weisel (1962) and many others have used similar techniques to reveal the spatial organization of the receptive fields of neurons in the visual cortex and thalamus (Bishop et al., 1971; Mullikin et al., 1984; Palmer and Davis, 1981; Shiller et al., 1976; Stevens and Gerstein, 1976).

Today, visual physiologists still apply many of the same principles of experimentation that originated with Kuffler 30 years ago. In the interim, however, the microcomputer arrived and forever changed the depth and scope of laboratory research. The computer allows visual physiologists a means for very precise control over stimulus conditions and the opportunity to quantify the responses of neurons. Using computers, it is possible to automate the process of presenting stimuli and monitoring neural discharges in real time. This makes it possible to do the following:

1. Quickly map the spatial distribution of excitation and inhibition in the receptive fields of visual neurons using spots of light.
2. Observe the time course of excitatory and inhibitory responses.
3. Use quantitative methods to analyze both the stimulus and response parameters.

Today's laboratory computers are inexpensive, fast, reliable, and versatile. When coupled with the latest laboratory interfaces they can quickly and easily be made to control equipment and consolidate information from monitoring devices. In this chapter we describe a complete visual physiology system suitable for the quantitative analysis of the receptive field organization of neurons in the visual brain.

System Hardware

Our system is currently designed to do the following:

1. Provide fingertip control over stimulus conditions.
2. Automatically produce on-line quantitative spatiotemporal maps of visual receptive fields.
3. Store and plot all data for future off-line analysis.

The system hardware is composed of four essential devices: an IBM XT personal computer with disk storage and graphics capabilities; an MI^2 laboratory interface (Modular Instruments Incorporated, Southeastern, PA); a Hewlett-Packard 7221A plotter; and an optical bench. The system software has been written using the C programming language.

Microprocessor

The IBM XT personal computer controls our system. This machine comes with a standard configuration that includes a 16-bit microprocessor (Intel 8088), 128 KB of read/write memory (RAM), and 40-KB read-only memory (ROM) that contains the easy to use BASIC language, two 5¼-inch disk drives (one for double-sided double density floppy disks and the other for a 10-MB fixed disk), eight bus slots for system expansion, and a flexible 83-key keyboard. System memory is addressable up to 20 bits or 1 MB. We have added the IBM color/graphics display adapter and the IBM color display monitor for full graphics capabilities. Three graphics modes are available:

1. text mode (25 lines with 40 or 80 characters per line; 256 characters in 16 foreground and 8 background colors);
2. medium-resolution mode (320 × 200 points, which allows a choice of three colors from one of two color palates as foreground and a fourth color for background); and
3. high-resolution black and white mode (640 × 200 points).

The color/graphics display adapter has 16 KB of on-board RAM, allowing us to use All Points Addressable graphics (APA) in both medium- and high-resolution modes.

Interface

The MI^2 laboratory interface is the hub of our system (see Figure 1). This interface provides a simple means of linking the microprocessor to all of our external devices. The MI^2 interface is a modular device that allows the user to combine 1–10 independent modules into a single package. This configuration affords the user total flexibility. In our system we have four modules to drive our optical bench and two modules to capture neural signals, which leaves four additional empty slots for future expansion. MI^2 offers a wide range of modules to suit many different applications. A description of the modules used in our system is given in our discussion of the optical bench and our methods for data acquisition. The modules include two ramp generators, which drive the x and y galvanometers, a level discriminator with window logic for spike detection, a pulse catcher that interrupts the computer following the occurrence of a spike, eight digital outputs to drive stepping motors, and an analog-to-digital converter (ADC) to receive signals from the remote joystick and potentiometers.

Optical Bench

Hubel and Wiesel (1962) reported that neurons in the visual cortex were excited by stationary or moving bars of light presented at a particular orientation. To generate such a stimulus, it is necessary to control the bar length and width, ori-

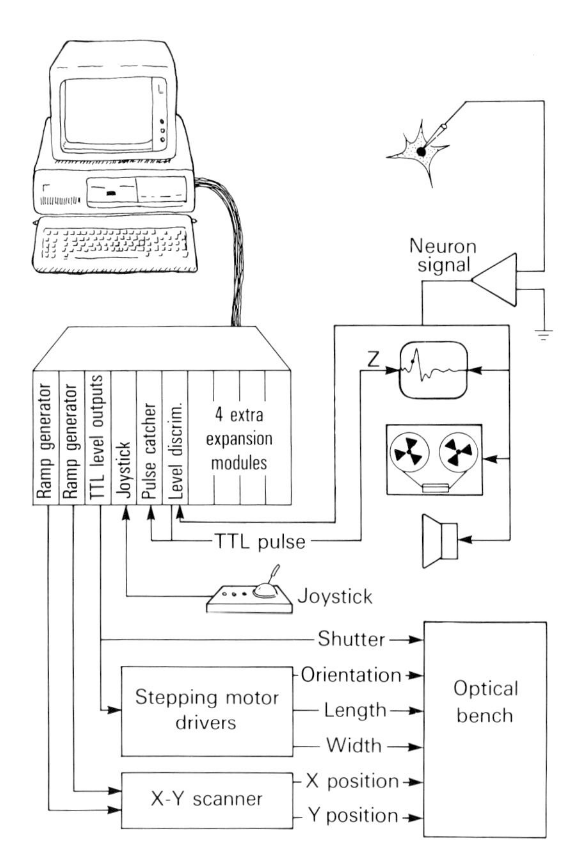

Figure 1. Illustration of system hardware for quantitative analysis of the receptive field organization of visual neurons.

entation, position, and velocity. These parameters are controlled by a computer-driven optical bench, designed and built by the Institute of Neurological Sciences machine shop at the University of Pennsylvania. Our optical bench, illustrated in Figure 2, consists of a light source, a dark source, a variable slit, stepping motors, a lens, a shutter, and two mirrors controlled by galvanometers. The length and width of the bar are controlled by a variable slit driven by two independent stepping motors. The variable slit is mounted in a holder that can be rotated by a third stepping motor to produce changes in orientation. The computer controls the movements of the three stepping motors by using six TTL-level outputs housed in one module of our interface (see Figure 1). Two TTL-level outputs are required for each stepping motor to produce movement in the forward and reverse directions. We can project a bar of light anywhere in the visual field by reflecting it off two mirrors that are mounted on the shafts of galvanometers. The galvanometers are positioned at right angles to one another, to give independent movement in horizontal and vertical directions, and are driven by two of the modules in our interface. These modules are general-purpose digital-to-analog (D/A) function generators with 12-bit accuracy. Each function generator houses a 4-KB on-board RAM that can be loaded with an arbitrary set of numbers to produce any signal. Once loaded, this RAM buffer can be dumped point by point at a user specified clock rate. For most of our experiments we fill the buffer with a ramp function to drift the stimuli. The clock rate sets the velocity of the drift. Once the function generator clock is enabled the buffered function is executed automatically, freeing the main computer for real-time data acquisition,

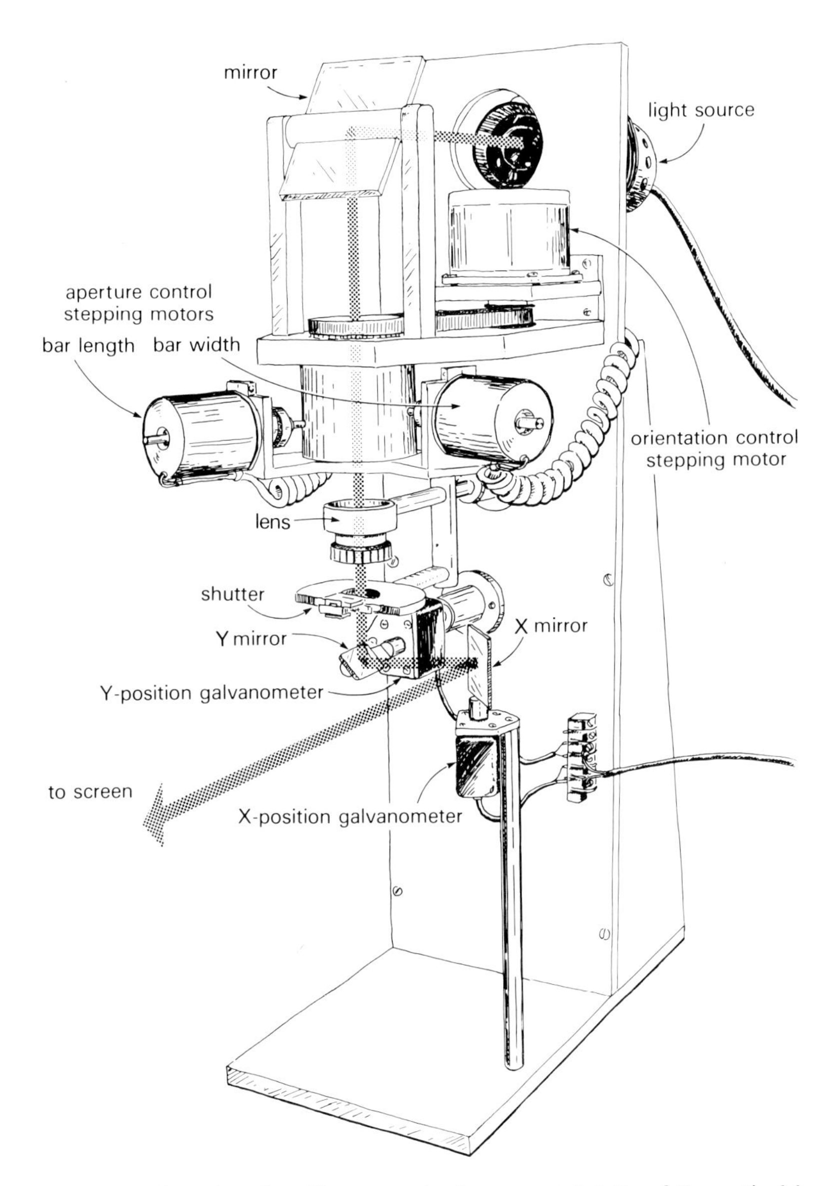

Figure 2. Drawing that illustrates the important details of the optical bench. Note light path.

graphics display, and other duties. Finally, we can turn the bar of light on or off using a shutter that is mounted between the variable slit and the galvanometers. The shutter is opened and closed by the computer using a single TTL-level output housed in the interface.

A joystick and three potentiometers provide complete remote control of the optical bench (i.e., x–y position, length, width, and orientation adjustments) from a single control panel. Voltages from the remote joystick and potentiometers are fed to an analog-to-digital (A/D) module in the interface. This module provides 12-bit accuracy in each of 16 A/D lines. For our purposes we use only five lines: two for the joystick to control x–y position and one for each pot to control the length, width, and orientation of our stimulus.

In summary, our optical bench provides a simple means for controlling the x–y position, length, width, and orientation of a bar of light. For visual physiology this system simplifies many of the problems associated with stimulus presentation and control. For example, when plotting the spatial organization of cortical receptive fields, it is customary to use oriented bars of light. Using this optical bench one can quickly scan the visual field and adjust the stimulus parameters for optimal stimulation.

Although our optical bench works well for our experiments, there are two limitations to the design. First, it is only possible to project light stimuli. In certain situations, dark stimuli may be preferred. Secondly, the optical bench is limited to projecting rectangular-shaped stimuli. It would be difficult if not impossible to modify the existing design so that more elaborate stimuli, such as sine or square wave gratings, could be projected. Innisfree (Cambridge, MA) manufactures an instrument that overcomes the limitations of our present optical bench. This instrument has the ability to present, on the face of an oscilloscope, a large repertoire of novel stimuli including both light and dark stimuli and sine or square wave gratings. We anticipate that the next generation of experiments will employ specialized devices like the Innisfree CRT image generator coupled with general-purpose machines like our computer and interface.

System Software

Language

When writing large programs it is advantageous to use "top–down" programming techniques that employ a structured, modular design. Before beginning, however, careful consideration should be given to the choice of a programming language. As a programmer, you should never adopt the attitude that you can rely upon a single programming language with which you have experience to solve *any* problem. Different languages offer different forms of expression that are more suited to particular tasks. The following considerations prompted us to write our system using the C programming language.

C is a highly consistent and structured language. Embodied in its syntax is a logic that is tailored to the problem of breaking down large program designs into many small pieces or modules that can be easily linked in any fashion. This is the primary objective in top–down programming. C is thoroughly documented in a book (Kernighan and Ritchie, 1978) that provides a formal definition of the language and thus encourages implementation that follows a standard syntax. This has contributed to the portability of programs written in C. The result is that programs written on one machine are easily transferable to other machines. A good C compiler will produce object or assembly code that performs efficiently and fast. Finally, it is relatively easy to combine your own assembly language routines and C when you require code running in real time, as our system requires.

Software Design

We have three primary objectives in designing the system software:

1. To design a routine that provided on-line graphic visualization and quantitative measurements of visual receptive fields.

2. To design a routine that enabled the user to have complete manual control over the optical bench using the remote joystick and pots.
3. To provide routines for storing data either on disks or on paper for future off-line analysis.

In addition to these objectives, we felt it equally important that the system implementation should be user friendly and easy to use in order to provide instant feedback of the program's status and simple methods for user interaction. Our overall software design is diagramed in Figure 3.

To produce on-line "pictures" of visual receptive fields we have adopted the peristimulus time response plane technique, first developed by Stevens and Gerstein (1976). With this technique we are able to visualize and quantify the spatiotemporal distribution of excitatory and inhibitory responses of visual neurons to a small spot of light turned on and off. The response plane technique is a logical extension of the technique employed by Kuffler (1953) but in our case the computer assumes control of the stimulus presentation sequence and averages the neural responses.

Conceptually, in order to generate a peristimulus time response plane, small bars of light are turned on and off by the computer in as many as 40 spatially adjacent but nonoverlapping positions within a visual receptive field. The bar of light is controlled so that it follows a path through the receptive field center perpendicular to the optimal stimulus orientation. Each time the computer presents the bar of light it captures and compiles, in the form of a histogram, the neural discharges that occur as a function of time for a given spatial position of the bar. These histograms are then stacked to form the response plane. The coordinates of the plane are as follow: peristimulus time on the x axis, spatial position of the bar on the y axis, and amplitude of response at each spatiotemporal coordinate represented as height above the space–time plane. In Figure 4 we illustrate this method as it was used to plot the receptive fields of two neurons from the dorsal lateral geniculate nucleus in the thalamus.

The software used to generate response planes is composed of three main routines. These include a data acquisition and graphics display routine, a parameter control routine, and an interrupt routine. Figure 3 illustrates the relationship of

Figure 3. Block diagram of system software.

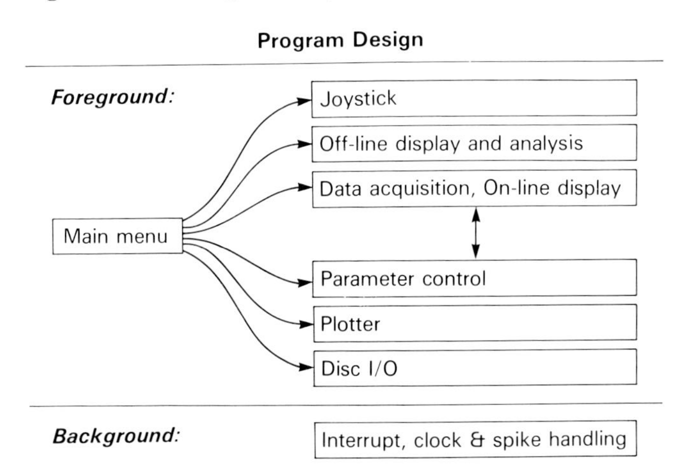

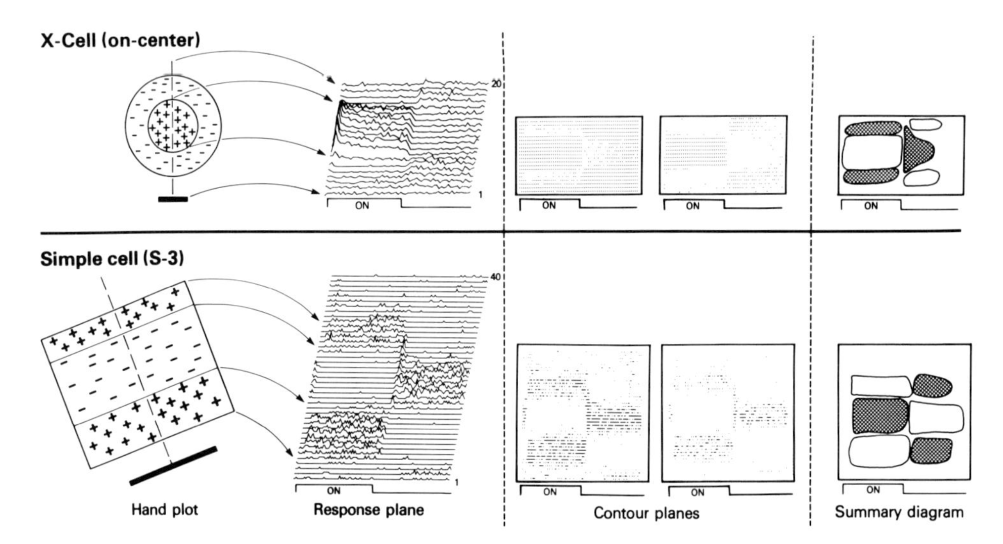

Figure 4. Illustration of the response plane technique as we have applied it to receptive fields in retina and striate cortex. *(Top)* Hand plot and response plane from a retinal *X*-cell. A small bright bar was turned on (640 msec) and then off (640 msec) at each of 19 positions on a path traversing the receptive field center. At each position, the unit discharge was accumulated into a PST histogram and the 19 histograms (plus 1 spontaneous) stacked to form a plane. Coordinates of the plane are time (left to right), space (vertically), and the probability of discharge is given by the height above the base at each space–time coordinate. Contour planes shown to the right are useful for detailed examination of the shapes of various response components. A summary diagram at the far right shows regions of relative excitation and inhibition in schematic form (excitation clear, inhibition shaded). *(Bottom)* Hand plot and response plane from a simple cell in layer IV of cat striate cortex. Exactly the same approach as with the retinal cell but great care was taken to use the optimal orientation. This was an S-3 cell since the response plane revealed three distinct, non-overlapping excitatory responses, one at light off and two at light on. In addition, three well-defined regions of relative inhibition may be observed.

these three software routines to the overall design of our system. Note in Figure 3 that the system is designed so that the bulk of the software operates in foreground while the interrupt routine operates in background. This design is well suited for data acquisition in real time. The foreground is reserved for the execution of software that does not critically depend upon time or special requests from devices that are external to the central processor. The background, on the other hand, is reserved for short routines that service external devices. Since our program must keep track of the time course of neural events following the presentation of a bar of light, we have placed special priority and emphasis on the execution of the interrupt routine by placing it in the background. It is essential that the interrupt routine is always ready to service a programmable external clock and keep track of time with millisecond accuracy. In addition, the interrupt routine is responsible for testing for the occurrence of neural events.

The data acquisition and graphics display routine is the heart of the response plane program. During foreground program execution, this routine is primarily responsible for continuously updating the response plane display and testing the keyboard for keystrokes. The background interrupt program keeps track of time and tests for neural responses in the form of action potentials. Upon initialization, this routine sets all of the data buffers to zero and prepares the optical bench for the beginning of the stimulus sequence. The stimulus sequence then follows. This sequence involves the following steps:

1. With the shutter closed, the bar of light is moved to the first spatial position by the galvanometers.
2. A programmable interrupt clock is enabled with 1 msec interrupts (this programmable clock comes as part of the MI^2 interface hardware).
3. The shutter opens for a period of N msec (default 640) during which time the foreground data acquisition routine continuously updates the response plane display, while the background interrupt service routine, executed with each tick of the clock, keeps track of time and tests for the occurrence of action potentials.

Action potentials are analyzed in the following manner. Amplified neural signals are fed to a *spike sorter* in the MI^2 interface. The output of the spike sorter, which is conditional on level crossings on the action potential, sets a bit in a flip-flop register housed in the pulse catcher interface module. The pulse catcher module interrupts the computer and summons the interrupt service routine, which records the event.

4. The shutter closes for a period of N msec (default 640). The foreground routine continues to update the display while the background routine keeps time and tests for action potentials.
5. The interrupt clock is disabled after the light-off period. With the shutter closed, the galvanometers reposition the bar of light to the next spatial position.

At this point the entire stimulus sequence is repeated from step 2 until all positions have been tested (up to 40). Once this is completed the sequence is automatically recycled, starting the new pass through the receptive field. With each new pass, the neural responses are added to the responses collected in the previous pass. In addition to these tasks, the data acquisition and graphics display routine are also responsible for monitoring the keyboard for commands, issued by the operator, to stop or suspend ("pause") data acquisition or alter the graphics display.

The parameter control routine allows the user to perform any last-minute adjustments to the length, width, or orientation of the stimulus before it is used for data acquisition. Furthermore, this routine allows the user to input parameters, such as the duration of the light on and off periods, the number of positions to be tested, and the extent of the visual field to be tested in degrees of visual angle. Other information such as the time, date, and any special comments is also handled here.

Before generating response planes, the stimulus must be located at the appropriate spot in the visual field. To accomplish this, visual receptive fields must be hand plotted by manually controlling the optical bench with the remote joystick and pots. A separate software routine is used for manual control of the optical bench. This routine is always executed before the acquisition of response planes. The joystick routine operates as follows. Twelve-bit information from the A/D samples of the remote joystick is routed to the 12-bit D/A function generators that control the galvanometers. Twelve-bit information is also routed from the pots to the TTL-level outputs that control the stepping motors. The consequence of this arrangement is that all of the stimulus parameters (i.e., length, width, orientation, and x–y position) are continuously monitored and can be stored at any time for use by the routines that generate the response planes. Because of this feature, it is not necessary to key in any of the stimulus parameters manually.

The rest of the system software is dedicated to storing and plotting actual data so that it can be analyzed in greater detail off-line. This support software is very standard and therefore requires no further discussion.

Creating a friendly "user environment" is as important as getting a program to do what you want it to do. We employ a number of commonsense devices to make our system friendly and easy to use. For example, the foreground program flow is designed to allow the user to enter any routine through one main branch point (see Figure 3). It is never necessary to follow some convoluted maze to enter and exit a part of the system. Messages, prompts, or menus are made clear but not overstated. If a user response is required, it usually can be entered by striking a single key. Most importantly, all routines provide continuous feedback of their status by updating displays in real time or producing sounds over a programmable speaker. We believe that continuous feedback is very important to the user. A program that does not appear to be doing something leaves the user with the same uneasiness that is felt when turning on a radio and hearing silence.

Conclusions

We have explained the principles of our computer controlled visual physiology system that enables us quickly to plot and quantify the spatiotemporal distribution of excitation and inhibition in the receptive fields of visual neurons. Our system provides the following:

1. on-line graphics visualization of receptive fields;
2. a simple means for controlling an optical bench either automatically or manually using a remote joystick and pots; and
3. support software to store and plot the data.

This and similar systems have been used successfully to provide detailed quantitative descriptions of the spatial organization of visual receptive fields in the retina (Stein et al., 1983), lateral geniculate nucleus (Bullier and Norton, 1979; Stevens and Gerstein, 1976), and visual cortex (Palmer and Davis, 1981; Mullikin et al., 1984).

Until a few years ago, such a system was beyond the reach of most visual physiology laboratories because its development required the demanding skills of electrical and mechanical engineering. Even if this talent had been available, the cost of building such a system was beyond the budgets of most small laboratories. Today all of this has changed. One can now purchase, at very reasonable prices, all of the electronic equipment required for building our system. Even the operation of the optical bench can be electronically simulated and enhanced using specialized devices like the Innisfree CRT image generator.

Although the hardware is readily available, specialized software for these applications has not been accessible. For this reason, many physiologists have spent too much of their valuable time designing and writing software to make their systems perform. Recently, however, MI^2 has made available a software interface language that is compatible with IBM personal computers and MI² laboratory interfaces. This software language augments the easy to use BASIC interpreter by providing support for the interface. As a result, many of the tedious steps involved in programming the interface can be easily condensed into a few simple statements, making programming time more cost effective. Clearly this represents an important initial step in development of industrial software to augment the advanced hardware available in the minicomputer market. It is now possible for software suppliers to focus their attention on making the operation of computers and computer peripherals easier.

We are deeply indebted to Michio Fujita, Fred Letterio, and Ed Shalna for technical matters. Credit for much of the original system design belongs to J. K. Stevens, G. L. Gerstein, and L. A. Palmer to whom we are indebted. We also wish to thank Robert Smith for freely giving his time to help trouble shoot the system during development.

This work was supported by the National Science Foundation Grant BNS 81-19839, Sloan Foundation Grant BR-2189, and National Eye Institute EY-04638.

T. L. Davis was a Sloan Foundation Research Fellow.

References

Bishop, P.O., J.S. Coombs, and G.H. Henry (1971) Responses to visual contours: Spatiotemporal aspects of excitation in the receptive fields of simple striate neurons. J. Physiol. (Lond.) 219: 625–657.

Bullier, J., and T.T. Norton (1979) *X* and *Y* relay cells in cat lateral geniculate nucleus: Quantitative analysis of receptive field properties and classification. J. Neurophysiol. 42: 244–273.

Hubel, D., and T.N. Wiesel (1962) Receptive fields, binocular interaction and functional architecture in the cat's visual cortex. J. Physiol. (Lond.) 160: 106–154.

Kernighan, B.W., and D.M. Ritchie (1978) The C programming language. Englewood Cliffs, NJ: Prentice-Hall.

Kuffler, S.W. (1953) Discharge patterns and functional organization of the mammalian retina. J. Neurophysiol. 16: 37–68.

Mullikin, W.H., J.P. Jones, and L.A. Palmer (1984) Receptive field properties and laminar distribution of *X*-like and *Y*-like simple cells in cat area 17. J. Neurophysiol. (in press).

Palmer, L.A., and T.L. Davis (1981) Receptive field structure in cat striate cortex. J. Neurophysiol. 46: 260–276.

Shiller, P.H., B.L. Finlay, and S.F. Volman (1976) Quantitative studies of single cell properties in monkey striate cortex. I. Spatiotemporal organization of receptive fields. J. Neurophysiol. 39: 1288–1319.

Stein, A., W.H. Mullikin, and J.K. Stevens (1983) The spatiotemporal building blocks of *X*-, *Y*-, and *W*- cat retinal receptive fields. Exp. Brain Res. 49: 341–352.

Stevens, J.K., and G.L. Gerstein (1976) Spatiotemporal organization of cat lateral geniculate receptive fields. J. Neurophysiol. 39: 213–238.

21

Ophthalmic Electrophysiology with a Microcomputer-Based System

Robert Massof, Bruce Drum, and Michael Breton

Introduction

Gross electrophysiological testing of the visual system is accomplished through three testing procedures: (1) electroretinogram (ERG), (2) visual evoked response (VER), and (3) electrooculogram (EOG). The ERG and EOG measure responses from the retinal elements of the eye, whereas the VER records potentials from the visual cortex of the brain.

The ERG has been described as a transient retinal potential that is initiated by a change in the light energy falling on the receptors (Armington, 1974). Armington points out that this apparently simple response (see Figure 1) is in fact exceedingly complex. Early analysis of waveform origins was limited by relatively slow and insensitive instrumentation. Following the pioneering work of Einthoven and Jolly (1908) and Piper (1911), Granit (1933) laid out an analysis of waveform components that is still valid today. Through a series of procedures, he was able to demonstrate the approximate physiological origins of three separate components (PI, PII, PIII) and show how these combined to produce the waveform observed in the clinical ERG. More recent analysis (Rodieck, 1972) has yielded a more detailed knowledge of the electrical properties of retinal layers, but has not contradicted Granit's basic findings.

The normal ERG has an amplitude of several hundred microvolts, and can thus be recorded using amplifiers of only moderate gain. Typical procedures use AC coupling to obtain a stable baseline and use an upper bandpass limit of about 1,000 Hz and a lower limit near 1 Hz. Before inexpensive microcomputers became available, recordings were typically displayed on an oscilloscope screen and photographed for archival storage. Waveform analysis was limited to measurements of the photographic image. In other words, progress in research and clinical use of the gross ERG was limited not by the recording technique and sensitivity, but by the lack of information processing capability.

The EOG was discovered very early in the history of electroocular recording (du Bois-Reymond, 1849; Dewar and M'Kendrick, 1873). Like the ERG, it depends on the integrity of retinal structures. Unlike the ERG, however, the EOG is the recording of a very slowly changing standing potential in the eye. This potential has an amplitude of several millivolts, with the cornea positive relative to the posterior pole. The EOG is typically recorded by placing electrodes near

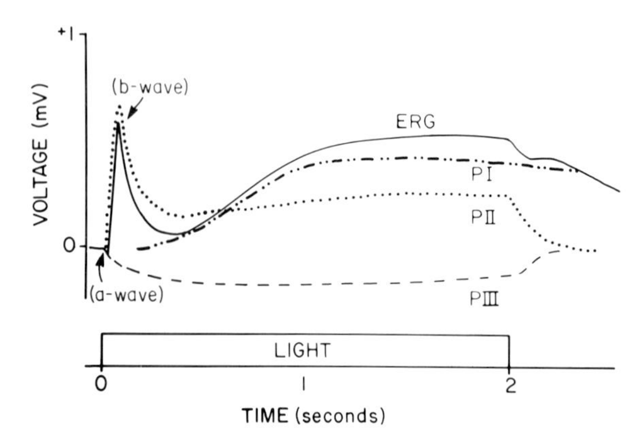

Figure 1. Component analysis of the cat ERG showing the *a*-wave and *b*-wave of the composite waveform and the three component potentials, PI, PII, and PIII, which disappear sequentially in ether narcosis. The *a*-wave and *b*-wave are the important clinical parameters. [After Granit (1933).]

the canthi on either side of the eye and recording the relative change of potential as the eye moves a prescribed lateral distance (Arden and Kelsey, 1962). The clinical significance of the EOG derives from the fact that the standing potential of the normal eye changes gradually in response to light and dark adaptation (see Figure 2). Diseases that affect the retinal pigment epithelium (RPE) and other retinal structures reduce the EOG responsiveness to light and dark adaptation.

Unlike either the ERG or the EOG, the VER is not easily recorded, even with sensitive amplifiers, and has a relatively short history of clinical use. The VER is a low-amplitude potential (typically 10–20 μV), recorded from the primary visual cortex through the scalp and skull by a surface electrode. Background electrical noise levels emanating from brain structures and other physiological sources are comparable in magnitude to the evoked signal. Clear recording of evoked signals was not easily accomplished until digital averaging techniques were introduced in the 1960s (see Figure 3). Early systems were relatively inflexible and required photographic records of display screens for archival storage and data analysis. However, more than in the other areas of electrophysiology, computer-aided

Figure 2. *(Top)* Saccadic eye movements recorded with the EOG. The alternation frequency of the fixation lights is 1.5 Hz. *(Bottom)* EOG amplitudes are plotted as a function of time for the right (●) and left (○) eyes of a normal individual. The normal EOG shows a minimum amplitude after 18 min of dark adaptation and a maximum amplitude 7 min after turning on the background light.

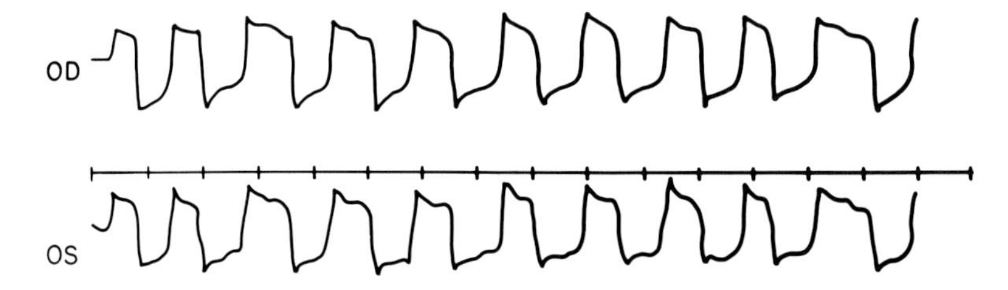

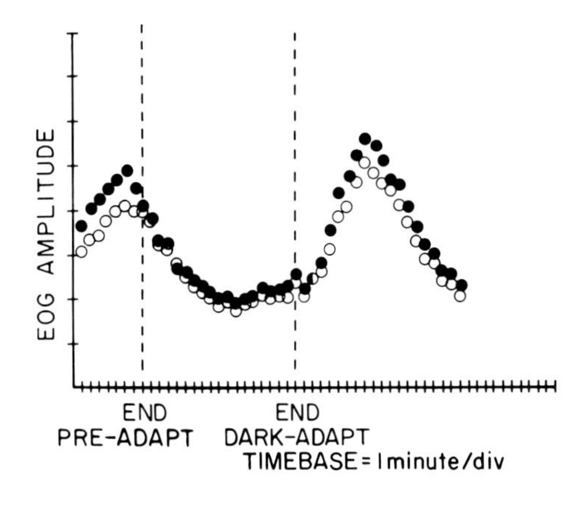

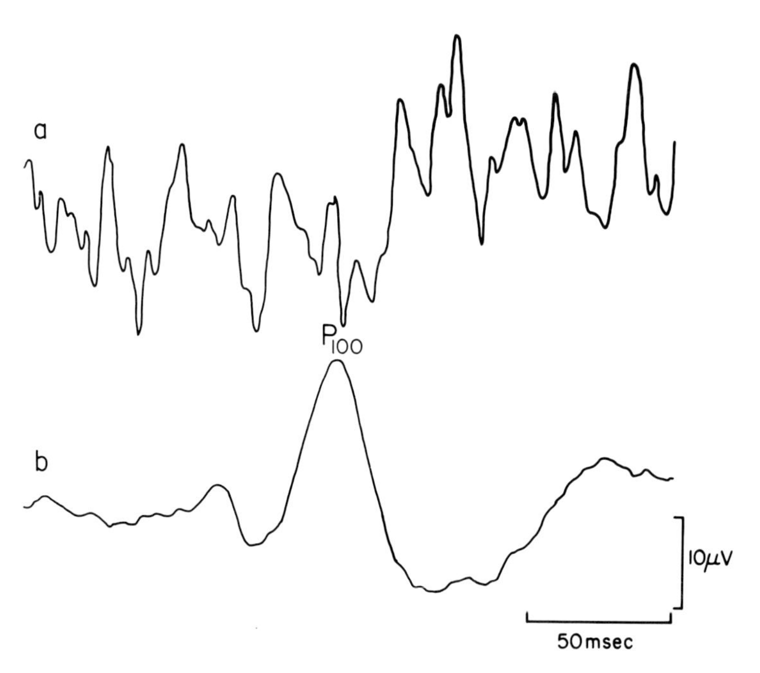

Figure 3. VER raw waveform **(a)** and average of 100 raw waveforms **(b)** to a checkerboard stimulus alternating at 2 Hz. The latency of P_{100} is the important clinical parameter.

analysis and stimulus control have been employed with the VER (see Bodis-Wollner, 1982).

Computer-Aided Recording

The introduction of inexpensive, general-purpose microcomputers has opened new possibilities for data handling in electrophysiology. Digital data handling may now include archival disk storage and later recovery, thus making possible sophisticated data analyses such as waveform additions and subtractions, waveform smoothing, Fourier transforms, and digital filtering. In addition, automatic stimulus control and data acquisition make it possible to specify entire test protocols with preset software. The use of manual overrides gives complete protocol flexibility for all critical program features.

Major advantages of a microcomputer-based system include:

1. complete digital data handling after acquisition;
2. archival digital storage of complete waveforms on disk media;
3. analysis of archived waveforms;
4. automatic stimulus presentation and protocol control;
5. manual keyboard override of protocol features;
6. ability to modernize obsolete procedures by updating software;
7. automatic calibration of stimulus conditions;
8. patient report preparation; and
9. modem transfer of data between systems.

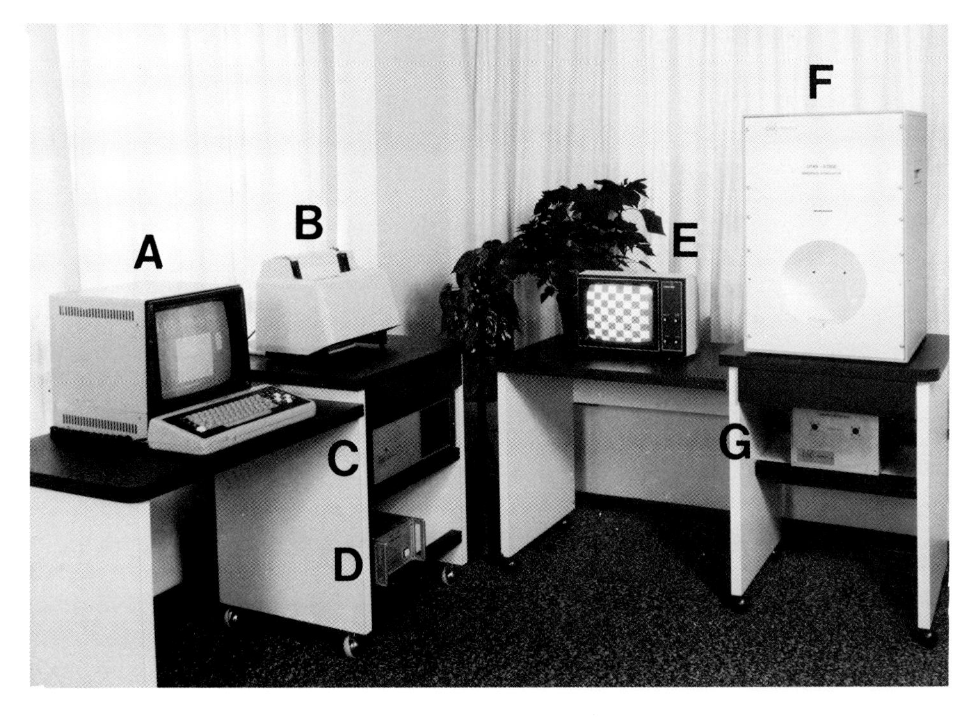

Figure 4. Main components of UTAS-E 1000 system: (**A**) CRT graphics display and keyboard, (**B**) dot-matrix printer (the printer shown is the IDS Paper Tiger, since replaced by the IDS Prism-80 described in the text), (**C**) microcomputer and disk drives, (**D**) power supply, (**E**) pattern stimulator, (**F**) Ganzfeld stimulator, and (**G**) interface box containing recording circuits, amplifiers, and ADCs.

This chapter describes a specific system known as the UTAS-E 1000* and gives the rationale behind the various features included in the system's design and operation.

Hardware Description

The UTAS-E 1000 (Universal Testing and Analysis System—Electrophysiologic 1000), shown in Figure 4, consists of a microcomputer, high-resolution graphics subsystem, dot-matrix printer, Ganzfeld and pattern stimulators, and four-channel recording apparatus. The Ganzfeld stimulator, amplifiers, and recording systems were designed and built by LKC Systems to meet specifications required for ophthalmic electrophysiologic recording. The peripheral devices were chosen to satisfy the following requirements:

1. Provide maximal display resolution and flexibility.
2. Make minimum demands on the microcomputer.

---

*UTAS-E 1000 is manufactured and distributed by LKC Systems, Inc. of Gaithersburg, MD, and was designed and programmed by the authors in cooperation with LKC. This chapter describes only the ophthalmologic capabilities of the UTAS system. The system can also be used for a variety of other neurologic tests, including auditory evoked potentials, somatosensory evoked potentials, auditory brainstem potentials, peripheral nerve conduction velocities, and electromyograms.

3. Be relatively inexpensive.
4. Be inexpensively serviceable by local computer dealers.
5. Have established reputations for reliability.

Computer and Peripheral Devices

Computer. The heart of the UTAS system is an 8-bit North Star Horizon computer with a Z80 microprocessor and 64 KB ($1K = 2^{10} = 1{,}024$) of random access memory (RAM). The Horizon was chosen because of its reputation for reliability and because it employs the S-100 bus, for which a wide variety of compatible circuit boards is available. The Horizon has two 350-KB (formatted) $5\frac{1}{4}$-inch quad-density floppy disk drives and controller, two RS232 serial ports, one input and one output 8-bit parallel port, a real-time clock communications baud-rate headers, and 80-bit device addressing on the bus. The Z80 microprocessor cycles at 4 MHz and the real-time clock provides vectored interrupts at 300 Hz.

Graphics System. High-resolution screen and printer graphics are controlled by a special S-100 board (Microangelo 512, Scion Corporation) that contains a Z80 coprocessor, 32 KB of RAM, a read-only memory (ROM) firmware package, and video graphics buffers and controllers. The video display terminal is a Ball monitor (RD 150) with a slow green phosphor (P39), and a maximum display resolution of either 480 vertical by 512 horizontal pixels or 40 lines of 85 alphanumeric characters. Graphics commands are sent by the North Star Z80 to the Microangelo board, which then handles all graphics functions. Also, the contents of the graphics display buffer on the Microangelo board can be read by the North Star Z80.

A 72 key parallel-input keyboard that is standard with the Microangelo subsystem is used for character input and system control commands. To facilitate program control, a separate matrix of 16 specially labeled command keys is provided in addition to the usual alphanumeric and control keys. The command keys generate ASCII codes that are redundant with other characters. Left and right arrows are also added for moving software-controlled cursors, and a "HELP" key is provided for calling user-training instructions.

Printer. To provide high-quality hard-copy alphanumeric and graphics output, an IDS Prism-80 graphics printer (Integral Data Systems, Inc.), also with its own Z80 coprocessor, 1.5-KB RAM buffer and ROM firmware, is interfaced to the Horizon Z80 through a serial port. The printer has a high-resolution print head that prints 150 characters per second (CPS) in normal mode and 200 CPS in lower-resolution draft mode on standard 8.5×11-inch paper. The firmware also provides a variety of other special printing features such as proportional spacing, margin justifications, and different character sizes.

Stimulators

The E-1000 is designed as a multiple-purpose machine capable of performing ERG, EOG, and VER testing. An automatic Ganzfeld (40-cm diameter integrating sphere) stimulator, shown in Figure 4F, is used to present global flashed stimuli for ERG and flashed VER testing. It also includes a central fixation light, and two eccentric lights mounted 15° to either side of center (i.e., 30° apart) for use as EOG fixation targets. Pattern VER and ERG stimulation are provided on a separate video display terminal (VDT), shown in Figure 4E.

Ganzfeld. Flashes in the Ganzfeld are produced by one of two light sources: (1) a standard Grass PS-22 photostimulator or (2) a Vivitar Photoflash photographic strobe for "bright-flash" requirements. The light sources are mounted to provide reasonably uniform illumination inside the sphere. The bright-flash unit is mounted on a carriage that rides on a track and is moved, when needed, into position over the aperture by a stepper-motor-driven pulley system.

Control over flash illuminance level is provided by neutral density filters mounted on two overlapping stepper-motor-driven filter wheels between the flash units and the aperture in the diffusing sphere. The two filter wheels provide for stimulus attenuations ranging from 0.0 to 4.8 log units in 0.2-log-unit steps. Since the Grass photostimulator nominally provides up to 3.4 cd-sec/m^2 and the bright-flash unit provides up to 3,400 cd-sec/m^2, the use of both sources and the neutral density filters allows intensity to be varied over a range of nearly 8 log units.

Ganzfeld background illumination is provided by a tungsten source that illuminates a diffusing glass over a second aperture at the top of the sphere. Maximum nominal background luminance is 170 cd/m^2 with a correlated color temperature of 2,800° K. Background luminance can be attenuated to 85, 34, or 17 cd/m^2 by changing aperture size with a stepper-motor-driven attenuator.

Calibrations of stimulus and background luminances are carried out automatically by a photodiode mounted at the base of the Ganzfeld diffusing sphere whenever the system is turned on. Three red light-emitting diodes (LEDs) are mounted on the posterior surface of the interior of the diffusing sphere. The central LED serves as a fixation point for ERG and VER testing, while two flanking LEDs may be alternately flashed by the computer to direct eye movements for EOG recording.

Pattern Stimulator. The VDT that provides pattern stimulation has a maximum luminance of 440 cd/m^2 and a maximum contrast of 93%. Patterns consist of alternating checkerboards or bars, and are generated by a two-page video graphics board (ALT-512, Matrox Ltd.) that plugs into the bus. Resolution is 256 × 256 pixels for each page of memory. The North Star CPU initially loads each page of memory with the proper bit pattern for the desired display and achieves pattern alternation by alternately displaying the two pages of memory synchronized with the VDT under software control. Software control of pattern generation allows great latitude of pattern types with the limitation that contrast cannot be manipulated (e.g., sine-wave gratings cannot be produced).

Recording Apparatus

Four physiological signal recording channels are provided. Electrodes are plugged into two cables that can be clipped to the patient's clothing. Each cable inputs two signals and a ground into the system amplifiers. The gains, low-pass filters, high-pass filters, and 60-Hz notch filters of the four amplifiers are set by the computer. The amplifier gain and filter settings are made in pairs; therefore channels 1 and 2 share the same amplifier parameters, as do channels 3 and 4. Within the software, each of the four amplifiers is called a *channel,* and each pair of amplifiers is considered to be a single amplifier for purposes of setting gain and filter parameters.

Each amplified signal is input into a different channel of an 8-channel multiplexed analog-to-digital converter (ADC) (LKC Systems). The ADC, which resides on the S-100 bus, has 12-bit resolution with a status bit for the data-ready signal, and each of the two bytes of data is separately addressed on the bus.

Program Operation

Development

All software for the UTAS-E 1000 was written in assembly language in order to obtain maximum speed and most efficient use of memory. However, during software development, the major components of the UTAS-E 1000 programs were first written in BASIC. This step facilitated the design and testing of display formats, program flow patterns, and hardware interfacing, as well as serving to simplify the task of developing and debugging algorithms for the analysis programs (e.g., fast Fourier transform). Once finalized, the BASIC programs were translated to assembly language.

The assembly language programs were written, assembled, and debugged under the CP/M operating system; however, the programs are all designed to execute under Northstar DOS. We chose Northstar DOS because it requires very little memory (100h–E00h),* whereas CP/M uses 00h–100h for BIOS and the top 20 KB for FDOS.

Design

The assembly language programs are organized into six separately loaded modules: (1) UTAS, (2) PROTOCOL, (3) ERGVER, (4) EOG, (5) ANALYSIS, and (6) REPORT. The memory allocations for these program modules and data buffers are illustrated in Figure 5.

The UTAS module contains subroutines for initialization, binary-to-ASCII and ASCII-to-binary conversion, binary multiplication and division, interrupt servicing, analog-to-digital conversion, input and output, graphics, interface control, disk read and write, and generation of option menus. The PROTOCOL module contains subroutines for collecting patient information, generating electrode maps on the graphics screen, generating protocol checklists, following protocol steps, processing protocol overrides, and controlling stimulus parameters. It also contains a text editor for the creation of REPORT files (see sections on Disk Files and Patient Reports below). The UTAS module and the first section of the PROTOCOL module remain in memory except when overwritten by the ANALYSIS or REPORT modules (see Figure 5).

The ERGVER module contains all routines required for data collection and buffering, signal averaging, waveform display, cursor operations on waveforms with corresponding amplitude and time alphanumeric displays, graphics printing, and command key processing. For ERG and VER recordings, all the data are

*h = hexidecimal.

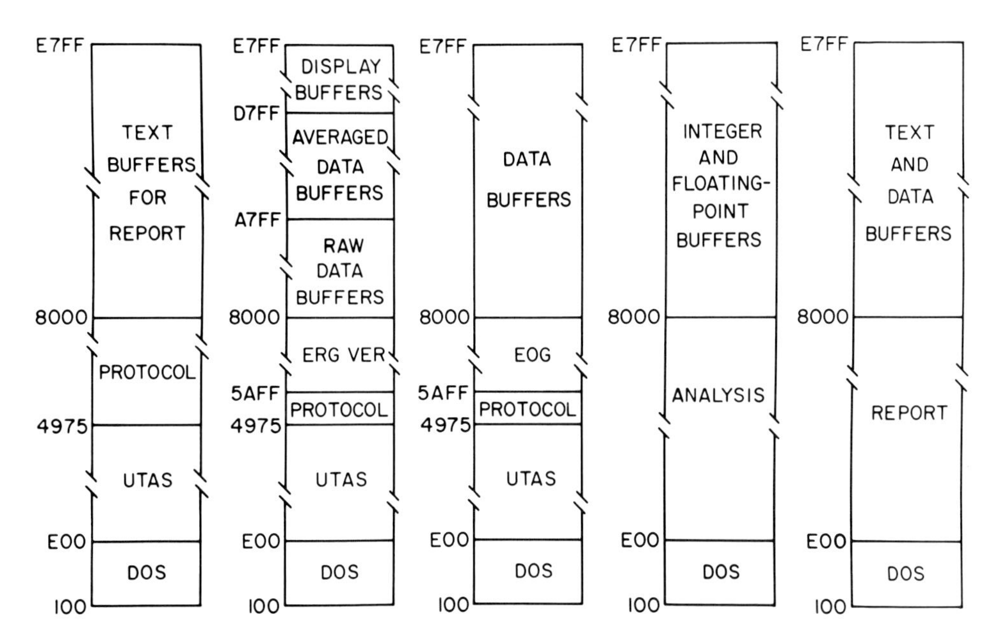

Figure 5. Memory maps for the six assembly language program modules of the UTAS-E 1000 system. The five columns, from left to right, represent the organization of computer memory for (1) start-up procedures and protocol selection, (2) ERG and VER recording, (3) EOG recording, (4) data analysis, and (5) report generation.

acquired and buffered before they are displayed on the graphics screen. This sacrifice of real-time data display is necessary because each data point must be transferred to the graphics board coprocessor through a hand-shaking routine. Therefore, the graphics display is slow relative to the data sampling rates. In the case of signal averaging, the user is given the option of suppressing the raw data display in order to shorten the time of data acquisition. However, the user is able to view an updated average at any frequency desired.

For ERG and VER recordings, the double-precision data from the ADCs are clocked-in using a calibrated timing loop. A timing loop is used because the real-time clock is too slow for the data rates normally employed (300 Hz to 2 KHz). However, interrupts from the real-time clock are used to control stimulus presentations (e.g., Ganzfeld flashes or pattern reversals). The raw A/D conversions from the four recording channels are stored in four double-precision 1*K* buffers (2 KB per buffer). All four channels are sampled and stored irrespective of the number of channels that are displayed. For signal averaging, the contents of the 1*K* double-precision raw data buffers are added to the contents of 1*K* triple-precision average data buffers. Thus, for ERG and VER recording the UTAS software reserves 8 KB of RAM for raw data storage and another 12 KB for average data storage.

During signal averaging the keyboard buffer is examined after every data sweep in order to trap manual interrupts. Upon interrupt, the user may implement one or more of the following options:

1. Examine the baseline.
2. Change averaging parameters.

3. View the updated average.
4. Change stimulus and/or recording parameters.
5. Store the most recent waveform on disk.
6. Change the artifact reject threshold.

If the stimulus and recording parameters have not been changed, signal averaging can continue from the point of interruption. Otherwise, the averaging must be restarted. If the artifact reject feature is selected, the minimum and maximum values are found in the raw data buffers before the raw data are added to the average data buffers. If the difference between the minimum and maximum values exceeds the user-defined threshold, the raw data buffers are discarded. When signals are rejected from the average, the graphics screen contrast is transiently reversed to alert the user.

The EOG module contains subroutines to record and display eye movements (EOG), to control stimulus parameters and presentation, to compute and graph average amplitudes, and to compute the Arden light/dark ratio. Also, there are special subroutines designed for examining and editing eye-movement data prior to the calculation of the Arden ratio.

For EOG recordings, the A/D values are sampled at 300 Hz using interrupts from the real-time clock. Software index counters advance during servicing of real-time clock interrupts and control events such as alternation of the fixation lights and timing of eye-movement and rest periods. The raw A/D values for each eye from 15-sec data collection intervals are stored in 4.5K double-precision RAM buffers. Thus, for EOG measures, 18 KB of RAM are dedicated to storage of the raw A/D values for the most recent eye movement recordings.

The UTAS-E 1000 is a turnkey system. This means that the UTAS software is automatically loaded and executed on system power-up or manual reset of the computer. The automatic start option of Northstar DOS was used to implement this turnkey feature. To return from the UTAS program to Northstar DOS, the user must either select "computer" from the first option menu (see below) or replace the UTAS system disk with a North Star system disk and manually restart the computer.

The UTAS software is designed to be sufficiently user friendly that a naïve user can collect data according to standard protocols without obtaining extensive prior training. Yet, it is sufficiently flexible that a more sophisticated user can override standard protocols or design new protocols.

A set of special command keys is largely responsible for the ease with which data can be collected, displayed, and processed. By pressing the appropriately marked command key, the user can do the following:

1. Change amplifier parameters.
2. Change the recording timebase (A/D sampling rate).
3. Display baseline responses.
4. Initiate recording of waveforms.
5. Adjust DC levels of displayed waveforms.
6. Initiate, interrupt, halt, or continue signal averaging.
7. Set artifact reject thresholds.
8. Erase displayed waveforms.
9. Call-up moveable cursors on displayed waveforms.
10. Print the current screen display on the graphics printer.

11. Store waveforms on disk.
12. Proceed to the next step in the protocol.

In addition, pressing the HELP key at any time interrupts program execution and displays a complete list of available options and their consequences.

Two examples serve to illustrate both the sophistication and the convenience of the command key operations. Waveform cursors can be used to show the amplitude and time delay of any point on a displayed waveform. When the CURSOR command key is pressed, two cursors consisting of vertical lines centered on the displayed data points are positioned on each waveform. An active cursor is designated by the user and moved by pressing right or left arrow keys. The active cursor is flashed by alternately erasing and refreshing the cursor line. As the cursor is moved along the waveform, the cursor line is erased and then refreshed at the next point (the waveform point is replaced after the cursor line is erased). If the shift key is pressed along with the arrow, the cursor slews 10 points along the waveform between refreshes. Buffer pointers track the cursor positions, and corresponding buffer values are displayed at the bottom of the screen as amplitude and time. The difference of the amplitude and time values for each cursor pair is also displayed.

When the STORE command key is pressed, the number of displayed waveforms is checked. If it is greater than 1, the user is prompted to identify which waveforms are to be saved. Depending on the type of waveform displayed, the entire raw data buffer (2 KB), averaged data buffer (3 KB), or floating-point data buffer (4 KB) is saved on disk along with the appropriate index and parameter information for the waveform. All disk bookkeeping is done by the software; the user must decide only how many and which waveforms to save.

Flow

Upon system start-up, the UTAS and PROTOCOL modules are read into RAM from disk (Figure 5) and a set of initialization routines is executed. The luminances of the Ganzfeld background light and the Grass photostimulator are automatically measured using a photodiode in the base of the diffusing sphere. The photodiode is connected to a system amplifier and read through the ADC. The measured luminances are displayed and the user is prompted for a command to advance to the next frame. The user is then prompted for the date and given the option to test the other system stimulators (EOG fixation lights, bright-flash stimulator, and pattern generator). Following these initial steps, all stepper-motor-driven devices are set to their home positions, the main option menu is displayed, and the user is asked to choose from the following options: electroretinogram, electrooculogram, visual evoked response, data analysis, report, or computer.

Following the choice of a specific option, the appropriate program module is read into RAM from disk. That is, if electroretinogram or visual evoked response is chosen, then the ERGVER module is read into RAM (Figure 5). A second level of menus is then displayed to permit the user to choose from among more specific options; for example, following the choice of electroretinogram the user then chooses from among standard, bright-flash, intensity-response, pattern, or operator-defined protocols.

For ERG, VER, and EOG recording, the user is prompted for alphanumeric patient information and then is walked through the respective protocol checklist. In addition, electrode placement maps are drawn on the graphics screen for the VER and EOG protocols. Following completion of the protocol checklist, the data display screen is created and the keyboard is polled for a command key response.

For ERG and VER recording, the command key options include recording and displaying either a sample baseline response (no stimulus presented) or a response to the stimulus. In either case, the waveform is recorded by sampling the ADC at a rate determined by the user-defined time base and all 1,024 double-precision values are stored in the raw waveform buffer. Once the entire waveform has been buffered, a display file is created that consists of every other data word in the raw waveform buffer, scaled to the range of the display screen, and the display file in turn is sent to the graphics board. All screen operations (e.g., DC adjust, scaling, erase feature) are performed on the display file through the use of command keys.

When a response to the stimulus is recorded, a trigger command is sent to the stimulator immediately prior to starting the A/D conversions. If the stimulus is periodic, the trigger commands are executed by real-time clock interrupt-driven subroutines. The raw or averaged waveform buffers for any or all of the currently displayed waveforms, along with their respective INDEX and PARAMETER file entries can be saved on disk by using the STORE command key. The STEP command key then advances the program to the next step in the protocol. If desired, however, the protocol can be overridden by using various control characters (e.g., control "C" to change the stimulus color).

For EOG recording, eye movement waveforms are constantly displayed in a scrolling window at the top of the screen (similar to Figure 2) while the alternating EOG fixation lights are presented to the patient. By pressing the RECORD command key the user begins the recording session, which consists of 15 sec of eye movement recording per minute for a period of approximately 30 min. If the STEP key is pressed, the elapsed time is displayed and the user is prompted for new adaptation period and stimulus presentation parameters. A 3.4-sec segment of the last eye movement record can be selected and stored on disk by moving a cursor-defined window to the desired location on the waveform and pressing the STORE command key. (Typically, eye movement waveforms are saved for the purpose of measuring saccadic velocities.) At the end of each 15-sec eye-movement recording period, the program finds the running minima and maxima in each record, saves them in the appropriate buffers along with the corresponding elapsed times, and computes the average EOG amplitude (peak-to-trough) for the last 10 sec of the record. (The first 5 sec are excluded to allow time for the patient's eye movements to synchronize with the alternating fixation lights.) The average amplitudes for each eye are plotted on a graph in the lower half of the screen (see Figure 2). At any time, the user can adjust the eye-movement display gain and DC levels. Also, a printout of the current screen, including eye-movement records, can be obtained by pressing the PRINT key.

Upon completion of the EOG session, the minimum amplitude during the dark period and the maximum amplitude during the light period are found for each eye and the Arden ratio (light maximum/dark minimum) is computed. The user can store on disk the buffers containing the maximum and minimum amplitudes for each minute of the session, and can recall and edit the buffers (i.e., remove artifacts) to compute new average amplitudes. Also, the user can position special cursors on the summary graph to choose new values of the minimum and maximum amplitude and then recompute the Arden ratio.

Storage Formats for Disk Files

The UTAS system provides floppy disk storage for digitized ERG and VER response waveforms, EOG summaries, results of waveform analyses, and detailed patient reports. This section describes the system of disk file organization and indexing that allows individual files of any type to be quickly identified and read from the diskette along with all related test and patient information.

The North Star disk operating system (DOS) divides the 350 KB of memory on each diskette into 700 consecutive 512-byte blocks numbered from 0 to 699. These blocks are physically organized as 70 concentric tracks (35 per side), each of which is divided into 10 equal sectors with one block per sector. The operating system reserves blocks 0–3 for a file directory that lists the file name, type, size, and starting block address of each file stored on the disk.

The UTAS program divides the remaining 348 KB of disk space into three main sections. Blocks 4–30 (13.5 KB) are reserved for a comprehensive file index of supplementary patient and test information for all the data and analysis files on the disk. Blocks 31–50 (10 KB) are reserved for a parameter file that contains numerical information about calibrations, stimulus settings, response parameters, and analysis parameters for each disk file. This leaves blocks 51–699 (324.5 KB) for up to 160 data and analysis files. Storage requirements for the various file types range from 1.5 to 8 KB, as described in the following paragraphs.

Waveform Files. Three main types of digitized waveforms can be stored: single (or "raw") responses, averaged responses, and analyzed waveforms. All three types contain 1,024 data points, but the storage formats differ for the individual points. Raw responses are the direct output of the 12-bit ADC, and thus can be stored as 2-byte integers. Averaged responses are actually sums of raw responses, and are stored as 3-byte integers to avoid loss of precision. Since all of the analysis routines use floating-point calculations, analyzed waveforms are stored as 4-byte floating-point numbers. Because of these different formats, raw waveform files require 2 KB of storage, averaged waveforms require 3 KB and analyzed waveforms require 4 KB.

EOG Files. As described in the above section on program design, the EOG generates too much data to store the digitized waveforms for an entire session. The maximum and minimum amplitudes of the EOG waveform are thus extracted by on-line analysis and stored along with the corresponding elapsed times. Normally, 10 maxima and 10 minima per minute are stored for a total duration of 30 min. The amplitudes and times are both stored as 2-byte integers, in parallel files requiring 1.5 KB each. Brief segments of the EOG waveform can also be saved as raw waveform files, in order to compute estimates of saccadic velocity (see Program Operation section and Data Processing and Analysis section).

Fourier Analysis Files. These files contain the Fourier frequency spectra of waveform files. The frequency functions are complex, and thus need twice as many elements as the source waveforms to express the real and imaginary components. Since the algorithm used for the fast Fourier transform uses floating-point arithmetic, 8 KB are required to store the transform of a 1,024-element digitized waveform.

Report Files. Report files are essentially "pages" of alphanumeric information entered by the system operator. They may contain anything that the operator cares to enter, normally comments related to the other files on the diskette. Examples might be patient histories, diagnostic evaluations, or supplementary information about test conditions. Individual report files may contain up to 3*K* characters (bytes), arranged in 36 85-character lines. Thus, an entire report file can be displayed at once on the screen, with enough space left at the bottom to display options and instructions. Although report files are numbered sequentially in order of their creation, it is possible to correct or revise an earlier report at any time.

Index File. The index file provides supplementary identifying information for all the data and analysis files on the diskette. Each 85-character entry contains a unique serial number between 1 and 160, the file name, the date, the patient's name, sex, and birthdate, and descriptive information about the test. The first 85 bytes of the index file are reserved for "bookkeeping information" related to access and updating of the index. For details on how the index file is used to find and retrieve files for display or analysis, see the Data Analysis section below.

Parameter File. In addition to the index file, each data diskette has a parameter file that contains a 63-byte list of numerical parameters for each data or analysis file on the disk. Some types of data file (e.g., EOG files and report files) do not fully utilize the parameter file. The first 3 bytes of the parameter file are reserved for bookkeeping information.

File Names. Each file is labeled with a descriptive 8-character file name that is listed both in the North Star DOS disk directory and in the more extensive index file (see above). For data files, the file name consists of a 5-letter file type followed by a 3-digit "counter" that functions as a serial number for that file type alone. For example, if a data disk contained 103 raw waveform (RWVFM) files and 26 averaged waveform (AWVFM) files, the most recently recorded waveforms would be named RWVFM103 and AWVFM026. File names for analysis files start with only a two-letter file type followed by either one or two three-digit integers that correspond to the *index numbers* of the source file(s) for the analysis. To illustrate, let waveforms RWVFM070 and RWVFM102 in the above example have disk serial numbers 94 and 146. A file containing the difference of these two waveforms would then be named SB094146 and the derivative of RWVFM070 would be named DE000094. To allow multiple smoothed versions of the same waveform, file names for smoothed waveforms have both a counter and a single index number; e.g., two smoothed versions of RWVFM070 might be named AM001094 and SM002094. The file type sections of the various possible file names are shown in Table 1.

Table 1. List of Possible File Types

Data files (integers and ASCII characters)		
1.	RWVFM	raw response waveform (2 KB)
2.	AWVFM	averaged response waveform (3 KB)
3.	EOGAM	EOG amplitudes (1.5 KB)
4.	EOGTM	EOG elapsed times (1.5 KB)
5.	SACVL	EOG waveform, used to compute saccadic velocities (2 KB)
6.	REPRT	patient report, contains 36 85-space lines (3 KB)
Analysis files (floating-point numbers)		
7.	SM	sum of two waveforms (4 KB)
8.	DF	differences of two waveforms (4 KB)
9.	IN	integral of waveform (4 KB)
10.	DE	derivative of waveform (4 KB)
11.	SW	smoothed waveform (4 KB)
12.	FT	fast Fourier transform (8 KB)
13.	FW	digitally filtered waveform (4 KB)
14.	AC	autocorrelation function of waveform (4 KB)
15.	CC	cross-correlation function of two waveforms (4 KB)

Data Processing and Analysis

The data analysis capabilities of the UTAS system are comprehensive and sophisticated, but are nonetheless extremely simple to use. Once a digitized waveform has been stored on disk, it can be easily retrieved, displayed on the screen in a variety of formats, printed in any screen format, measured by a flexible system of cursors, and subjected to a large repertoire of numerical and statistical analyses. This section first gives a brief overview of the program operation and then describes the capabilities of specific analysis options.

The analysis program is entered initially from the main program menu. The operator can then direct the program flow with a series of single-keystroke commands. The operator first displays the disk file index on the screen and selects up to four files for display by typing their INDEX numbers. At any time, it is possible to either print out a hard copy of the screen display or print out a detailed parameter list for each displayed file. Individual files can easily be removed from the display or replaced by other files.

When a file (or files) has been selected for analysis, a list of numerical and statistical analysis options is displayed and the desired option can be called by a single-keystroke command (e.g., "F" for Fourier analysis). When any particular analysis is completed, the operator can save the result on disk (unless an equivalent file is already on disk) before returning to the main program menu.

Manipulating the Screen Display

The vertical scale of a displayed file can be easily increased or decreased, and the waveform can be raised or lowered to any position on the screen.

Most waveforms are normally displayed at half resolution because of the limitations of the display monitor. However, the "window" between any pair of cursors can also be displayed at full resolution, and waveforms that extend beyond the screen can be scrolled leftward or rightward as needed.

Four pairs of waveform cursors are available for measuring ordinate and abscissa values on the displayed waveforms. Each waveform is allotted one cursor pair if three or four waveforms are displayed, two pairs if two waveforms are displayed and all four pairs if only one waveform is displayed. The bottom of the screen contains a readout of the x and y position of each cursor, plus the differences between x and y positions for each cursor pair.

Numerical Waveform Analyses

Addition and Subtraction of Waveforms. Any two waveforms with common axes and the same number of data points can be added to or subtracted from each other. Previously analyzed waveforms are acceptable, making it possible, for example, to compute differences of differences.

Integration. The discrete integral, $G(t) = \Sigma_{u=0}^{+}(g(u) - C) \cdot \Delta u$, can be computed for waveform $g(t)$. Since the shape of the integrated function is strongly affected by the DC response level C, the operator is given the option of setting the zero baseline with a vertical cursor. If this option is not exercised, the default baseline is the average amplitude of the 6th through the 25th points of the source waveform.

Differentiation. The discrete derivative, $F(t) = f(t)/\Delta t$, can be computed for any well-behaved (i.e., reasonably continuous) waveform $f(t)$. Previously analyzed waveforms are acceptable. It is usually desirable to smooth noisy waveforms before differentiating, in order to avoid excessive spiking in the derivative. A special application of differentiation is the computation of saccadic velocity functions from digitized EOG waveforms.

Waveform Smoothing. The smoothing program replaces each point in a waveform with a weighted average of points in the surrounding region. A symmetrical triangular weighting function is used. That is, if the weighting function spans $2S + 1$ points, then the weight is $(S - |k| + 1)/[\Sigma_{i=-S}^{S}(S - |i| + 1)]$ at a distance of k points from the center of the function (for $0 \leq k \leq S$). The width parameter S can be selected by the operator, and can range from one to 60 points on either side of the central point. The default value of S is 1. For points near the beginning or end of the source waveform, parts of the weighting function that extend beyond the edge of the waveform are truncated.

Fourier Analysis and Digital Filtering. The fast Fourier transform can be computed for any function of time, including analyzed waveforms. Implementation of the Cooley–Tukey algorithm (see Brigham, 1974) has been optimized for the Z80 microprocessor and the IEEE standard format for binary floating-point numbers. In addition, execution time has been significantly shortened by omitting certain computations when their result can be predicted with certainty. With these modifications, a 1,024-sample transform can be completed in less than a minute.

Once the transform has been computed, the operator may choose to simply store the resulting frequency functions. Alternatively, the four cursor pairs may be used to set the amplitudes to zero in as many as four adjustable frequency bands, after which the frequency functions are back-transformed to retrieve the filtered waveform. If a previously transformed FT file is selected for analysis, the filtering procedure can be initiated immediately.

Normally, only the amplitude spectrum is displayed. If desired, however, the amplitude and phase spectra can be alternately displayed by successive presses of the STEP command key. Cursors may be used to measure either amplitudes or phases, but frequency bands can be erased from the amplitude spectrum only.

The phase spectrum display is intended to show true phase rather than the principal value only. Perfect accuracy cannot be guaranteed, however, because to the phase computations assume that no phase differences greater than or equal to π occur between adjacent points of the spectrum.

Statistical Analyses

Correlation Coefficient. Correlation coefficients can be computed either between two entire waveforms or between two cursor-selectable "windows" of equal width. The two windows can be either on different waveforms or on the same waveform. The results can be printed, but they cannot be stored on disk.

Correlation Functions. The autocorrelation function of a waveform or the cross-correlation function between two waveforms is computed by multiplying the Fourier transform of one function with the complex conjugate of the Fourier transform of the other and computing the inverse transform of the result (Brigham, 1974). Because of memory limitations, these computations are made only on the even-numbered points of the source waveform(s). In most cases, the results do not differ significantly from those that include every point.

Patient Reports

Report Formats. Information from data files, analysis files, and report files can be coordinated and printed out in formal patient reports. The procedures for generating reports are designed to allow the user maximum flexibility in selecting and organizing the material to be included.

Two report formats are available for ERG, VER, and analyzed waveforms. Choice of format is a trade-off between the resolution of displayed waveforms and the number of waveforms on a page. The high-resolution format allows the positioning of up to four waveforms on a screen or page with full recorded resolution. Waveforms are read from the diskette sequentially by typing the waveform number listed in the index at the bottom of the index display screen. Following simultaneous display of the desired waveforms, cursors may be positioned on important waveform features. Pressing the PRINT command key then copies the screen to the printer and prints patient information, cursor positions, and cursor amplitude and time differences in an alphanumeric block below the graphics printout (Figure 6).

A second format allowing up to 10 waveforms per screen or page is designed for summary reporting of complete ERG or VER studies. Waveforms for the 10 report positions are chosen in sequence from the diskette index, which is scrolled as needed in a constantly visible screen segment. All chosen waveforms are then displayed in two parallel columns. Cursors may be placed as in the high-resolution format except that only four sets of cursors are actively positioned at a time.

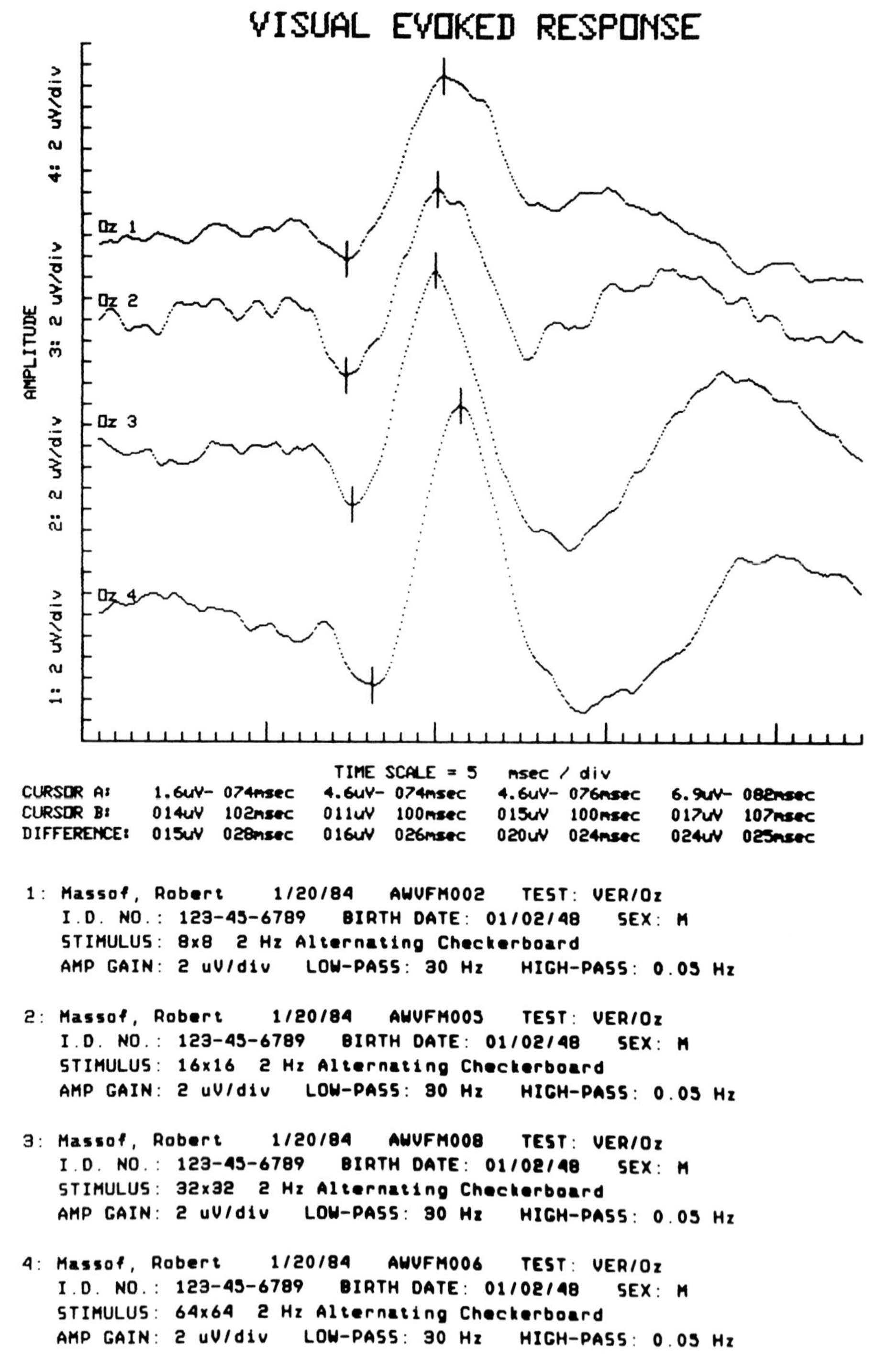

Figure 6. Example of a patient report printed in high-resolution format. The VER waveforms were recalled from disk and were positioned on the screen using the SIZE command key. The cursors were then set to the relevant peaks and the entire report was printed out by pressing the PRINT key. Cursor amplitudes and latencies for waveforms 1–4 are printed from left to right under the graph. Cursor *A* always has the smaller latency.

Once all four cursor pairs are set, the values are stored and the active cursors are moved to the next set of four waveforms until cursor settings are satisfactory for all waveforms. Cursor information may then be displayed along with previously entered patient information on a second report page by pressing the STEP control key. Subsequent presses of the STEP key switch back and forth between the waveform and alphanumeric displays.

In addition to high-resolution and summary waveform reports, single-page EOG summary reports similar to the on-line output generated at the time of recording can be automatically constructed from EOG disk files. REPORT disk files also can be created or modified, and can subsequently be displayed and printed as desired. (REPORT files can also be created using the PROTOCOL program module, but existing files cannot be recalled from disk.) These four different formats can be printed individually or combined in any order, allowing the user great freedom to compose reports as simple or as complex as the occasion dictates.

Modem Operations. As part of the REPORT option, the user is able to communicate by modem with other UTAS-E 1000 systems. For modem communications, we have chosen the Auto Dial 212A (U.S. Robotics, Inc.), operating at 1200 baud with 11-bit/character data transmission (1 start bit, 8 data bits, 1 parity bit, 1 stop bit). We are using the 11-bit format instead of the standard 10 bits per character in order to transmit a full byte of data in each character (the 10-bit format uses the most significant data bit as the parity bit). ASCII characters are transmitted in the same way as data bytes. The data transmission software allows users to toggle between computer and voice communication over the same link.

To transmit waveforms, we first send identifying information indicating the type of file to be transmitted, then the INDEX and PARAMETER file information, and finally the buffered data. In order to minimize the time of data transmission, every other data point in the buffer is transmitted first and displayed, and then the odd points are transmitted during interrupts to complete the data set. This allows the user at the receiving end to examine and run cursors over the waveform while the balance of the data is being transmitted and all data are being verified.

To verify data, the receiving computer echoes characters back to the sending computer. The sending computer performs a logical AND between the original and echoed data. If the AND produces a zero, the data are retransmitted.

Conclusions

Electrophysiology has developed in parallel with recording, display, and, most recently, data processing technology. Advances in inexpensive digital technology have had their greatest impact on methods of data storage and analysis. The electrophysiology system described in this chapter has been designed to take advantage of the inherent flexibility in combining a general-purpose microcomputer with special-purpose peripheral subunits. The major system subunits are a Z80-based microcomputer, a high-resolution graphics system, a dot-matrix printer, an automated Ganzfeld for flash presentation, a TV-based pattern stimulator, and a custom amplifier and interface unit. Subunits were selected for high performance-to-price ratio, high reliability, and inexpensive service availability.

All system software is written in assembly language for greater speed and efficient use of computer memory. The operating program is modularized into six subunits that are called from floppy disk storage into CPU memory as needed. The six modules contain the programming code necessary to operate the graphics system, to acquire, average, display, and store data, to analyze results, and to produce reports. Standard operating protocols simplify use of the system, while built-in manual overrides allow continuous, documented change of the standard procedures through the use of specially marked command keys. The filing system for data storage and retrieval reserves disk space according to data type and required precision, and maintains special index and parameter files that hold identifying information for all data and analysis files stored on a diskette. Powerful arithmetic and statistical analysis capabilities are also provided.

A system such as this defies obsolescence by its capability for change through modification of its operating program. The system may thus evolve with new developments in both hardware capability and scientific knowledge.

References

Arden, G.B., and J.H. Kelsey (1962) Changes produced by light in the standing potential of the human eye. J. Physiol. 161:189–204.

Armington, J.C. (1974) The Electroretinogram. New York: Academic Press.

Bodis-Wollner, I. (ed.) (1982) Evoked potentials. Ann. NY Acad. Sci. 388:1–738.

Brigham, E.O. (1974) The Fast Fourier Transform. Englewood Cliffs, NJ: Prentice-Hall.

Dewar, J., and J.G. M'Kendrick (1873) On the physiological action of light. Proc. R. Soc. Edinb. 8:100–104, 110–114, 179–182.

du Bois-Reymond, E. (1849) Üntersuchungen über theirische Electricitat, vol. 2. Berlin: Reimer, pp. 256–257.

Einthoven, W., and W.A. Jolly (1908) The form and magnitude of the electrical response of the eye to stimulation by light at various intensities. Q.J. Exp. Physiol. 1:373–416.

Granit, R. (1933) The components of the retinal action potential in mammals and their relation to the discharge of the optic nerve. J. Physiol. 77:207–239.

Piper, H. (1911) Über die Netzhantströme. Arch. Anat. Physiol. Abt. (Leipzig): 85–132.

Rodieck, R.W. (1972) The Vertebrate Retina. San Francisco: W.H. Freeman and Co.

Index

I

J

K

L

M

S